Reliability Analysis for Asset Management of Electric Power Grids

Reliability Analysis for Asset Management of Electric Power Grids

Robert Ross
IWO, Ede, the Netherlands
TU Delft, Delft, the Netherlands

Registered Offices
John Wiley & Sons, Inc., 111 River Street, Hoboken, NJ 07030, USA
John Wiley & Sons Ltd, The Atrium, Southern Gate, Chichester, West Sussex, PO19 8SQ, UK

Editorial Office
The Atrium, Southern Gate, Chichester, West Sussex, PO19 8SQ, UK

For details of our global editorial offices, customer services, and more information about Wiley products visit us at www.wiley.com.

Wiley also publishes its books in a variety of electronic formats and by print-on-demand. Some content that appears in standard print versions of this book may not be available in other formats.

Library of Congress Cataloging-in-Publication Data

Names: Ross, Robert, author.
Title: Reliability analysis for asset management of electric power grids /
    Prof Dr Robert Ross, IWO, Ede, the Netherlands; TU Delft, Delft, the
    Netherlands.
Description: Hoboken, NJ : Wiley, [2019] | Includes bibliographical
    references and index. |
Identifiers: LCCN 2018031845 (print) | LCCN 2018033843 (ebook) | ISBN
    9781119125181 (Adobe PDF) | ISBN 9781119125198 (ePub) | ISBN 9781119125174
    (hardcover)
Subjects: LCSH: Electric power distribution–Reliability.
Classification: LCC TK3001 (ebook) | LCC TK3001 .R65 2018 (print) | DDC
    621.319–dc23
LC record available at https://lccn.loc.gov/2018031845

Cover Design: Wiley
Cover Image: © iStock.com/yangphoto; Equations courtesy of Robert Ross

Set in 10/12pt WarnockPro by SPi Global, Chennai, India

Printed in Singapore by C.O.S. Printers Pte Ltd

10 9 8 7 6 5 4 3 2 1

In remembrance of my parents and dedicated to my dear wife, children and sister.
R. R.

Contents

Preface

Reliability Analysis for Asset Management of Electric Power Grids aims to provide understanding and skills for analysing data in order to assess the reliability of components and systems. The understanding and skills support not only asset management and maintenance, but also incident management. The latter deals with unexpected failures that need to be evaluated to assist in decision-making.

The structure of the book is presented in Table 1 below. After an introduction (Chapter 1) that pictures asset management and incident management in qualitative terms, seven chapters follow. The subjects of these chapters are: the basics of statistics (Chapter 2), measures to quantify (Chapter 3), a range of statistical distributions with their aims and properties (Chapter 4), graphical analysis of data (Chapter 5), distribution parameter estimation (Chapter 6), system and component reliability (Chapter 7) and, finally, system states with their reliability, availability and redundancy (Chapter 8). These provide an arsenal of techniques that form a foundation for statistical analysis in asset and incident management. These eight chapters form the core of the course in reliability analysis.

The final two chapters (Chapters 9 and 10) aim to provide deeper insight and may be used in parallel with Chapters 1–8. Chapter 9 discusses a range of practical cases from asset management and incident management while using the techniques as explained in the previous chapters. Per case, it is indicated from which section the information is taken. Elements of the sections can be used for illustration during the course to complement the teaching from other chapters. Interested readers from the electric power industry may choose to start with Chapter 9 and select the aspects of reliability analysis that they would like to study more deeply, then follow the references.

Chapter 10 also aims at providing deeper insight, not so much by treating practical cases, but rather by studying a range of subjects in more depth. Depending on the courses given, the topics from this chapter may be added to lectures on Chapters 1–8.

The book covers some relatively new subjects and approaches, such as:
- The difference in the meaning of statistical distribution and probability, as discussed and followed throughout the text.
- Child mortality and the bath tub model, discussed with two different meanings. Statistically often associated with a declining hazard rate, in practice the meaning of child mortality is often encountered as a weak subpopulation which does not necessarily mean a declining hazard rate at all. The two models are discussed.

Table 1 Overview of the subjects treated in the book.

Qualitative introduction on:	**Chapter 1**

- asset management; maintenance styles
- incident management

↓

Basics of statistics, addressing:	**Chapter 2**

- concept outcomes, sample space, events, distribution, probability
- statistical functions F, R, f, h, H; combinations of distributions and processes; two bath tub models depending on child mortality type
- concept of ageing dose, power law and accelerated ageing

↓

Measures in statistics:	**Chapter 3**

- expected values; conditional values and Bayes' theorem
- moments; mean, median, mode, variance, standard deviation
- covariance, correlation, similarity index and compliance

↓

The purpose, characteristics and use of various specific distributions:	**Chapter 4**

- uniform, beta, Weibull, exponential, normal, lognormal, binomial, Poisson, hypergeometric and multinomial

↓

Graphical data analysis and representations of distributions:	**Chapter 5**

- parameter-free graphs, confidence intervals
- parametric plots: Weibull, exponential, normal and lognormal, Duane and Crow/AMSAA

↓

Parameter estimation:	**Chapter 6**

- bias, efficiency, consistency and small data sets
- maximum likelihood, least squares and weighted least squares
- application to Weibull, exponential and normal distributions
- asymptotic behaviour, power function and unbiasing
- beta distribution-based and regression-based confidence limits

↓

System and component reliability:	**Chapter 7**

- block diagrams
- series systems and competing processes, parallel systems and redundancy, combined systems and common-cause failure
- analysis of complex systems

↓

System states in terms of working versus down:	**Chapter 8**

- states and transitions; failure and repair, absorbing down-states
- Markov chains and Laplace transforms
- mean time to first failure and mean time between failures
- availability and steady states

(Continued)

Table 1 (Continued)

⇅

Practical applications to asset management and incident management: **Chapter 9**

- period-based, corrective, condition-based, risk-based maintenance
- health index, risk index and combined health index
- testing and quality with small test sets and accelerated ageing
- failure cases and probability forecast of next failures

⇅⇅

Miscellaneous background subjects: **Chapter 10**

- combinatorics and the gamma function
- power functions and asymptotic behaviour
- regression analysis and regression-based confidence intervals
- sampling, Monte Carlo and random number generators
- hypothesis testing

Graph template and data tables **Appendices**

- The power law concept associated with the ageing dose concept and used for discussing accelerated ageing and testing.
- Asymptotic behaviour of bias and variances, described with a three-parameter power function. This approach leads to an elegant unbiasing method in parameter estimation.
- The power function also used to approximate the normal distribution.
- The similarity index, introduced to compare distributions, which is useful for evaluating whether two distributions are the same and for estimating the number of failures yet to come. Determining the significance is discussed, and various examples are elaborated.
- Consistency between graphical analysis and parameter estimation, which means that the best fit in a graph is identical to the best fit from parameter estimation.
- Comparable views on the confidence limits based on random sampling (beta distribution) versus linear regression.
- The relation between Monte Carlo simulations and sampling from the ranked cumulative distribution space. Numerical integration and mapping the ranked F-space is discussed. The effect of quality control testing on the resulting ranked F-space and confidence limits is demonstrated.
- Much attention is paid to analysing small data sets. It is acknowledged that large data sets are necessary for accurate statistics. On the other hand, data are often scarce, with incident management and timely decision-making required. While conclusions may not be very accurate, for decision-making after unexpected failures they may be good enough and – more important – can support timely decision-making.
- The statistical models related to maintenance models, like corrective, period-based, condition-based and risk-based maintenance, as well as models like the health index, risk index and combined health index.

As shown in Table 1, a range of distributions and plotting methods are discussed, including the Poisson distribution and Crow/AMSAA plots. Five distributions stand out in the discussions:

- Weibull, because it is the asymptotic distribution for the weakest link in the chain, which applies to many failure incidents.
- Exponential, because maintained components and systems tend to have a more or less constant hazard rate, which is a property of the exponential distribution.
- Normal, because it is the asymptotic distribution for the mean and standard deviation. It helps in evaluating the accuracy of regression analysis.
- Beta, for confidence limits and ranked sampling.
- Uniform, which is fundamental to random sampling.

Additional distributions are discussed due to their peculiar properties.

The book is a considerable extension of the manuscript used for courses on system reliability at the Netherlands Royal Institute for the Navy. It is extended with experience gained during failure investigations and forensic studies, as well as asset management of grids and maintenance.

Finally, the book contains more than 30 years of experience and, moreover, stands on the shoulders of generations of experts before me. It is the sincerest wish of the author to match the needs in this field of students and colleagues in charge of strategic infrastructures, and the electric power industry in particular. This book also aims to contribute to the lively discussions that accompany the further development of asset management and incident management. The subject is not considered a finished topic with this book. Many experts contribute with questions, ideas and research daily, which makes this a still maturing field. The author welcomes any suggestions for improvements in order to maximize the contribution to the field of reliability analysis.

Robert Ross
IWO, Ede, the Netherlands
TU Delft, Delft, the Netherlands

Acknowledgements

I gratefully acknowledge the reviews, reflections and comments of Sonja Bouwman (AVANS), Remko Logemann (TenneT), Marijke Ross (New Perspective), Robbert Ross (Rijkswaterstaat) and Peter Ypma (IWO and TU Delft). In addition, I would like to thank Aart-Jan de Graaf (HAN), René Janssen (NLDA) and various colleagues at TenneT for fruitful discussions on selected topics in this book.

List of Symbols and Abbreviations

∞	Infinity
$\pm$	Plus or minus; e.g. $x = \pm 1$ means x can be $+1$ or -1
$\neq$	Is not equal to
$\approx$	Is approximately equal to; e.g. $\pi \approx 3.14159$
$\sim$	Is similar to

- with quantities: same order of magnitude; e.g. 4~6
- with probability: has a similar probability distribution
- with functions: is asymptotically equal to; definition $f(x) \sim g(x)$ $(x \to a)$ means $\lim\limits_{x \to a} \dfrac{f(x)}{g(x)} = 1$; e.g. $x^2 + 2x + 3 + \frac{4}{x} \sim x^2$ $(x \to \infty)$

$\cong$ Isomorph; having the same measures or being congruent (identical measures)

- let ABC and A′B′C′ be triangles. ABC and A′B′C′ are isomorphic (congruential) if and only if their shape (i.e. angles) and distances are equal. They may be oriented differently

$*$ Complex conjugate

- if $z = (x + iy)$ with $z \in \mathbb{C}$ then $(x + iy)^* = (x - iy)$ with $x,y \in \mathbb{R}$ and i the imaginary unit, i.e. $i^2 = -1$

$\propto$ Is proportional to

- $x \propto y$ means: $x = c.y$ with c being a constant, i.e. a linear relationship

$\equiv$ Equal by definition; congruence

- with definitions: $x \equiv y$ means x by definition is equal to y
- with congruence: $a \equiv b$ (mod c) means $a - b$ is divisible by c; e.g. $7 \equiv 4$ (mod 3)

$\sqrt{}$ Square root

- $\sqrt{x}$ means the non-negative number whose square is equal to x

$!$ Factorial; e.g. $n! = 1 \cdot 2 \cdot \ldots \cdot n$, and $n! = \Gamma(n + 1)$, the gamma function

$<$	Is smaller than
$\leq$	Is smaller than or equal to
$\ll$	Is much smaller than
$>$	Is larger than
$\geq$	Is larger than or equal to
$\gg$	Is much larger than
$<x>$	Expected value of x. This is the mean value of x as the average of all possible x values
$\Rightarrow$	Implication
	• with statements: implies that; e.g. A $\Rightarrow$ B: if A is true then B is true
$\Leftarrow$	Reverse implication
	• with statements: implies that; e.g. A $\Leftarrow$ B: if B is true then A is true
$\Leftrightarrow$	Equivalence, mutual implication
	• with statements: is equivalent to; e.g. A $\Leftrightarrow$ B: A is true if and only if B is true
$\in$	Is an element of
	• with set theory: is an element of; e.g. $n \in \mathbb{Z}$ means n is an integer
$\notin$	Is not an element of
	• with set theory: is not an element of; e.g. $\pi \notin \mathbb{Z}$ means π is not an integer
$\emptyset$	Empty set; e.g. $A \cap A^c = \emptyset$ means the intersection of a set A and its complement is an empty set
$\cap$	Intersection; e.g. $x \in A \cap B$ means x is an element of both set A and set B
$\cup$	Union; e.g. $x \in A \cup B$ means x is an element of the union of set A and set B
$\subset$	Is a proper subset of; i.e. not an empty set or equal set; e.g. $A \not\subset A$
$\subseteq$	Is a subset of; includes proper sets and the complete set; e.g. $A \subseteq A$
$\backslash$	Excluding; e.g. $\mathbb{N}\backslash\{0\}$ means the set of natural numbers without 0, i.e. $\{1, 2, \ldots\}$
$\wedge$	And; e.g. $x > 0 \wedge x < 1$ means $0 < x < 1$, i.e. both statements are true
$\vee$	Or; e.g. $x < 0 \vee x > 1$ means x outside range $[0, 1]$, i.e. either statement is true

$\|\ldots\|$	Absolute value

- with real number x: absolute value of x; i.e. if $x \geq 0$, $|x| = x$; if $x < 0$, $|x| = -x$
- with complex number $x = a + i \cdot b$ where i is imaginary number and a and b are real numbers: $|x| = \sqrt{|a|^2 + |b|^2}$, with $|a|$ and $|b|$ absolute real numbers
- with vector $X = (x_1, \ldots, x_n)$: the Euclidean length; i.e. $|X| = \sqrt{x_1^2 + \ldots + x_n^2}$
- with matrix A: determinant of the matrix: $|A|$
- with set $X = \{x_1, \ldots, x_n\}$: the cardinality, i.e. number of elements. $|X| = n$

$\|\ldots\|$	Rounded off; $\|x\|$ is the nearest integer to x; e.g. $\|1.5\| = 2$; $\|1.4\| = 1$

$\binom{n}{k}$	Binomial coefficient, also called n-over-k; see also factorial $n!$

$$\binom{n}{k} = \frac{n!}{k! \cdot (n-k)!} = \frac{n \cdot (n-1) \cdot \ldots \cdot (n-k+1)}{k \cdot (k-1) \cdot \ldots \cdot 1}; \text{ e.g.}$$

$$\binom{5}{2} = \frac{5!}{2! \cdot 3!} = \frac{5 \cdot 4}{2 \cdot 1} = 10$$

$\lfloor \ldots \rfloor$	Floor function of real number x: largest integer n less than or equal to x; i.e. $n + 1 > \lfloor x \rfloor \geq n$
$\lceil \ldots \rceil$	Ceiling function of real number x: smallest integer n greater than or equal to x; i.e. $n - 1 < \lceil x \rceil \leq n$
$A(t)$	Availability at time t
A_∞	Availability long term, i.e. for infinite period of time or the limit $t \to \infty$
$A\%$	In context of distributions: failed fraction of population at specific reliability; e.g. 50% median where 50% of the population failed
	In context of availability: up-time as a percentage of total time (i.e. up-time + down-time)
A^c	Complement of

- with set theory: the set A^c contains all elements of the total sample space Ω that are not in set A

α	Lowercase alpha

- with two- or three-parameter Weibull distribution: scale parameter

$\forall$	For all; $\forall t : t \cdot t = t^2$, i.e. for all variables t the product of t and itself is the square of t
a.u.	Arbitrary unit; i.e. unit not specified
AM	Asset management
$B(i,j)$	Beta distribution

β	Lowercase beta
	• with two- or three-parameter Weibull distribution: shape parameter
$\mathbb{C}$	The set of complex numbers
	• the sum of a real and an imaginary number; e.g. $z = x + iy$ with $z \in \mathbb{C}$ and $x,y \in \mathbb{R}$ and i the imaginary unit, i.e. $i^2 = -1$
CM	Corrective maintenance, i.e. action after failure (usually replacement)
CBM	Condition-based maintenance
Cigré	Conseil International des Grands Réseaux Électriques, Eng.: International Council on Large Electric Systems
D	Ageing dose in terms of power law
$D_{\%}$	Relative (ageing) dose; i.e. $D_{\%} = D/D_{tot}$
D_{tot}	Total or fatal (ageing) dose
Δ	Delta; e.g. Δt (usually small) step or variation in time t
$f(t)$	Distribution density; derivative of cumulative failure distribution
$F(t)$	Cumulative failure distribution
$\varphi_X(t)$	Characteristic function
$G_X(t)$	Moment generating function
γ	Lowercase gamma
	• Euler's constant $\gamma \approx 0.5772156649\ldots$
	• with three-parameter Weibull distribution: threshold or location parameter
Γ	Uppercase gamma; e.g. used for gamma function (see below)
$\Gamma(n)$	Gamma function; if n is a natural number: factorial of $(n-1)$, i.e. $\Gamma(n) = (n-1)!$
$h(t)$	Hazard rate
$H(t)$	Cumulative hazard rate over range $[0,t]$
$H(t,\Delta t)$	Cumulative hazard rate over range $[t, t+\Delta t]$
i	Lowercase i
	• often used as index to distinguish an arbitrary element from a group; e.g. x_i ($i = 1, \ldots, 10$) to represent any element from the group $\{x_1, x_2, x_3, x_4, x_5, x_6, x_7, x_8, x_9, x_{10}\}$
	• in case of imaginary number i: the square root of -1; i.e. $i^2 = -1$
IAM	Institute for Asset Management (UK)
IEC	International Electrotechnical Commission (CH)
IEEE	Institute for Electronic and Electrical Engineers (USA)

j	Lowercase *j*

- often used as index to distinguish an arbitrary element from a group; e.g. x_j ($j = 1, ..., 10$) to represent any element from the group $\{x_1, x_2, x_3, x_4, x_5, x_6, x_7, x_8, x_9, x_{10}\}$
- in case of imaginary number *j*: the square root of -1; i.e. $j^2 = -1$

λ	Lambda

- with exponential distribution: inverse characteristic time
- with systems: hazard rate

$L(t)$	Likelihood function; product of failure probability densities (uncensored case)
M	As suffix: denoting median
μ	Lowercase mu

- with distributed variable X: expected mean, i.e. $\mu = <X>$
- with systems: repair rate

Max(*a,b*)	Function that selects the largest of two values *a* and *b*
Min(*a,b*)	Function that selects the smallest of two values *a* and *b*
MTBF	Mean time between failure
MTTFF	Mean time to first failure
$\mathbb{N}$	The set of natural numbers; i.e. $\mathbb{N} = \{0, 1, ...\}$
Ω	Uppercase omega

- total sample space

π	Lowercase pi

- fundamental constant $\pi \approx 3.1415926536...$

Π	Uppercase pi; symbol denoting product

- $\prod\limits_{i=1}^{n} a_i$ means: $a_1 \cdot a_2 \cdot ... \cdot a_n$

PBM	Period-based maintenance
PM	Periodic maintenance; also preventive maintenance; both used as synonym for PBM
$\mathbb{Q}$	The set of rational numbers

- numbers that can be derived from ratios of integer numbers; i.e. $\mathbb{Q} = \{p/q: p, q \in \mathbb{Z}, q \neq 0\}$; e.g. $3\frac{14}{100} = 3.14 \in \mathbb{Q}$, but $\pi \notin \mathbb{Q}$

$\mathbb{R}$	The set of real numbers

- the set of rational and irrational numbers; e.g. $3\frac{14}{100} = 3.14 \in \mathbb{R}$ and $\pi \in \mathbb{R}$

$R(t)$	Reliability, also survivability

RBM	Risk-based maintenance
RCM	Reliability-centred maintenance
s	Lowercase s

- with averaging: scatter or standard deviation of (population) sample
- with Laplace transform: variable of Laplace-transformed function

σ	Lowercase sigma

- with Gaussian or normal distribution: scatter or standard deviation

Σ	Uppercase sigma; symbol denoting summation

- $\sum_{i=1}^{n} a_i$ means: $a_1 + a_2 + \ldots + a_n$

τ	Lowercase tau; measure of time
θ	Lowercase theta

- mean lifetime
- with exponential distribution: characteristic time, i.e. scale parameter

$\mathbb{Z}$	The set of integer numbers; i.e. $\mathbb{Z} = \{0, \pm 1, \pm 2, \ldots\}$

About the Companion website

This book is accompanied by a companion website:

www.wiley.com/go/ross/reliabilityanalysis

The website includes:

Figures and tables from the book in PowerPoint format, answers to the questions and collected comments by readers.

Scan this QR code to visit the companion website.

1

Introduction

This chapter provides an introduction to electric power grids as strategic infrastructures (Section 1.1); introduces asset management (AM) as an approach to optimize performance versus efforts (Section 1.2); provides an overview of maintenance concepts with their underlying philosophies and under which conditions they are preferred (Section 1.3); and finally discusses incident management (IM) (Section 1.4). The purpose of this chapter is to describe the context of the statistical methods discussed in this book.

1.1 Electric Power Grids

In 1880 Joseph Wilson Swan started to provide electricity to a residence in Rothbury, UK and in 1882 Thomas Edison's Pearl Street Station power plant generated electricity for homes in Manhattan. Edison envisaged: 'After the electric light goes into general use, none but the extravagant will burn tallow candles' [1]. This statement evolved into: 'We will make electricity so cheap that only the rich will burn candles'. Interestingly, these statements refer to three themes that are important to AM: performance (electric light versus candles), general use (utility) and price (extravagant, rich).

In many countries electricity lived up to this expectation and became a dominant energy carrier, not just for lighting, but also for industrial machines, transportation, many household functionalities and more. Electric power networks grew, were integrated and expanded to continental grids and larger. Interconnections have been developed, such as the 580-km NorNed submarine cable that enables hydropower reservoirs and consumption in Norway to act as energy storage for, for example, a surplus of wind power generated out of the German coast. That energy can be transmitted through the Netherlands' grid and transmitted through this NorNed cable or connected to other grids in Europe and beyond. The 130-kW power plant of 1882 has been succeeded by GW power plants installed by State Grid, China. Nowadays, developed countries depend to a great extent on a reliable supply of electric power, which makes electric power grids very strategic infrastructures. The growing environmental awareness and focus on sustainable energy led to initiatives like the German 'Energiewende' (in English: energy transition, namely the shift from fossil and nuclear energy to sustainable energy), which only adds to the importance of electric power grids with electricity as a convenient energy carrier.

Reliability Analysis for Asset Management of Electric Power Grids, First Edition. Robert Ross.
© 2019 John Wiley & Sons Ltd. Published 2019 by John Wiley & Sons Ltd.
Companion website: www.wiley.com/go/ross/reliabilityanalysis

A power grid consists of many assets, but it is much more than a collection of components. The assets depend on each other to such an extent that they form clusters and connections which can be treated as units. For instance, a cable system consists of a cable termination and cable sections that are linked with cable joints. In addition, this cable system combined with primary components like switchgear, measuring transformers (current, voltage), a power transformer and a rail forms a circuit. Two parallel circuits can form a redundant connection, and so on. The components and connections are often controlled and protected by secondary systems. All those systems together form the network or grid.

Utilities are responsible for the performance of these networks. The assets that together form the grids need to be designed for their function, supervised, maintained, repaired and finally replaced. The network as a whole must fulfil its duty of collecting, conducting and supplying energy. It must be robust and cost-effective, and moreover must be safe and blend well into the environment. Planning and decision-making to meet these requirements is the core of AM. Mature AM is based on lessons learnt from the past and present philosophies. It works with assumptions and predictions to optimize decision-making. There are different styles of AM and of decision-making with IM, but they all take great benefit from statistical techniques. This book aims at providing the essential background for applied statistics to be used in AM and IM.

1.2 Asset Management of Electric Power Grids

Since electrical energy started being transmitted and distributed through power networks, the electric power supply has changed modern society, but society in turn has also changed the electric power supply. Starting out as more or less a gadget offered by inventors, entrepreneurs and technical engineers, many countries nowadays have ministries of energy or economic affairs to develop strategies for energy supply. Political choices set the targets of the performance and conditions that utilities have to meet. The expectations have to be fulfilled through an infrastructure composed of many components and connections that form the networks or grids. The physical components are the tangible assets that together shape this infrastructure. The question is how to reach the requested performance and added values at the lowest price. Value and price do not necessarily concern merely financial or economic concepts, but can also involve other business values like safety, power quality, security of supply, and so on.

AM is the collective term for the structured decision-making and execution of plans to reach an optimized balance between performance, efforts and risk with the utilization of assets. This includes all aspects that are relevant to the performance of the grid, such as the strategy, operations, finance, regulations, information, value and more. An AM system is an organized set of systematic and coordinated activities to stay in control.

Standards have been developed for common practice in AM. The Institute of Asset Management (IAM), together with the British Standards Institute (BSI), developed the Publicly Available Specification 55 (PAS 55) [2]. This standard led to the ISO standards 55000, 55001 and 55002 entitled 'Asset Management — Overview, Principles and Terminology [55000]; Requirements [55001]; and Guidelines on the Application of ISO 55001 [55002]', which are generally applicable for the purpose of managing physical assets [3–5]. ISO 55000 defines an asset as 'an item, thing or entity that has potential or actual value to an organization. The value will vary between different organizations and their stakeholders, and can be tangible or intangible, financial or non-financial'. AM then translates the organization's objectives into asset-related decisions, plans and activities, using a risk-based approach. Effective AM is not just an activity, but rather an integrated system that must be implemented organization wide and combine all related disciplines including managerial, financial and technical in order to create optimum value during the complete asset lifecycles.

Considering all aspects there will be conflicting interests, which urges organizations to find optimum balance. For instance, keeping assets in service as long as feasible is a way to postpone investments, but also increases the risks associated with unplanned outage by failing assets. The optimum between postponing investments and preventing damage due to failures touches the interest of not only various organizational departments, but also of stakeholders including employees, clients, investors, legislation and even people passing by a substation (for safety). AM therefore requires a comprehensive system to provide an integral approach. The treatment of such a system is beyond the scope of the present book, but the matter is very well introduced by a free document provided by the IAM (see also 'Asset Management – An Anatomy' [6]). Various AM subgroups (Table 1.1) are distinguished. Most of these subgroups use data that is processed to provide information. This involves descriptive statistics as well as inferences to analyse scenarios and carry out forecasting that is used for planning.

Section 1.3 elaborates further on the essential aspects of AM, namely maintenance styles (part of 'Lifecycle delivery activities' in Table 1.1) in relation to failure behaviour

Table 1.1 AM subject groups (for a detailed description, see [6]).

	AM subject group	
1.	AM strategy and planning	Core activities to develop AM policy into plans. Based on organizational strategy, organization and people enablers and AM decision-making, giving input to the lifecycle delivery activities.
2.	AM decision-making	Framework for decision-making on the optimum blend of activities to achieve specific objectives. It is based on asset knowledge enablers as well as organization and people enablers. Balancing cost and value as well as strategies related to ageing assets are part of this subject.
3.	Lifecycle delivery activities	The lifecycle comprises acquisition, operation, maintenance and disposal of assets. It is based on AM strategy and planning, asset knowledge enablers as well as risk and review. Incident response is part of this subject.
4.	Asset knowledge enablers	Data, information (data in a context) and knowledge (experience, values, information, insight combined as basis for decision-making). The knowledge comprises a SWOT analysis (strengths, weaknesses, opportunities and threats) related to the assets. This subject also concerns what and how information is collected and used in the organization.
5.	Organization and people enablers	Concerns contract and supplier management, as well as leadership, structure, culture, competence and behaviour within the organization.
6.	Risk and review	Among others, assessment of criticality and risk management of performance monitoring and accounting.

versus preventive measures. In the light of these maintenance styles, asset health and risk are also discussed. The purpose is to provide a focus on the relationship between these concepts and statistical failure behaviour.

1.3 Maintenance Styles

Managing the grid assets requires decision-making about the whole lifecycle of assets. This starts with the design and specifications of the local grid and the way compliance is tested. It ends with disposal of the assets when operation of that part of the grid is no longer needed. In between, the assets are part of the operational grid. As the performance, condition and properties of the assets that build the grid may change over time, maintenance is required to keep the grid fit for purpose. There are various policies that can be followed when managing (systems of) assets to remain in operation. In the following sections, the maintenance activities are divided into:

- inspections
- servicing
- replacement.

Inspections are activities to check the functionality and other properties of the asset. In case of a transformer, this can be visual inspection of the housing to check for oil leakage, testing the quality of bushings, monitoring the properties of the oil, and so on. Servicing the asset concerns all preventive or corrective activities that aim at prolonging operational life and improving the condition of an asset, like replacing contacts in switchgear or cleaning insulator surfaces. Repair, refurbishment and overhaul also fall into this category. Replacement means ending the operation of one asset or subsystem and installing another asset or subsystem to take over the required functionality. The collective term of 'maintenance style' is used for the combination of inspections, servicing and replacement.

Maintenance styles are rooted in the AM system. AM decisions will generally balance various aspects. Examples of such aspects are listed in Table 1.2, where the aspects are grouped in the categories 'Licence to Operate', 'Resources' and 'Financial'. In reality, the

Table 1.2 Examples of aspects to be balanced with asset management.

Licence to operate	Resources	Financial
• Being in control • Safety • Security of supply • Power quality • Politics • Permits • Public relations • Image of social responsibility • In compliance with law • ...	• Reliability and availability of grid assets • Access to expertise • Availability of qualified and skilled personnel and contractors • In-time delivery of services and parts • Stock of spare parts • Local circumstances • ...	• Capital expenses (new equipment) • Operational expenses (maintenance costs, etc.) • Compensation (damage, non-delivery, right of way) • Hiring personnel • Income through tariffs • Public investment • ...

AM systems embrace many more aspects and categories, but Table 1.2 illustrates the variety of aspects that are balanced with AM. For instance, the utility earns its licence to operate the grid by adequate performance, which includes securing the power supply and power quality (a set of characteristics of the voltage like amplitude and frequency stability, low disturbance in the form of harmonics and transients, etc.) in a safe and affordable way.

In order to earn this licence, the utility needs access to resources that enable the construction, operation and control of the grid. Resource aspects concern reliability and availability, including personnel, parts and market partners like manufacturers, contractors, and so on. It may require that reliable equipment can be bought from at least two different manufacturers in order to secure the supply of products in case one manufacturer faces problems in production. Power supply is important, but must also be financially balanced. Financial expenses are generally divided into capital expenses (CAPEX) and operational expenses (OPEX). CAPEX involves all investments in the construction of new and/or replacement assets; OPEX refers to all costs of operating assets such as inspections and servicing.

Historically, the maintenance style developed with the role of electric power in society, characteristics of assets, political views, technological breakthroughs, environmental circumstances and the role of information technology, to mention a few. The role of electric power developed from a luxury article replacing candles to a commodity article that significantly improved health and working conditions (cf. [7–9]). The importance of reliability and availability of the electric power supply shifted from it being a luxury item to basically a necessity of life. Whatever the degree of dependence, the power system must fulfil its task. This is the role of AM, and the AM boundary conditions will determine what maintenance styles are preferred.

Though many variations in maintenance styles exist, four main styles can be distinguished, as described in the following sections:

1. corrective maintenance (CM)
2. period-based maintenance (PBM)
3. condition-based maintenance (CBM)
4. risk-based maintenance (RBM).

For each style, implications are discussed with respect to inspections, servicing and replacement, and when this style is typically applied. The term 'hazard rate' is used in the discussion on maintenance styles to express the danger that an asset fails. For the moment, this qualitative concept does not need to be explained further. However, Chapter 2 discusses the hazard rate as one of the basic statistical functions (see Section 2.4.5).

1.3.1 Corrective Maintenance

CM aims at mitigating situations with defective assets (i.e. failed, insufficient functioning or unacceptably damaged). No preventive action is carried out until the moment the asset becomes defective. Another name for this style is run-to-fail. This may look like a rather careless type of maintenance, but there can be very good reasons to apply this style. Figure 1.1 shows the concept of CM.

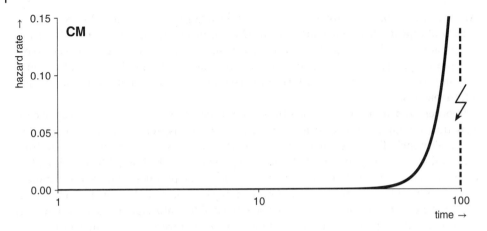

Figure 1.1 The concept of CM: the asset is left to operate without preventive maintenance. Repair or replacement is carried out after failure.

A familiar example of domestic CM is generally applied with lighting. Lamps at home will be used without maintenance until failure and are then replaced.

1.3.1.1 CM Inspections

If inspections are carried out, this is mainly to check whether the asset is still functioning well. It may also include determining whether the local environment is suitable. For example, it may be concluded that vegetation should be cut if it threatens the asset. If an asset has a task to fulfil in one way or another, it must be considered how it will become apparent that corrections must be carried out (i.e. when the asset has stopped working).

1.3.1.2 CM Servicing

CM avoids servicing. If the asset under consideration is not able to fulfil its task, it must be repaired or replaced. Therefore, CM restricts maintenance to corrective actions after an asset fails. Usually it is evaluated whether repair or replacement of the asset is preferred.

1.3.1.3 CM Replacement

Replacement means removing the old asset and installing a new one. The CM maintenance style will probably influence the specification of the asset. For instance, if an asset will be employed in a remote or otherwise hard-to-access place where servicing requires unreasonable efforts, then CM may be the only realistic option. In order to optimize the performance, the specification of the asset may prescribe a long life without servicing. In contrast, another maintenance style may be preferred over CM in accessible locations where convenient servicing procedures exist. The serviceability would then most likely appear in the specifications.

1.3.1.4 Evaluation of CM

CM is particularly applicable in situations where:

(a) OPEX is considerably larger than CAPEX (i.e. it is significantly cheaper to replace than carry out preventive maintenance). Large parts of cable networks are managed by this style, although monitoring and diagnostic systems are increasingly employed.

(b) Maintenance initiates failure behaviour. In such cases the relevant phrase is 'as long as it works, don't touch it'. A typical example used to be generators, which had a bigger chance of failing after maintenance and recommissioning than when they ran without human interference. This changed when sensors were developed that could indicate the adequate moment for maintenance (CM then turns into CBM, see Section 1.3.3).

(c) The asset life cannot be extended by maintenance, because of its properties. A lamp is a typical example; with most lamps there is very little that can be done to improve their condition by servicing (other than sometimes, e.g. if dirt can be removed to avoid overheating).

(d) The asset cannot be maintained because it is out of reach, which may apply to remote unmanned stations, satellites in space, assets in harsh environments, and so on.

CM is sensible in various situations, but it needs to be checked whether situations allow CM and indeed, if not, whether the situation can be adapted to become suitable. For instance, it may be unacceptable that a complete system goes down after failure of a single asset in the system. A solution to that situation is to introduce redundancy (see Section 7.4) and emergency plans. Examples are double lights on a car, quick repair (e.g. due to quick access to spare parts – lights in a house, cable splices and terminations in a storage facility; see also Section 8.5) or providing an alternative like allowing a manual override if a control system itself is subject to CM.

An important issue mentioned above is whether a failed asset is detected in a timely manner. An insulation failure results in a breakdown which is usually readily observed, but if the contacts of a circuit breaker or disconnector are welded together on closing the contacts, malfunction may only appear at the moment that the circuit breaker needs to break a current but is unable to do so. Again, redundancy is the general solution to that, which translates into switching equipment in series with circuit breakers.

Overall, CM has the advantage of low OPEX and taking advantage of the full technical lifetime of the asset. But it is not always true that the expenses of CM mainly consist of CAPEX. A failure or interruption of service may also be solved by refurbishment. Keys to successful CM are timely detection of malfunction, fast response, flexible workforce and robustness of the grid after asset failure and/or matching asset lifetimes with system lifetimes (e.g. with non-serviceable systems).

1.3.2 Period-Based Maintenance

PBM is also referred to as 'time-based maintenance', 'use-based maintenance' or 'periodic maintenance'. This style aims at preventing asset failure by timely preventive actions with fixed intervals (periods) in between (see Figure 1.2). These intervals can be periods in terms of calendar times, but can also be operational hours, number of switching operations, and so on. So, 'period' does not necessarily refer to time. For that reason, the term 'period-based maintenance' is preferred over 'time-based maintenance'. Typical examples are routine visual inspections at a substation (calendar periods), replacement of turbine blades after fixed running hours (usage or operational hours), servicing switchgear after a fixed number of switching actions, or maintaining

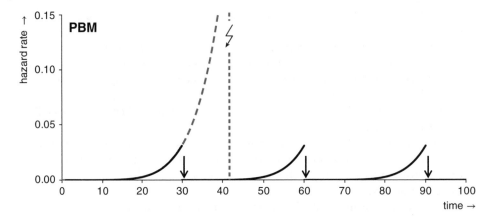

Figure 1.2 The concept of PBM: the asset is periodically maintained to reduce the hazard of failure expressed in terms of the hazard rate (Section 2.4.5). The dotted line shows the development without PBM and the solid lines show maintenance with PBM to extend the operational life of the asset. *Note*: In practice, the reset of the hazard rate is not perfect (cf. Section 9.1.1.4).

a car after a prescribed number of driven kilometres/miles. Traditionally, this style has been (and still is) used in grids to keep the system in optimum technical condition. Because of its high costs, however, it is increasingly replaced by CBM (see e.g. Sections 1.3.3 and 9.1.1).

1.3.2.1 PBM Inspections
Inspections with PBM are according to a plan with fixed periods in between, irrespective of the asset condition. As mentioned before, these periods can be defined in terms of calendar time, but also running hours, number of rotations, switching actions or lightning strikes, and so on. Various inspection cycles may be employed simultaneously (e.g. visual inspections at a higher frequency and more labour-intensive deeper investigations at a lower frequency). Because the planning will be comprehensive, there is a high probability that all assets are inspected in due time and a good, up-to-date overview of the asset states can be available at any moment. Inherent to PBM is that it tends to be labour-intensive and most inspections are carried out without the need for follow-up actions.

1.3.2.2 PBM Servicing
Servicing with PBM involves planned cleaning, overhaul or refurbishment. Whereas CM acts on asset failure, PBM plans to prevent these failures. This means carrying out servicing before failure might occur. Overall, a grid that is maintained by the PBM style tends to be in very good technical shape at relatively high cost. The advantage is that the grid suffers fewer failures, faces fewer emergencies and becomes more predictable. The disadvantage is that equipment may be serviced too often, equipment may be more frequently planned out of service and resources may be unnecessarily spent. Some assets may be vulnerable to excessive servicing, with the danger of introducing child mortality (i.e. unexpectedly early failures) into the equipment caused by maintenance, which produces the reverse of what is aimed for. However, these disadvantages may be overcome by employing a larger period/lower frequency for PBM.

1.3.2.3 PBM Replacement

Applied to replacement, PBM means replacing assets after a fixed period, which could be calendar time, operational time or other measures of deployment. The advantages of PBM are high reliability of the assets in use due to timely replacement, having time to wait for the right price to acquire assets and receiving assets in time. The disadvantage is that valuable remaining life of the assets is sacrificed (cf. Section 9.1.3). In some particular cases, child mortality may appear in the new assets, which does not increase but rather decreases grid reliability. Another concern is that frequent replacement means taking parts of the grid temporarily out of operation, which limits the operational flexibility in the grid. Although PBM may give insight into the development of the early stages of ageing, it also prevents the build-up of experience with actual asset lifetimes because every asset is replaced before failing. This effect can be reduced by analysis of removed assets to evaluate the likelihood that the asset would have failed in the near future or by testing assets on remaining life after taking them out of the grid.

1.3.2.4 Evaluation of PBM

PBM requires a planning, register and progress monitoring system. It will have relatively high OPEX due to frequent preventive actions, irrespective of the asset condition, and also relatively high CAPEX because of early replacements to keep the failure probability low. On the other hand, the periodic servicing may extend the operational life of assets.

Situations where PBM is particularly applicable are where:

(a) Assets are of such strategic value that relatively high OPEX and high CAPEX are justified.
(b) OPEX is a worthwhile investment compared to CAPEX, when servicing is known to prolong operational life and/or is more cost-effective than a higher level of redundancy. So, the depreciation of the assets is lower (e.g. regular visual inspection of a (sub)station is a fairly cheap way to detect problems such as oil leaking out of transformer coils or surface deterioration of insulators). Being able to detect problems at an early stage enables timely plan servicing. This helps to prevent early failure and replacement.
(c) The ability to plan maintenance and replacement provides significant profits compared to the required flexibility for ad-hoc mitigation with CM.

PBM is particularly favoured in a technology-driven environment and where uninterrupted performance is required. However, the assets must be suitable for periodic check-ups. If inspections require the asset to be temporarily taken out of operation, the system as a whole must allow that interruption. Again, redundancy may enable that situation.

Overall, PBM is usually associated with a high degree of reliability, but also with relatively high OPEX for frequent inspections and maintenance and high CAPEX for timely (early) asset replacement. Keys to successful PBM are assets with a predictable lifetime, readily observable quality, good planning and access for inspections, servicing and replacement.

1.3.3 Condition-Based Maintenance

CBM or also reliability-based maintenance aims at a compromise between CM and PBM by using smart technologies and knowledge rules. The concept is that preventive

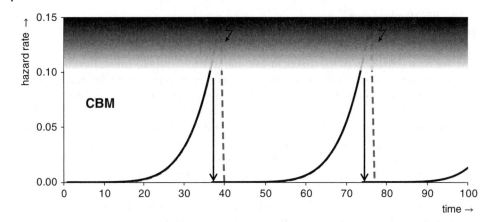

Figure 1.3 The concept of CBM [11]: the hazard of failure can be detected by diagnostics and expert rules on remaining life. This is depicted by the dark band. The dotted line indicates that a failure would occur if no preventive action is taken. The arrows show that maintenance or replacement reduces the hazard rate, which is the essence of CBM.

actions are avoided until the final stages of ageing reveal themselves through a diagnostic technique, in combination with adequate failure forecast of remaining life based on knowledge rules. CBM leans heavily on the ability to detect the early stages of failure or a (side) effect that correlates well with the hazard of failure (see Figure 1.3), as well as on the correct interpretation of the observations by expert rules that translate the observations into an adequate diagnosis. With proper diagnostics and expert rules, CBM aims at timely interventions when the hazard of failure becomes significant. Expert rules are also called knowledge rules.

For instance, cables and cable terminations may break down due to electrical treeing that precedes the actual breakdown. Depending on the specific situation, like materials and circumstances, this electrical ageing process develops in weeks up to even years and is driven by partial discharge (a kind of sparking within the insulation without being a full breakdown). These partial discharges can be detected by sensors and signal processing systems, that may even allow the partial discharge location to be pinpointed. A condition monitoring system (CMS) may be in place to monitor the occurrence of partial discharge. If experience is built up about the correlation of partial discharge and remaining life, partial discharge monitoring can generate alerts for imminent breakdown. Preventive measures can then be taken in time [10]. This is a typical CBM approach.

The style gained importance after utilities were forced to operate at lower cost. Cutting back on PBM was a first reflex, but generally the lack of maintenance did not meet the expected cost reductions and – even worse – jeopardized the security of supply. As mentioned above, PBM is planned irrespective of the asset condition, but that does not mean that all maintenance can be reduced without damage. The efficiency of maintenance could be improved if inspections and servicing could target assets that require attention, while leaving healthy assets unattended.

Investments in cost-effective, smart technologies became essential. The success of CBM was partly due to new sensitive detectors, but also due to knowledge rules that effectively translated observations and expert opinions into meaningful alerts.

A domestic example is a smoke sensor that reveals the early stages of fire. Another example is inflating bicycle tyres after they appear too soft and the hazard of damaging the tyre by riding the bike is deemed unacceptable (or too uncomfortable).

1.3.3.1 CBM Inspections

One of the first achievements of CBM was to reduce the number of periodic inspections by using inspection results to forecast whether the next planned inspection might or might not be skipped. For instance, if insulators are found to be in a very good state on inspection and the local circumstances are mild to the equipment, it may be practically certain from past experience that the next inspection also (as planned with PBM) would not urge the crew to take any corrective actions. So those inspections might be skipped as well. In the line of CBM the frequency of inspections can be reduced, which reduces OPEX without increasing CAPEX. If, on the other hand, the asset does not require immediate attention, but the condition seems to deteriorate, then the next inspection should be carried out as planned or even earlier. Such smart planning and alerts require expertise to be laid down in knowledge rules. With the automated processing of inspection findings, development of sensors and smart monitoring techniques, the inspection frequency can thus be optimized. Of course, there has to be a sound balance between investing in CMSs and reduced inspection costs.

1.3.3.2 CBM Servicing

CBM servicing involves methods to determine the condition and then be able to timely plan overhaul and repair the asset (just) before it fails. An alert that an asset is reaching possibly dangerous levels of degradation can be obtained from diagnostic programs or inspections (by CBM or PBM). Often inspections (even with PBM) contain elements of CBM, because if unacceptable situations are observed, then also with PBM such situations are normally mitigated immediately. The CBM style responds to triggers that call for servicing just-in-time.

1.3.3.3 CBM Replacement

Condition-based replacement aims at deploying assets until their last stages of operational lifetime. In that way asset depreciation is minimized and most benefit is taken from CAPEX. The definition of end of life may not just be time until technical failure, but alternatively the time until end of economic life. For instance, once the rate of inspection and servicing costs exceed the depreciation of new equipment, the conclusion can be that economic life is over and it is more cost-efficient to replace the asset than to keep it in operation. This interpretation of condition and operational life tends towards RBM, but stays close to the focus of managing the condition of assets and could therefore be regarded as CBM.

1.3.3.4 Introduction to the Health Index

A method that enables a convenient overview of the condition of the assets in a grid is provided by the so-called health index (HI). It is particularly helpful in evaluating populations of assets and can be extended towards groups of different assets such as connections consisting of linked assets and (sub)grids. The concept rates the condition of each asset in terms of the HI. It is an indicator for the urgency to service or replace an asset. The HI concept is presented in Figure 1.4 (cf. [12]) and explained below.

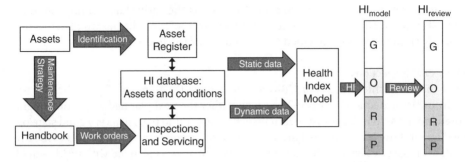

Figure 1.4 The concept of the health index [12]. Based on the asset type (static data) and observations (dynamic data), a HI model calculates the HI_{model}, which may be checked and adjusted to yield the HI_{review}. The cell heights in the columns (right side of the flow chart) reflect a percentage of the asset population. The codes in the cells mean: G = green (OK, no action required); O = orange (fair, but action required to return to G); R = red (suspect, investigation required, plan for repair or replacement); P = purple (urgent attention required).

Assets with their identification (ID) and characteristics like location, year of installation, and so on are stored in an asset register. These are the so-called static data. The assets are subject to a certain maintenance strategy, which is described in a handbook (which can also be an electronic document or interactive website) and induces orders for inspections (condition assessment) and servicing. The results from the condition assessment and servicing are observations which are translated into condition indicators (CIs). These findings are the so-called dynamic data. The static and dynamic data are stored in a HI database and used as input to the HI model.

This model consists of knowledge rules that are used to assign a HI_{model} based on the static and dynamic data. The HI_{model} reflects the probability that the asset will fail in a defined upcoming period. This HI_{model} is generally a score on a certain scale (which can be 1–10 or 1–100, etc.). This mark is often also translated into some colour code (as in Figure 1.4). Four colours may be used, but three or two are also practiced. Typically, the colour scheme follows the traffic-light convention, possibly extended with an additional colour (green = good; orange/yellow = fair, but action required and possible to return to green; red = suspect, plan for replacement or refurbishment; and purple = urgent attention required). A group of assets will have a certain percentage of each colour, which gives the HI_{model} colour distribution shown in Figure 1.4. As the HI_{model} is calculated by a model and the actual situation may be somewhat different, the results should be checked, which leads to a HI_{review}. Ideally, the HI_{model} and HI_{review} will be the same, but in the stage where the knowledge rules still need to be developed there may be significant differences. This is a learning cycle that should improve diagnostics, knowledge rules and the HI model, and possibly lead to considering other CIs. Even after many years, issues with new information may come up that urge updates of the HI method. Some scatter will usually also occur due to differences in assets, observations and circumstances.

The HI_{review} of the individual assets is input to planning of servicing or replacement. The collective HI_{review} gives insight into the condition of the total asset base in the grid and can be used to plan maintenance and replacement of assets. The HI of assets that together form a connection or even (sub)grid can be integrated into a combined health index (CHI), which can be used to evaluate the state for those integrated systems.

After discussing statistical techniques to evaluate the state of assets and grids, the HI is discussed in greater depth in Section 9.2 and further.

1.3.3.5 Evaluation of CBM

CBM assumes the ability to fully assess the condition and adequately alert for imminent failure. This full assessment is of course an ideal situation, but in an imperfect reality, CBM can still be very effective. CBM is built on diagnostic methods and knowledge rules that provide an estimate of failure moments based on the observed and perceived condition of the asset. The quality of the forecasts leans heavily on the quality of the observations and of course on the validity of the knowledge rules. These are the greatest challenges within CBM.

Situations where CBM is applicable:

(a) On the one hand, assets are of such value and/or failure has such an impact that CM is not acceptable but on the other hand, effective PBM is too expensive. CBM is a good compromise granted that condition assessment is possible, reliable and sufficiently cost-effective.
(b) Condition assessment techniques are feasible and cost-effective.
(c) The organization is able to work with a flexible planning, keep an overview of the system and has access to both human and material resources for timely servicing and replacement.

CBM is particularly favoured in an advanced technological environment where automated systems can process monitoring data and planning can be adjusted according to the needs of the moment and near future. Besides adequate diagnostics and knowledge rules, key to successful CBM is organizational flexibility to respond to alerts in a timely manner. Of the three discussed maintenance styles, CBM requires the most expertise on signals that correlate with failure mechanisms.

1.3.4 Risk-Based Maintenance

The previous three maintenance styles – CM, PBM and CBM – focused on the assets and their condition. RBM takes into account that events (e.g. failures) impose risks to many more aspects than just technical functionality. In daily conversation the term 'risk' is often used for the probability that a certain danger will become reality. However, risk has another meaning in RBM. Risk comprehends both the likelihood of an event occurring as well as the impact of the event with respect to business values like safety and security of supply. Risk is therefore often defined as the product of probability and impact or effect:

$$risk = probability \times impact \tag{1.1}$$

A similar definition is:

$$risk = occurrence\ rate \times consequence \tag{1.2}$$

Whether risk should be defined exactly as the mathematical products in Eq. (1.1) or (1.2) is debatable, nevertheless AM systems often employ a so-called risk index (RI) as a measure of risk for prioritization that scales with both probability and effect on both occurrence rate and consequence.

For example, some cable terminations may build up pressure and subsequently fail with a blast. If such a termination is next to a place where people reside, the problem of an operational interruption after a blast may be secondary to the safety issues involved. The failure of the very same type of termination in a remote or shielded area will cause the same operational nuisance, but safety may not be an issue in that case. The failure probability is the same, but the risk with respect to safety is much higher in the first case.

The concept of RBM is that the mitigation of situations with a higher risk level gets a higher priority. It may mean that a high risk urges an asset to be taken out of service earlier than would be done with CBM. In that case the loss of operational life is weighed against the violation of other values. On the other hand, if a local grid is sufficiently redundant (so a single failure will not interrupt energy supply) and other consequences of failure are acceptably small, then mitigation urgency is lacking and RBM may lead to postponing replacement until a convenient moment for action is found. RBM aims at optimizing the total performance with respect to the various corporate business values (CBVs).

The difference between CBM and RBM is that CBM focuses on the integrity of the asset and RBM focuses on the performance of the utility in all its aspects (see Figure 1.5).

1.3.4.1 Corporate Business Values and the Risk Matrix

RBM evaluates the risk of events by combining the occurrence frequency and the impact with respect to business values. It depends on the utility with which values will be involved in risk assessments and how each of the risks is weighed.

Some commonly used CBVs are:

(a) performance
(b) safety
(c) finances
(d) reputation

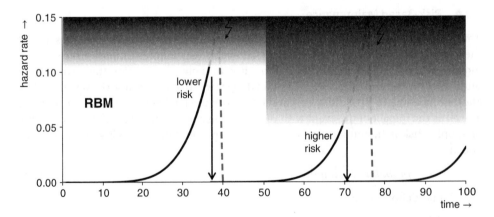

Figure 1.5 With RBM not the asset condition, but the risk in terms of corporate business values determines the priority of mitigation: the higher the risk of violating corporate business values, the lower the hazard rate that triggers treatment of an asset.

(e) customer satisfaction
(f) environment
(g) compliance
(h) social responsibility
(i) …

These CBVs will normally be defined quantitatively by a reference level. For instance, for CBV 'safety' the standard might be: 'nobody gets hurt'. If an incident occurs where someone gets bruised or his arm broken or worse, then the CBV is violated. The levels of violation are normally categorized according to their impact gravity.

In case of an undesired event, one or more CBVs may be violated. For instance, if a termination or a bushing fails with an explosion the value 'performance' of the grid is locally jeopardized, but the blast and the bits that are propelled from the termination, as well as the possible fire following the failure, will also endanger people in the near vicinity of the termination. So, the CBV 'safety' can also be violated. Another example is oil leaking from a transformer or oil-pressure cable, where the CBV 'environment' can be violated.

With RBM risk analysis is used for prioritizing the mitigation of issues. There are various challenges to overcome. One challenge is how to evaluate the risk based on the likelihood of an event and the impact it imposes with multiple risks. Another challenge is how to compare the violation levels of different CBVs.

A traditional way is to translate impacts into money, which may lead to the awkward duty of estimating the financial value of human life, of pricing operating within the law versus damage to the environment. From the viewpoint of simplicity, it is appealing to just calculate the cost of damage compensation, but the CBVs may be valued much higher than financial compensation alone. The frequency of CBV violation can be established through statistical inference, as will be shown in the coming chapters. Since inflation and deflation influence the value of money, the option of translating CBVs into money must be continuously adapted with the financial markets. However, even financial impacts themselves may be weighed differently. For instance, the same amount in CAPEX, penalties to be paid, interest, asset book value or loss of income may have a very different weight. Probably formulas can be drafted that overcome such differences, but in the end it may be concluded that the financial balance is not such a convenient value after all.

An alternative to translating CBVs into only money is provided by a risk plane or more popular, in its discrete form, the so-called risk matrix (RM). This is illustrated in Figures 1.6 and 1.7. Here shades of grey are used, but normally colours indicate the acceptability of the risk (which inspired the cover design of this book). Those colours do not relate directly to the colour scheme of the HI (see Section 1.3.3.4). The idea of a risk plane is to scale the impact with respect to CBVs by a number that can be translated into terms of 'minor' to 'extreme' impact. In this way most of the CBVs can also be evaluated regardless of the cost. CBVs with a financial impact can be adjusted to the corporate financial strength independent of non-financial CBVs like safety. Again, the frequency of CBV violation can be established through statistical inference. The combination of frequency and impact leads to two-dimensional coordinates in a risk plane, as in Figure 1.6.

Usually not a scale but a range of impact categories is designed through interviews with the stakeholders, including the management board. Also, the frequency may be

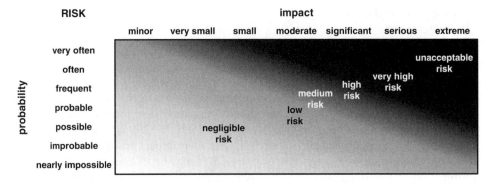

Figure 1.6 Risk is a combination of the quantitative occurrence and impact of the event. The risk for an evaluated business value has coordinates in this risk plane. A strategy can be to mitigate all risks from and above a certain level (e.g. the medium risk level).

RISK	impact						
	minor	very small	small	moderate	significant	serious	extreme
very often	low	medium	high	very high	unacceptable	unacceptable	unacceptable
often	negligible	low	medium	high	very high	unacceptable	unacceptable
frequent	negligible	negligible	low	medium	high	very high	unacceptable
probable	negligible	negligible	negligible	low	medium	high	very high
possible	negligible	negligible	negligible	negligible	low	medium	high
improbable	negligible	negligible	negligible	negligible	negligible	low	medium
nearly impossible	negligible	negligible	negligible	negligible	negligible	negligible	low

probability

Figure 1.7 Risk is often discretely categorized, which yields the risk matrix. The risk for each evaluated CBV is assigned a cell in this risk matrix [12].

categorized. This leads to the discrete representation of the evaluation results in a risk matrix (see Figure 1.7). With the discretization some information gets lost, but it may be easier to harmonize risk evaluations of different inspectors and reduce scatter.

The HI may not be a good indicator for value-violating frequencies. The frequency of an event (like failure with an explosion) is not necessarily the frequency that belongs to a certain risk (like putting people in danger). For instance, if at some point a termination is estimated to fail once per 5 years, the frequency for interrupted power supply will be 0.2 per year (in case of no redundancy). If the termination is properly shielded, or people rarely come into the vicinity, say 1% of operational time, the estimated frequency of human exposure to a termination explosion may then be 1% of 0.2 per year, so 0.002 per year (the assumption here is that the explosion and the presence of people are independent). Following that line, the frequency for CBV safety is also one-hundredth of the frequency for performance. So, in this case the same event yields different frequencies for different CBVs.

For each CBV the risk evaluation leads to coordinates in the risk plane or a cell in the risk matrix. The risk plane and risk matrix replace Eqs (1.1) and (1.2). One method is to use the equations for each business value and then take the highest as the criterion for mandatory mitigation. After a positive decision to add the issue to the portfolio of issues (the 'to-do list'), normally still a single risk index like the weighed risk is constructed to allow an unambiguous priority ranking.

Another method is to first weigh each CBV risk, compose a weighed risk index (cf. Section 9.1.4.2) and then apply a criterion to add the issue to the portfolio or not. The difference between the two methods is that in the first case each single CBV can trigger action, while in the second case a good or bad score on certain CBVs may reduce respectively advance the mitigation on the other CBVs. If all CBVs are weighed equally, the above listed CBVs would each add 12.5% to the risk index, but usually different CBV weighing factors are used in priority ranking for risk mitigation (see also Section 9.1.4).

One may wonder why the weighing is applied while the risk matrix might be adapted as well. However, the risk matrix is used to qualify events for the mitigation portfolio, while the weighed risk index is used for prioritizing only the issues that made it to the portfolio.

1.3.4.2 RBM Inspections

RBM applied to inspections means that the inspection regime is tuned to the identified risks. For example, not only the asset condition may be inspected, but also whether the surroundings changed such that the risk increased. For example, inflammable products may be placed in the near vicinity of the assets, which could increase the impact in case of a failure. Particularly the frequency of inspections or monitoring of specific aspects may be intensified if the circumstances require so. RBM may or may not lead to lower OPEX than CBM, but the aim is primarily to prevent violations of the utility CBVs. It requires not only the smart planning of CBM, but also an evaluation method to estimate the impact on CBVs. As this is applied to all assets, many risk evaluations have to take place and a practical rather than a very precise risk evaluation system may be developed. For example, a geographical information system may be used to evaluate whether assets are close to a populated area that would make the asset more likely to be a safety risk. Assets in an environmentally sensitive area like a nature park, in a populated area or of extraordinary strategic value may be inspected more frequently with RBM.

1.3.4.3 RBM Servicing

RBM servicing, similar to RBM inspections, is based on estimated risks, which requires knowledge of both condition and impact with respect to the CBVs. Depending on the estimated risks and anticipated scenarios, the preferred option can also be to replace rather than carrying out maintenance. Similar to risk-based inspections, assets in an environmentally sensitive area like a nature park, in a highly populated area or of extraordinary strategic value may be serviced more frequently because of the RBM policy.

1.3.4.4 RBM Replacement

RBM replacement aims at replacing assets when the risk in terms of one or more of the CBVs becomes too high and preserving the CBVs can be directly reflected in the investment plan. The overall performance of the utility is thus optimized in this way. Again, assets in an environmentally sensitive area like a nature park, in a populated area or of extraordinary strategic value may be replaced earlier with RBM.

1.3.4.5 Evaluation of RBM

RBM not only assumes the ability to assess the condition and to forecast imminent failure like CBM, but also assumes the ability to assess the impact that asset failure has on CBVs. It requires more complicated evaluations, but the overall performance of the utility may benefit from this approach.

Situations where RBM would typically be used are where:

- The focus is on utility performance measured in CBVs rather than only asset survivability. For establishing the likelihood of failure events, the prerequisites for CBM have to be fulfilled.
- RBM may be selectively applied to replacement and investments while any other maintenance style might be applied to inspections and/or servicing.

RBM is particularly favoured in an advanced technological environment and setting where utilities are evaluated for their performance in a broader sense. RBM may resemble CBM for forecasting events, but it aims at better tempering the adverse effects where the stakes are higher. Keys to successful RBM are diagnostics, knowledge rules that can adequately forecast imminent failure, organizational flexibility to respond to alerts in a timely manner, methods to measure CBVs are in place and the ability exists to quantify the risks. Of the discussed maintenance styles, RBM has the broadest scope.

1.3.5 Comparison of Maintenance Styles

The maintenance styles discussed in previous sections are summarized in Table 1.3. The first rows show the differences in strategy, focus and consideration that lead to the respective maintenance style. In short, CM in its most extreme form leaves the asset unattended until failure. This style employs assets to the very last moment. PBM aims at a prescheduled lifecycle, which makes this style the most predictive in CAPEX as long as the scheduled lifetime is shorter than the technical lifetime. However, the technical life may be prolonged with dedicated servicing efforts. The periods between servicing should be optimized. CBM aims at keeping track of the asset's condition. This alternative to PBM usually employs diagnostics and knowledge rules that must then adequately forecast imminent failures. The focus of CBM is on the asset. RBM primarily aims at the best performance in terms of CBVs, rather than focusing on the asset alone. Of course, failure ends operation of each asset and RBM will most probably never produce a longer lifetime than CBM, but it may prevent the worst violations of CBVs from happening.

The two following rows in Table 1.3 show typical asset features and action. The maintenance styles are split into inspection, servicing and replacement. Inspection consists of checks, servicing consists of preventive measures or refurbishments, and replacement consists of removing the existing (possibly failed) asset and installing another one instead.

The answer to the question of which maintenance style is the most cost-effective very much depends on the balance between various aspects like CAPEX, OPEX, required operational life, predictive value of condition monitoring and the weight of CBVs (safety, compliance, etc.). Each maintenance style can be the best option depending on the circumstances.

In many cases the maintenance styles are not used in their purest form, but rather blended. For instance, inspection may follow a PBM regime for ease of planning, servicing may follow a CBM regime, while replacement may follow an RBM regime. If

Table 1.3 Characteristics of maintenance styles.

Maintenance style aspect	CM	PBM	CBM	RBM
Strategy	Run to fail	Periodic preventive action	Timely action based on asset condition	Timely action based on CBVs
Focus	Maintenance-free	Scheduled plan	Asset condition	Utility performance
Consideration	Solving failure outvalues preventive actions	Preventive action outvalues solving failure	Condition monitoring outvalues periodic preventive and corrective action	Utility performance based on managing risks in terms of CBVs
Typical asset feature	Non-serviceable or disposable	Cost-effective servicing or very strategic	Able to be monitored	Able to be monitored and fail impact to be estimated
Typical action	Inspection and/or maintenance impossible or unwanted	Cost-effective preventive servicing	Cost-effective condition monitoring	Measures to control mixed and weighed risks
Inspection	Ability to detect failure	Periodic as scheduled	Indicated by perceived condition	Indicated by risk in terms of CBVs
Servicing	Not preventive, possible alternative to replacement after failure	Periodic as scheduled	Indicated by perceived condition	Indicated by risk in terms of CBVs
Replacement	After detected failure	After scheduled operational life; before significant failure probability	Indicated by perceived condition	Indicated by risk in terms of CBVs
Typical needed statistical information	Average lifetime Warranted minimum lifetime CAPEX per lifetime Asset operational availability	See CM + OPEX per action Optimum cycle	See CM + OPEX per action Success rate of monitoring Predictive window of monitoring	See CBM + Utility performance Likelihood and estimated impact

maintenance costs are relatively high and the risks associated with failure are low, CM may still be a good option in the early stages of operational life, while replacement may still follow an RBM regime.

The last row in Table 1.3 shows typical information that is used for decision-making and evaluation of AM performance.

1.4   Incident Management

Maintenance styles aim at managing assets in normal operation. Asset managers should, however, be prepared for the fact that usually unusual things happen as well. This is IM. Of particular interest is the evaluation of the extent to which an incident can develop into an emergency situation.

Although IM has various aspects in common with the CM style, there are differences. CM can be a deliberate choice based on a balance between saving OPEX due to not servicing the asset and willingly increasing CAPEX due to earlier replacement, since no effort is undertaken to prolong operational life. Also with CM there will be an expectation of average operational life after which replacement is necessary and with the choice for CM this expectation is apparently acceptable or preferable over other maintenance styles.

However, if assets fail considerably faster than expected, then not only corrective action is necessary. The asset manager may most likely have to evaluate whether the incident is unfortunate, but not worrying on the one hand or on the other hand the situation requires adequate measures to prevent (or contain) an emergency situation and eliminate the possibility of similar failures in the future. One possible reason for early failures is that assets suffer from errors in production and/or installation. Another possible reason is that the operational conditions differ from the specification – like a salty environment that appears more harmful to the assets than anticipated. Repair or replacement will probably not improve the situation if the new assets suffer the same problems or if the condition problems are not solved.

The following issues may typically call for IM:

(a) *Early failures*. If a group of assets is commissioned, a few unexpected early failures may occur. The question is raised of whether or not these failures are statistically unavoidable and should be accepted. The answer may be challenging because at that stage there are too few events to draw firm conclusions. Forensic studies, in combination with unbiased statistics, may be applied to take optimum benefit from the small data set to evaluate the likelihood of these early failures considering the specifications (see also Section 9.4 and following).

(b) *Non-compliance*. Early failures may still fall in an acceptable range, but their trend may suggest that the next failures will be too early and that the average lifetime will be considerably shorter than specified. A question is raised over the certainty with which a batch will not comply with the specifications, and when measures must be taken. Again, forensic studies and statistics will be employed in the evaluation (Section 3.6.5 and 3.7).

(c) *Repair-or-replacement decisions*. Failures may occur near the end of the specified minimum life. Many systems are repairable, but after several repairs asset replacement may be more appropriate. The questions is thus whether the latest failures are incidents that can be resolved with repair or whether a wave of failures can be expected, and replacement of the entire batch is appropriate (see also Section 9.4.2).

(d) *Adjustment of maintenance style*. Maintenance styles should be selected based on circumstances, but if the circumstances deviate from expectations, then the maintenance style may have to be adjusted. This may be the cause when failures encountered with CM could be prevented with minor preventive servicing indicated by

PBM or CBM. Another example is that servicing based on CBM assumes successful forecasting, but if this appears not possible, then a PBM style may be a better approach. Aspects of maintenance styles are also discussed in Section 9.1.

Statistical description, together with forensic studies, often provides insight into the actual situation and statistical inferences are often used for prognosis and decision-making.

1.5 Summary

Electrical energy systems started as luxury features to replace candles by electric light. Within a century, electric power supply gained a dominant position in modern societies. The role of electric power supply is still growing. Transmission and distribution networks consist of a large number of assets like transformers, cables that are integrated into subsystems like circuits and stations which are integrated into the networks that build the grid. Utilities are responsible for the performance of these networks.

AM is the collective term for the structured decision-making and execution of plans to reach an optimized balance between performance and efforts with the utilization of the assets. This includes all aspects that are relevant to the performance of the grid, such as the strategy, operations, finance, regulations, information, value and more. AM also determines the way the assets are maintained. The maintenance activities are subdivided into inspection, servicing and replacement.

Though many variations in maintenance style exist, four styles are described here: CM, PBM, CBM and RBM. Each aims at optimizing the balance between OPEX, CAPEX and performance including risk control.

CM aims at mitigating situations with defective assets while avoiding preventive inspections and servicing. The advantage is low OPEX, but since no effort is put into prolonging asset life, CAPEX may be higher. PBM is based on planned actions. Preventive activities take place whether or not the asset requires attention. The disadvantage is that OPEX is generally relatively high. Life may be prolonged by preventive actions, but asset replacement is also planned, whether or not operation could have been continued. CBM aims at both avoiding unnecessary preventive actions and still carrying out maintenance in a timely manner based on the asset condition. CBM leans heavily on the ability to detect the early stages of failure or a (side) effect that correlates well with the hazard of failure. CM, PBM and CBM all focus primarily on asset functionality.

RBM resembles CBM in assessing the likelihood of failure events, but additionally assesses the impact of events as violation of a set of CBVs (like safety and compliance with regulations). Risk is defined by a combination of event likelihood and impact on CBVs. Prioritization is based on the (usually weighed) risk index. RBM is not necessarily cheaper than the other maintenance styles, but aims at a better overall performance in terms of all CBVs involved. The focus of RBM is therefore on utility performance rather than merely technical performance of the assets.

Two methodologies in support of maintenance are the HI and the RM that may produce an RI. The HI enables a convenient overview of the condition of the assets in a grid. It rates the confidence in the asset health. A colour code reflects the HI and the urgency to undertake preventive or corrective actions. The RM is a plane or matrix that is useful to assess the risk for violation of CBVs. On the vertical axis the occurrence frequency is rated in line with the HI. On the horizontal axis the impact of a failure event on the CBVs is rated. The position in the RM determines the gravity of the risk. For instance, an event that will happen very often in combination with an extreme impact would be an unacceptable risk, while a nearly impossible hazard in combination with a minor impact is a negligible risk. Other combinations of occurrence frequency and impact would be risks that rate between negligible and unacceptable.

Apart from the organized AM and maintenance styles, unusual things usually happen as well. This is taken care of with IM. Both AM and IM require fact-finding and decision-making. Descriptive statistics is used in fact-finding and inferential statistics is used for exploring scenarios and making decisions. The coming chapters discuss methodologies for both types of statistic.

Questions

1.1 Categorize the following cases in maintenance activities (inspection, servicing, replacement) and styles (CM, PBM, CBM, RBM). Explain your choices.

a. New transformers are installed. The utility wishes not to rely merely on commissioning tests and decides to install equipment that detects hydrogen gas. This gas is associated with fast degradation. These systems are kept in place for 5 years and then removed. Up to 30 years of age, visual inspections and oil analysis is regularly carried out annually. After 30 years, equipment for on-line gas-in-oil analysis is installed on the transformers in order to detect various gases that signal imminent failure.

b. Transformers are installed in a remote area. Annually the transformers are checked visually and oil quality is measured. After 15 years the nearby city expands and the station becomes located in a newly developed urban area. Soon residents start to complain about transformer noise. The utility starts to monitor the acoustic noise from the transformer, which appears to be within legal limits. Additional research is carried out that points to the possibility of constructive interference (i.e. the houses are built in such a way that at some place the noise is amplified, which leads to complaints). Although compliant with the regulations, the utility wants to protect its reputation of social responsibility and takes measures to reduce noise and schedules installing new, silent transformers.

c. Cable terminations are components that can be built in a ceramic housing. Normally such components hardly require maintenance, but occasionally the oil level is checked. In theory these components can build up pressure and ultimately explode, propelling sharp porcelain fragments over a distance of tens of metres. After an actual incident with an exploding termination the utility decides to install screens near terminations to block possible fragments flying from exploding terminations. Furthermore, in residential areas and near roads, the utility starts to exchange such terminations by terminations with polymer housing, because these are safer due to less violent rupture and absence of sharp fragments. Later the rest of the terminations are exchanged as well.

1.2 Recommend maintenance strategies based on the provided information for the following assets and cases. If information is lacking, state a reasonable assumption and proceed.

a. A cable circuit consists of two accessible terminations (one at the beginning and one at the end), various lengths of direct buried underground cable that are connected by joints. The cable is planned to last for at least 40 years. Three failure breakdown modes are considered by their initiation:

   i. External damage mainly to cable and joints due to digging activities of third parties. Except for regulations for third parties to consult, there are no countermeasures taken.

   ii. Discharge breakdown (i.e. electrical insulation failure after various partial discharge phenomena, e.g. kinds of internal or external electric sparking but not yet a full breakdown); at the time of installation no on-line monitoring system had been developed for that voltage class. Incidental partial discharge measurements are possible at the cost of about 1% of the cable system price.

   iii. Thermal breakdown (i.e. electrical insulation failure after overheating of the insulation). This cable system contains a glass fibre sensor to monitor the temperature of the complete cable length with a resolution of typically a few metres. This can be used as a warning for imminent thermal breakdown, after which the load can be reduced in a time manner to prevent actual failure. The fibre may deteriorate after 20 years and cannot be replaced. Additional costs for installing a thermal monitoring system and communications amount to about 2% of the cable system, with 5% operating costs per year.

b. Surge arrestors capture overvoltages on the high-voltage systems (e.g. after lightning strikes). They are used amongst other countermeasures to protect transformers that typically cost a hundred times more than a surge arrestor and are regarded as strategic in the grid. The surge arrestors can withstand a certain number of strikes and transients. The arrestors do not have a remote monitoring system, but a counter is attached to keep track of the history. The counters occasionally appear not to work well.

c. Various types of high-voltage equipment at an open-air substation near the ocean coast are placed on insulators. These insulators are to prevent flash-over from the high-voltage conductors to earth and must therefore remain in good condition for about 30 years. However, sea salt may be deposited on the insulator surface and may lead to flash-overs after typically 2 years. After 10 years of operation there are two options to be considered: continue the annual cleaning of the insulators at a cost of 2% of the equipment or exchange the insulators for new insulators featuring self-cleaning surfaces. With annual cleaning there is still a 5% probability that equipment will fail. Doubling the cleaning frequency would reduce the failure probability to 1%. Exchange of the equipment would mean a depreciation of 20 years.

2

Basics of Statistics and Probability

The field of statistics is often divided into descriptive statistics and inferential statistics. Descriptive statistics concerns collecting and representing facts and interpretations. It usually results in an inventory to provide insight into the characteristics of a situation or population (e.g. an overview of the age of grid components, value of assets, condition of equipment, etc.). This is the type of statistics used in annual reports and performance dashboards. Descriptive statistics usually also explores correlations that may be postulated, investigated and tested in order to get a deeper understanding of the facts.

Inferential statistics aims at decision-making based on scenarios and probabilities (e.g. deciding between replace or repair after an incident in a cable, planning a replacement programme, setting priorities when resources are limited or weighing the odds that a system can be energized and relied on). This requires the ability to estimate the likelihood and impact of events that may happen and to make decisions that influence the course of events. Inferential statistics aims at forecasting and what scenarios to be prepared for.

Both types of statistics are used in asset management, but inferential statistics is often the biggest challenge, because it requires forecasting an unknown future and acting on it. Only the future may tell whether the decision was right, but inferences justify the decision based on the known facts and past experience. Factual data or lack thereof is generally evaluated by statistical methods. Each of the various maintenance styles described in Chapter 1 addresses its own questions on the design, operation, maintenance, repair and replacement of assets. The performance of each style can be evaluated statistically, whether it concerns the data, prognoses or testing of hypotheses. Whether it is about descriptive or inferential statistics, both build on the fundamentals of statistics, which are the subject of the present chapter.

First, fundamental concepts like outcomes, sample space and events are discussed (Section 2.1). Next, the concept of probability (Section 2.2) and the relation of and difference between probability and distribution are explained (Section 2.3). The fundamental appearances of a distribution – be it cumulative distribution F, reliability R, distribution density f, mass function f, hazard rate h or cumulative hazard function H – together with the bath tub model are presented (Section 2.4). Subsequently, mixed distributions either due to competing processes or inhomogeneous populations are addressed (Section 2.5). The discussion includes the conditional distribution according to the Bayes theorem (Section 2.5.2.1). In addition to the bath tub model for competing processes, Section 2.5.3 introduces the bath tub model for inhomogeneous asset populations with

Reliability Analysis for Asset Management of Electric Power Grids, First Edition. Robert Ross.
© 2019 John Wiley & Sons Ltd. Published 2019 by John Wiley & Sons Ltd.
Companion website: www.wiley.com/go/ross/reliabilityanalysis

an ill-performing fraction. Finally, multivariate distribution and accelerated ageing are discussed with the power law and the concept of ageing dose (Section 2.6).

2.1 Outcomes, Sample Space and Events

If a breakdown experiment is conducted, an impedance measurement performed, lifetime data collected or any other observations made, there will be outcomes. In statistics an outcome of an observation is called an event.

The event can be a binary outcome like a component passing or failing a certain test. The event can also be a discrete number, like the number of surviving components of a batch after some period of exposure to a load or a test. The event can be a rational number, like the fraction of objects that survive a test period. Or, the event can be a real number like the lifetime of an object in operation or in an ageing test. But an event can also be the outcome of an evaluation of the surface material of which an insulator is made (e.g. silicon rubber, epoxy resin or porcelain). An event can also be a subset like non-porcelain.

The collective possible outcomes form a set. A single outcome of this set is called an elementary event or element. A selected outcome is called a sample. So, if a dice is rolled, the elementary events are throwing 1, 2, 3, 4, 5 and 6 (see Example 2.1). The outcome 3 is a sample and throwing with outcome 3 is also an (elementary) event. Throwing an odd number is also an event, but not an elementary event as it consists of the outcomes 1, 3 and 5. If it is known that the three insulator types mentioned above are all options, and the material of one insulator is checked, then the three elements are: silicon rubber, epoxy resin and porcelain. If the lifetime of a cable is determined, there are an infinite number of elementary events, namely lifetimes ranging from zero to infinity.

The complete set of outcomes is called the sample space, indicated with the symbol Ω. Within the space Ω, various sets or subsets of elements may exist. Example 2.1 shows that the sample space of throwing a dice includes the set of odd numbers and also a set of prime numbers. The intersection of these two sets forms a subset of outcomes that are both odd and a prime number, which are the elements 3 and 5. Important relationships are shown in Table 2.1 and Figure 2.1.

From the space Ω, one or more elements can be taken, which is called a sample. The number of elements in the sample is called the sample size. Rolling a dice seven times

Table 2.1 Notations in set theory: A, B and C are sets of elements in the sample space Ω.

Notation	Name	Meaning
$A = \Omega$	Sample space	A is equal to the sample space (i.e. contains all elements)
$A = \emptyset$	Empty set	A is an empty set (i.e. contains no elements)
$A \subseteq B$	Subset	Every element of A is an element of B
$A \subset B$	Proper subset	Every element of A is an element of B, but $A \neq B$ and $A \neq \emptyset$
$C = A \cup B$	Union	C contains all elements of A and B
$C = A \cap B$	Intersection	C contains only the elements that A and B have in common
$C = A \backslash B$	Set difference	C contains all elements of A excluding elements of B
$A^C = \Omega \backslash A$	Complement	A^C contains all elements of Ω excluding elements of A (A^C not to be confused with power C of A)

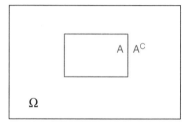
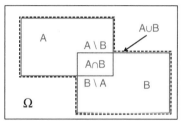

Figure 2.1 Left: Event *A* and its complement in the complete sample space Ω. Right: Events *A* and *B* with their intersection and union in the complete sample space Ω.

produces a set of seven outcomes and the sample size is seven. The action of taking a sample is called sampling. For determining the number of possible combinations, it is relevant to consider whether sampling is done with or without replacement (i.e. whether elements taken from the sample space return to the sample space or not). Rolling a dice does not take away the elements from the sample space and every outcome is eligible again, which means rolling a dice is sampling with replacement. Destructive testing of insulators takes elements out of the sample space without replacement. Their times to breakdown will not be measured again. Also, taking assets from a stock of assets is sampling without replacement as the stock will get exhausted.

A Venn diagram is a graphical representation of sample spaces with subsets that visualizes the logical relations between events. A Venn diagram illustrating the various relationships in Table 2.1 is shown in Figure 2.1.

Example 2.1 *Set Theory Applied to Throwing a Dice* A dice is a cube with numbered sides used to generate random numbers from 1 to 6. The six elements in the sample space when throwing a dice are: 1, 2, 3, 4, 5 and 6. The total sample space associated with throwing a dice is therefore:

$$\Omega = \{1, 2, 3, 4, 5, 6\} \tag{2.1}$$

There are various subsets one can think of, like odd and even outcomes, or the outcome being a prime number (remember 1 is not a prime number). Intersections of sets form subsets which can be empty or not. A range of events and (sub)sets are shown in Table 2.2. The range of relationships or operations is explained in Table 2.1.

Table 2.2 Various events for the case of throwing a dice.

Event	(Sub)set	Outcome set	Relationships (see Table 2.1)
Elementary	Ω	{1,2,3,4,5,6}	Contains all elements and sets
Odd number	*A*	{1,3,5}	$A \subseteq \Omega$
Even number	*B*	{2,4,6}	$B \subseteq \Omega; A \cap B = \emptyset; A \cup B = \Omega$
Prime number	*C*	{2,3,5}	$C \subseteq \Omega$
Non-prime number	*D*	{1,4,6}	$D \subseteq \Omega; C \cap D = \emptyset; C \cup D = \Omega$
Even prime number	*E*	{2}	$E \subseteq \Omega; E = B \cap C$
Odd non-prime number	*F*	{1}	$F \subseteq \Omega; F = A \cap C$

Note: 1 is not a prime number.

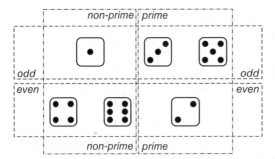

Figure 2.2 Venn diagram illustrating the events listed in Table 2.2. The figure shows the six elementary events and also the events that are a collection of one or more elementary events (like odd, even, prime and non-prime number outcomes). Such events may or may not have an empty space as cross-section (e.g. odd ∩ even is empty, odd ∩ prime is not empty).

Although the Venn diagram is a very helpful tool to show relationships between subsets, it is noteworthy that a Venn diagram is a two-dimensional representation of a sample space, while the events themselves may be one-, two-, three-dimensional and up to much higher dimensions. So, the dimensions of the Venn diagram and the events may not match.

For instance, the outcomes of throwing a dice are one-dimensional, namely any integer from 1 to 6. The Venn diagram in Figure 2.2 shows the outcomes scattered in a two-dimensional square, which is practical for showing the various subsets, but drops the sense of outcome dimension. Another example is measuring the three-dimensional vertex (corner) coordinates of a unity cube centred at (0,0,0). The vertex coordinates are eight three-dimensional vectors $(\pm\frac{1}{2}, \pm\frac{1}{2}, \pm\frac{1}{2})$ (i.e. $(-\frac{1}{2}, -\frac{1}{2}, -\frac{1}{2})$, $(-\frac{1}{2}, -\frac{1}{2}, +\frac{1}{2})$, $(-\frac{1}{2}, +\frac{1}{2}, -\frac{1}{2})$, etc.). These eight three-dimensional vectors are samples from a three-dimensional space, but can be put in a Venn diagram as eight entities in a two-dimensional square. So, the two-dimensional representation by the Venn diagram is not related to the dimension of the sample space itself. Nevertheless, it may provide a good map to reflect relationships between sample sets. In Example 2.2 this is applied to a test of five insulators.

Example 2.2 *Difference Between Venn Diagram and Sample Space Dimensions*
Five insulators are put to an accelerated ageing test to determine their times to breakdown. It is evaluated how many of these insulators have a breakdown time larger than a required breakdown time τ. The test lasts until all insulators have failed.

After the test there will be a series of five breakdown times which forms a five-dimensional vector of variables: $(t_1, t_2, t_3, t_4, t_5)$. Therefore, the sample space Ω consists of five-dimensional vectors. In this test the times are ranked such that $t_1 \leq t_2 \leq t_3 \leq t_4 \leq t_5$.

A Venn diagram is a two-dimensional representation of Ω. With respect to the required minimum breakdown time τ, six sets can be defined as an example:

- Set A: all insulators have a breakdown time larger than τ.
- Set B: at least one insulator has a breakdown time smaller than τ.
- Set C: at least two insulators have a breakdown time smaller than τ.
- Set D: at least three insulators have a breakdown time smaller than τ.
- Set E: at least four insulators have a breakdown time smaller than τ.
- Set F: all insulators have a breakdown time smaller than τ.

This is shown in Figure 2.3.

Figure 2.3 Five insulators are subject to a test which results in a five-dimensional vector of breakdown times. The two-dimensional Venn diagram can represent the relationships between sets of five-dimensional breakdown time vectors. Note: $t_1 \leq t_2 \leq t_3 \leq t_4 \leq t_5$.

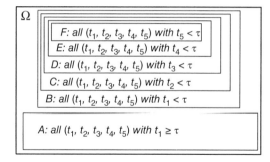

Being able to categorize events, the next question is how to determine the probability of events. This is the subject of the next section. After that, the relation between probability and the statistical distribution is discussed.

2.2 Probability of Events

Given a sample space with all events and a set of elementary events, each event has a probability of occurring. The notation for the probability P of an event X, meaning an event with an outcome in the subset X, is $P(X)$. If all elementary events are equally likely to occur (which is assumed unless stated otherwise), then the probability of an event out of the set X can be calculated as:

$$P(X) = \frac{\text{Number of outcomes in } X}{\text{Number of outcomes in } \Omega} \tag{2.2}$$

A number of rules apply to probabilities. For each event X:

$$0 \leq P(X) \leq 1 \tag{2.3}$$

As each elementary event belongs to the complete sample space Ω, and vice versa, space Ω contains all possible elementary events. Therefore the probability for Ω is 1:

$$P(\Omega) = 1 \tag{2.4}$$

An empty set does not contain any of the possible events and therefore the probability that an event from an empty set will happen is 0:

$$P(\emptyset) = 0 \tag{2.5}$$

These statements can be checked easily in the case of a discrete number of outcomes in the sample space Ω, as shown in Example 2.3.

Example 2.3 *Probability of Events When Rolling a Fair Dice Once* When rolling a dice once, six elementary events are possible which have an equal probability of 1/6 of occurring. Additionally, various sets can be defined (see Example 2.1, Table 2.2 and Figure 2.2). Using Eq. (2.2), the probabilities can be calculated for various (sets of) events (see Table 2.3).

In case of an infinite number of possible outcomes Eq. (2.2) still holds, but relates to the density of outcomes in certain ranges. This applies, for instance, to determining

Table 2.3 Examples of probabilities.

Event	Calculated from Eq. (2.2)	Mathematical notation
Throwing six	1/6	$P(6) = 1/6$
C: throwing a prime number	3/6	$P(C) = 3/6$
B: throwing an even prime	1/6	$P(B \cap C) = 1/6$
$B \cap D$: throwing an even non-prime	2/6	$P(B \cap D) = 2/6$
Throwing any element of {1,2,3,4,5,6}	6/6	$P(\Omega) = 6/6 = 1$
Throwing none of {1,2,3,4,5,6}	0/6	$P(\emptyset) = 0/6 = 0$

Note: For the event labels, refer to the (sub)sets of Table 2.2.

breakdown times, electric strength or other performances. The next section discusses the relationship between probability and statistical distributions.

2.3 Probability versus Statistical Distributions

Reliability studies usually concern a relationship between probability and a measurable performance indicator. This is a quantity like lifetime, breakdown strength, number of switching actions, and so on. Most examples will be in terms of time, but performance can be any measure of performance. A typical statement could be that there is a 90% probability that an asset will survive 40 years of operation, where the performance is operational lifetime. Another statement could be that 10% of a batch of insulators may fail at $20\,\text{kV mm}^{-1}$ or lower in a ramp voltage test (i.e. a test where the test voltage rises linearly with time), which means that the performance is an electric field. The probability of passing or failing a performance level plays an important role in the evaluation of reliability of assets.

In Section 2.1, sampling from a distribution with or without replacement was mentioned. In the case of sampling with replacement, like rolling a dice, the initial situation is restored and the set will not be exhausted. In the case of sampling without replacement, the number of elements is diminishing. In the following, it is assumed that sampling is without replacement unless stated otherwise.

Often, failure data from practice and/or from tests are gathered to describe the failure behaviour of assets. As an illustration, consider the case of Example 2.4.

Let a series of assets be manufactured as in Table 2.4 and these assets be put in operation until failure. After all assets have failed, it is possible to rank the assets in increasing order of their performance (here, lifetime) and to assign an index i to them (like in Table 2.4 and Figure 2.4). Then, it is possible to quantify which fraction F of the asset population failed within a certain period of lifetime and to quantify which percentage survived. It is also possible to quantify how many assets or which fraction fails per unit time at any moment. Example 2.4 shows how such a relationship between F and T emerges. This is called the statistical failure distribution.

Table 2.4 Example of a batch of assets, labelled according to installation and ranked by lifetime.

Label	Lifetime T (hr)	Index i	Ranked lifetime T (hr)	Failed fraction F (%)	Surviving fraction R (%)	Original label
		—	0	0	100	
L1	288	1	87	9	91	L7
L2	875	2	182	18	82	L4
L3	1386	3	288	27	73	L1
L4	182	4	405	36	64	L11
L5	539	5	539	45	55	L5
L6	2485	6	693	55	45	L9
L7	87	7	875	64	36	L2
L8	1792	8	1099	73	27	L10
L9	693	9	1386	82	18	L3
L10	1099	10	1792	91	9	L8
L11	405	11	2485	100	0	L6

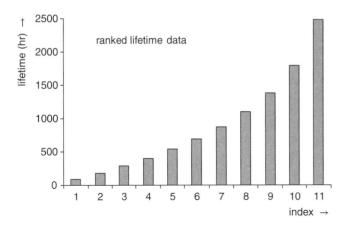

Figure 2.4 Example of ranked lifetime data of belonging to the case of Table 2.4.

As long as assets cannot be distinguished by properties that are characteristic of the failure behaviour, they are regarded as forming a homogeneous population of assets. Even though post-mortem analysis (i.e. forensic studies after failure) may reveal why one asset performed better or worse than others, at the start it was not possible to predict for any given time $T = t$ which asset would belong to the failed fraction F or the surviving fraction R, as shown in Example 2.4. Provided assets are indistinguishable with respect to their failure behaviour (whether fact or mere assumption), this leads to the statement that the probability of a randomly selected unit having lifetime T shorter than t is equal to the fraction $F(t)$:

$$P(T < t) = F(t) \tag{2.6}$$

It should be noted that the relationship between F and T is an observable fact. However, the equality between probability and failed fraction leans heavily on the assumption that there is no prior knowledge about individual times to failure. It is noteworthy that CBM aims at breaking this assumption. Successful diagnostics are able to predict which assets will, and which will not, fail in a well-defined period of time. This will change the estimated failure probability $P\,(T < t)$ for each individual asset, even though the overall $F(T)$ remains the same.

For instance, if at the start a technique would be able to indicate imminent failure within the next 250 hours, units L4 and L7 could be distinguished from the other nine units as the units that are about to fail. If that technique exists and is applied, it is no longer true that a randomly selected unit would have a probability of 18% of failing. Instead, L4 and L7 would have a 100% probability of failing within the next 250 hours and the other nine would have a 0% probability of failing. This is an example of Bayesian statistics that will be elaborated on in Section 2.5.2.1. However, on average and/or when the technique is not applied, the probability of a random unit failing within the next 250 hours is still 18%.

With the application of the diagnostic, the distribution $F(T)$ still holds, but the relationship in Eq. (2.6) is broken not only for the diagnosed units, but also for the others. It is noteworthy that the distribution F and the probability P are different concepts that may be equal under certain conditions like random selection.

Example 2.4   *Performance, Ranking, Fraction, Probability*   Let a batch of 11 units be installed. These assets have unique labels L1, L2, …, L11 (see Table 2.4 column 'Label'). It is assumed that the units cannot be distinguished except for these labels at the start. It cannot be predicted how each unit will perform in practice. After all units have been put into operation and ultimately failed, the lifetimes T appear as in the second column of Table 2.4, 'Lifetime'. Clearly, the lifetimes of the units are not the same but scattered. No correlation between the labels and the lifetimes is apparent.

In the right-hand part of Table 2.4 the lifetimes are ranked in increasing order of magnitude (column 'Ranked lifetime T (hours)') and an index is added (column 'Index i'). After this ranking it is possible to quantify the fraction of units that perform better or worse than a given lifetime T. For instance, at $T = 1200$ hours, 8 out of 11 units (i.e. 73% of the batch) failed and 3 out of 11 units (i.e. 27% of the batch) survived. So, there is a relationship between the performance of the 'lifetime' T and the failed fraction F that can be noted as a function $F(T)$.

It may be true that a post-mortem analysis can explain why unit L6 ($i = 11$) belongs to the 27% fraction with a lifetime $T > 1200$ hours and why L1 belongs to the 73% fraction with a lifetime $T < 1200$ hours, but this could not be predicted at the start because the units were initially indistinguishable, as stated above. As long as no relationship is known between the labels and the lifetimes, at the start L1 might very well be guessed to belong to the 27% strong fraction and/or L6 could have been guessed to belong to the 73% weaker fraction. If tests like these have been done before, then $F(T)$ may have become known. Randomly selecting a unit without knowing its individual lifetime, there is a 73% probability that this randomly selected unit belongs to the group of units that fails at or before $T = 1200$ hours.

2.4 Fundamental Statistical Functions

In the previous section the concepts of statistical distribution and probability were discussed. This section discusses the statistical distribution in more detail, by its various appearances.

The relation between performance (usually time) on the one hand and the population fraction that meets the performance on the other hand is described by a statistical distribution. The performance is an outcome from the sample space and is a variable. In case the sample space has a finite number of outcomes, the variable is discrete. If the sample space is a continuum, the variable is continuous.

A distribution can be expressed in various ways, namely in terms of a so-called cumulative distribution function F, reliability R, distribution density f, hazard rate h and the cumulative hazard function H. The fundamental functions are all appearances of the very same distribution and can be translated into each other. The performance (variable) depends on the context and can refer to all kinds of entities like operational time, calendar time, pressure, voltage, number of rotations, and so on. However, for ease of discussion, the performance will often be described in terms of lifetime T.

It should be noted that the variable 'lifetime' throughout this book can be replaced by other applicable entities. Also, lifetime is a positive quantity in the range 0 to $+\infty$, but a performance in terms of voltage might in principle run from $-\infty$ to $+\infty$, while other quantities may even have other ranges. Where appropriate, the boundaries of mathematical operations like integrals should therefore be adjusted if required.

2.4.1 Failure Distribution *F*

The cumulative distribution function $F(x)$ is a function of a variable X (e.g. time) and describes the population fraction for which an event has taken place at a value of $X \leq x$. In reliability statistics this event is usually the time T to failure of assets being smaller or equal to t (i.e. the event that the asset failed at or before t). The cumulative distribution function is also called the cumulative failure probability function. The term probability can be misleading if the proper assumptions are not met. It refers to the probability $P(T < t)$ that a sample from the population will have a shorter lifetime T than t. As stated before, two assumptions are made:

1. The test object is selected randomly from the population.
2. Prior to the test, the samples are indistinguishable with respect to their lifetime (i.e. they may be recognizable, for example by a label, as long as this label does not carry information about the lifetime or reveal the lifetime).

In case the assumptions are met, the probability $P(T < t)$ that a particular test object will fail before t is (repeating Eq. (2.6)):

$$P(T < t) = F(t) \tag{2.7}$$

The assumptions do not hold if the appearance of the sample does reveal the failure behaviour to some extent. For example, assets that are about to fail may start to produce alarming signals like smoke, smell, noise, discharges, discoloration, leakage or other phenomena that distinguish the soon failing assets from the others. If there is prior knowledge about the failure of individual assets, then $F(t)$ is not a good measure for the probability (see Section 2.3). Also, the term 'cumulative probability' function becomes misleading then, but the interpretation of 'cumulative (failure) distribution' that indexes the assets by their failure sequence still applies and is therefore preferred. Expertise to distinguish the failure behaviour of one asset from that of another asset is the foundation of a knowledge rule that allows a failure forecast with a greater precision than using $F(t)$. Such rules are the key to successful CBM (see Section 1.3.3).

For the failure distribution $F(t)$ with $0 \leq t < \infty$, the following holds:

$$F(0) = 0 \tag{2.8}$$

$$\lim_{t \to \infty} F(t) = 1 \tag{2.9}$$

If the domain boundaries are t_{start} and t_{end}, then Eqs (2.8) and (2.9) become:

$$F(t_{start}) = 0 \tag{2.10}$$

$$F(t_{end}) = 1 \tag{2.11}$$

In case other performance variables than time apply, like voltage, electric field, mechanical strength, and so on, Eqs (2.3) and (2.4) still apply, but in some cases the minimum and maximum boundaries of the variable may have to be adjusted.

2.4.2 Reliability R

The reliability function or survivability R is the complement of F (i.e. the reliability function $R(x)$ describes the population fraction that an event is characterized by a variable X with a value greater than x). In terms of time to failure, it describes the population fraction that survives the specific moment t. As the events of failure and survival exclude each other, it follows that:

$$R(t) = 1 - F(t) \tag{2.12}$$

For reliability, the following holds:

$$R(0) = 1 \tag{2.13}$$

$$\lim_{t \to \infty} R(t) = 0 \tag{2.14}$$

Again, other boundaries like t_{start} and t_{end} and variables like voltage may apply. Under the same assumptions of Section 2.4.1, the reliability can be used to calculate the probability that a test object from the population will have an equal or longer lifetime T than t (i.e. will fail at or after t). This probability is:

$$P(T \geq t) = R(t) \tag{2.15}$$

Similar to the discussion in Section 2.3, once test objects can be distinguished from other objects by a diagnostic technique that allows a failure forecast with greater precision, this opens the door to successful CBM.

2.4.3 Probability or Distribution Density *f*

The probability density, distribution density or relative failure frequency $f(t)$ is the portion of events per unit variable T that happen at $T = t$. In failure statistics, $f(t)$ describes which population fraction dF fails per time fraction dt. It is the increase of F and decrease of R per unit time:

$$f(t) = \frac{dF(t)}{dt} = -\frac{dR(t)}{dt} \tag{2.16}$$

The term 'probability density' again refers to the interpretation of a randomly selected test object, whose probability of failure is equal to $F(t)$. The term 'distribution density' avoids this probability interpretation and is preferred. Distribution density remains a valid term, even if knowledge is available that influences the assessment of the probability of failure of the individual assets.

At the starting moment $T = 0$, no object will have failed (Eq. (2.8)) and with increasing t a larger fraction of the population $F(T)$ fails:

$$\int_0^T f(t)dt = F(T) - F(0) = F(T) \tag{2.17}$$

Ultimately, with infinitely long time, the complete population will have failed:

$$\int_0^\infty f(t)dt = \lim_{t \to \infty} F(t) = 1 \tag{2.18}$$

Likewise, R will diminish from 1 to 0 and the integrated distribution density f can also be calculated in terms of R, which yields the same as in Eq. (2.18):

$$\int_0^\infty f(t)dt = -\lim_{t \to \infty} R(t) - (-R(0)) = R(0) - \lim_{t \to \infty} R(t) = 1 \tag{2.19}$$

That is, the integral of the distribution density is the fraction of the complete population that has failed and is also the fraction that has ceased to survive.

2.4.4 Probability or Distribution Mass *f*

If the variable T is discrete, then the derivative with infinitesimal dt cannot be used. The discrete approach involves differentials and summations. For the discrete case the frequency of occurrence of the outcomes t_i is represented by the probability or distribution mass function f. The possible variable values T can be indexed as T_i with $i \in \mathbb{N}$. The sum of the frequencies of all possible outcomes is 1:

$$\sum_i f_i = \sum_i f(t_i) = 1 \tag{2.20}$$

For example, the outcomes t_i ($i = 1, \ldots, 6$) of rolling a dice are: 1, 2, 3, 4, 5, 6. The frequency of a fair dice is 1/6 for each outcome t_i. Distribution density and mass differ, because the first needs integration over an interval in order to produce a distribution or population fraction, whereas the latter produces values that can be used as distribution or population fractions directly.

2.4.5   Hazard Rate *h* and the Bath Tub Model

The hazard rate or failure rate *h* describes what survivor fraction will fail in the coming time fraction. It is a very important function to describe the failure behaviour of assets. The surviving fraction at $T = t$ equals $R(t)$ (Eq. (2.12)):

$$h(t) = \frac{1}{R(t)} \frac{dF(t)}{dt} = -\frac{1}{R(t)} \frac{dR(t)}{dt} = \frac{f(t)}{R(t)} \tag{2.21}$$

Because $R(t)$ is always equal to 1 or less, $h(t)$ is always equal to $f(t)$ or more.

Whereas the (failure) distribution density $f(t)$ concerns the fraction of the original population that fails per time unit, the hazard rate $h(t)$ relates this failure frequency only to the surviving fraction. To illustrate the importance of the two:

- The (failure) distribution density or failure frequency $f(t)$ is of interest for a repair crew or fire brigade that needs to know how many failures occur per unit time (e.g. for planning resources). An example in the energy sector is how many joints may fail during a week and must be replaced.
- The hazard rate $h(t)$ is of interest for operators who depend on the reliability of the assets that are in operation and need to estimate the likelihood that they will be successful. An example in the energy sector is the likelihood that an operational circuit may have a problem in the coming week.

Another illustrative example is the hypothetical situation that 10 assets are put to service and every day one of them appears to fail (so after 10 days no asset is available anymore). At the beginning, 100% of the population is available and consists of 10 assets. The failure distribution and probability density $f(t)$ is 10% per day and constant: each day 1 of the original 10 fails. If one has to work with an asset, one is interested in the performance of the assets that still work up to that moment, which is the hazard rate $h(t)$. This hazard rate increases steadily in this example from 10% on the first day (i.e. 1 out of 10 fails), to 11% on the second day (i.e. 1 out of 9 fails), 12.5% on the third day (i.e. 1 out of 8 fails), 20% on the fifth day (i.e. 1 out of 5 fails), and so on until 100% on the 10th and last day (i.e. 1 out of the only remaining 1 fails). So, $f(t)$ is related to the original batch and $h(t)$ is related to the surviving fraction of the original batch.

The derivative of the hazard rate is used to categorize perceived types of failure. The hazard rate can be decreasing, constant or increasing. This means that the failure probability of survivors decreases, remains stable or increases. This is summarized in Table 2.5 and Figure 2.5. These types of failure behaviour are indicated by names:

- *Child mortality, infant mortality, or (a milder expression) child disease.* This refers to a process where the hazard rate is high during the early stages of operational life, after which the hazard rate reduces. This means that assets surviving the early stages of operational life have less probability of failing per unit time than assets from the original batch. It is justified to have an increasing confidence in the assets that survive the early stages. Typical examples are batches of products that grow in reliability (like resins, which are curing (hardening)). Note that in daily life the term 'child mortality' is also used for products that are weak due to, for example, imperfections in manufacturing or installation. Such assets may just age faster than proper products and have an increasing hazard rate, and would be regarded

Table 2.5 Three classes of hazard rate types.

Range	Hazard rate $h(t)$	Process type	Characteristic
$\dfrac{dh}{dt} < 0$	Decreasing	Child mortality	High initial hazard rate
$\dfrac{dh}{dt} = 0$	Constant	Random failure	Flat hazard rate
$\dfrac{dh}{dt} > 0$	Increasing	Wear out	High hazard rate at the end

Note: It should be noted that the term 'child mortality' is generally adopted as above but a common wider meaning of child mortality is discussed in Section 2.5.3, which includes the unexpected fast wear out of a deficient subpopulation.

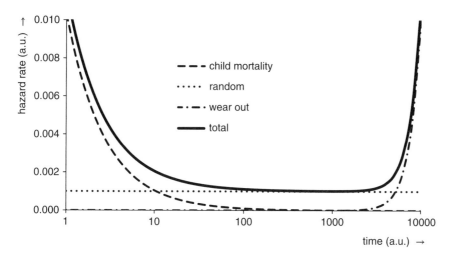

Figure 2.5 The bath tub model: total hazard rate as the sum of hazard rates due to three concurrent failure processes [11].

as a wear-out phenomenon (cf. Section 2.5.3). Generally, quality control aims at removing child mortality assets that would otherwise lead to early failure in practical application (see also Example 2.10 in Section 2.6.2).

- *Random failure.* This refers to a process with no increase or decrease in hazard rate (i.e. *h* is constant). As an example, this is typically found with radioactive isotopes. The decay of a uranium atom does not depend on its age. Another example is assets that fail due to external causes that are not related to the quality of the assets. For example, if assets fail due to extreme weather there may hardly be a correlation with the lifetime or quality of the asset.
- *Wear-out failure.* This refers to an increasing hazard rate that extinguishes the asset population. That is, the increasing hazard rate announces the end of life of what is left of the batch. This means that the population should be replaced once this process is recognized. An example is a batch of transformers in which the paper insulation degrades to such an extent that more and more are failing.

The hazard rate is usually treated as a function of a continuous variable, but it can also be used for a discrete variable with values t_i. In the discrete case it can be defined as:

$$h(t_i) = \frac{f(t_i)}{R(t_{i-1})} = \frac{f(t_i)}{1 - F(t_{i-1})} = \frac{f(t_i)}{1 - \sum_{j=1}^{i-1} f(t_j)} \tag{2.22}$$

2.4.6 Cumulative Hazard Function H

The integral of $h(t)$ is the cumulative hazard rate H or hazard. If the cumulative hazard rate is taken from the starting point at $t = 0$, it can be written as a single-variable function $H(t)$:

$$H(t) = \int_0^t h(x)dx \tag{2.23}$$

The hazard can also be expressed in terms of the reliability R:

$$H(t) = \int_0^t h(x)dx = -\int_0^t \frac{1}{R}\frac{dR}{dx}dx = -\int_{R(0)}^{R(t)} \frac{1}{R}dR = -\int_{\ln R(0)}^{\ln R(t)} d\ln R = -\ln R(t) \tag{2.24}$$

The last step in Eq. (2.24) uses the fact that $R(0) = 1$ and $\ln(1) = 0$. The reliability $R(t)$ and failure distribution $F(t)$ can also be expressed in terms of $H(t)$:

$$R(t) = \exp[-H(t)] \tag{2.25}$$

$$F(t) = 1 - \exp[-H(t)] \tag{2.26}$$

The hazard can also be calculated over a period from t_1 to t_2. The hazard function then takes the form of a two-variable function:

$$H(t_1, t_2) = \int_{t_1}^{t_2} h(x)dx = H(t_2) - H(t_1) \tag{2.27}$$

In this way the danger of breakdown over periods of various assets can be compared. Unlike the case with $f(t)$, the integral of $h(t)$ over the complete time range will not be equal to 1. This becomes obvious if H is expressed in terms of R and integrated over the complete time domain. For $t \to \infty$, the value of $-\ln(R(t))$ becomes infinite:

$$H(0, \infty) = H(\infty) = -\int_{\ln R(0)}^{\ln R(\infty)} d\ln R \to \infty \tag{2.28}$$

The cumulative hazard rate $H(t)$ will increase to infinity over the complete time domain. In case of a variable T with other boundaries than 0 and ∞, the integral has to be adapted accordingly.

2.5 Mixed Distributions

The previous sections discussed the fundamental functions in which distributions can be expressed. In practice, distributions rarely consist of a single distribution, but usually appear to be a combination or mix of distributions. This section discusses the mix of mechanisms and (sub)populations from the viewpoint of the underlying processes that build up the resulting distribution. Section 2.6 then discusses multivariate distributions in the light of ageing conditions. These subjects do overlap, but as these viewpoints may both be encountered with asset management, both are discussed here.

A combination of distributions arises from building systems by combining assets. For instance, circuits, station bays and grids consist of combinations of assets, each of which may consist of various parts. As a result, the failure distribution of the system will be a combination of all those distributions. Some assets themselves may be regarded as a system consisting of parts. This type of combined distribution can be treated with system reliability in Chapter 7. This includes groups of assets that feature competing failure modes within each asset, like simultaneous child mortality, random and wear-out failure. Assets (or systems) with competing processes are discussed in Section 2.5.1.

Another type of combined distribution is caused by inhomogeneous populations of assets. For instance, the failure behaviour of transformers in a grid may be investigated. This transformer population may consist of various transformers with different failure distributions. The overall distribution will be a mix of the underlying distributions. Such inhomogeneous asset populations are discussed in Section 2.5.2.

A special case of inhomogeneous populations is a group of assets of which only a part of the batch features serious child mortality failure due to production or installation errors, while the rest of the batch only features random and wear-out failure. This leads to a dedicated bath tub model, as discussed in Section 2.5.3.

2.5.1 Competing Processes

Assets may fail to various processes (e.g. as in Section 2.4.5, Figure 2.5). In case the assets themselves can be regarded as systems, the discussion with system reliability provides an alternative approach (Chapter 7). But in general, if various failure mechanisms compete within assets and the failure modes cannot be assigned to distinct parts, the following applies.

Assume two failure modes exist simultaneously in each asset. For instance, an insulator may fail due to internal breakdown or by flash-over. The first process means the insulation is punctured with a short-circuit, while the second means a short-circuit along its outer surface. The two processes may each have their distinctive distribution, characterized by indices 1 and 2, say F_1 and F_2, respectively. The combined distribution will be indicated without an index (e.g. F).

The failure mode that strikes first makes the asset fail. In other words, an asset only survives if it survives both failure modes. The following relationships hold for reliability R, failure function F, probability density f and hazard rate h:

$$R = R_1 \cdot R_2 \tag{2.29}$$

$$F = 1 - ((1 - F_1) \cdot (1 - F_2)) = F_1 + F_2 - F_1 \cdot F_2 \tag{2.30}$$

$$f = f_1 \cdot R_2 + f_2 \cdot R_1 \tag{2.31}$$

$$h = h_1 + h_2 \tag{2.32}$$

$$H = H_1 + H_2 \tag{2.33}$$

This type of mixed distribution is also referred to as multiplicative distribution, which draws its name from Eq. (2.29).

This approach can be extended to multiple competing failure modes. In Section 2.4.5 three classes of hazard rate were introduced: child mortality, random and wear-out failure. With a batch of assets, these three types of failure may compete in each asset. If that is the case, the resulting hazard rate is the sum of the hazard rates of all processes. Figure 2.5 shows the case of three processes: child mortality, random and wear-out. This is the so-called bath tub model.

This model is often used to describe the stages of an asset population. In the early stages after commissioning, some assets may fail due to the child mortality process. After some time the importance of the child mortality process declines (though the process is still active) and random failure dominates the failure behaviour. At some point wear out will take over and will ultimately extinguish the complete population. In this form the bath tub model assumes the three processes to be present in all assets. Therefore, the three hazard rates may be added, similar to Eq. (2.32). In Section 2.5.3 another bath tub model is shown that concerns the case where only a fraction of the assets suffer early failures.

In graphical representations, often a chart is constructed with scales such that the failure distribution shows as a straight graph of $F(t)$ versus t. Examples are exponential plots, Weibull plots, and so on. In such a representation the existence of two (or more) different processes can be recognized by a graph of the total distribution that is no longer straight but curves towards an increasingly steep graph (see Figure 2.6). The steepest graph is exhausting the population, like the wear-out process in Figure 2.5. The example of Figure 2.6 can be a random process competing with a wear-out process.

If failure data are plotted in a graph that should produce a straight line, but instead an upward curve is found, this can indicate two competing processes. Other possible indications are that the scaling was unsuccessful due to an invalid assumption about the type of distribution or that a wrong plotting position was used. In order to investigate the first indication, the asymptotes provide information about the distinctive processes. These are indicated by dotted lines in Figure 2.6, which coincide with the solid line in the regions where the respective distributions dominate.

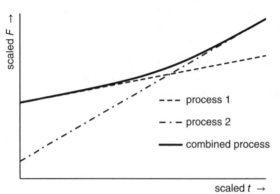

Figure 2.6 Effect of mixing competing processes within each asset. The axes are scaled in such a way that a single-failure distribution shows as a straight line. Process 1 has more random or 'child mortality' behaviour than process 2. The graph shows that in the initial stages, process 1 dominates while process 2 takes over later.

--- process 1

— ·— process 2

—— combined process

scaled F ↑

scaled t →

2.5.2 Inhomogeneous Populations

AM usually deals with asset populations that consist of varied assets. For instance, the lifetimes of underground cables can be discussed, but a closer look at this asset population may reveal that there are many types of cable. Furthermore, these cable types are used at different voltage classes. It is possible that the failure distributions of these cable types differ.

With a failure distribution Γ of an inhomogeneous population, a first step is to find out how the population can be split into its fractions. Next, the failure distribution of each failure mechanism for each population fraction is to be evaluated.

Example 2.5 *Inhomogeneous Population of High-Voltage Cables* The alternating current grid of a utility consists of various types of high-voltage (HV) cable. In a certain area all cable types listed in Table 2.6 are found. The first two columns show that this population of cables actually consists of four cable types: the two HV classes 150 and 380 kV, where three technology types are found for 150 kV. The third column shows two main types of insulation, namely paper and cross-linked polyethylene (XLPE), but paper can be combined with oil or mass (which is a kind of fatty substance, i.e. thicker than oil). The fourth column shows an ageing mechanism that ultimately may lead to breakdown of a cable. If interface discharge occurs, breakdown can follow within months. This process mainly (if not merely) occurs with polymer-insulated cables. Waxing is generally a slow process, but is typical for paper with oil or mass. As the timescales of these processes usually differ significantly, it makes sense to evaluate the failure behaviour separately. A Venn diagram may be used to group relevant sets.

As long as the necessary information is not available at the time of evaluation or is ignored, the cables have to be analysed as one population. But if subpopulations can be distinguished, then conditional statistics can be used. For instance, the breakdown times for HV cables that as a condition have paper insulation provide information about the waxing phenomenon. As an additional condition, the cable technology may be involved to study the relevance of the difference between oil and mass. As for the interface discharging process, the condition of XLPE insulation can be used to select the cables that bear information to study that phenomenon. The voltage level may be used as an additional condition to evaluate whether it correlates with interface discharging problems. Conditional statistics are called Bayesian statistics and help to unravel the effect of inhomogeneous populations.

Table 2.6 Overview of 150 and 380-kV HV cables in a given grid with technologies, applied insulation and two typical ageing processes.

Voltage (kV)	Technology	Electrical insulation	Possible ageing process
150	Oil pressure cable	Paper and oil	Waxing
	External gas pressure cable	Paper and mass	Waxing
	Polymer-insulated cable	XLPE	Interface discharge
380	Polymer-insulated cable	XLPE	Interface discharge

Note: XLPE is the polymer insulation.

Fractions of this HV cable population may be viewed in terms of circuit lengths when the impact of the ageing processes on cable reliability is studied. The background lies in the fact that the probability of finding a weak spot in a longer length of cable is larger than that in a shorter length of cable. This so-called 'volume effect' will be discussed further in Section 7.3.

An example is that, having prior knowledge that a certain cable is a 380-kV cable, the posterior (i.e. after using the prior knowledge) probability that waxing will be found in this cable is 0%, because all 380-kV cables in that grid have polymer insulation according to Table 2.6. However, if there were no prior knowledge about the voltage level, then the probability that waxing occurs is non-zero, because it could be a 150-kV cable with paper and oil or paper and mass insulation that might suffer from waxing.

If the subpopulations cannot be or are not distinguished, then the respective distribution functions differ from those in Section 2.5.1. Let F_i, R_i, f_i and h_i be the fundamental functions of subpopulation i. Let the number of assets that belong to subpopulation i be n_i. Let F, R, f, h and n apply to the total population (i.e. the sum of all subpopulations i). Then the fraction p_i of subpopulation i at the start ($t = 0$) is:

$$p_i = \frac{n_i}{\sum_i n_i} = \frac{n_i}{n} \tag{2.34}$$

The fundamental functions applying to the complete population are:

$$R = \sum_i p_i \cdot R_i \tag{2.35}$$

$$F = \sum_i p_i \cdot F_i = 1 - R \tag{2.36}$$

$$f = \frac{dF}{dt} = \sum_i p_i \cdot f_i \tag{2.37}$$

$$h = \frac{f}{R} = \frac{\sum_i p_i \cdot f_i}{\sum_i p_i \cdot R_i} = \sum_i p_i \cdot \frac{f_i}{R} \tag{2.38}$$

Note the difference between Eqs (2.32) and (2.38). As long as the subpopulations are not distinguishable, the fractions p_i may remain unknown. If the subpopulations can be distinguished, then they may be treated and analysed as separate populations.

2.5.2.1 Bayes' Theorem

In order to analyse inhomogeneous populations, it can be very helpful to incorporate conditions in calculations. In Example 2.5, it was shown how setting a condition could enable a specific ageing process to be studied that otherwise might be obscured by non-relevant data. In terms of set theory (see Section 2.1), the correct subset must be selected in the sample space Ω. Thomas Bayes (1701–1761) studied conditional statistics and formulated an important theorem.

Assume that there are two subsets A and B of the sample space Ω. These do not need to be complements, but may partly overlap as on the right side of Figure 2.1. Let the probability of an event from any set X be $P(X)$. It is also assumed that $P(A)$ and $P(B)$ are independent. If the outcome belongs to one subset, then it may or may not belong to the other subset too. If the outcome does belong to both subsets, then it belongs to the intersection $A \cap B$. This intersection can also be defined as a condition in two ways, namely:

- $A \mid B$ means the outcome belongs to subset A under the condition (i.e. *given* it belongs to the subset B).
- $B \mid A$ means the outcome belongs to subset B under the condition (i.e. *given* it belongs to the subset A).

The probability that the outcome belongs to the intersection $A \cap B$ is $P(A \cap B)$. Following the sequence in the first definition this can be written as the product of the probability that the event belongs to subset A under the condition that it belongs to subset B times the probability that the event actually belongs to subset B (i.e. the probability that the condition is met indeed):

$$P(A \cap B) = P(A|B) \cdot P(B) \tag{2.39}$$

Similarly, using the sequence in the second definition of the intersection:

$$P(A \cap B) = P(B|A) \cdot P(A) \tag{2.40}$$

Combining Eqs (2.39) and (2.40) yields:

$$P(A|B) \cdot P(B) = P(B|A) \cdot P(A) \tag{2.41}$$

Bayes' theorem follows from this equation and is stated as:

$$P(A|B) = \frac{P(B|A) \cdot P(A)}{P(B)} \tag{2.42}$$

Bayes' theorem is widely applied in various types of statistics, where conditional probabilities apply. The use of these equations is illustrated in Example 2.6 for calculating the probability of selecting a certain bushing from an inhomogeneous population. More examples of Bayesian statistics are discussed in Sections 9.4.1.3 and 9.4.2.2.

Example 2.6 *Bayesian Statistics for an Inhomogeneous Population of Bushings*
Bushings are components that are used to let conductors with high voltage cross walls or earthed housings. They basically consist of a conductor surrounded by an insulator that separates the HV from earth (see Figure 2.7). They are mounted on transformers, a wall of a station, a compensation coil, and so on. A long-established design for bushings in the HV grid applies a porcelain housing. These have a good track record, but if pressure is built up inside, they may explode and cause safety issues if people are nearby. One of the alternatives is to install bushings with glass fibre-reinforced epoxy with silicone rubber protection. Such a bushing is called a polymer bushing. These have significantly less impact on their environment.

Imagine a station with four transformers and a compensation coil. On one transformer the porcelain bushings have been replaced by polymer bushings. The bushings are installed as in Table 2.7. Assume partial discharge diagnostics have been carried out on all bushings and one bushing is suspect. Assume that at the time of evaluation it is not known on which transformer the polymer insulators are mounted (it should be known, but for the sake of the example assume it is not known). Say, there is good reason to believe that the suspect bushing is of porcelain. What is the probability that it is

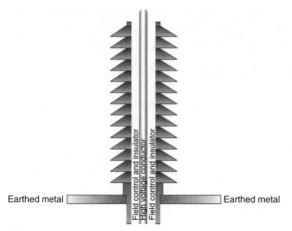

Figure 2.7 Typical geometry of a bushing: a high-voltage conductor runs vertically, surrounded by an insulator that consists of an insulating housing and insulating oil inside.

Earthed metal

Earthed metal

Table 2.7 A hypothetical case of a considered set of bushings that are mounted on transformers (set *A*) or coils (set *A^c*) and may consist of porcelain (set *B*) or polymer (set *B^c*).

Equipment		A: transformers	A^c: coils	Total
Number		4	1	5
Bushings outer protection				
B:	Porcelain	18	3	21
B^c:	Polymer	6	0	6
	Total	24	3	27

Note: For the sake of simplicity, the transformers are supposed to have only primary and secondary terminals (six in total) and the coils three terminals.

Table 2.8 The distribution of bushings in Table 2.7 translated into probabilities for porcelain transformer bushing.

Equipment		A: transformers	A^c: coils	Total
Number		4	1	5
Bushing probabilities				
B:	Porcelain	$P(A \mid B) = 18/21$	...	$P(B) = 21/27$
		$P(B \mid A) = 18/24$		
		$P(A \cap B) = 18/27$		
B^c:	Polymer	...	...	...
	Total	$P(A) = 24/27$	...	$P(A) + P(A^c) = 27/27$
				$P(B) + P(B^c) = 27/27$

Note: The remaining table cells are left for the questions.

mounted on a transformer? Table 2.8 shows the results. $P(A|B)$ is 18/21 (among the 21 porcelain bushings it is one of the 18 on a transformer). It can be checked that Eq. (2.42) is indeed valid.

2.5.2.2 Failure Distribution of an Inhomogeneous Population

This section demonstrates how the failure distribution of an inhomogeneous population can be analysed. An inhomogeneous population consists of two or more subpopulations which form the sets A_k. It is assumed that any element belongs to only one population. The number of elements in a subset A_k is indicated by n_k. The sum of the subset numbers equals the total number of elements n:

$$\sum_k n_k = n \tag{2.43}$$

The probability $P(A_k)$ of randomly selecting an element (whether failed or surviving) from A_k equals the (combined failed and surviving) fraction of the set A_k of the total sample space Ω:

$$P(A_k) = \frac{n_k}{\sum_k n_k} = \frac{n_k}{n} \tag{2.44}$$

Each population has its own failure distribution, which can be expressed in terms of the functions F_k, R_k, f_k and h_k (see Sections 2.4.1–2.4.3, 2.4.5). Similarly, the overall failure distribution of the inhomogeneous population can be written as:

$$F = \frac{\sum_k n_k F_k}{n} = \sum_k P(A_k) \cdot F_k \tag{2.45}$$

$$R = \frac{\sum_k n_k R_k}{n} = \sum_k P(A_k) \cdot R_k \tag{2.46}$$

$$f = \frac{\sum_k n_k f_k}{n} = \sum_k P(A_k) \cdot f_k \tag{2.47}$$

$$h = \frac{f}{R} = \frac{\sum_k P(A_k) \cdot f_k}{\sum_k P(A_k) \cdot R_k} = \frac{\sum_k P(A_k) \cdot h_k \cdot R_k}{\sum_k P(A_k) \cdot R_k} \tag{2.48}$$

Example 2.7 *An Inhomogeneous Population of Two Subsets* Assume $n = 15$ insulators are installed on a new station. Of these insulators, nine are of Type I and six are of Type II. The two types together form an inhomogeneous insulator with sample size $n = 15$. This population consists of two sets, namely set I of Type I insulators with $n_I = 9$ and set II of Type II insulators with $n_{II} = 6$. In this case these sets are each other's complement:

$$I^C = II \Leftrightarrow II^C = I \tag{2.49}$$

The probability of randomly selecting an arbitrary insulator from a list is equal for each insulator, namely $1/n = 1/15$. However, the probability of selecting an arbitrary insulator that belongs to either subgroup depends on the relative subgroup size. The population fraction of Type I insulators is $P(I)$, while that of Type II insulators is $P(II)$:

$$P(I) = \frac{n_I}{n_I + n_{II}} = \frac{n_I}{n} \tag{2.50}$$

$$P(II) = \frac{n_{II}}{n_I + n_{II}} = \frac{n_{II}}{n} \tag{2.51}$$

Each set has its own failure distribution, denoted as F_I and F_{II}, while the total distribution has distribution F. Functions R, f and h follow the same indexing rules. After a certain time T, three out of nine original Type I insulators failed and four out of six Type II insulators failed. The original inhomogeneous population falls into two sets, namely the eight surviving insulators (set R) and seven failed insulators (set F). Again, these sets are complements:

$$F^C = R \Leftrightarrow R^C = F \tag{2.52}$$

The sets are shown in Figure 2.8. This is just a snapshot in time. With increasing time, more insulators will move from set R to set F.

If, for each insulator type, the failure distributions F_I and F_{II} are known, then the resulting distribution F is also known:

$$F = \frac{n_I F_I + n_{II} F_{II}}{n_I + n_{II}} \tag{2.53}$$

Filling in the values for n_I, n_{II}, F_I and F_{II} yields:

$$F = \frac{9 \cdot \dfrac{3}{9} + 6 \cdot \dfrac{4}{6}}{9 + 6} = \frac{7}{15} \tag{2.54}$$

Then the fraction of Type I insulators in the set of failed insulators is $P(I|F)$, and can be calculated according to Eq. (2.42):

$$P(I|F) = \frac{P(F|I) \cdot P(I)}{P(F)} = \frac{F_I \cdot P(I)}{F} = \frac{F_I \cdot \dfrac{n_I}{n}}{F} \tag{2.55}$$

Substituting the numerical values yields:

$$P(I|F) = \frac{\dfrac{3}{9} \cdot \dfrac{9}{15}}{\dfrac{7}{15}} = \frac{3}{7} \tag{2.56}$$

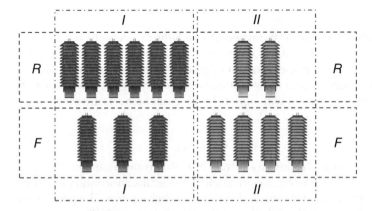

Figure 2.8 Venn diagram of inhomogeneous populations of Type I and Type II insulators (set I and set II, respectively). After a certain time T, the population partly failed and can also be divided into two sets: the surviving set R and the failed set F.

From Figure 2.8 it can be checked that indeed three out of seven failed insulators are Type I insulators.

In Example 2.7 the distribution was described after a fixed time T, but Eq. (2.55) also applies to failure distributions as functions of time t. Writing $P(I|F)$ as $P_{I|F}$ in order to avoid confusion with time, the relationship becomes:

$$P_{I|F}(t) = \frac{F_I(t)}{F(t)} \cdot \frac{n_I}{n_I + n_{II}} \tag{2.57}$$

If n_I and n_{II} are known (or guessed) and both $F(t)$ and $P_{I|F}(t)$ are continuously kept track of by checking the failures as well as to which insulator type the failures belong, then $F_I(t)$ can be derived by rewriting:

$$F_I(t) = P_{I|F}(t) \cdot F(t) \cdot \left(1 + \frac{n_{II}}{n_I}\right) \tag{2.58}$$

This can be further generalized to the case of multiple subsets A_k, as at the start of this section.

A typical example of two sets A and B that together form an inhomogeneous population is shown in Figure 2.9. Assume set A to consist of assets that age faster than the majority of set B. The graph is scaled such that the cumulative failure probability of a single set shows as a straight line as in Figure 2.6. The cumulative failure probability of the total population is a synthesis of the set A and set B functions and appears as a z-shaped curve. The asymptotes approach the two individual functions.

2.5.3 Early Failures Interpreted as Child Mortality

In Table 2.5, child mortality was denoted as a function with a decaying hazard rate. However, often any early failures are indicated as child mortality even when the hazard rate of the early failures is not truly decaying, but may rather be a fast wear-out process of a small part of the population. This is a wider interpretation than in Table 2.5. So, early failures may not only be due to a competing process with decaying hazard rate, but also due to a (hopefully) small population fraction that exhibits imperfections leading to accelerated ageing and consequently early failures. This is elaborated in Example 2.8.

Example 2.8 *A Fast Wearing Out Fraction of an Inhomogeneous Population*
Assume 20 assets in a batch of 1000 assets (so 2%) have a severe defect that leads to fast

Figure 2.9 Typical example of two sets A and B that together form the total population. Set A dominates F at the start, because set A ages considerably faster than set B. After set A is extinguished, set B dominates F.

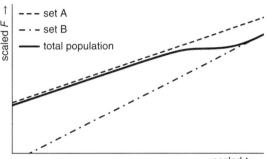

wear out. In this subset of 20 assets an additional fast wear-out process is competing with the normal ageing that also occurs in the other 980 assets, such as random failure due to external influences. Therefore, this is an inhomogeneous population of which a fraction features an accelerated, competing process.

The case is shown in Figure 2.10. If these 20 assets can be recognized and treated as a separate batch, it would reveal that their hazard rate does not decrease but increases instead, albeit on a shorter term than the other 98%. However, if these 20 assets cannot be distinguished from the other 980 assets, then the 20 assets may give the false impression that all 1000 assets have a child mortality process, because just before the 20th failing asset this type of failure became exhausted and the total hazard rate decreased.

In Figure 2.10 the increasing hazard rate of the 2% subpopulation is visible in the top left corner, but if the scale maximum of the hazard rate plot were chosen less than 0.007 and the time remains plotted from 1 a.u., then the early increase of h would not show in the graph. Then the rapid decay at about 3 a.u. would give the false impression that the early h only decays.

The case of a rapid wear-out process of a subpopulation is generally also referred to as a child mortality case. A forensic study may be able to tell the two asset types apart. Subsequently, two hazard rate plots can be drawn that both appear as wear-out phenomena but on different timescales. So, an appearing bath tub curve with child mortality can be due to an accelerated wear-out phenomenon if the fast ageing mechanism only applies to a subpopulation. This mixture of subpopulations and processes is quite common. Quality control at the manufacturer and testing regimes aim to resolve the matter by detecting such assets and preventing them being delivered and installed. Forensic studies usually aim to detect abnormal wear out.

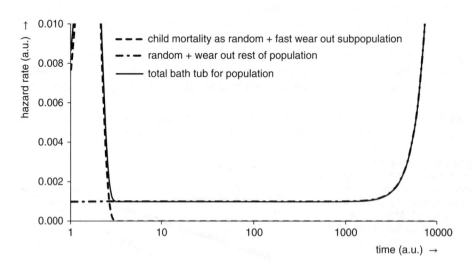

Figure 2.10 A bath tub appearance created by an inhomogeneous population: all assets feature a random failure. The 98% majority also features a wear-out mechanism at about 10 000 a.u., but 2% have a wear-out mechanism that extinguishes this subpopulation at about 3 a.u. (Eq. (2.38) was applied).

At this stage there are two types of child mortality: the first type is defined by having a decreasing hazard rate and is discussed in Section 2.4.5; the second type is defined as a subpopulation with a fast wear-out mechanism (i.e. an increasing hazard rate) and is discussed in the present section. In the literature, the first is usually introduced with Weibull statistics (Section 4.2.1.1), but in practice the second is often encountered. Both are subjects of this book, and it will be stated which type is being discussed.

2.6 Multivariate Distributions and Power Law

The previous section on mixed distributions led to a resulting failure distribution that was built up from the separate original distributions. The resulting distribution is an example of a multivariate distribution, which is a distribution that depends on more than one variable.

This applies to the failure distribution of assets consisting of more than one failure mechanism. For instance, an asset may consist of multiple parts that each can fail, but the first failing part causes the asset to fail. For instance, overhead lines consist of towers and conductors. The conductors are attached to the towers with insulators. There are various failure modes of such line insulators, like mechanical malfunction or electrical breakdown. With mechanical malfunction an insulator may break and the line is dropped, while with electrical breakdown a flash-over may have occurred that leaves the insulator not fit for purpose anymore. Both mechanisms will most likely have their own failure distribution, but they may be independent phenomena. The resulting distribution describing the lifetimes of insulators is a multivariate distribution. Section 2.5 applies where the ageing mechanisms can be considered independent.

Two other cases of failure mechanism are firstly those that consist of multiple degradation stages and secondly those that depend on ageing mechanism conditions.

An example of the first case is a phenomenon called water treeing that degrades the polyethylene insulation of underground cables. Water trees consist of nanoscale traces of impurities and damaged polyethylene that may eventually bridge the complete cable insulation. The exact properties of water trees depend on the circumstances during ageing. During the first phase of degradation, water treeing already lowers the breakdown strength of the cable insulation even before bridging the complete insulation. The second stage of failure is the electric breakdown of the cable insulation.

More information about degradation of polymers and stochastic models can be found in [13].

The second case of ageing mechanism conditions concerns the influence of ageing parameters (which may be variables themselves) and is the subject of the following sections. The case is treated in terms of the power law and introduces the concept of ageing dose.

2.6.1 Ageing Dose and Power Law

If items are tested for an ageing mechanism that is relevant to their lifetime, then the time to failure usually depends on the applied stress in an endurance test. For instance: the time to breakdown will depend on the applied electric field strength and frequency; the rate of photo-oxidation of a polymer will depend on the oxygen concentration and/or the intensity and wavelength of UV radiation; the thermal degradation of transformer oil-and-paper insulation will depend on the temperature, and so on. The lifetime of test objects usually depends on one or more stress parameters, which together form the set of ageing conditions.

The distribution and expected performance in the examples above are not fixed but will vary with the ageing conditions. That is, the ageing conditions have an impact on the failure distribution. If the stress on an asset deviates from the nominal stress, the time to failure can be considerably different. The dependency of a distribution on both time and stress can be interpreted as a distribution with a variable and a parameter. But if the stress can vary as well, then time and stress may both be interpreted as variables. Often a power relationship is found between the lifetime τ (i.e. time to failure) and the applied stress B (see Figures 2.11 and 2.12).

As an empirical finding, a so-called power law may apply as shown in Figure 2.11 and Eq. (2.59), where the lifetime τ multiplied by the stress B to a given power p appears constant:

$$\tau \cdot B^p = D_{tot} \tag{2.59}$$

Here, power p appears, after which the power law is named. The constant D_{tot} in the equations is called the damage or ageing dose at failure. Each asset has its own D_{tot} which is linked directly to its time to failure τ. The power p and the dose D_{tot} are the constants in the power law model defining the relation between lifetime τ and stress B for each asset. These constants p and D_{tot} have to be determined from case to case.

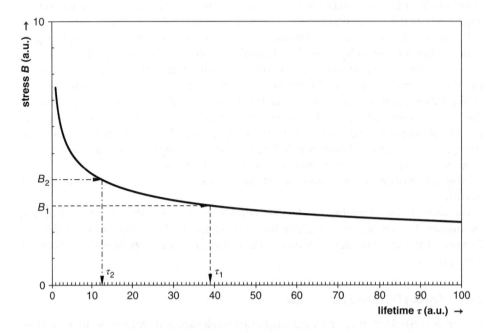

Figure 2.11 A power law relation between stress B and time to failure τ on linear–linear scale. The more severe stress B_2 leads to a shorter lifetime τ_2 than τ_1 at stress B_1.

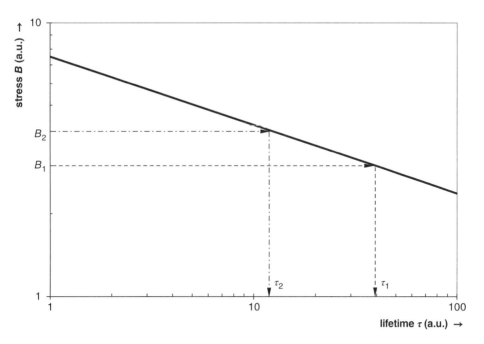

Figure 2.12 A linear relation between the logarithms of stress B and lifetime τ often appears. The more severe stress B_2 leads to a shorter lifetime τ_2 than τ_1 at stress B_1.

In linearized form this relation shows as in Figure 2.12 and can be expressed as:

$$\log B = \frac{1}{p} \log D_{tot} - \frac{1}{p} \log \tau \tag{2.60}$$

The dimension of D_{tot} is time multiplied by the stress dimension to a power p (which does not need to be an integer). The power law can also be written in a relative form to obtain a dimensionless expression of the relative ageing dose $D_\%$ at time τ:

$$\frac{\tau \cdot B^p}{D_{tot}} = D_\%(\tau) = 1 \tag{2.61}$$

If the stress B is fixed, the relative ageing dose $D_\%(t)$ can be defined for $\theta \leq \tau$ as:

$$\frac{\theta \cdot B^p}{D_{tot}} = D_\%(\theta) \leq 1 \tag{2.62}$$

This expresses that when the ageing time t of the asset reaches τ, the time to failure (i.e. the maximum ageing time), the relative ageing dose $D_\%$ becomes 1 and the asset fails. $D_\%$ can be interpreted as the percentage of the maximum ageing dose (i.e. the ageing dose D_{tot} at failure). If the ageing history period θ is the sum of k periods θ_j with load B_j with $j = 1, \ldots, k$, then $D_\%$ can be written as:

$$\frac{\sum_{j=1}^{k} \theta_j \cdot (B_j)^p}{D_{tot}} = D_\%((\theta_1, B_1), \ldots, (\theta_k, B_k)) \leq 1 \tag{2.63}$$

In integral form the relative ageing dose of an asset with varying stress and ageing history time $\theta \leq \tau$ is written as:

$$\frac{\int_0^\theta (B(t))^p dt}{D_{tot}} = D_\%(\theta; B(t)) \leq 1 \tag{2.64}$$

Again, it should be noted that the asset fails at $D_\% = 1$ (or 100% relative damage dose) in the power law model and that $D_\%(\theta)$ strongly depends on the stress profile $B(t)$.

In some cases the power law from the failure distribution is a theoretically correct description. In other cases the power law may apply only in good approximation and possibly only in a limited range, while outside this range lifetime and stress have a different dependency.

2.6.2 Accelerated Ageing

The power law is often used to translate the lifetimes τ_i from one stress level to another. If two stress levels B_1 and B_2 are considered and the power law is assumed to apply, then Eq. (2.59) with constant D_{tot} links lifetimes and stresses:

$$\tau_1 \cdot B_1^p = D_{tot} = \tau_2 \cdot B_2^p \tag{2.65}$$

This can be rewritten as:

$$\tau_1 = \tau_2 \cdot \left(\frac{B_2}{B_1}\right)^p \tag{2.66}$$

A practical case is worked out in Example 2.9 below.

Example 2.9 *Accelerated Ageing Test of a Capacitor* Assume a capacitor has a nominal operational stress B_1 of 500 V. Over time such capacitors age and break down electrically. A capacitor is tested at an enhanced stress B_2 of 1 kV, which is known to accelerate the ageing process. The test result at enhanced stress level B_2 is a lifetime τ_2 of 6 hours. Also assume that from a range of experiments it is already known that the power law applies and that the power $p = 10$. From the time to failure τ_2 in the test and knowledge of the stresses B_1 and B_2, the equivalent lifetime τ_1 if the asset was aged at stress B_1 can be estimated from Eq. (2.66):

$$\tau_1 = \tau_2 \cdot \left(\frac{B_2}{B_1}\right)^p = 6 \cdot \left(\frac{1\,\text{kV}}{500\,\text{V}}\right)^{10} \text{hr} = 6 \cdot 1024\,\text{hr} = 6144\,\text{hr} \tag{2.67}$$

The stresses B_1 and B_2 differ by a factor 2, but their lifetimes τ_1 and τ_2 differ by a factor 1024. A power p in the range 7–14 is not unusual. It shows that maltreating (i.e. applying – even relatively briefly – a much more severe than nominal stress to an asset can lead to much faster ageing and failure). It also shows that testing at elevated stress levels consumes lifetime. Particularly with older assets, the stress level at a repeated test must be considered in the light of possibly wasting remaining life.

The power law and relative ageing dose $D_\%$ as in Eqs (2.63) and (2.64) also provide a means to translate the life of assets with varying stresses into the equivalent life θ_0 at nominal stress B_0. For the case that the ageing history time θ of the asset consists of k

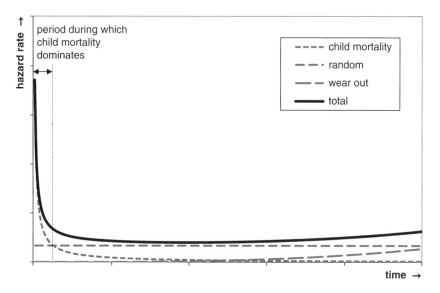

Figure 2.13 A case where a child mortality process dominates the hazard rate. Accelerated ageing aims at consuming the child mortality lifetime until the hazard rate drops to an acceptable level. After that, the delivered products are largely cleared from the high initial hazard rate.

periods θ_j with stresses B_j (see Eq. (2.63)), the corresponding estimated lifetime θ_0 at nominal stress B_0 becomes:

$$\theta_0 = \frac{D_\% \cdot D_{tot}}{(B_0)^p} = \frac{\sum_{j=1}^{k} \theta_j \cdot (B_j)^p}{(B_0)^p} \leq \tau_0 \tag{2.68}$$

And in terms of a continuously varying stress $B(t)$:

$$\theta_0 = \frac{D_\% \cdot D_{tot}}{(B_0)^p} = \frac{\int_0^\theta (B(t))^p dt}{(B_0)^p} \leq \tau_0 \tag{2.69}$$

It should be noted that the total ageing history in these equations is θ, which is not necessarily equal to the estimated lifetime θ_0 at nominal stress B_0.

The value of these equations is that the lifetimes of assets with a different, but known, ageing profile can now be compared under the assumption that a power law applies and the power p is known. An illustrative case is elaborated in Section 9.4.2.2.

Example 2.10 *Preventing Early Failures After Delivery* Dominating child mortality processes can be burnt out of a product batch by accelerated ageing as a method of quality control. This approach applies to both the infant mortality cases of Section 2.4.5 (decreasing hazard rate) and Section 2.5.3 (fast ageing subpopulation). After a period of accelerated ageing, the significance of the child mortality process for the hazard rate is reduced and the remaining part of the batch can be delivered at higher quality. Figure 2.13 shows an example of the first case.

The disadvantage is that some life is already consumed before delivery and the time until the onset of wear out is reduced. Therefore, the accelerated ageing in the pre-delivery period must not be too long and not at too severe ageing conditions.

It must be considered to what level the stress can be enhanced in order to burn out the child mortality efficiently. As three processes are present, a stress enhancement will most likely accelerate all processes. If the power model is applicable to all of them, they may still have different powers. The wear-out mechanism may have a higher or lower power than the child mortality process. It must therefore be analysed in practice how effective accelerated ageing is, and this analysis may result in a range of useful acceleration factors. Forensic studies may reveal which processes have caused the individual test objects to fail.

Example 2.11 shows how accelerated ageing provides a way to assess the power p and the ageing dose D_{tot} at failure.

Example 2.11 *Determining the Power of the Power Law* As mentioned in Section 2.6.1, the power p and dose D_{tot} in the power law model have to be determined from case to case. This can be achieved by endurance tests at various stresses, but another way is to vary the stress B with time (i.e. to vary B as a function of time t). A standard way to do this is a linear relationship with various stress rates (see Figure 2.14). Due to the linear relationship at breakdown for each stress rate c:

$$B(t) = c \cdot t = D_{tot} \tag{2.70}$$

As the stress varies continuously (or with a small adaption in steps), the following holds at any time θ:

$$D_{tot} = \int_0^\theta (B(t))^p dt = c^p \int_0^\theta t^p dt = \frac{c^p \cdot t^{p+1}}{p+1}\bigg|_0^\theta = \frac{c^p \cdot \theta^{p+1}}{p+1} \tag{2.71}$$

In logarithmic terms:

$$\log B = \frac{1}{p} \log D_{tot} - \frac{1}{p} \log \theta + \frac{1}{p} \log(p+1) \tag{2.72}$$

This resembles Eq. (2.60), but differs in the term $(1/p)\log(p+1)$. The background is that in Eq. (2.60) the stress was constant during the lifetime, while in Eq. (2.72) the stress was lower during most of the lifetime and less ageing dose was accumulated.

The power p can be found by repeating destructive testing at various stress rates c. From the slope of $\log(B)$ as a function of $\log(\theta)$ the power can be obtained according to Eq. (2.72), as shown in Figure 2.15. With known stress rate c, the ageing dose at failure D_{tot} can subsequently be calculated through the relationship in Eq. (2.71).

2.6.3 Multi-Stress Ageing

The power law as defined in Eq. (2.59) can be extended for multi-stress ageing (i.e. ageing where the number m of stresses B_j ($j = 1, ..., m$) can vary and independently follow the power law with power p_j). For instance, with electric testing the electric field, frequency,

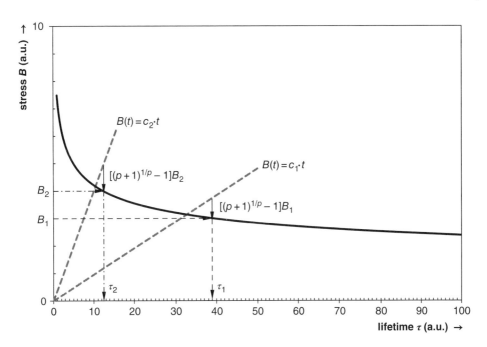

Figure 2.14 A way to determine the power relation is to increase the load with time (e.g. linearly). Failure occurs when time θ reaches lifetime τ.

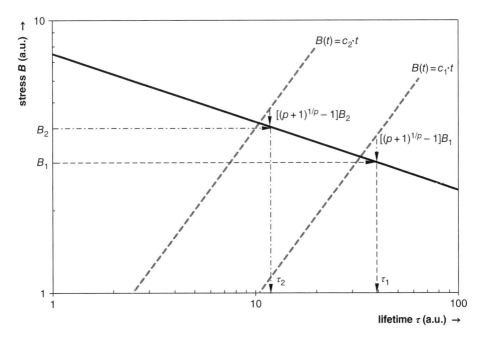

Figure 2.15 A way to determine the power relation is to increase the stress with time (e.g. linearly). Failure occurs at lifetimes τ_i.

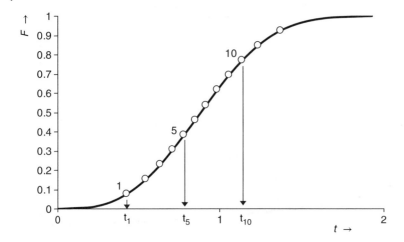

Figure 2.16 Example of cumulative distribution of 12 assets and their breakdown times. The index of the assets is shown for the first, fifth and 10th asset to fail.

mechanical stress and temperature may vary, which could all follow a power law. If these stresses can be regarded as independent indeed, then the power law model becomes:

$$\tau \cdot \prod_{j=1}^{m} B_j^{p_j} = D_{tot} \tag{2.73}$$

If all stresses are kept constant except for one single type of stress, then the product of the constant stresses can be put to the other side of the equation and Eq. (2.73) reduces to Eq. (2.59).

In case the stresses cannot be regarded as independent, their relationship must be put into the equation if known.

As for temperature as an ageing stress, thermal excitation may accelerate degradation processes [14]. The remaining life is then considered to relate to an activation energy ΔH according to the Arrhenius law. The remaining lifetime is influenced by temperature as:

$$t_{\text{remaining}} = A \cdot \exp(\Delta H/kT) \tag{2.74}$$

Here, A is a (time) constant, T is the absolute temperature in Kelvin and k is the well-known Boltzmann constant. Accelerated ageing is possible by increasing the temperature (the stress in this case), but it has an exponential dependency on temperature rather than a power law.

2.6.4 Cumulative Distribution, Ageing Dose and CBM

The power law is applied to means and medians, but can be applied to individual assets as well. Then, Eq. (2.59) applies to each single asset in use. In this approach there is a direct link between the cumulative distribution, variable T and dose D. This is illustrated by an example of 12 assets represented by their cumulative distribution and breakdown times (see Figure 2.16).

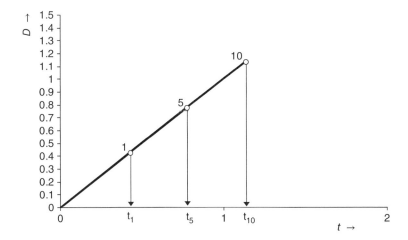

Figure 2.17 Example of ageing dose build-up assets and their breakdown times.

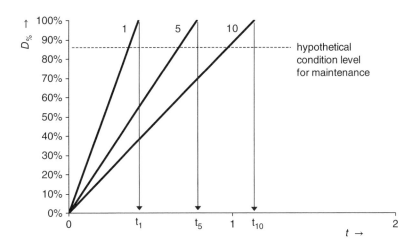

Figure 2.18 Example of relative ageing dose build-up assets and their breakdown times. A hypothetical alert level for CBM is indicated.

In terms of Eq. (2.59), each of the assets builds up an ageing dose D. The weakest asset fails first (at $t = t_1$) followed by the other assets. Figure 2.17 shows the build-up of the ageing dose for 3 of the 12, namely assets 1, 5 and 10, to fail.

Instead of against time, the cumulative distribution can also be plotted against the ageing dose using the linear relationship of Eq. (2.59) and the time t could be replaced by the ageing dose D in Figure 2.16. The same stress on all the assets causes a faster ageing in the weaker assets than in the stronger. This is expressed in Eq. (2.62), where the ageing dose is relative to the individual fatal ageing dose $D_{100\%}$ of each asset. Figure 2.18 shows the build-up of the relative dose for the same assets as in Figure 2.17.

The aim of CBM is to repair or replace assets in a timely manner. The key to successful CBM is the ability to detect on time that the ageing dose D is approaching the

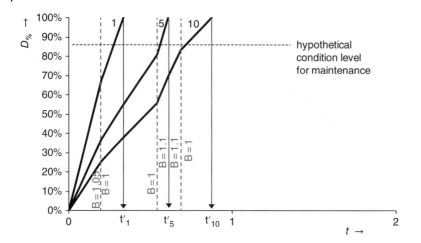

Figure 2.19 Example of relative ageing dose build-up for assets and their breakdown times with varying stress *B* as indicated in the figure. Again, the hypothetical alert level for CBM is indicated.

fatal level D_{tot}, or equivalently that the relative ageing dose $D_\%$ is approaching the $D_{100\%}$ level. Some ageing mechanisms are accompanied by processes that can be detected by diagnostic measurements. For example, electrical treeing leads to breakdown eventually. Measurement of the breakdown has no predictive value (except maybe for locating the repair site), but the development of an electrical tree normally involves partial discharging. This phenomenon consists of sparks that show as current pulses and can be picked up by various methods. The fingerprint of partial discharges may predict the extent of degradation and provide an alert level for maintenance. Figure 2.18 also shows a hypothetical level that might be detected by inspections and announces nearing the end of technical life.

If the stress *B* varies over time, the rate of the ageing doses changes according to Eq. (2.64). As a consequence, the breakdown times t'_i at varying stress may differ from the breakdown times t_i with the reference (i.e. design) stress. Higher stresses shorten the breakdown times and lower stresses lengthen the breakdown times. Figure 2.19 shows the effect of varying stresses *B* on the accumulation of the relative ageing doses $D_\%$.

Some degradation phenomena consist of sequential ageing mechanisms. For instance, XLPE insulation of underground cables can be deteriorated by an ageing phenomenon called water treeing, which lowers the breakdown strength. The exact lowering of the breakdown strength depends on the combination of ageing conditions. An electrical tree can be generated with a rate that depends on the growth characteristics of the water trees. Because there are two sequential ageing processes, two sequential ageing times may be treated with two power laws.

The power law model aims at lifetime and ageing damage. It provides a method to account for different levels of stress *B*. In practice, uncertainties can be expected due to variance in *B*, external influences (if not taken into account in the model), uncertainties with the assessment of the condition, and so on. Therefore, it is not stated that the power model applies exactly to all cases of accelerated ageing, but even if not exact, it may still be a very useful model, both conceptually and quantitatively.

2.7 Summary

This chapter discussed probability and statistical distributions: the basics of set theory (Section 2.1), the probability of events (Section 2.2), overlap and difference in meaning between probability and distributions (Section 2.3), fundamental distribution functions (Section 2.4), mixed distributions (Section 2.5) and multivariate distributions (Section 2.6).

As for the basics of set theory (Section 2.1), in terms of statistics an event is an outcome of an experiment. A single outcome is called an elementary event or element. Groups of possible outcomes form a set. The set of all possible outcomes is called the sample space Ω. A group that does not contain an element is called an empty set. Terms referring to parts and combinations of sets are listed in Table 2.1. A Venn diagram is a two-dimensional representation of the coherence of sets.

As for the probability of events (Section 2.2), if all elementary events are equally likely to occur (which is assumed unless stated otherwise), then the probability of an event out of the set X can be calculated as $P(X) =$ [number of outcomes in X]/[number of outcomes in Ω] (Eq. (2.2)). The probability that the event belongs to the entire sample space Ω is 1 and that it belongs to an empty space is 0. $P(X) \in [0,1]$.

As for overlap and difference in meaning between probability and distributions (Section 2.3), a distribution describes how the outcomes of experiments are distributed. In particular, a failure distribution $F(t)$ describes how the sequence and number of failures are distributed in time (or another performance indicator). This does not change with knowledge about the individual test objects. A probability $P(T < t)$ is a measure for the likelihood that an event takes place before $T = t$. If no knowledge about an individual object is used (i.e. an object is randomly selected), then $P(T < t)$ equals $F(t)$. However, if prior knowledge about the condition exists, $P(T < t)$ may change, but whether prior knowledge is available or not, $F(t)$ remains the same. Distribution and probability are therefore different concepts.

As for the fundamental distribution functions (Section 2.4), the different appearances of a distribution are discussed. These are:

- *Cumulative distribution.* $F(t)$ describes the population fraction that failed before t.
- *Reliability function.* $R(t)$ describes the population fraction that survived at least up to t. R is the complement of F (i.e. $R = 1−F$) (Eq. (2.12)).
- *Distribution density.* $f(t)$ describes the population fraction dF that fails per period dt at t. $f(t)$ is the derivative of F and the negative derivative of R: $f = dF/dt = −dR/dt$ (Eq. (2.16)) and the integral of f over the entire range of t is 1.
- *Probability mass function.* $f_i = f(t_i)$ applies to discrete outcomes t_i and describes the fraction ΔF_i that occurs at t_i.
- *Hazard rate.* $h(t)$ describes the *surviving* population fraction dF that fails per period dt at t. The hazard rate $h(t)$ is equal to $f(t)$ scaled to the surviving fraction $R(t)$ (i.e. $h = f/R$) (Eq. (2.22)).
- *Cumulative hazard rate.* $H(\Delta t)$ describes the accumulated failure hazard over a period $\Delta t = [t_{start}, t_{end}]$ and therefore is a function of two variables: t_{start} and t_{end}. If the start of the period is taken as $t_{start} = 0$, then it is customary to write H as a single-variable function $H(0, t = t_{end}) = H(t)$. H is an integral of h: $H = \int h \, dx$ (Eqs (2.27) and (2.23)).

The hazard rate $h(t)$ is used in the bath tub model that describes three ageing types that dominate the lifecycle of assets: 'child mortality' (early failures), random failure (normal operating life) and wear out (end of life). These three are usually indicated as $dh/dt < 0$, $dh/dt = 0$ and $dh/dt > 0$. It is noteworthy though that early failures are commonly called 'child mortality' cases, but these do not necessarily have a decreasing hazard rate $dh/dt < 0$.

As for mixed distributions (Section 2.5), the consequences of competing processes and inhomogeneous populations (i.e. mixed, differently behaving subpopulations) are discussed. Competing processes are also the foundation of the bath tub model, whereas mixed subpopulations lead to an alternative bath tub model (Section 2.5.3). The theorem of Bayes is about conditional statistics and states that the probability $P(A \mid B)$ that an outcome of set A occurs while it is previously known that set B already applies can be determined by $P(B \mid A) \cdot P(A)/P(B)$ (Eq. (2.42)). This is particularly useful if $P(B \mid A)$, $P(A)$ and $P(B)$ can be determined.

As for multivariate distributions and the power law (Section 2.6), the focus is on accelerated ageing. In this respect test parameters (i.e. applied stresses) are regarded as variables that have an impact on the failure distribution. The concept of ageing dose is introduced in Section 2.6.1 as a measure of damage. The power law is discussed and used to translate consumed lifetime τ_1 at one stress level B_1 to the equivalent consumed lifetime τ_2 at another stress level B_2. $\tau_1 = \tau_2 \cdot (B_2/B_1)^p$ (cf. Eqs (2.59) and (2.67)). This is elaborated for situations where stress levels vary during the lifetime of an asset. It may be noted that accelerated ageing does not always follow a power law (cf. thermal excitation in Section 2.6.3), but for many processes the power law provides a useful method to compare ageing at different stress levels.

Questions

2.1 With reference to Section 2.1: what set is $\Omega \backslash A \cup B$?

2.2 With reference to Section 2.4.5: derive the equation for the total hazard rate h_{tot} in the bath tub model from the hazard rates of child mortality, random failure and wear-out failure (see Figure 2.5).

2.3 With reference to Section 2.5.1: derive the equations for the cumulative distribution F (see Eq. (2.30)), the distribution density f (see Eq. (2.31)) and the hazard rate h (see Eq. (2.32)) from the equation of the reliability R in Eq. (2.29).

2.4 With reference to Section 2.5.2.1:
   a. Consider Table 2.8 and fill in the other cases.
   b. Check the validity of Eqs (2.39)–(2.42).
   c. What if $P(A)$ and $P(B)$ are not independent in those cases?

2.5 With reference to Example 2.7 in Section 2.5.2.2: how, from knowledge of n_I, n_{II}, F and $P(II|F)$, can F_I also be calculated?

2.6 With reference to Section 2.5.2.2: derive an equation similar to Eq. (2.58) for any $F_k(t)$ being the failure distribution of set A_k. What is the minimum required information for calculating $F_k(t)$?

2.7 With reference to Section 2.6.2: in case of burning out child mortality assume that the domination of early failures lasts 25 days in operational conditions. Assume all processes have the same power $p = 9$. The stress is increased by a factor 3. Determine the required period of accelerated ageing to burn out dominant child mortality.

3

Measures in Statistics

The concepts of probability and statistical distribution were discussed in Chapter 2. We defined the relationship between the failure distribution F and a performance variable X, and how ageing conditions influenced the results. Once the distributions are known, conclusions can be drawn about the distributed outcomes. For instance, if the distribution of breakdown times is determined in an endurance test, the data can be analysed and conclusions drawn on questions like:

- What is the average breakdown time? This is about the mean and expected values (Section 3.1).
- How much do the breakdown times scatter? This is about the standard deviation and the variance (Section 3.1.8).
- At what time will a given percentage of assets have failed? This is about medians and quantiles (Section 3.2).
- When will the highest failure frequency occur? This is about the mode (Section 3.3).
- How do the mean, median and mode compare? This is discussed in Section 3.4.

  And if various distributions have been determined:

- How well do the variables and covariables correlate? This is about covariance and correlation (Section 3.5).
- How similar are distributions? How well does an observed distribution comply with a specified distribution? This is about the similarity index (Section 3.6) and compliance (Section 3.7).

These questions are all about measures or quantities that characterize the distribution of experimental outcomes. In Chapter 4, these measures will be used to characterize specific distributions like the Weibull, exponential, normal, and more. Chapter 5 deals with graphic representations of distributions and Chapter 6 with distribution parameter estimation, where the measures will prove important again. But as a start, the above questions will first be answered in this chapter by discussing various measures that characterize distributions.

3.1 Expected Values and Moments

The mean or arithmetic mean $<x>$ represents the centre of gravity (or centre of mass) of a distribution variable X. The expected value of a distributed variable X is written as

Reliability Analysis for Asset Management of Electric Power Grids, First Edition. Robert Ross.
© 2019 John Wiley & Sons Ltd. Published 2019 by John Wiley & Sons Ltd.
Companion website: www.wiley.com/go/ross/reliabilityanalysis

$E(X)$ and is equal to the mean. In the discrete case of a finite number n of all variable values x_j (with index $j = 1,2,\ldots,n$), the mean $<X>$ of a discrete uniform distribution is calculated as:

$$E(X) = \langle X \rangle = \frac{1}{n} \sum_{j=1}^{n} x_j \tag{3.1}$$

If variables with the same values are grouped and/or values x_j can have different frequencies, the probability mass function $f(x_j)$ will describe the weight each variable value has in the mean. The expected value $E(X)$ is then given by:

$$E(X) = \langle X \rangle = \frac{\sum_{j=1}^{n} f(x_j)\,x_j}{\sum_{j=1}^{n} f(x_j)} \tag{3.2}$$

In case of a continuous distribution F, the mean $<x>$ can be calculated as:

$$E(X) = \langle X \rangle = \frac{\int_0^1 x(F)dF}{\int_0^1 dF} = \int_0^1 x(F)dF \tag{3.3}$$

As $f(x)$ is the derivative of $F(x)$ (see Eq. (2.16)), rewriting Eq. (3.3) yields:

$$E(X) = \langle X \rangle = \int_0^1 x(F)dF = \int_{-\infty}^{\infty} x \cdot \frac{dF}{dx}\,dx = \int_{-\infty}^{\infty} x \cdot f(x)\,dx \tag{3.4}$$

A special case of determining the expected value concerns the expected value of the cumulative distribution itself (see Example 3.1).

Example 3.1 *Expected Value of the Failure Distribution* An example of such an expected value is the expected value of F itself, so in this case X is defined as the identity function I:

$$X(F) = I(F) = F \tag{3.5}$$

The expected value $<F>$ is:

$$\langle F \rangle = \int_0^1 F\,dF = \frac{1}{2}F^2 \Big|_0^1 = \frac{1}{2} \tag{3.6}$$

An important application of an expected value is the mean lifetime θ with failure distribution $F(t)$. The lower integration boundary is then 0 if t is taken as $0 \le t < \infty$:

$$\theta = \langle t \rangle = \int_0^1 t(F)\,dF \tag{3.7}$$

In the form of Eq. (3.4), this becomes:

$$\theta = \langle t \rangle = \int_0^{\infty} t \cdot f(t)\,dt = \int_0^{\infty} R(t)\,dt \tag{3.8}$$

The right-hand side of Eq. (3.8) is not trivial. It is derived by partial integration of the third part of the equation under the condition that the product $t \cdot R(t)$ approaches 0 with increasing t.

The discrete analogue of Eq. (3.4) is:

$$E(X) = \langle X \rangle = \sum_{j=1}^{n} x_j f_j \tag{3.9}$$

where f_j equals $1/n$ if all x_j are different (i.e. weigh equally). The probability mass function can have different values if, for instance, the set of possible x_j values contains doubles (or multiples). Instead of having the doubles all in the set, these could also be taken singly with a double (or multiple) density. In Eqs (3.4) and (3.9) the mean is calculated by integration over x with $f(x)$, respectively f_j, as weighing function.

The mean of a function $G(t)$ can be determined in a similar way as Eq. (3.3):

$$\langle G \rangle = \int_0^1 G(t)\,dF(t) = \int_0^\infty G(t)\frac{dF(t)}{dt}dt = \int_0^\infty G(t)f(t)dt \tag{3.10}$$

3.1.1 Operations and Means

After the expected value (i.e. mean) of a variable is determined, follow-up calculations may be necessary. For instance, the mean time per failure may have been determined and next the failure frequency is calculated as the reciprocal of the time to failure. The reciprocal of the time to failure θ is generally not equal to the mean failure frequency ϕ.

Example 3.2 *Mean Time to Failure and Mean Failure Frequency* Assume two insulators are tested in an accelerated ageing test. The first fails after 1 hour, the second after 4 hours. The mean time to failure θ is:

$$E(T) = \theta = \frac{1\,\text{hr} + 4\,\text{hr}}{2} = 2.5\,\text{hr} \tag{3.11}$$

The failure frequency based on mean time to failure is $1/\theta = 0.4\,\text{hr}^{-1}$. The individual failure frequencies are $1\,\text{hr}^{-1}$ and $0.25\,\text{hr}^{-1}$. Therefore, the mean failure frequency ϕ is:

$$E\left(\frac{1}{T}\right) = \varphi = \frac{1\,\text{hr}^{-1} + 0.25\,\text{hr}^{-1}}{2} = 0.625\,\text{hr}^{-1} = \frac{5}{8}\,\text{hr}^{-1} \tag{3.12}$$

The time to failure based on the mean failure frequency is $1/\phi = 1.6$ hours. Clearly, θ and ϕ are not each other's reciprocal (i.e. the mean is not a preserved property under the reciprocal operation).

Example 3.2 shows that it is important to consider which mean is taken and what is actually the expected value. The rules for the preservation of the mean are as follows.
Summation of X with constant A:

$$\langle X + A \rangle = \langle X \rangle + \langle A \rangle = \langle X \rangle + A \tag{3.13}$$

Product of X and constant A:

$$\langle A \cdot X \rangle = A \cdot \langle X \rangle \tag{3.14}$$

Summation of two variables X and Y:

$$\langle X + Y \rangle = \langle X \rangle + \langle Y \rangle \tag{3.15}$$

Product of two independent variables X and Y:

$$X, Y \text{ independent}: \langle X \cdot Y \rangle = \langle X \rangle \cdot \langle Y \rangle \tag{3.16}$$

Example 3.3 illustrates the calculation of various means.

Example 3.3 *Calculations with Means* Let the domain of variable X be $[0;1]$ (i.e. $0 \leq x \leq 1$). The reciprocal function is non-linear. The reciprocal of the mean $1/<X>$ and the mean of the reciprocal $<1/X>$ are compared:

$$\langle X \rangle = \int_0^1 x \; dx = \frac{1}{2}x^2 \Big|_0^1 = \frac{1}{2} \Rightarrow \frac{1}{<X>} = 2 \qquad (3.17)$$

$$\left\langle \frac{1}{X} \right\rangle = \int_0^1 \frac{1}{x} \; dx = \ln x |_0^1 \to \infty \Rightarrow \frac{1}{<X>} \neq \left\langle \frac{1}{X} \right\rangle \qquad (3.18)$$

Similarly, the means of X and its square X^2 can be compared and are not the same:

$$\langle X \rangle = \int_0^1 x \, dx = \frac{1}{2}x^2 \Big|_0^1 = \frac{1}{2} \Rightarrow (<X>)^2 = \frac{1}{4} \qquad (3.19)$$

$$\langle X^2 \rangle = \int_0^1 x^2 \, dx = \frac{1}{3}x^3 \Big|_0^1 = \frac{1}{3} \Rightarrow \langle X^2 \rangle \neq (\langle X \rangle)^2 \qquad (3.20)$$

Of course, occasionally the boundaries and functions can be such that in this special case non-linear operations do yield the same mean, but in general the mean is not preserved with non-linear operations.

On the other hand, the mean is preserved under summation. For example:

$$\langle X + X^2 \rangle = \int_0^1 (x + x^2) \, dx = \int_0^1 x \, dx + \int_0^1 x^2 \, dx = \langle X \rangle + \langle X^2 \rangle \qquad (3.21)$$

In case G is a function of variables X and Y:

$$\langle G(X, Y) \rangle = \int_{-\infty}^{\infty} \int_{-\infty}^{\infty} G(x, y) f(x, y) dx dy \qquad (3.22)$$

The discrete analogue is:

$$\langle G(X, Y) \rangle = \sum_x \sum_y G(x, y) f(x, y) \qquad (3.23)$$

It should be noted that the mean of a function of a variable is not necessarily equal to the function of the mean of the variable. So, except for special cases:

$$\text{generally} : \langle g(X) \rangle \neq g(\langle X \rangle) \qquad (3.24)$$

The next section discusses the moments of a distribution which are examples of means of functions, where the rules above will be applied.

3.1.2 Bayesian Mean

If an asset is in operation without failure for a period of time t_{op}, then it certainly does not belong to the population part that failed before $t = t_{op}$. With the knowledge that the life will exceed t_{op}, the expected life of such assets will also be larger than the original θ of all assets. As the expected lifetime of a reduced part of the population has to be assessed, the lower integral boundary has to be adjusted for both averaging over time as well as the population part:

$$\theta|_{t>t_{op}} = \langle t \rangle |_{t>t_{op}} = \frac{\int_{t_{op}}^{\infty} t \cdot f(t) \, dt}{\int_{t_{op}}^{\infty} f(t) \, dt} \qquad (3.25)$$

Up to this stage it is assumed that no other knowledge about this asset is available than the minimal lifetime. If a better estimation of the lifetime can be made because information about the asset condition is available, then this will have an effect as well. For instance, if a certain condition would lead to the understanding that a probability function $g(t)$ is valid rather than $f(t)$ and this is established at moment $t = t_{op}$, then the expected time θ_{cond} becomes:

$$\theta_{cond}|_{t>t_{op}} = \langle t \rangle_{cond,t>t_{op}} = \frac{\int_{t_{op}}^{\infty} t \cdot g(t)\, dt}{\int_{t_{op}}^{\infty} f(t)\, dt} \qquad (3.26)$$

The suffix 'cond' means that from the specific condition it is concluded that $g(t)$ is valid for the asset that still survives up to that moment (so both the condition and the minimum survival time do hold). The integral is therefore also over the part of the population that survives up to at least $t = t_{op}$.

In the following sections it is assumed that only the distribution is known (i.e. no other information that would justify a Bayesian approach).

3.1.3 The Moments of a Distribution

The moments are a series of measures that can be used to characterize a distribution, like a fingerprint. The moments μ_k are the mean values of k-powers of the performance variable X. The kth moment $<X^k>$ of the distribution for the discrete, respectively, continuous case is defined as:

$$\mu_k = \langle X^k \rangle = \sum_{j=1}^{n} (x_j)^k \cdot f_j \qquad (3.27)$$

$$\mu_k = \langle X^k \rangle = \int_0^1 x^k\, dF = \int_{-\infty}^{\infty} x^k f(x)\, dx \qquad (3.28)$$

The first moment $\mu = \mu_1$ is the mean $<x>$ as defined in Eqs (3.1), (3.3), (3.4) and (3.9).

3.1.4 EXTRA: Moment Generating Function

Sometimes it is more convenient to compute the moments with the aid of an auxiliary function rather than through Eqs (3.27) and (3.28). Such a function is the exponential of $t{\cdot}X$, where X is the variable of the distribution and t is the variable of this so-called moment generating function $G_X(t)$. The expected value of the derivatives produces the moments. In the case of a discrete distribution (characterized by $f(x)$):

$$G_X(t) = \langle e^{t \cdot X} \rangle = \sum_j e^{t \cdot x_j} f(x_j) \qquad (3.29)$$

The first moment is then found by a single differentiation (provided this is possible):

$$\frac{d}{dt} G_X(t) = \frac{d}{dt} \langle e^{t \cdot X} \rangle = \sum_j x_j e^{t \cdot x_j} f(x_j) \qquad (3.30)$$

$$\frac{d}{dt} G_X(t)\Big|_{t=0} = \langle X \rangle \qquad (3.31)$$

Comparison with Eq. (3.27) shows that this equals the first moment $<X>$ for $t = 0$. Likewise, the kth moment $<X^k>$ of the distribution is derived from the kth derivative and setting $t = 0$:

$$\frac{d^k}{dt^k} G_X(t) = \frac{d^k}{dt^k} \left\langle e^{t \cdot X} \right\rangle = \sum_j x_j^k e^{t \cdot x_j} f(x_j) \tag{3.32}$$

$$\frac{d^k}{dt^k} G_X(t) \bigg|_{t=0} = \left\langle X^k \right\rangle \tag{3.33}$$

Similarly, for the continuous case a moment generating function $G_X(t)$ can be defined:

$$G_X(t) = \left\langle e^{t \cdot X} \right\rangle = \int e^{t \cdot x} f(x) \, dx \tag{3.34}$$

Again, the moments can be found by taking the kth derivative ($k \geq 1$):

$$\frac{d^k}{dt^k} G_X(t) = \frac{d^k}{dt^k} \left\langle e^{t \cdot X} \right\rangle = \int x^k e^{t \cdot x} f(x) \, dx \tag{3.35}$$

Again, the moments are found by setting $t = 0$ as in Eq. (3.33).

As a prerequisite, the series and integrals must converge in order to reach a finite solution. If all moments exist, then the moment generating function $G_X(t)$ can be expanded using the Taylor expansion (for further explanation, see Section 10.2.1):

$$\lim_{t \to 0} G_X(t) = \lim_{t \to 0} \sum_{k=0}^{\infty} \frac{t^k}{k!} \frac{d^k}{dt^k} G_X(t) = \lim_{t \to 0} \sum_{k=0}^{\infty} \frac{t^k}{k!} \left\langle X^k \right\rangle \tag{3.36}$$

If this series converges for $t > 0$ then $G_X(t)$ is uniquely defined by the moments $<X^k>$ (cf. [71]).

3.1.5 EXTRA: Characteristic Function

Convergence problems with the moment generating function may occur, but can often be avoided by the complex characteristic function $\varphi_X(t)$, which in case of a continuous distribution is close to the inverse Fourier transform of the distribution density function $f(x)$:

$$\phi_X(t) = \left\langle e^{it \cdot X} \right\rangle = \int_{-\infty}^{\infty} e^{it \cdot x} f(x) \, dx = \int_0^1 e^{it \cdot X(F)} \, dF \tag{3.37}$$

where t is real and the domain of x may have to be adjusted depending on the exact distribution. The function $X(F)$ is the inverse of the cumulative distribution $F(x)$. It may be noted that the difference with the standard Fourier notation is a factor of -2π, which can be regarded as a matter of scaling. Mathematical treatment of the characteristic function is beyond the scope of the present book, but some important properties are summarized here.

If the random variable X has a moment generating function $G_X(t)$, with $t \in \mathbb{R}$, then the domain of the characteristic function can be broadened to the complex numbers (i.e. $t \in \mathbb{C}$). Then, $G_X(t)$ and $\varphi_X(t)$ are related as:

$$\phi_X(-i \cdot t) = G_X(t) \tag{3.38}$$

If the moments of variable X exist up to at least kth order, then again a Taylor expansion for $t \to 0$ can be carried out. The moments are related to the derivatives of $\varphi_X(t)$ as:

$$\langle X^k \rangle = (-i)^k \frac{d^k}{dt^k} \phi_X(t) \Big|_{t=0} \tag{3.39}$$

The characteristic function of a real numbered variable always exists. If two random variables X_1 and X_2 have distributions $F_1(x)$ and $F_2(x)$, then their characteristic functions $\varphi_{X1}(t)$ and $\varphi_{X2}(t)$ are the same if and only if the distributions are the same:

$$F_1 = F_2 \quad \Leftrightarrow \quad \phi_{X1} = \phi_{X2} \tag{3.40}$$

The characteristic function of the product of a variable X and a constant A is:

$$\phi_{AX}(t) = \langle e^{it \cdot AX} \rangle = \langle e^{iAt \cdot X} \rangle = \phi_X(At) \tag{3.41}$$

The characteristic function of the sum of two independent variables X_1 and X_2 is:

$$\phi_{X1+X2}(t) = \phi_{X1}(t) \cdot \phi_{X2}(t) \tag{3.42}$$

An important result is that the characteristic function of the mean of independent variables X_j with $j = 1, \ldots, n$ (n a natural number) is:

$$\phi_{\frac{1}{n} \sum_j X_j}(t) = \prod_j \phi_{X_j} \left(\frac{t}{n} \right) \tag{3.43}$$

If the variables X_j are the same, then Eq. (3.43) becomes:

$$\phi_{<X>}(t) = \left[\phi_X \left(\frac{t}{n} \right) \right]^n \tag{3.44}$$

Finally, granted the distribution density exists, it can be derived through the inverse Fourier transform when the characteristic function $\varphi_X(t)$ is known. As mentioned under Eq. (3.37), a scaling correction has to be carried out in order to retrieve $f(x)$ through the inverse Fourier transform:

$$f(x) = \frac{1}{2\pi} \int_{-\infty}^{\infty} e^{it \cdot x} [\phi_X(t)]^* \, dt \tag{3.45}$$

Here the asterisk denotes the complex conjugate. Even when the variable has no density function, the characteristic function can still be regarded as the Fourier transform of a measure of the variable.

3.1.6 Central Moments of a Distribution

The central moments are defined as the moments of X minus its mean μ. The mean bears information about the position of the centre of mass of the population and the central moments bear information about the distribution shape of the population around this centre of mass. The kth central moment of the distribution for the discrete (respectively continuous) case is defined as:

$$\langle (X - \mu)^k \rangle = \sum_{j=1}^{n} (x_j - \mu)^k \cdot f_j \tag{3.46}$$

$$\langle (X - \mu)^k \rangle = \int_0^1 (x - \mu)^k dF = \int_{-\infty}^{\infty} (x - \mu)^k f(x) dx \tag{3.47}$$

The central moments can be expressed in the moments. In general, the kth central moment can be written in terms of the moments μ_i and as:

$$\left\langle (X - \mu)^k \right\rangle = \left\langle \left[\sum_{i=0}^{k} \binom{k}{i} X^i \cdot (-1)^{k-i} \cdot \mu^{k-i} \right] \right\rangle = \sum_{i=0}^{k} \binom{k}{i} \mu_i \cdot (-1)^{k-i} \cdot \mu^{k-i}$$

(3.48)

3.1.7 The First Four Central and Normalized Moments

The first four central moments are often used to characterize the distribution, like a fingerprint. The central moments can be expressed in terms of the moments. The first central moment $<X - \mu>$ simply reflects the fact that the first moment $<X>$ equals μ and therefore the first central moment becomes zero:

$$\langle X - \mu \rangle = \langle X \rangle - \mu = \mu - \mu = 0$$

(3.49)

The second central moment is the variance of the distribution. The variance is a measure of the (squared) scatter of variable X around its mean μ. The variance is:

$$\text{var}(X) = \left\langle (X - \mu)^2 \right\rangle = \left\langle X^2 \right\rangle - 2 \cdot \langle X \rangle \cdot \mu + \mu^2 = \mu_2 - \mu^2$$

(3.50)

Here, μ_2 is the second moment (see Section 3.1.3). The (positive) square root of the variance is called the standard deviation σ and is a measure of the scatter. The advantage of the standard deviation is that it has the same dimension as the mean. The standard deviation σ is:

$$\sigma = \left| \sqrt{\langle X^2 \rangle - \langle X \rangle^2} \right| = \left| \sqrt{\mu_2 - \mu^2} \right|$$

(3.51)

Because of this definition, the variance $\text{var}(X)$ is often represented by σ^2:

$$\text{var}(X) = \sigma^2$$

(3.52)

Almost trivially, if normalized with the squared standard deviation, the normalized variance becomes equal to 1:

$$\frac{\text{var}(X)}{\sigma^2} = \frac{\left\langle (X - \mu)^2 \right\rangle}{\sigma^2} = \frac{\mu_2 - \mu^2}{\sigma^2} = 1$$

(3.53)

The third central moment is a measure of the asymmetry of the distribution, but not always conclusive. If the third moment is normalized with the standard deviation, it is called the skewness of the distribution. The skewness γ_1 is:

$$\gamma_1 = \frac{\left\langle (X - \mu)^3 \right\rangle}{\sigma^3} = \frac{\mu_3 - 3\mu_2\mu + 2\mu^3}{\sigma^3}$$

(3.54)

If the skewness is not equal to zero, then the distribution of X is not symmetrical around μ. A skewness equal to zero, however, does not imply that X is distributed symmetrically around its mean. So, the skewness is only decisive if it is non-zero. The skewness is positive if the distribution has a longer right tail (i.e. is skewed to the right). The skewness is negative if the distribution has a longer left tail (i.e. is skewed to the left).

The fourth central moment is a measure of the peakedness and the heaviness of the tails of the distribution. If it is normalized with the standard deviation, it is called the kurtosis of the distribution. The kurtosis γ_2 is therefore:

$$\gamma_2 = \frac{\left\langle (X - \mu)^4 \right\rangle}{\sigma^4} = \frac{\mu_4 - 4\mu_3\mu + 6\mu_2\mu^2 - 3\mu^4}{\sigma^4}$$

(3.55)

The Gaussian or normal distribution has a kurtosis of 3. This distribution (although the domain may be quite different) is often taken as a reference and the 'excess kurtosis' $\gamma_{2,e}$ is defined as:

$$\gamma_{2,e} = \frac{\langle (X - \mu)^4 \rangle}{\sigma^4} - 3 \qquad (3.56)$$

Unfortunately, in the literature kurtosis and excess kurtosis are often not distinguished very well, which may cause confusion. Even more, the term γ_2 is used in the literature for both kurtosis and excess kurtosis depending on the source. A zero excess kurtosis distribution is called mesokurtic and the most important example is the Gaussian distribution. A distribution with a negative excess kurtosis is called platykurtic and a distribution with a positive excess kurtosis is called leptokurtic.

Mathematically, there is also another reason for using the excess kurtosis, namely skewness, and (excess) kurtosis can also be related to the so-called cumulants κ_i. In this book it is explicitly indicated which kurtosis is being discussed.

3.1.8 Mean, Standard Deviation, and Variance of a Sample

The variance and standard deviation as defined in Eq. (3.50), respectively Eq. (3.51), describe the mean deviation of variable X from its mean $<X>$. Instead of a complete population with infinite sample size (or large N), in practice often only a limited sample of n observations is available with values x_j ($j = 1, \ldots, n$; $n < N$). It is assumed that this is a random selection. This section describes the relation of the average and the variance of this sample and compares it with the population mean and variance.

Having only the sample and in principle no knowledge of the exact population, at best the average $\bar{x}$ is known rather than the population mean $<X>$:

$$\bar{x} = \frac{1}{n} \sum_{j=1}^{n} x_j \qquad (3.57)$$

Note that this case is different from the case in Eq. (3.1) or Eq. (3.2), where n is the sample size of the complete distribution. In that case the average and the mean are the same. But here, n is not the population size but the (smaller) sample size. So, generally, the sample average $\bar{x}$ is not the population mean $<X>$. Furthermore, the observations x_j are not ranked, which implies that each x_j is a random sample x and for each j it holds that the mean $<x_j>$ equals the population mean $<X>$.

As for the first moment of the average $\bar{x}$, the expected value $<\bar{x}>$ also equals $<X>$:

$$\langle \bar{x} \rangle = \left\langle \frac{1}{n} \sum_{j=1}^{n} x_j \right\rangle = \frac{1}{n} \sum_{j=1}^{n} \langle x_j \rangle = \frac{n}{n} \langle X \rangle = \langle X \rangle \qquad (3.58)$$

The difference between an observation x_j and the expected or mean $<X>$ can be regarded as an observation deviation or error e_j from the expected value $<X>$:

$$e_j = x_j - \langle X \rangle \qquad (3.59)$$

Note that the expected observation error $<e_j> = <e>$ is zero:

$$\langle e_j \rangle = \langle x_j - \langle X \rangle \rangle = \langle X \rangle - \langle X \rangle = \langle e \rangle = 0 \qquad (3.60)$$

Likewise, the error between the average and the expected value is defined as an error E in the average. This appears equal to the average observation error:

$$E = \bar{x} - \langle X \rangle = \frac{1}{n} \sum_{j=1}^{n} x_j - \langle X \rangle = \frac{1}{n} \sum_{j=1}^{n} e_j \tag{3.61}$$

The mean square of the observation error $<e^2>$ equals the variance σ^2 (cf. Eq. (3.52)):

$$\langle e^2 \rangle = \langle e_j^2 \rangle = \langle x_j^2 \rangle - 2 \langle x_j \cdot \langle X \rangle \rangle + \langle X \rangle^2$$
$$= \langle X^2 \rangle - \langle X \rangle^2 = \sigma^2 \tag{3.62}$$

The mean square of the average error $<E^2>$ equals:

$$\langle E^2 \rangle = \langle (\bar{x} - <X>)^2 \rangle = \left\langle \left(\frac{1}{n} \sum_{j=1}^{n} x_j - <X> \right)^2 \right\rangle$$

$$= \frac{1}{n^2} \left\langle \sum_{j=1}^{n} e_j^2 \right\rangle - \left\langle \sum_{\substack{k=1 \\ j \neq k}}^{n} \sum_{j=1}^{n} e_j e_k \right\rangle = \frac{1}{n^2} \left\langle \sum_{j=1}^{n} e_j^2 \right\rangle - \frac{1}{n^2} \left\langle \sum_{j=1}^{n} e_j \right\rangle^2$$

$$= \frac{1}{n} \langle e^2 \rangle - \langle e \rangle^2 = \frac{1}{n} \langle e^2 \rangle$$

$$= \frac{1}{n} \sigma^2 = \sigma_n^2 \tag{3.63}$$

The j,k cross-term in Eq. (3.63) cancels out because the summations are independent and the expected value $<e> = 0$ according to Eq. (3.60). The standard deviation of the population is σ and the standard deviation of the average $\bar{x}$ of an n-sized sample drawn from the population is σ_n.

The difference between observation x_j and $\bar{x}$ is called the residual d_j:

$$d_j = x_j - \bar{x} \tag{3.64}$$

The average squared residual $\overline{d_j^2}$ is called the variance s^2 of the sample x_j ($j = 1, \ldots, n$):

$$s^2 = \frac{1}{n} \sum_{j=1}^{n} d_j^2 = \frac{1}{n} \sum_{j=1}^{n} (x_j - \bar{x})^2 \tag{3.65}$$

The relationship with the population variance σ^2 can be found by inserting $0 = <X> - <X>$ and taking the mean $<s^2>$ of the sample variance s^2 in Eq. (3.65):

$$\langle s^2 \rangle = \left\langle \frac{1}{n} \sum_{j=1}^{n} (x_j - \langle X \rangle - \bar{x} + \langle X \rangle)^2 \right\rangle = \left\langle \frac{1}{n} \sum_{j=1}^{n} (e_j - E)^2 \right\rangle$$

$$= \left\langle \frac{1}{n} \sum_{j=1}^{n} e_j^2 - 2E \frac{1}{n} \sum_{j=1}^{n} e_j + E^2 \right\rangle = \langle e^2 \rangle - \langle E^2 \rangle \tag{3.66}$$

From Eqs (3.62) and (3.63) it follows that:

$$\langle s^2 \rangle = \langle e^2 \rangle - \langle E^2 \rangle = \sigma^2 - \sigma_n^2 = \frac{n-1}{n} \sigma^2 \tag{3.67}$$

Therefore, the variance of the population σ^2 and the expected variance of an n-sized sample $<s^2>$ are related as:

$$\sigma^2 = \frac{n}{n-1} \cdot \langle s^2 \rangle = \frac{1}{n-1} \cdot \left\langle \sum_{j=1}^{n} (x_j - \bar{x})^2 \right\rangle \tag{3.68}$$

Furthermore, the variance σ_n^2 of an n-sized sample drawn from a population with variance σ^2 is:

$$\sigma_n^2 = \frac{1}{n}\sigma^2 = \frac{1}{n-1} \cdot \langle s^2 \rangle = \frac{1}{(n-1)^2} \cdot \left\langle \sum_{j=1}^{n} (x_j - \bar{x})^2 \right\rangle \tag{3.69}$$

In practice, $<s^2>$ may not be known and only the sample x_j ($j = 1, \ldots, n$) may be available.

In that case, s^2 can serve to estimate σ^2 and σ_n^2 as:

$$\sigma^2 \approx s_X^2 = \frac{1}{n-1} \cdot \sum_{j=1}^{n} (x_j - \bar{x})^2 = \frac{n}{n-1} \cdot s^2 \tag{3.70}$$

$$\sigma_n^2 \approx \frac{1}{n \cdot (n-1)} \cdot \sum_{j=1}^{n} (x_j - \bar{x})^2 = \frac{1}{n-1} \cdot s^2 \tag{3.71}$$

The term $n - 1$ is also considered as the sample size n minus one degree of freedom, which comes from taking the mean in Eq. (3.63) and its role in Eq. (3.66). The range of n residuals is dependent because its sum is model-bound, namely equal to 0 as stated explicitly in Eq. (3.60). In other situations, a model exists that takes two degrees of freedom and then a factor $n - 2$ appears. An example of that is the discussion on variances related to the best-fit linear regression in Section 10.3 (cf. Eqs (3.70) and (10.91) in that respect), which requires two parameters.

Such estimates play an important role in assessing the accuracy of measured quantities, which are often expressed in terms of standard deviations. Standard deviation s quantifies the scatter in a sample and can then be used to estimate the standard deviations σ and σ_n^2 of the population, respectively, of n-sized sample averages.

3.2 Median and Other Quantiles

An alternative to the mean value is selecting a (ranked in order of magnitude) value that marks a given percentage of the population. For instance, the median value is the value at 50% of the population. But other percentages are named as well.

The cumulative distribution $F(x)$ runs from 0 up to 1. This range can be divided into equal parts. If the number of such parts is Q, then there are $(Q - 1)$ cumulative distribution values that mark the boundary between two successive parts. The parts can be numbered j, with $j = 1, \ldots, Q$. The boundaries between these sections are at:

$$F_{Q,j} = \frac{j}{Q} \qquad (1 \leq j \leq Q - 1) \tag{3.72}$$

So, if the range of F is split into two equal parts, then $F_{2,1} = 0.5$ is the boundary between the two parts. Sometimes it is more convenient to use the ratio in Eq. (3.72) as a subscript

Table 3.1 Quantiles $x_{Q,j}$ with corresponding boundaries $F_{Q,j}$ between distribution parts and their alternative name.

Boundaries between distribution parts $F_{Q,j}$	Q-quantile	Name of $x_{Q,j}$
0.5	2-quantile	Median
0.333, 0.667	3-quantiles	Tertile
0.250, 0.500, 0.750	4-quantiles	Quartile
0.200, 0.400, 0.600, 0.800	5-quantiles	Pentile
0.167, 0.333, 0.500, 0.667, 0.832	6-quantiles	Sextile
0.143, 0.286, 0.429, 0.571, 0.714, 0.857	7-quantiles	Septile
0.125, 0.250, 0.375, 0.500, 0.625, 0.750, 0.875	8-quantiles	Octile
0.111, 0.222, 0.333, 0.444, 0.555, 0.666, 0.777, 0.888	9-quantiles	Nonile
0.1, 0.2, 0.3, 0.4, 0.5, 0.6, 0.7, 0.8, 0.9	10-quantiles	Decile
0.01, 0.02, 0.03, …	100-quantiles	Percentile
0.001, 0.002, 0.003, …	1000-quantiles	Permille

and the notation becomes $F_{50\%}$ instead of writing $F_{2,1}$. It will be indicated where this is appropriate.

The value of the X variable that corresponds to the boundary is called the jth Q-quantile $x_{Q,j}$. Again, a similar subscript notation can be used as with $F_{Q,j}$. By definition, the following holds for the $x_{Q,j}$:

$$F(x_{Q,j}) = \frac{j}{Q} \tag{3.73}$$

In other words, $x_{Q,j}$ could be regarded as the inverse function F^{inv} with variable F:

$$x_{Q,j} = F^{inv}(F_{Q,j}) = F^{inv}\left(\frac{j}{Q}\right) \tag{3.74}$$

Some quantiles have special names, like the percentiles and the permilles (see Table 3.1). Another important quantile is the two-quantile, which is better known as the median. The median is the value x below which 50% of the variables X lie:

$$x_{median} = x_{Q,j} = x_{2,1} = x_{50\%} = F^{inv}\left(\frac{1}{2}\right) \tag{3.75}$$

If an operation is applied that preserves the ranking of the elements (and does not introduce discontinuities at the median), then the median is preserved under the operation. Unlike the case with the mean (see Section 3.1.1), this does not require the operation to be linear. The condition may even be widened as for the median it is also allowed to reverse the ranking. The median of a continuous cumulative distribution and the median of the corresponding reliability function are the same.

With populations having a finite number of elements n (i.e. population size n), the elements can be ranked and indexed i, with $i = 1, …, n$. The population size n can be odd or even. If n is odd, then the index $i_{50\%}$ that corresponds to 50% of the ranked population is $i_{50\%} = (n + 1)/2$. In that case the median is also preserved with operations as mentioned above.

If n is even there is no index $i_{50\%}$. The median is then calculated by averaging the elements with $i = n/2$ and $i = (n+2)/2$. In that case the median is not preserved under the operations mentioned above, because this again is a mean which is only preserved with the rules mentioned in Section 3.1.1, albeit the mean of only two elements is taken instead of all elements. Generally, the median of a population with even size n is likely better preserved than the mean, with the exception of special cases.

If the probability $f(x)$ is a symmetrical function, then the median is equal to the mean. Many distributions related to ageing have a time or load variable that runs from 0 to ∞, and they are not symmetrical.

3.3 Mode

The mode of a continuous distribution is the value x, where the distribution density is highest. If the distribution consists of discrete values, then the mode is the x value that occurs most frequently.

Where the cumulative distribution $F(x)$ rises fastest and $R(x)$ declines fastest, by definition most failures occur per unit time. This is the maximum of the distribution density function $f(x)$.

Figure 3.1 shows the measures mode, median and mean for a continuous distribution that is not symmetrical. If the distribution or probability density function is symmetrical, the three measures become the same (provided it has a single peak at its median and mean). The mean is calculated with Eq. (3.3).

If a distribution density function has a single maximum, then the mode x_{mode} is found by locally resolving:

$$\frac{df(x)}{dx} = 0 \quad \wedge \quad \frac{d^2f(x)}{dx^2} < 0 \tag{3.76}$$

The second inequality is added to ensure this is a maximum and not a minimum.

3.4 Merits of Mean, Median and Modal Value

As shown in Figure 3.1, the mean (or expected), median and modal value characterize a distribution, but generally differ if the underlying distribution density is not symmetrical around its maximum. In various situations a single representative value is required, which raises the question of whether the mean, median or modal value might be preferred. They are all frequently used. In short, the meaning of these values is:

- *Mean*: expected value. If many tests are run, the average result approximates the mean or expected value.
- *Median*: the value that splits the total group of results equally into two subgroups, the 50% lower results and the 50% higher results. The median is the value that forms the border between the 50% lower and 50% higher values. A randomly selected result has equal probabilities of being lower or higher than the median.
- *Mode*: the result that occurs most frequently.

The advantage of the median is that it is conserved with non-linear operations, given that the function associated with the operation increases or decreases monotonically.

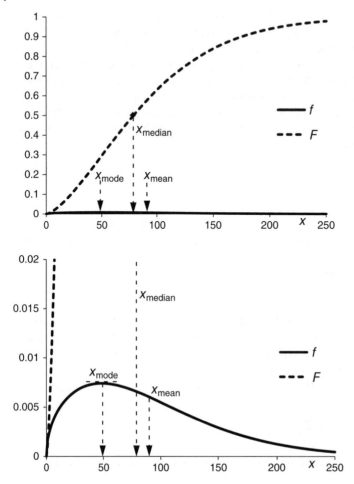

Figure 3.1 Mode, median and mean of a distribution. The top and bottom graphs represent the same cumulative distribution $F(x)$ and probability density $f(x)$, but are scaled differently along the vertical axis in order to show the origin of median and mode.

For example, if x_M is the median of a set of positive real numbers X, and the variable is squared (which is a non-linear operation), then $(x_M)^2$ is the median value of the set X^2.

A disadvantage of the median and mode is that extreme values have little impact, while extreme values do have an impact on expected values. For instance, if a supplier must pay a €1000 fine for each late delivery and only 80% of the deliveries are in due time, then in fact a €200 fine is paid per delivery on average. If the supplier keeps track of the mean delivery fine, the goal can be set to reduce this €200 average fine. However, if the company keeps track of the median fine, then it is found that this is €0 because more than 50% of the deliveries are in due time. The median fine will not provide useful information if the supplier's performance needs to be improved. Likewise, the modal fine is also €0 and no trigger for improvement is generated there either.

Each of the measures mean, median and mode have merits, but it must be evaluated per situation which serves the purpose of the exercise best.

3.5 Measures for Comparing Distributions

In the previous sections, measures of a single distributed variable were explored. These measures can be used to characterize a distribution or make forecasts related to a single variable. This section discusses measures to compare different variables (or the variable with itself) in order to find relationships (i.e. covariances and correlations). This indicates to what extent the variables are independent.

3.5.1 Covariance

If the relationship between two variables X and Y is investigated, two measures are often used: the covariance and the correlation (see Section 3.5.2). Both are measures of linear dependence of the two variables (i.e. to what extent they jointly vary linearly around their means μ_X, respectively μ_Y).

The covariance σ_{XY} is defined as:

$$\sigma_{XY} = \langle (X - \mu_X) \cdot (Y - \mu_Y) \rangle = \langle X \cdot Y \rangle - \mu_X \cdot \mu_Y \tag{3.77}$$

There is a strong resemblance with the variance, which in fact is the covariance of a variable X with itself. The latter equals the square of its standard deviation σ_X of that variable:

$$\sigma_{XX} = \langle (X - \mu_X)^2 \rangle = \langle X^2 \rangle - \mu_X^2 = \sigma_X^2 = \text{var}(X) \tag{3.78}$$

The covariance is a measure for linear dependence, which can be a positive or negative number or equal to zero. The sign indicates whether X and Y increase jointly or have a reverse relationship. If X and Y are independent, then σ_{XY} is 0. However, the reverse is not true, as Eq. (3.82) shows in Example 3.4.

Example 3.4 *Covariance of Linear and Quadratic Relationships* Suppose X is a variable representing the real numbers from 0 to 1 (i.e. $0 \leq X \leq 1$ and $Y = 3 \cdot X$; see Figure 3.2). Clearly, the relationship between X and Y is linear. The covariance σ_{XY} can be calculated as:

$$\sigma_{XY} = \left\langle \left(X - \frac{1}{2} \right) \cdot \left(Y - \frac{3}{2} \right) \right\rangle = 3 \cdot \left\langle \left(X - \frac{1}{2} \right)^2 \right\rangle = \frac{1}{4} \tag{3.79}$$

Suppose X represents the real numbers from 0 to 1 (i.e. $0 \leq X \leq 1$ and $Y = 1 - X$; see Figure 3.3). The relationship is negatively linear. The covariance σ_{XY} can be calculated as:

$$\sigma_{XY} = \left\langle \left(X - \frac{1}{2} \right) \cdot \left(Y - \frac{1}{2} \right) \right\rangle = \left\langle X \cdot Y - \frac{1}{2}(X + Y) + \frac{1}{4} \right\rangle = -\frac{1}{12} \tag{3.80}$$

Suppose X again represents the real numbers from 0 to 1 (i.e. $0 \leq X \leq 1$ and $Y = X^2$; see Figure 3.4). Obviously X and Y do not have a linear relationship, but if X increases, so will Y. The covariance σ_{XY} can be calculated as:

$$\sigma_{XY} = \left\langle \left(X - \frac{1}{2} \right) \left(X^2 - \frac{1}{3} \right) \right\rangle = \left\langle \left(X^3 - \frac{1}{2}X^2 - \frac{1}{3}X + \frac{1}{6} \right) \right\rangle = \frac{1}{12} \tag{3.81}$$

Suppose X represents the real numbers from -1 to 1 (i.e. $-1 \leq X \leq 1$ and $Y = X^2$; see Figure 3.5). Note the difference in domain compared to the cases above. Obviously X

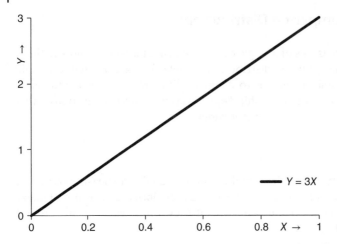

Figure 3.2 Variables X and Y with $Y = 3X$.

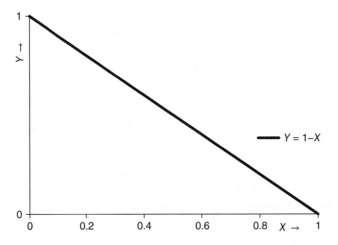

Figure 3.3 Variables X and Y with $Y = 1 - X$.

and Y do not have a linear relationship. Now X^2 increases with the absolute value of X. The covariance σ_{XY} can now be calculated as:

$$\sigma_{XY} = \left\langle (X - 0)\left(X^2 - \frac{1}{3}\right)\right\rangle = \left\langle \left(X^3 - \frac{1}{3}X\right)\right\rangle = 0 \tag{3.82}$$

This is an example of a well-defined relationship between X and Y, but nevertheless the mean relationship appears neutral in terms of the covariance.

The covariance can be derived for a set of n variable pairs (x_i, y_i) $(i = 1, ..., n)$ in a similar fashion as for the standard deviation and variance in case of a sample (see Section 3.1.8). For instance, at a series of n times x_i a quantity (e.g. travelled distance) y_i can be measured, after which the covariance is one of the measures of interest. The

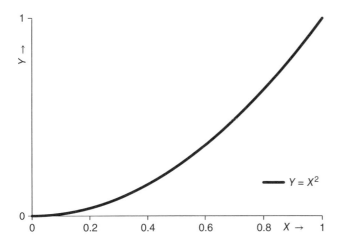

Figure 3.4 Variables X and Y with $Y = X^2$.

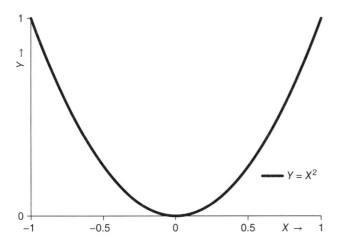

Figure 3.5 Variables X and Y with $Y = X^2$.

covariance σ_{XY} based on such a sample is:

$$\sigma_{XY} = \frac{1}{n-1} \left\langle \sum_{i=1}^{n} ((x_i - \bar{x}) \cdot (y_i - \bar{y})) \right\rangle \tag{3.83}$$

In case of a single sample the expected value remains unknown and σ_{XY} is estimated from the sample (x_i, y_i) $(i = 1, \ldots, n)$ as s_{XY}:

$$\sigma_{XY} \approx \frac{1}{n-1} \sum_{i=1}^{n} ((x_i - \bar{x}) \cdot (y_i - \bar{y})) = s_{XY} \tag{3.84}$$

3.5.2 Correlation

Correlation is closely related to covariance (see Section 3.5.1). The measure of correlation is the correlation coefficient ρ_{XY}, which is the covariance of X and Y normalized with the standard deviations σ_X and σ_Y (see Eq. (3.51)):

$$\rho_{XY} = \frac{\sigma_{XY}}{\sigma_X \sigma_Y} \tag{3.85}$$

The correlation coefficient is normalized such that its value ranges from -1 to 1. The extremes are only reached with a fully linear dependence. For non-zero correlation coefficients between -1 and 1, the variables have a more or less linear relationship. If the correlation coefficient is 0, the variables are called uncorrelated. If X and Y are independent, then ρ_{XY} is always 0. However, the reverse is not true. As with the covariance, a well-defined relationship may still result in a correlation coefficient of 0 (see Example 3.5, Eq. (3.89)).

The questions accompanying this chapter elaborate further on other relationships with variance and correlation.

Example 3.5 *Correlation Coefficient of Linear and Quadratic Relationships* In this example the correlation coefficients of the variables in Example 3.4 are determined.

Suppose X is a variable representing the real numbers from 0 to 1 (i.e. $0 \le X \le 1$ and $Y = 3 \cdot X$; see Figure 3.2). Clearly, the relationship between X and Y is linear. The correlation coefficient ρ_{XY} becomes 1 due to the fully and positive linear relationship:

$$\rho_{XY} = \frac{\left\langle \left(X - \frac{1}{2} \right) \cdot \left(3X - \frac{3}{2} \right) \right\rangle}{\sqrt{\langle X^2 \rangle - \frac{1}{4}} \cdot \sqrt{\langle 9X^2 \rangle - \frac{3}{4}}} = \frac{\frac{1}{4}}{\sqrt{\frac{1}{12}} \cdot 3 \cdot \sqrt{\frac{1}{12}}} = 1 \tag{3.86}$$

Suppose X again represents the real numbers from 0 to 1 (i.e. $0 \le X \le 1$ and $Y = 1 - X$; see Figure 3.3). The relationship is negatively linear. The correlation coefficient ρ_{XY} becomes -1 due to the fully and negative linear relationship:

$$\rho_{XY} = \frac{\left\langle \left(X - \frac{1}{2} \right) \cdot \left(Y - \frac{1}{2} \right) \right\rangle}{\sqrt{\langle X^2 \rangle - \frac{1}{4}} \cdot \sqrt{\langle Y^2 \rangle - \frac{1}{4}}} = \frac{-\frac{1}{12}}{\sqrt{\frac{1}{12}} \cdot \sqrt{\frac{1}{12}}} = -1 \tag{3.87}$$

Suppose X again represents the real numbers from 0 to 1 (i.e. $0 \le X \le 1$ and $Y = X^2$; see Figure 3.4). Obviously X and Y do not have a linear relationship, but if X increases, so will Y. The correlation coefficient ρ_{XY} merely approaches 1 due to the not fully linear relationship:

$$\rho_{XY} = \frac{\left\langle \left(X - \frac{1}{2} \right) \left(X^2 - \frac{1}{3} \right) \right\rangle}{\sqrt{\langle X^2 \rangle - \frac{1}{4}} \cdot \sqrt{\langle X^4 \rangle - \frac{1}{9}}} = \frac{\frac{1}{12}}{\sqrt{\frac{1}{12}} \cdot \sqrt{\frac{4}{45}}} \approx 0.97 \tag{3.88}$$

Suppose X represents the real numbers from -1 to 1 (i.e. $-1 \le X \le 1$ and $Y = X^2$; see Figure 3.5). Note the difference in domain compared to the cases above. Obviously X and Y do not have a linear relationship. Now X^2 increases with the absolute value of X.

The correlation coefficient ρ_{XY} becomes 0 due to the negative and positive relationships that compensate each other over the full range:

$$\rho_{XY} = \frac{\left\langle (X-0)\left(X^2 - \frac{1}{3}\right)\right\rangle}{\sqrt{\langle X^2\rangle - 0} \cdot \sqrt{\langle X^4\rangle - \frac{1}{9}}} = \frac{0}{\sqrt{\frac{1}{3}} \cdot \sqrt{\frac{8}{45}}} = 0 \tag{3.89}$$

This is an example of a well-defined relationship between X and Y, but nevertheless the mean relationship appears neutral in terms of the correlation coefficient.

Section 3.1.8 discussed the mean, standard deviation and variance for a limited sample x_i drawn from a population and at the end of Section 3.5.1 a similar situation was discussed for a sample in the form of observation pairs (x_i, y_i) with $i = 1, \ldots, n$. The correlation coefficient can be used for such pairs as well. Using the approximation in Eq. (3.84) for σ_{XY} and in Eq. (3.70) for σ_X and σ_Y, an approximation for the correlation coefficient is the empirical correlation coefficient r_{XY}:

$$\rho_{XY} = \frac{\sigma_{XY}}{\sigma_X \cdot \sigma_Y} \approx \frac{\sum\limits_{i=1}^{n}((x_i - \bar{x}) \cdot (y_i - \bar{y}))}{\sqrt{\sum\limits_{i=1}^{n}(x_i - \bar{x})^2} \cdot \sqrt{\sum\limits_{i=1}^{n}(y_i - \bar{y})^2}} = r_{XY} \tag{3.90}$$

It may be noted from Eq. (3.85) that the correlation coefficient itself is not an expected value, but a ratio of the expected value σ_{XY} and the product of σ_X and σ_Y (which latter two are square roots of expected values). Whereas the means of the approximations of σ_{XY}, σ_X^2 and σ_Y^2 yield the expected (co)variances, it is not true that the mean of the approximation in Eq. (3.90) yields ρ_{XY}. In general, $<r_{XY}> \neq \rho_{XY}$. The sequence of averaging is relevant when calculating the correlation coefficient from a series of experiments. First the covariance and the variances must be averaged, after which the combined correlation coefficient can be determined.

3.5.3 Cross-Correlation and Autocorrelation

The above examples showed a measure for a linear relationship of variables X and Y. Such variables can be functions $X(t)$ and $Y(t)$ of a variable T with values t. This variable can be time, for instance. The covariance and correlation of these functions can be explored as described above. In some cases correlation exists, but one of the two functions can be shifted by a so-called lag or delay τ. Depending on the field of work, the definitions vary, but essentially the cross-correlation is a covariance in combination with a delay.

The cross-correlation $R_{XY}(\tau)$ can be defined as:

$$R_{XY}(\tau) = \langle (X(t) - \mu_X) \cdot (Y(t + \tau) - \mu_Y)\rangle \tag{3.91}$$

If the means $R\mu_X$ and μ_Y are both 0, then $R_{XY}(\tau)$ reduces to:

$$\mu_X = \mu_Y = 0 \Rightarrow R_{XY}(\tau) = \langle X(t) \cdot Y(t + \tau)\rangle \tag{3.92}$$

The cross-correlation $(X^*Y)(\tau)$ of two signals X and Y, of which the means $R\mu_X$ and μ_Y are both 0 and Y lags X by τ, is similarly defined as:

$$(X * Y)(\tau) = \int X^*(t) \cdot Y(t + \tau)dt \tag{3.93}$$

X^* is the complex conjugate of X, which for real values becomes:

$$X \text{ real}: \quad (X * Y)(\tau) = \int X(t) \cdot Y(t + \tau)dt \tag{3.94}$$

Examples are time functions with a phase shift like $\sin(\omega t)$ and $\cos(\omega t)$, but also functions like distributions with difference in the mean of the variable.

The cross-correlation may be normalized like the correlation coefficient.

Autocorrelation is the cross-correlation of a variable or function with itself. The autocorrelation $R_X(\tau)$ can be defined as:

$$R_X(\tau) = \langle (X(t) - \mu_X) \cdot (X(t + \tau) - \mu_X) \rangle \tag{3.95}$$

It is assumed that the mean does not vary with τ, otherwise μ_X must be written as a function of τ.

Again, if the mean is 0, then $R_X(\tau)$ reduces to:

$$\mu_X = 0 \Rightarrow R_X(\tau) = \langle X(t) \cdot X(t + \tau) \rangle \tag{3.96}$$

The autocorrelation can be used to evaluate the repetitiveness of $X(t)$; that is, whether $X(t)$ is a periodic function like a sine wave. If τ is taken to be 0, then the autocorrelation becomes the variance:

$$R_X(0) = \langle (X(t) - \mu_X)^2 \rangle = \sigma_X^2 = \text{var}(X) \tag{3.97}$$

3.6 Similarity of Distributions

This section discusses a measure to determine how similar two sets of data or even two distributions are. The background lies in being able to tell whether the experienced failure behaviour complies with the specifications of the expected failure behaviour. Firstly, the similarity of two discrete groups is discussed. Secondly, the similarity index of two discrete distributions and next of two continuous distributions is explained. Finally, the significance and some mathematical considerations are addressed. Non-similarity can point at a lower quality, but alternatively at a higher quality as well. For that reason, a measure of compliance is also discussed in Section 3.7. Sections 9.4.3 and 9.4.4 present two cases where this similarity index is used to estimate the expected number of future failures after some faults occurred.

3.6.1 Similarity of Counting in Discrete Sets

Similarity is a quality that two groups feature mutually. It is a symmetrical property; if one is similar to the other, the other is similar to the first.

The similarity or similarity index S_{fg} is introduced as a normalized measure to evaluate the similarity of two sets of outcomes F and G. The similarity S_{fg} is 1 if the two sets are identical and 0 if they have nothing in common. With reference to Section 2.1 for set theory definitions, F and G are subsets of their union $F \cup G$. The intersection $F \cap G$ is formed by the outcomes that belong to both F and G and is therefore a subset of both sets.

As an example, a group of diverse insulators is studied. They differ in rated voltage and material of the housing. Two subgroups F and G are identified. Group F consists of insulators with a rated voltage of 150 kV. Group G consists of insulators where the housing material is a polymer. The total group of insulators falls partly in F, G, $F \cap G$ or even outside $F \cup G$ (see Table 3.2). The logic variables f_i and g_i test whether object i belongs to group F (respectively G). The outcomes f_i and g_i are represented in a Venn diagram (Figure 3.6). The question is how similar distributions F and G are.

The actual elements in the sets are the outcomes of presence in a group (i.e. the values of f_i and g_i). If groups F and G were identical, then all 150 kV-rated insulators would be polymer and vice versa all polymer insulators would be rated 150 kV. In that case the set of outcomes f_i would be identical to g_i. However, in this example that is clearly not the case. In total, nine outcomes (f_i, g_i) differ. The question is now how to determine to what extent the groups F and G are similar (i.e. given that an insulator belongs to at least F or G). It may be noted that the question is not to what extent the outcomes f_i and g_i

Table 3.2 Example for determining similarity.

Object index i	Rated voltage (kV)	Insulator material	Logic outcome f_i	Logic outcome g_i	Element of set
01	150	Porcelain	1	0	F
02	150	Porcelain	1	0	F
03	150	Porcelain	1	0	F
04	150	Porcelain	1	0	F
05	150	Porcelain	1	0	F
06	150	Porcelain	1	0	F
07	150	Polymer	1	1	$F \cap G$
08	150	Polymer	1	1	$F \cap G$
09	150	Polymer	1	1	$F \cap G$
10	150	Polymer	1	1	$F \cap G$
11	150	Polymer	1	1	$F \cap G$
12	150	Polymer	1	1	$F \cap G$
13	220	Porcelain	0	0	$\Omega \backslash (F \cup G)$
14	220	Porcelain	0	0	$\Omega \backslash (F \cup G)$
15	220	Porcelain	0	0	$\Omega \backslash (F \cup G)$
16	380	Polymer	0	1	G
17	380	Polymer	0	1	G
18	380	Polymer	0	1	G

Note: A collection of diverse insulators is studied. The outcome for each insulator is whether it belongs to a group. Two sets F (rated voltage 150 kV) and G (material polymer) are defined. The logic outcomes (f, g) reflect the presence in the sets F and G (i.e. $f_i = 1$ means object i is present in set F and $g_i = 1$ means object i is present in set G).

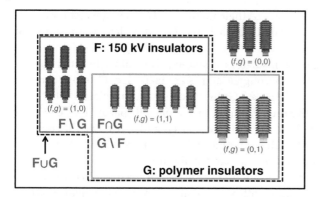

Figure 3.6 Two sets F and G describe the populations of 150-kV insulators (respectively polymer insulators) among a larger sample space of insulators. A Venn diagram shows that F and G are not identical, but are not completely different either. There is a limited similarity.

are identical, which would include objects $i = 13$–15 of Table 3.2. Here the similarity of F and G within their union $F \cup G$ is studied.

In the literature, various similarity indices are defined. A famous definition is provided by Jaccard [15]. It is often applied to counting presence in groups:

$$S_{fg} = \frac{|F \cap G|}{|F \cup G|} = \frac{\text{number of elements } F \cap G}{\text{number of elements } F \cup G} \tag{3.98}$$

where the notation $|A|$ means the number of elements in set A. It may be noted that summing the outcomes over the subsets yields the same results as summing over the full union, because f_i and g_i are logic variables that are 0 outside the borders of their set. In terms of the outcomes f_i and g_i, this similarity index becomes:

$$S_{fg} = \frac{\displaystyle\sum_{F \cap G} f_i g_i}{\displaystyle\sum_F f_i + \sum_G g_i - \sum_{F \cap G} f_i g_i} = \frac{\displaystyle\sum_{F \cup G} f_i g_i}{\displaystyle\sum_{F \cup G} f_i + \sum_{F \cup G} g_i - \sum_{F \cup G} f_i g_i} \tag{3.99}$$

The subtraction of $|F \cap G|$ in the denominator prevents the elements in the union $|F \cup G|$ from being counted double. In the case of Figure 3.6, the similarity of sets F and G is therefore:

$$S_{fg} = \frac{6 \cdot (1 \cdot 1)}{12 \cdot 1 + 9 \cdot 1 - 6 \cdot (1 \cdot 1)} = \frac{6}{15} = 0.4 \tag{3.100}$$

It is noteworthy that with counting presence, the representation is a dimensionless number. However, as soon as the outcomes in a set are expressed in terms of a unit (e.g. 'gram' or even 'item'), then the denominator in Eq. (3.100) becomes improper due to summing terms with different dimensions. Using '*item*' as the unit for a count:

$$S_{fg} = \frac{6 \cdot (1 \, item \cdot 1 \, item)}{12 \cdot 1 \, item + 9 \cdot 1 \, item - 6 \cdot (1 \, item \cdot 1 \, item)} = \frac{6 \, item^2}{21 \, item - 6 \, item^2} \tag{3.101}$$

The Jaccard index therefore yields an improper expression if dimensions are involved. Realizing that the square of a logic (binary) variable is identical to the variable itself, an alternative expression for Eq. (3.99) can be formulated which is proper with respect to dimensions. This is called the similarity index and it is elaborated to be applied not only to counting presence, but also to a series of outcomes, to continuous distributions and even to distribution functions themselves in the following. In discrete form the similarity index S_{fg} is defined firstly as:

$$S_{fg} = \frac{\sum\limits_{F \cap G} f_i g_i}{\sum\limits_{F} f_i f_i + \sum\limits_{G} g_i g_i - \sum\limits_{F \cap G} f_i g_i} = \frac{\sum\limits_{F \cup G} f_i g_i}{\sum\limits_{F \cup G} f_i f_i + \sum\limits_{F \cup G} g_i g_i - \sum\limits_{F \cup G} f_i g_i} \qquad (3.102)$$

As long as counts are considered dimensionless, the Jaccard approach in Eq. (3.98) and the similarity index in Eq. (3.102) yield the same result as the (dimensionless) original, but the similarity index of Eq. (3.102) is now consistent with respect to dimensions:

$$S_{fg} = \frac{6 \cdot (1 \cdot 1)}{12 \cdot 1 \cdot 1 + 9 \cdot 1 \cdot 1 - 6 \cdot (1 \cdot 1)} = \frac{6}{15} = 0.4 \qquad (3.103)$$

Or explicitly, using the unit '*item*' for a count again:

$$S_{fg} = \frac{6 \cdot (1\ item \cdot 1\ item)}{12 \cdot 1\ item \cdot 1\ item + 9 \cdot 1\ item \cdot 1\ item - 6 \cdot (1\ item \cdot 1\ item)} = \frac{6\ item^2}{15\ item^2} = 0.4 \qquad (3.104)$$

In case the set sizes are not finite, but infinite, it may still be possible to calculate a similarity index based on counting provided the ratio of the sample sizes remains finite.

3.6.2 Similarity of Two Discrete Distributions

In the previous section the similarity between sets was described in terms of counting elements in the sample space as counts of presence. However, one might not only be interested in the similarity of counts, but also in the similarity of other outcomes like measurements on two different occasions. For instance, think of a group of n insulators at a substation that are diagnosed for their partial discharge level on two different occasions. These levels may be indexed as i. The test results are f_i and g_i; the mass functions (Section 2.4.4) that indicate the occurrence of the levels i. The question can be to what extent the two sets of results are similar.

Therefore, the similarity of the outcomes themselves is explored. It is assumed that there is an equal number n of events (diagnostic results), but two different sets F and G with outcomes f_i and g_i ($i = 1, ..., n$). Note that f_i and g_i are not necessarily presence counts any longer, but can be any set of outcomes like partial discharge level, oil acidity, and so on. If i is a parametrization that indexes the outcomes and can be used in the summations, then f_i and g_i can be thought of as two functions of i.

The concept of similarity as in Eq. (3.102) is now generalized to functions on discrete sets by defining the similarity index as:

$$S_{fg} = \frac{\sum\limits_{i} f_i g_i}{\sum\limits_{i} f_i f_i + \sum\limits_{i} g_i g_i - \sum\limits_{i} f_i g_i} = \frac{\sum\limits_{i} f_i g_i}{\sum\limits_{i} f_i^2 + \sum\limits_{i} g_i^2 - \sum\limits_{i} f_i g_i} \qquad (3.105)$$

A numerical case is elaborated in Example 3.6.

Table 3.3 The results of throwing a dice 30 times.

Event	1	2	3	4	5	6
Counts in experiment	4	7	5	6	4	4
Observed f_i	4/30	7/30	5/30	6/30	4/30	4/30
Theoretical g_i	1/6	1/6	1/6	1/6	1/6	1/6

The first row shows the possible events, namely throwing 1, 2, 3, 4, 5 or 6. The second row shows how often each event occurred in the first experiment of throwing 30 times. The third row shows the observed probability mass f_i, which is equal to the counts divided by the 30 throws. The last row shows the theoretically expected probability mass g_i of a fair dice.

Example 3.6 *Similarity of Discrete Sets of Two Probability Mass Functions* A particularly interesting application concerns two experiments with a fixed number of outcomes. Assume as the first experiment a dice like in Example 2.1 is thrown 30 times. The possible events are $\{1,2,3,4,5,6\}$, as shown in the first row of Table 3.3. The observed number of these six possible events is shown in the second row of Table 3.3. The second experiment is the calculation of the theoretically expected outcomes for a fair dice. The third and fourth rows show the observed and theoretical mass functions f_i (respectively g_i). These characterize two distributions, namely the distribution of the experimental observations and the theoretical distribution. The question is to what extent they are similar. The similarity index of these two is determined with Eq. (3.105), as elaborated in Eq. (3.106).

The similarity index can now be employed as a measure of similarity between these two sets. This could be used as an indication of how similar the dice in the experiment is compared to a fair dice. The similarity index through Eq. (3.105) of these two sets thus becomes:

$$S_{fg} = \frac{\frac{4}{180} + \frac{7}{180} + \frac{5}{180} + \frac{6}{180} + \frac{4}{180} + \frac{4}{180}}{\frac{16+49+25+36+16+16}{900} + \frac{6 \cdot 1 \cdot 1}{36} - \left(\frac{4}{180} + \frac{7}{180} + \frac{5}{180} + \frac{6}{180} + \frac{4}{180} + \frac{4}{180}\right)} = \frac{30}{31.6} = 0.95$$

(3.106)

As a conclusion, from 30 throws the investigated dice appears to have 0.95 similarity with a fair dice, which is close to 1 (i.e. fully similar). A follow-up question is then how significant this finding is and when a dice is deemed fair enough. That is the topic of Section 3.6.4.

Equation (3.105) can be regarded as the ratio of inner products of functions or vectors $f = (f_1, \ldots, f_n)$ and $g = (g_1, \ldots, g_n)$. Dividing all sums in Eq. (3.105) by the number of events n expresses the similarity index in terms of the means following Eq. (3.1). As the division by this number n appears in both the numerator and the denominator, they cancel each other and therefore Eq. (3.105) can also be written as:

$$S_{fg} = \frac{\langle f \cdot g \rangle}{\langle f \cdot f \rangle + \langle g \cdot g \rangle - \langle f \cdot g \rangle}$$

(3.107)

In case of probability mass functions, Eq. (3.105) can also be interpreted as follows. As the cumulative functions F and G, as well as the mass functions f and g, are functions of i, the mass functions can also be expressed in terms of F and G, which provides two alternative parametrizations. Means $\langle f_F \rangle$, $\langle g_G \rangle$, $\langle f_G \rangle$ and $\langle g_F \rangle$ can be calculated as (note that the sum of frequencies is 1; cf. Section 2.4.4, Eq. (2.20)):

$$\langle f_F \rangle = \frac{\sum\limits_{i=1}^{n} f_i f_i}{\sum\limits_{i=1}^{n} f_i} = \sum\limits_{i=1}^{n} f_i^2 \tag{3.108}$$

$$\langle g_G \rangle = \frac{\sum\limits_{i=1}^{n} g_i g_i}{\sum\limits_{i=1}^{n} g_i} = \sum\limits_{i=1}^{n} g_i^2 \tag{3.109}$$

$$\langle f_G \rangle = \frac{\sum\limits_{i=1}^{n} f_i g_i}{\sum\limits_{i=1}^{n} g_i} = \sum\limits_{i=1}^{n} f_i g_i = \frac{\sum\limits_{i=1}^{n} f_i g_i}{\sum\limits_{i=1}^{n} f_i} = \langle g_F \rangle \tag{3.110}$$

In terms of these expected values or means, the similarity index can also be expressed as:

$$S_{fg} = \frac{\langle f_G \rangle}{\langle f_F \rangle + \langle g_G \rangle - \langle f_G \rangle} = \frac{\langle g_F \rangle}{\langle f_F \rangle + \langle g_G \rangle - \langle g_F \rangle} \tag{3.111}$$

3.6.3 Similarity of Two Continuous Distributions

The logical step following the similarity of discrete sets is the similarity index of continuous sets. The discrete distributions were parametrized by a discrete variable i. For a continuous distribution a continuous variable t is used for parametrization.

In integral form the similarity index S_{fg} of Eq. (3.105) becomes:

$$S_{fg} = \frac{\int f(t) \cdot g(t)\, dt}{\int f^2(t)dt + \int g^2(t)dt - \int f(t) \cdot g(t)\, dt} \tag{3.112}$$

with f and g distribution density functions. This can again be interpreted as a ratio of means (cf. Eq. (3.111)):

$$S_{fg} = \frac{\langle f_G \rangle}{\langle f_F \rangle + \langle g_G \rangle - \langle f_G \rangle} = \frac{\langle g_F \rangle}{\langle f_F \rangle + \langle g_G \rangle - \langle g_F \rangle} \tag{3.113}$$

with:

$$\langle f_G \rangle = \langle g_F \rangle = \int f(t) \cdot g(t)\, dt \tag{3.114}$$

$$\langle f_F \rangle = \int f(t) \cdot f(t)\, dt = \int f^2(t)\, dt \tag{3.115}$$

$$\langle g_G \rangle = \int g(t) \cdot g(t)\, dt = \int g^2(t)\, dt \tag{3.116}$$

With distribution density functions $f(t)$ and $g(t)$, the similarity index S_{fg} for two distributions can now be calculated with Eq. (3.112) or Eq. (3.113). In a wider approach, the similarity between other functions such as hazard rates might also be elaborated, but density functions provide a meaning with means. In a sense, the means $\langle f_G \rangle$ and $\langle g_F \rangle$ can be regarded as a measure of overlap or intersection of the distributions (Section 2.1). The sum of means $\langle f_F \rangle + \langle g_G \rangle - \langle f_G \rangle$ is the union. Expressing the densities in terms of the cumulative functions with an alternative expression of Eqs (3.114)–(3.116) is possible, as shown in Figure 3.8 (but often not convenient):

$$\langle f_G \rangle = \int f(G)\, dG \tag{3.117}$$

$$\langle g_F \rangle = \int g(F)\, dF \tag{3.118}$$

$$\langle f_F \rangle = \int f(F)\, dF \tag{3.119}$$

$$\langle g_G \rangle = \int g(G)\, dG \tag{3.120}$$

Interestingly, the similarity index can compare distributions of different families. For instance, it is possible to compare a lognormal distribution $f(t)$ with a Weibull distribution $g(t)$ or other combinations. In this way we can explore how significant the exact choice of a given distribution is compared to alternative single or mixed distributions.

Similarity is a symmetrical relationship. It is noteworthy that the definition of similarity also fulfils the requirement of symmetry:

$$S_{fg} = S_{gf} \tag{3.121}$$

If the distributions F and G (respectively f and g) are identical, then the similarity is 1 indeed:

$$f(t) = g(t) \quad \Rightarrow \quad S_{ff} = \frac{\int f^2(t)dt}{\int f^2(t)dt + \int f^2(t)dt - \int f^2(t)dt} = 1 \tag{3.122}$$

Also, if the distributions have nothing in common, the similarity is 0:

$$\forall t : \quad f(t) \cdot g(t) = 0 \quad \Rightarrow \quad S_{fg} = \frac{0}{<f^2> + <g^2> - 0} = 0 \tag{3.123}$$

The evaluation of the similarity can also be restricted to a certain interval, for instance the similarity between $t = 0$ and the time $t = t_{10,1}$ of F (i.e. the first 10-quantile of F, where $F(t_{10,1}) = 0.1$; see Section 3.2). The similarity index could then be defined as a function $S_{fg}(t_1, t_2)$ of two variables, namely an integration interval start value t_1 and end value t_2:

$$S_{fg}(t_1, t_2) = \frac{\int_{t_1}^{t_2} f(t) \cdot g(t)\, dt}{\int_{t_1}^{t_2} f^2(t)dt + \int_{t_1}^{t_2} g^2(t)dt - \int_{t_1}^{t_2} f(t) \cdot g(t)\, dt} \tag{3.124}$$

This is particularly useful if a group of assets suffers early failures and there is a need to evaluate the similarity between the observed and reference (specified) distribution during a prescribed early period of operation. For instance, it might be interesting to evaluate the similarity during the first year. In that case, $t_1 = 0$ year and $t_2 = 1$ year. The similarity index can thus provide a means in quality control.

Example 3.7 *Similarity of Two Distributions* A series of assets is maintained and supposed to have a cumulative failure distribution $F_E(t)$ defined by:

$$F_E(t) = 1 - \exp\left(-\frac{t}{\theta}\right) \tag{3.125}$$

with $\theta = 35$ years. However, several years of experience lead to the conclusion that the actual failure distribution is $G_W(t)$:

$$G_W(t) = 1 - \exp\left[-\left(\frac{t}{a}\right)^b\right] \tag{3.126}$$

with $a = 31.5$ year and $b = 1.3$. The scale parameters θ and a differ and $G_W(t)$ has an additional parameter b which changes the appearance of the distribution as well. The scale parameter is reduced by 10% (from 35 to 31.5 years). We are asked to quantify the similarity of these functions. How similar are they?

This is done using the similarity index S_{fg}. A numerical integration for this case yields:

$$S_{fg} \approx 0.933 \tag{3.127}$$

which indicates that the behaviour is 0.067 short of 1. Whether this is dramatic depends on the required level of similarity.

As a check, the hazard rate (see Section 2.4.5) of both distributions is calculated. The hazard rate $h_{FE}(t)$ of the exponential distribution F_E is a constant:

$$h_{F_E}(t) = \frac{f(t)}{1 - F_E(t)} = \frac{1}{\theta} \approx 0.0286 \ \text{yr}^{-1} \tag{3.128}$$

and the hazard rate $h_{GW}(t)$ increases with time (see also Table 3.4, four rightmost columns):

$$h_{G_W}(t) = \frac{g(t)}{1 - G_W(t)} = \frac{b \cdot t^{b-1}}{a^b} \tag{3.129}$$

Since $b > 1$, $h_{GW} > 1$ and the failure mechanism has the character of a wear-out mechanism (see Section 2.4.5).

At the early stages the assets appear to perform better under G_W than under the supposed distribution F_E. When the operational time approaches and exceeds the scale parameter values, the ruling failure mechanism of G_W reveals its wear-out behaviour with an increasing hazard rate. The similarity S_{fg} indicated an almost 7% decay of the similarity over the entire time range $[0,\infty)$. The deviation in R is of the same order of magnitude. The similarity tends to indicate the behaviour of the bulk and particularly the extremes might therefore be studied separately (using a limited interval as in

Table 3.4 Reliability and hazard rate for the two distributions F_E and G_W in this example.

	R(5)	R(20)	R(31.5)	R(35)	R(40)	h(5)	h(20)	h(31.5)	h(35)	h(40)
F_E	0.867	0.565	0.407	0.368	0.319	0.0286	0.0286	0.0286	0.0286	0.0286
G_W	0.913	0.575	0.368	0.318	0.256	0.0238	0.0360	0.0413	0.0426	0.0443
Ratio (%)	105	102	90	86	80	83	126	144	149	155

The ratio means dividing the value for G_W by that for F_E in the same column.

Table 3.5 Similarity varies with the analysed interval.

Period [0,t] (yr)	5	10	20	31.5	35	40	60	80	100	150
$S_{fg}(0,t)$	0.81	0.90	0.93	0.93	0.93	0.93	0.93	0.94	0.93	0.93

The values apply to the similarity study of the defined F_E and G_W in this example.

Eq. (3.124)). The similarity of F_E and G_W depends on the period that is used for evaluation (see Table 3.5). If the similarity is lower over a certain period, it means that the distributions differ more in that interval. Whether this is good or bad depends on the specifications. In this example, G_W actually provides a more favourable failure rate at small t than the originally anticipated F_E.

Similarity is not a measure of compliance per se, and lack of similarity is not necessarily bad. Dissimilarity may be welcome in the range where assets perform better than specified.

To elaborate a little further on the cross-terms in the S_{fg} definition, Figure 3.7 shows an example of two functions $F(t)$ and $G(t)$ (top) with their distribution density functions $f(t)$ and $g(t)$ (bottom) that are used for determining S_{fg}. Visual evaluation shows a limited overlap in this particular case, which suggests that there is a non-zero similarity index, but S_{fg} is certainly not 1 because there is also a significant non-overlap.

The density functions $f(t)$ and $g(t)$ can be determined at each value of F and G (see Figure 3.8, top), yielding $f(F)$ and $g(G)$ as in Eqs (3.117)–(3.120). As shown, it is also possible to express f in G and g in F. Though $<f_G>$ and $<g_F>$ are equal according to Eq. (3.114), $f(G)$ and $g(F)$ are not the same. It may also be noted that $<f_F>$ and $<g_G>$ are not the same. The more peaked a density function is, the more peaked the squared density function is and the higher the mean of its square tends to be.

In Figure 3.8 (bottom) the surface under the dotted lines divided by the surface under the solid line yields the similarity index. If f and g (and thus F and G) are the same, then the dotted and solid lines lie on top of each other.

3.6.4 Significance of Similarity

If n assets are installed and early failures occur, it may be checked to what extent the observed failure data match with the reference data (the specified or expected behaviour). The two distribution densities of observed data $g(t)$ and reference data $f(t)$ can be compared through the similarity index S_{fg}.

The more failures are observed, the more accurately the observed distribution $g(t)$ is defined and the better the similarity S_{fg} between observed and reference distribution can be determined. However, in case of doubt about the quality of the assets, such analysis will be carried out long before many failures occurred. Timely decision-making requires this in order to prevent the situation from escalating out of control with too many (if not all) n assets failing too early. Such studies are normally carried out on limited data sets of r failure times t_i ($i = 1, \ldots, r$) with r often considerably smaller than n. The question is how accurate the similarity index S_{fg} is with a limited set of observed failure data t_i ($i = 1, \ldots, r$).

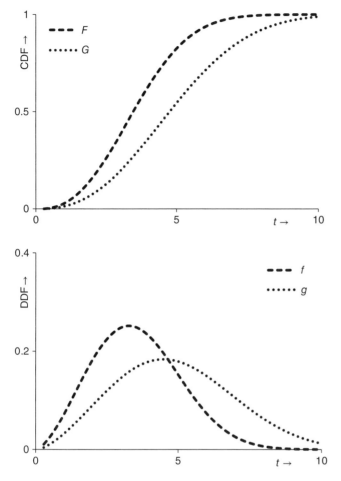

Figure 3.7 On the top, two distributions $F(t)$ and $G(t)$ with their density functions $f(t)$ and $g(t)$ on the bottom.

There are no equations available as yet. A study on the similarity between observed and expected ageing behaviour and its significance was presented in [16]. Its method is explained below and further examples can be found in that paper.

The idea is to estimate how likely the found S_{fg} is under the hypothesis that the observations are truly drawn from the population with distribution $f(t)$. This is done with simulations in which distributions $g_{gen}(t)$ are generated based on random samples drawn from the population. Next, these $g_{gen}(t)$ are used to study how their similarity indices with $f(t)$ are distributed. Finally, it is checked where the observed S_{fg} ranks among the simulated similarity indices. From this ranking it is decided how likely it is that the hypothesis is justified.

In more detail, the significance (or confidence limits) of the similarity are investigated by carrying out a Monte Carlo simulation (see also Section 10.4.4) to generate data sets based on the reference distribution $f(t)$. These sets simulate a variety of observed data. In the simulation these sets can be censored either in time t_r or in number r of

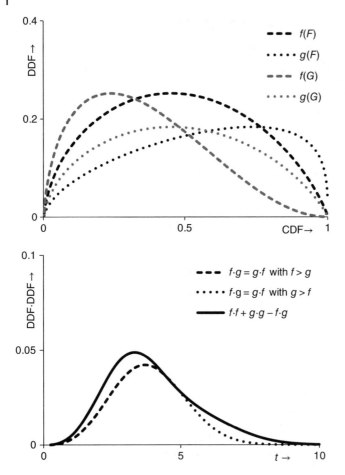

Figure 3.8 On the top, the distribution density functions (DDFs) of the two distributions as functions of *F* and *G*. On the bottom, the DDF products are shown.

observations, which means that sets of *n* failure times are generated and only the *r* smallest are analysed with the knowledge that there are *n* − *r* hidden (i.e. 'censored' or 'suspended') greater failure times. For each of these generated data sets the distribution $g_{gen}(t)$ with its parameters is determined (by parameter estimation techniques to be discussed in Chapter 6) and thus a set of observation distributions $g_{gen}(t)$ is generated. Due to the random selection of small, censored data sets, the estimated parameters of the $g_{gen}(t)$ will scatter about the parameters of the true reference distribution $f(t)$.

Next, all similarity indices $S_{f,ggen}$ are determined between the reference distribution *f* and each distribution g_{gen}. This yields a distribution $F_S(S_{f,ggen})$ of similarity indices $S_{f,ggen}$. This can be done at various moments and yields a graph as in Figure 3.9, which is based on data in [16]. Note that $F_S(S_{f,ggen})$ is very different from the distribution of the times (i.e. $F(t)$). Furthermore, it is noteworthy that $F_S(S_{f,ggen})$ and a graph like Figure 3.9 can be produced with simulations once a reference $f(t)$ is defined. They do not depend on the observed $g(t)$.

Subsequently, the ranking among the $S_{f,ggen}$ of the actually observed similarity index S_{fg} is determined. If S_{fg} ranks among, say, the top 90% of $S_{f,ggen}$ then the observed set can

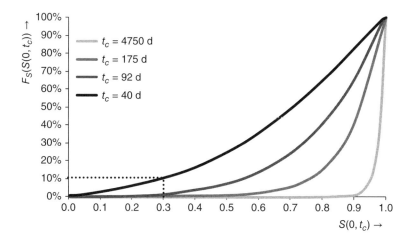

Figure 3.9 Significance of the similarity index after [16]. The graphs represent the distribution of similarity indices after a specified interval $[0,t_c]$ (i.e. at the evaluation moment t_c). The figure shows that after 40 days a similarity index $S(0,t_c = 40$ days) of 0.3 ranks about 10%. This means that at $t_c = 40$ days, 90% of samples randomly drawn from the reference population would have a similarity index higher than 0.3 and the probability of wrongful rejection is about 10%.

be regarded to fall within acceptable scatter of events, however if S_{fg} ranks among the bottom 10% of $S_{f,ggen}$ then it may be concluded that the result is abnormal.

This can be regarded as a test. Having a distribution such as $F_S(S_{f,ggen})$ over the interval $[0,t_c = 40$ days] in Figure 3.9 makes it possible to test the hypothesis. The hypothesis H_0 is that the failed objects do belong to the population that is distributed as the distribution $f(t)$. A low similarity index would then be a matter of scatter. One approach is that H_0 is rejected if the observed similarity index S_{fg} drops below a given S_L. Section 10.5 discusses hypothesis testing and defines the error probabilities α and β that H_0 is wrongfully rejected (respectively wrongfully accepted). If $S_L = 0.3$ were chosen to reject the hypothesis if $S_{fg} < S_L$, then 10% of randomly drawn samples from the population will have $S_{f,ggen} < S_L$ according to Figure 3.9. So, there is a probability of $\alpha = 10\%$ that H_0 is wrongfully rejected.

Another use would be to determine S_{fg} and evaluate how it ranks among the simulated $S_{f,ggen}$. If the critical value $S_L = S_{fg}$, then the ranking would be the maximum probability of Type 1 error (i.e. wrongfully rejecting H_0).

As an example, Figure 3.10 shows a case that was discussed in [11]. The total number of assets is 87, of which 6 failed within 40 days. The reference distribution with confidence intervals (a range in which the observation may fall with a given certainty, here (10%, 90%)) is plotted in grey and the six observed failures with their estimated distribution plus confidence intervals are plotted in black. The details about graphical representation and distribution parameter estimation will be discussed in Chapter 5 (respectively Chapter 6), but for now the focus is on the concept of the similarity index itself.

The first six fall within the confidence intervals of the reference distribution and because of that there may be no reason to doubt the quality of the assets. However, extrapolation of the observed distribution gives a very worrying image. The characteristic lifetime at $F_t = 63\%$ of the extrapolated observed distribution is almost a factor 10 too low compared to the reference. On the other hand, it is possible that the next observation will fall within the confidence intervals too and that the observed

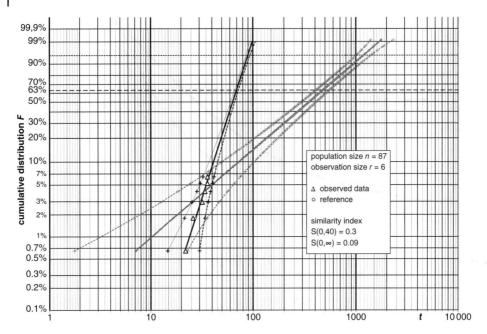

Figure 3.10 Example of an observed distribution based on six observations and the reference distribution of 87 assets. The six observations fall within the [10%, 90%] confidence intervals drawn around the reference distribution, but the extrapolation of the observed distribution is very worrying. The similarity index is calculated for the observation period yielding S(0,40) = 0.3 and for the extrapolation yielding S(0,∞) = 0.09.

distribution will start to resemble the reference better and better. There is uncertainty about the quality and how this can be expressed.

The similarity index can be calculated for given intervals with Eq. (3.124). The similarity index for the observed period [0,40] appears to be 0.3, which is considerably lower than 1 and reflects the worries quite well. The extrapolated distribution covers the period [0,∞] and the similarity index over this complete range is 0.09, which is dramatically lower than 1. $S(0,40) = 0.3$ may be enough reason to discuss the matter between customer and supplier, and be prepared for the scenario that the extrapolation comes true. For instance, plans may be agreed upon if the similarity index drops any further.

How significant is this analysis? Actually, the earlier discussed Figure 3.9 applies to this case, as described in [16]. So, in this case $S(0,40) = 0.3$ ranks among the 10% lowest simulated similarities, as shown in Figure 3.9. This suggests that this is a rather abnormal case of 1 out of 10. However, it may be required that the probability of a Type I error (i.e. wrongfully rejecting the hypothesis H_0 that observations still agree with the reference) should be smaller than $\alpha = 10\%$.

If an error $\alpha = 1\%$ were required, then the S level for rejecting H_0 would be $S_L = 0.03$. Then, having $S_{fg} = 0.3$ after 40 days is not low enough to reject H_0. If, after 92 days, the graph of observed data proceeded as in the first 40 days, the similarity S_{fg} would decrease to 0.085 as reported in [16], while a chosen $S_L = 0.25$ would correspond to about $\alpha = 1\%$. So, after 92 days the hypothesis H_0 could be rejected with a small Type I error. The similarity index can thus be used to detect and quantify deviant behaviour at an early stage.

3.6.5 Singularity Issues and Alternative Similarity Indices

The continuous similarity index reads (Section 3.6.3, Eq. (3.112)):

$$S_{fg} = \frac{\int f(t) \cdot g(t)\, dt}{\int f^2(t)dt + \int g^2(t)dt - \int f(t) \cdot g(t)\, dt} \tag{3.130}$$

As reported in [16], functions f and g may exist that become infinite (i.e. have singularities). For instance, the distribution density $f(t)$ of a Weibull distribution $F(t)$ with a shape parameter $\beta < 1$ has a singularity at $t = 0$. The cross-term $f \cdot g$ of two Weibull distribution densities f and g in Eq. (3.130) will have a singularity if the sum of their shape parameters β_f (respectively β_g) is smaller than 2: $\beta_f + \beta_g \le 2$.

Not every singularity is an issue. For instance, the integral of a Weibull distribution density $f(t)$ from 0 to ∞ yields $\int f(t)dt = 1$. However, when $\beta_f + \beta_g \le 1$ a discontinuity occurs. The similarity index will still have a value in the range $[0,1]$, but jumps to 0 if $\beta_f \ne \beta_g$ if either $\beta_f < \frac{1}{2}$ or $\beta_g < \frac{1}{2}$. If $\beta_f = \beta_g$, then the similarity index yields 1 as it should.

In Monte Carlo experiments issues may appear even when $\beta_f + \beta_g > 1$ and similarity indices are calculated for randomly selected samples. Drawing from a population will also contain samples with estimated shape parameter $b < \beta$, as will be shown in Section 10.4.3. Depending on parameters like β, the similarity index may suffer singularities. This is also the main source: employing distributions with $\beta < 1$ or close to 1 with Monte Carlo simulations.

A question is whether the similarity index should and might be adapted. The similarity index can be written as Eq. (3.107) in terms of inner products:

$$S_{fg} = \frac{\langle f \cdot g \rangle}{\langle f \cdot f \rangle + \langle g \cdot g \rangle - \langle f \cdot g \rangle} \tag{3.131}$$

This might be regarded as a next generalization of the similarity index. The f and g may no longer need to be the distribution densities, but could be any of the other fundamental functions F, R, h or H (see Section 2.4 and further). The hazard rate h behaves in a similar way as f. The integral forms F, R and H have no singularities, except that H becomes infinite with increasing time. The use of the distribution density f stays close to the original meaning of counting representation in subpopulations (see Sections 3.6.1 and 3.6.2).

In practice, the similarity index has been used mostly for wear-out phenomena including child mortality cases with fast ageing subpopulations (Section 2.5.3). Such distributions have shape parameters $\beta > 1$ and since $\beta_f + \beta_g > 1$, singularities are not an issue.

In the cases discussed in Sections 9.4 and further, the discrete similarity index is used for investigating the similarity between distributions. In those cases, usually no failures are observed at $t = 0$, but always some (at least a little) time after energizing. This generally prevents singularities to occur. The discrete singularity index is (cf. Eq. (3.105)):

$$S_{fg} = \frac{\sum_i f_i g_i}{\sum_i f_i f_i + \sum_i g_i g_i - \sum_i f_i g_i} = \frac{\sum_i f_i g_i}{\sum_i f_i^2 + \sum_i g_i^2 - \sum_i f_i g_i} \tag{3.132}$$

As discussed in Sections 9.4 and further, the f_i and g_i are approximated by the reciprocal values of the times between failure. The method is explained in more detail in Section 9.4.3.4. An encountered issue is the sensitivity of the use of reciprocal times between failures for fluctuations in those time intervals. The similarity may prove weak then, but the similarity index adequately produces a low score (see Section 9.4.4.4).

If singularity issues occur with the similarity index applied to Weibull distributions, it seems to concern discontinuous jumping between a non-zero value and zero for severe child mortality cases in the sense that $\beta < 1$ while Monte Carlo experiments sample in that region. The similarity index, however, can still be calculated. Use of the discrete similarity index based on observed data may be an adequate solution.

3.7 Compliance

Compliance in asset management means conformity with specifications, directives and/or regulations. The compliance of assets with respect to reliability may be specified in terms of measures of minimum required quality. This means that the asset failure times must be equal to or larger than the specified minimum values. As mentioned above, similarity is not a good measure of compliance per se, because a much better behaviour than specified will also cause non-similarity and consequently the similarity index will drop significantly below 1. So, another measure is needed, which is compliance.

Assume in all cases that $\Phi(t)$ is the specified distribution and $F(t)$ is the observed distribution.

If the distribution $F(t)$ is fully compliant with the specified $\Phi(t)$, then the failed part of the population must be equal to or less than the distribution $\Phi(t)$:

$$\forall t : F(t) \leq \Phi(t) \tag{3.133}$$

One consideration is that the two functions may cross (i.e. that times $t = t_{X,j}$ exist where $\Phi(t) = F(t)$). If this is the case, then the range can be split up into subranges where $F(t)$ performs better than $\Phi(t)$ (meaning that $F(t)$ is smaller than $\Phi(t)$) and subranges where $F(t)$ performs worse than $\Phi(t)$ (meaning that $F(t)$ is greater than $\Phi(t)$).

Compliance can be elaborated as a difference or as a ratio. Difference compliance C_{diff} is concluded for each t or range $[t_1, t_2]$ where the difference between $\Phi(t)$ and $F(t)$ is larger than or equal to 0. Pointwise difference compliance $C_{F,diff}(t)$ can be defined as:

$$C_{F,diff}(t) = \Phi(t) - F(t) \tag{3.134}$$

The value of $C_{F,diff}(t)$ falls in the interval $[-1,1]$, where: -1 means that the observed failure behaviour finished the complete population before the specified ageing behaviour even kicked off (i.e. the population is fully non-compliant over the full range of t); 0 means that the observed and specified behaviour are completely in line (i.e. the population is compliant); 1 means that no failures have taken place before the specified failure behaviour has finished the complete population (i.e. the population is fully overcompliant).

Range compliance $C_{F,diff}(t_1, t_2)$ over a limited interval $[t_1, t_2]$ is then defined as:

$$C_{F,diff}(t_1, t_2) = \int_{t_1}^{t_2} [\phi(t) - f(t)]dt = \Phi(t_1) - \Phi(t_2) - F(t_1) + F(t_2) \tag{3.135}$$

It may be noted that the observed distribution density $f(t)$ can be larger than the specified distribution density $\varphi(t)$ and yet, the observations may comply because averaging occurs in Eq. (3.135) and a seemingly non-compliant part is then compensated by an overcompliant part of the batch. Compliance over a range $[t_1,t_2]$ in terms of Eq. (3.135) requires that the difference between $\Phi(t)$ and $F(t)$ is non-negative, that is:

$$C_{F,diff}(t_1,t_2) = \int_{t_1}^{t_2} [\phi(t) - f(t)]dt = \Phi(t_1) - \Phi(t_2) - F(t_1) + F(t_2) \geq 0 \qquad (3.136)$$

If the range $[t_1,t_2]$ is taken to be $[0,\infty)$, then $C_{F,diff}$ is zero. Compliance can also be defined more precisely by investigating in which ranges $[t_1,t_2]$ Eq. (3.133) applies.

Evaluating the difference compliance is also useful over a finite range, where the specification of $\Phi(t)$ is useful (e.g. $[t_1,t_2] = [0,t_{100,99}]$ where $t_{100,99}$ is the 99th percentile of $\Phi(t)$ (see Section 3.2). However, since early failures are often of most interest, $C_{F,diff}$ is usually evaluated for $[t_1,t_2] = [0,t_{10,1}]$ with $t_{10,1}$ the first decile of $\Phi(t)$ (see Section 3.2).

Ratio compliance C_{rat} evaluates the ratio of $\Phi(t)$ and $F(t)$. It is observed for each t where:

$$C_{F,rat} = \frac{\Phi(t)}{F(t)} \geq 1 \qquad (3.137)$$

It is assumed that the functions $F(t)$ and $\Phi(t)$ have the same domain (i.e. the full set of variable values t). As both functions are also bijective (i.e. a specific function value is associated with each value t), there is also a bijective relation between the two functions $F(t)$ and $\Phi(t)$.

A measure of ratio compliance $C_{F,rat}$ over a range can be defined as:

$$C_{F,rat}(F; \Phi_1, \Phi_2) = \int_{\Phi 1}^{\Phi 2} \frac{1}{F(t)} d\Phi = \int_{t1}^{t2} \frac{\phi(t)}{F(t)} dt \qquad (3.138)$$

where indices 1 and 2 again denote the start and end of the integration interval. Instead of comparing cumulative functions, the reliability $R(t)$ can also be used for measures of compliance.

3.8 Summary

This chapter discussed measures: expected values and moments (Section 3.1); median and quantiles (Section 3.2); mode (Section 3.3); merits of mean, median and mode (Section 3.4); covariance and correlation of data sets (Section 3.5); similarity of distributions and compliance (Section 3.6).

As for expected values and moments (Section 3.1), means or expected values are determined for both discrete and continuous variables as well as functions thereof. In case of discrete variables, some values may have a higher frequency than others and therefore they can be weighed with the probability or distribution mass function (Eq. (3.2)). The mean is preserved in sums (i.e. $<X + Y> = <X> + <Y>$; Eq. (3.15)) and provided X and Y are independent also in products $<X \cdot Y> = <X> \cdot <Y>$ (Eq. (3.16)). The moments are means of powers of variables. The kth moment of X is $\mu_k = <X^k>$ (Eqs (3.27) and (3.28)). As determining the kth moment can be complicated, it can be more convenient to use a moment generating function (Section 3.1.4) or a characteristic function (Section 3.1.5).

The central moments are the moments of a variable minus its mean (i.e. the kth central moment equals $<(X - \mu)^k>$; Eqs (3.46) and (3.47)). The first four central moments are often used to characterize a distribution (Section 3.1.6): the first is trivial, because it is

0; the variance is a measure of data scatter; the skewness is a measure of asymmetry; the (excess) kurtosis is a measure of peakedness or the heaviness of tails.

If the entire population is not tested, but just a random sample from it with sample size n, the expected value or mean of the population and of random samples are the same. The population variance as the second central moment with the sample will be estimated as a factor $n/(n-1)$ smaller than the true population variance.

The median is the variable at which the cumulative distribution $F = 1/2$. This and other fractions are called quantiles (Section 3.2). The mode is the variable value at which a single peaked distribution density reaches its maximum (Section 3.3). The merits of mean, median and mode are compared. It depends on the purpose of use which measure might be preferred (if a single one has to be chosen). Median and mode are preserved under many operations. The mean takes the influence of extremes into account.

Covariance (Section 3.5.1) and correlation (Section 3.5.2) are measures for the linear relationship between two variables. If these are 0 there is no overall linear relationship, but still another relationship (e.g. quadratic) may exist nevertheless. The covariance of X with itself is the variance. The correlation coefficient ρ_{XY} is the covariance of X and Y normalized with their standard deviations σ_X and σ_Y. The correlation may be larger if one of the variables is delayed (like sine and cosine, or two distributions with different means). This can be measured with the cross-correlation and the autocorrelation (Section 3.5.3).

As for the similarity of distributions and compliance (Section 3.6), the concept of the similarity index S_{fg} is introduced (Section 3.6). It measures to what extent two data sets or particularly two distributions f and g are the same. This is generally used to evaluate whether two distributions match, whether it makes a difference which distribution is used and particularly to quantify the similarity between observed and expected failure behaviour as a measure of quality control. The latter often requires timely decision-making, which implies the analysis of an often small data set. The significance of the similarity index for such small data sets is discussed in Section 3.6.4. Finally, measures of compliance are discussed in Section 3.7.

Questions

3.1 With reference to Section 3.1.7:
   a) Prove Eq. (3.54).
   b) Prove Eq. (3.55).

3.2 With reference to Section 3.1.7:
   a) Give the general expression for the kth central moment *normalized* with σ.

3.3 With reference to Section 3.1.7:
   a) Calculate the first four central moments, the skewness, the kurtosis and the excess kurtosis for the case of Table 2.4.

3.4 With reference to Section 3.2:
   a) Prove that the mean and the median have the same value for a symmetrical distribution.

3.5 With reference to Section 3.5.2:

a) Assume variables $X_1, \ldots, X_n$ are independent. Prove: $\mathrm{var}\left(\sum_{i=1}^{n} X_i\right) = \sum_{i=1}^{n} \mathrm{var}(X_i)$.

3.6 With reference to Section 3.5.2:

a) Show that $\mathrm{var}(X + Y) = \mathrm{var}(X) + \mathrm{var}(Y) + 2\rho_{XY} \cdot \sqrt{\mathrm{var}(X) \cdot \mathrm{var}(Y)}$.

b) Elaborate $\mathrm{var}(X - Y) = \ldots$

3.7 With reference to Sections 3.6.3 and 3.7:

a) Find the median $t_{FE,M}$ for distribution F_E in Example 3.7.

b) Determine $S_{fg}(0, t_{FE,M})$.

c) Determine the difference compliance $C_{diff}(0, t_{FE,M})$ (note that F_E is the reference and G_W is the observed distribution).

d) Determine the ratio compliance $C_{rat}(0, t_{FE,M})$ (note that F_E is the reference and G_W is the observed distribution).

4

Specific Distributions

This chapter describes a range of distributions that are used for specific purposes. These distributions are not only of interest for variables that follow these distributions by nature. Various of these distributions are also so-called asymptotic distributions. This means that under certain conditions other distributions approach such an asymptotic distribution. As a consequence, such distributions are generally used in fair to good approximation.

These categories and distributions will be studied in the following sections. For each distribution the purpose or context is explained and the parameters and characteristics are elaborated. The present chapter provides a kaleidoscope of distributions. Rather than reading through all of them, the reader may select the interesting distribution(s) and relevant sections from Table 4.1.

4.1 Fractions and Ranking

When managing assets, important variables are population fractions that failed or survived and their ranking (i.e. the sequence of failing assets). A part of the analysis does not require knowledge about the statistic behaviour in time, but merely how asset groups can be shuffled.

In Section 2.4.1 one of the interpretations of the cumulative distribution function F was a measure for the failed fraction or index of failure sequence. Every asset can be identified with its own F value between 0 and 1. The cumulative distribution ranks the assets by their sequence of failing. The F value can be regarded as a variable in the domain [0,1] which can be characterized by measures like the mean, median, variance, and so on. With random sampling each asset and therefore each F value has the same probability of occurring (i.e. the F values are uniformly distributed). Sampling assets by their F value is an application of the uniform distribution, which is discussed in Section 4.1.1.

If a series of assets are selected, then the number of assets in a series is called the sample size n. In service or tests the assets can be indexed by their increasing strength, which is called ranking. The behaviour of the first (or any other ranking index) out of n assets can be studied with their measures. This is discussed with the beta distribution in Section 4.1.2.

Reliability Analysis for Asset Management of Electric Power Grids, First Edition. Robert Ross.
© 2019 John Wiley & Sons Ltd. Published 2019 by John Wiley & Sons Ltd.
Companion website: www.wiley.com/go/ross/reliabilityanalysis

Table 4.1 Statistics and distributions in the present chapter.

Section	Statistics	Distribution	Domain
4.1	**Fractions and ranking statistics**		
4.1.1	Constant probability density (continuous variable) or constant probability mass (discrete variable)	Uniform	$x \in [a,b] \subseteq \mathbb{R}$ or $\mathbb{Z}$
4.1.2	Ranked probability x from a uniform distribution	Beta	$x \in [0,1] \subset \mathbb{R}$
4.2	**Extreme value statistics**		
4.2.1, 4.2.2, 4.2.3	The smallest extreme value x (weakest link of chain)	Weibull, other[a]	$x \in [0,\infty) \subseteq \mathbb{R}$
4.2.4	Random failure at x (i.e. constant hazard rate h)	Exponential	$x \in [0,\infty) \subseteq \mathbb{R}$
4.3	**Mean and variance statistics**		
4.3.1	The mean of a variable x with domain $(-\infty,\infty)$	Normal or Gaussian	$x \in (-\infty,\infty) \subseteq \mathbb{R}$
4.3.2	The logarithmic mean of a variable t with domain $[0,\infty)$	Lognormal	$t \in [0,\infty) \subseteq \mathbb{R}$
4.4	**Frequency and hit statistics**		
4.4.1	Number z of two elementary outcomes in n trials (sampling with replacement)	Binomial	$z \in [0,n] \subseteq \mathbb{N}$
4.4.2	Number z of two elementary outcomes in n trials without knowing how may trials happened	Poisson	$z \in [0,n] \subseteq \mathbb{N}$
4.4.3	Number of two elementary outcomes in n trials (sampling without replacement)	Hypergeometric	$z \in [0,n] \subseteq \mathbb{N}$
4.4.4	Frequency that one out of two outcomes occurs with a large number n of trials	Normal	$x \in (-\infty,\infty) \subseteq \mathbb{R}$
4.4.5	Numbers z_k of $k > 2$ elementary outcomes in n trials (sampling with replacement)	Multinomial	$z_k \in [0,n] \subseteq \mathbb{N}$
4.4.6	Number of more than two elementary outcomes in n trials (sampling without replacement)	Multivariate hypergeometric	$z_k \in [0,n] \subseteq \mathbb{N}$

a)   Other domains: Gumbel with domain $(-\infty,\infty)$, Fréchet with domain $(-\infty,0]$.

4.1.1   Uniform Distribution

The uniform distribution is a very simple but essential distribution. The term 'uniform' refers to the feature that each variable value has the same probability of occurrence. This is also called a homogeneous distribution. The domain is generally taken to be $[a,b]$. The variable X can be continuous or discrete. Both are discussed in the following sections.

Very important applications of the uniform distribution are random sampling assets from a group (i.e. picking an asset at random). This translates into sampling values from the cumulative distribution if the assets can be ranked by a measure of performance such as time to breakdown, and so on. Some applications of uniform distributions were already discussed in previous sections, like rolling a dice (Section 2.2) and random selection from a list of insulators (Section 2.5.2.2). Sampling a uniformly distributed variable is also the foundation of many Monte Carlo simulation experiments.

4.1.1.1 Continuous Uniform Distribution Characteristics

In its continuous form the uniform distribution has two parameters, a and b, which mark the boundaries of the domain. The uniform distribution is abbreviated as $U(x;a,b)$ or $U(x)$ for short. As all variable values have the same probability of occurring, the probability density function $u(x) = u$ is a constant and therefore the cumulative distribution U has a linear relationship with variable X:

$$\frac{dU(x)}{dx} = u \quad \Rightarrow \quad U(x) = U(a) + \int_a^x u\,dx = u \cdot (x - a) \tag{4.1}$$

$U(a) = 0$ because a is the lower limit of the domain.

As the integral of $u(x)$ over the full domain must equal 1 (see Section 2.4.1) and the domain is finite, all fundamental functions (see Section 2.4) are defined. These are shown in Figure 4.1. The main characteristics of the uniform distribution are summarized in Table 4.2.

Random number generators are often used in Monte Carlo simulations. A good random number generator produces numbers with an equal probability within a defined range (i.e. each number has the same probability of occurring). This can be done by sampling uniformly distributed numbers from the range [0,1] and treating these as values of the cumulative distribution function F. In this way the cumulative distribution function F itself is regarded as an example of a uniform distribution with variable $X = F$ and parameters $a = 0$ and $b = 1$. This makes $U(x;0,1)$ the identity function on [0,1]. The characteristics of $U(x;0,1)$ are presented in Table 4.3. Random number generators should be tested (cf. Section 10.4.4).

Another example is measuring the position of the hands of an analogous clock at random time. For example, the results of reading the minute hand are from the sample space [0,60) — that is, the readings are real numbers ranging from (and including) 0 up to (and not including) 60. If the time of reading is random indeed, the results are continuous and uniformly distributed.

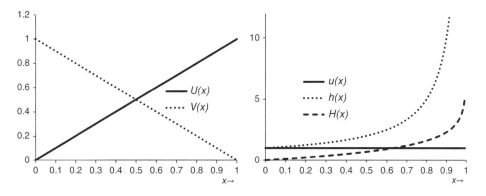

Figure 4.1 Fundamental functions of the continuous uniform distribution $U(x)$ with $a = 0$ and $b = 1$.

Table 4.2 Characteristics of the continuous uniform distribution $U(x)$.

Parameters	a, b with $a < b$	a: start of domain b: end of domain
Fundamental functions		
Domain	$a \leq x \leq b$	outside domain:
Cumulative distribution U	$U(x) = \dfrac{x - a}{b - a}$	$x < a$: $U(x) = 0$ $x > b$: $U(x) = 1$
Reliability V	$V(x) = \dfrac{b - x}{b - a}$	$x < a$: $V(x) = 1$ $x > b$: $V(x) = 0$
Distribution density u	$u(x) = \dfrac{1}{b - a}$	$x < a$: $u(x) = 0$ $x > b$: $u(x) = 0$
Hazard rate h	$h(x) = \dfrac{1}{b - x}$	$x < a$: $h(x) = 0$ $x > b$: $h(x) \to \infty$
Cumulative hazard H	$H(x) = -\ln(b - x)$	$x < a$: $H(x) = 0$
Measures		

Mean	Variance	Skewness	Excess kurtosis	Median	Mode
$\dfrac{1}{2}(a + b)$	$\dfrac{1}{12}(b - a)^2$	0	$-\dfrac{6}{5}$	$\dfrac{1}{2}(a + b)$	n.a.

Table 4.3 The cumulative distribution U for F value as a uniformly distributed variable x.

Parameters	$a = 0 \quad \wedge \quad b = 1$	a start of domain b end of domain
Fundamental functions		
Domain	$0 \leq F \leq 1$	outside domain:
Cumulative distribution U	$U(F) = F$	$F < 0$: $U(F) = 0$ $F > 1$: $U(F) = 1$
Reliability V	$V(F) = 1 - F$	$F < 0$: $V(F) = 1$ $F > 1$: $V(F) = 0$
Distribution density u	$u(F) = 1$	$F < 0$: $u(F) = 0$ $F > 1$: $u(F) = 0$
Hazard rate h	$h(F) = \dfrac{1}{1 - F}$	$F < 0$: $\quad h(F) = 0$
Cumulative hazard H	$H(F) = -\ln(1 - F)$	$F < 0$: $\quad H(F) = 0$
Measures		

Mean	Variance	Skewness	Excess kurtosis	Median	Mode
$\dfrac{1}{2}$	$\dfrac{1}{12}$	0	$-\dfrac{6}{5}$	$\dfrac{1}{2}$	n.a.

4.1.1.2 Discrete Uniform Distribution Characteristics

In its discrete form the uniform distribution again has the two parameters a and b that mark the boundaries of the domain. A third parameter is the number n of elements in the variable space. So, there are n variables X_i with index $i = 1,\dots,n$. Again, this information is sufficient to determine the fundamental statistical functions. These are presented in Table 4.4. If the index does not start at 1, then the starting index value can be regarded as a fourth parameter.

Throwing a dice is an example of a uniform distribution with $a = 1$, $b = 6$ and $n = 6$. The probability that an elementary event occurs per throw is 1/6. This is an example of

Table 4.4 Characteristics of the discrete uniform distribution $U(x)$ with n variable values x_i.

Parameters	a, b with $a < b$	a: start of domain b: end of domain n: number of variable values
Domain	$a \le x_i \le b;$ $x_i = a + \dfrac{i-1}{n-1} \cdot (b-a)$	outside domain:
Fundamental functions		
Cumulative distribution U	$x \in [x_i, x_{i+1}): \ U(x) = \dfrac{i}{n}$	$x < a: \ U(x) = 0$ $x \ge b: \ U(x) = 1$
Reliability V	$x \in [x_i, x_{i+1}): \ V(x) = \dfrac{n-i}{n}$	$x < a: \ V(x) = 1$ $x \ge b: \ V(x) = 0$
Distribution density u	$u(x_i) = \dfrac{1}{n}$	$x < a: \ u(x) = 0$ $x > b: \ u(x) = 0$
Hazard rate h	$h(x_i) = \dfrac{1}{n-i}$	$x < a: \ h(x) = 0$
Cumulative hazard H	$x \in [x_i, x_{i+1}): \ H(x) = \sum\limits_{j=1}^{i} \dfrac{1}{n-j}$	$x < a: \ H(x) = 0$
Measures		

Mean	Variance	Skewness	Excess kurtosis	Median	Mode
$\dfrac{1}{2}(a+b)$	$\dfrac{1}{12}((b-a+1)^2 - 1)$	0	$-\dfrac{6}{5} \cdot \dfrac{n^2+1}{n^2-1}$	$\dfrac{1}{2}(a+b)$	n.a.

sampling with replacement. With every round the same event has the same probability of occurring.

Randomly selecting an insulator from a stock for testing or installing takes the selected insulator out of the stock. It cannot be selected anymore. This is sampling without replacement. As a consequence, the number of variable values n decreases. The probability for each rendering event grows, because fewer variable values remain.

For each round with or without replacement, the fundamental functions are as in Table 4.4 given that the appropriate number of variables is used with each round. The basic functions of the discrete distribution are shown in Figure 4.2.

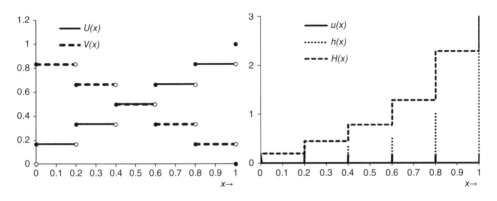

Figure 4.2 Basic functions of the discrete uniform distribution $U(x)$ with $a = 0$, $b = 1$ and $n = 6$.

Table 4.5 Moment generating function G of the uniform distribution U.

Variable	Moment generating function	Distribution parameters
Continuous	$t \neq 0:$ $\quad G_X(t) = \dfrac{e^{tb} - e^{ta}}{t \cdot (b-a)}$ $t = 0:$ $\quad G_X(0) = 1$	a, b with $a < b$
Continuous	$t \neq 0:$ $\quad G_X(t) = \dfrac{e^t - 1}{t}$ $t = 0:$ $\quad G_X(0) = 1$	$a = 0 \quad \wedge \quad b = 1$
Discrete	$G_X(t) = \dfrac{e^{ta} - e^{t(b+1)}}{n \cdot (1 - e^t)}$	a, b with $a < b$ $n:$ number of variable values

4.1.1.3 EXTRA: Moment Generating Function and Characteristic Function

The moment generating functions $G_X(t)$ (Section 3.1.4) for the uniform distribution are shown in Table 4.5. The kth moment $<X^k>$ can be found by taking the kth derivative of $G_X(t)$ and setting $t = 0$:

$$\frac{d^k}{dt^k} G_X(t) \bigg|_{t=0} = < X^k > \tag{4.2}$$

The Taylor series of the exponential function for $t \to 0$ (i.e. the Maclaurin's series) can be used to calculate the kth derivatives of $G_X(t)$ at $t = 0$:

$$\lim_{t \to 0} e^{cx} = \sum_{j=0}^{\infty} \frac{t^j}{j!} \frac{d^j}{dt^j} e^{ct} = 1 + c \cdot t + \frac{c^2 t^2}{2} + \frac{c^3 t^3}{6} + \frac{c^4 t^4}{24} + \dots \tag{4.3}$$

where c is a constant to be replaced by the domain boundaries a and b here.

The characteristic functions $\varphi_X(t)$ (Section 3.1.5) for the uniform distribution are shown in Table 4.6. The kth moment $<X^k>$ can be found with the kth derivative of $\varphi_X(t)$ as follows:

$$< X^k > = (-i)^k \frac{d^k}{dt^k} \phi_X(t) \bigg|_{t=0} \tag{4.4}$$

4.1.2 Beta Distribution or Rank Distribution

One of the applications of the uniform distribution (see Section 4.1.1) concerns the cumulative distribution F values assigned to each asset as a variable itself. The F value is uniformly distributed. In a related application a series of n assets may be taken from a population of assets and tested. This will produce a series of n observations which may be time to failure T (or any other measure of performance). These test outcomes can be ranked from small to large, yielding a ranked series of n observations t_i with $i = 1,\dots,n$ and $t_1 < \dots < t_n$. Provided the cumulative distribution $F(t)$ is known, the assets can also each be associated with an F_i value, again with $i = 1,\dots,n$ and $F_1 < \dots < F_n$. Repeated tests on n assets will produce yet other series of n observed and ranked F_i values, again

Table 4.6 Characteristic function φ of the uniform distribution U.

Variable	Characteristic function	Distribution parameters
Continuous	$t \neq 0: \quad \phi_X(t) = \dfrac{e^{itb} - e^{ita}}{it \cdot (b-a)}$ $t = 0: \quad \phi_X(0) = 1$	a, b with $a < b$
Continuous	$t \neq 0: \quad \phi_X(t) = \dfrac{e^{it} - 1}{it}$ $t = 0: \quad \phi_X(0) = 1$	$a = 0 \quad \wedge \quad b = 1$
Discrete	$\phi_X(t) = \dfrac{e^{ita} - e^{it(b+1)}}{n \cdot (1 - e^{it})}$	a, b with $a < b$ n: number of variable values

with $i = 1,\ldots,n$ and $F_1 < \ldots < F_n$. The ranked F_i observations are distributed according to a beta distribution, of which the parameters α and β are linked to i and n.

So, a single sample from the space of cumulative distribution F values is a uniformly distributed variable. A series of n ranked F_i values are each beta distributed with parameters that depend on i and n. This application is important in the statistical analysis of failure data and inferences.

The beta distribution can be much more widely applied than for the purpose above, and can actually be determined for any parameter set of positive real numbers. The beta distribution is very useful for estimating confidence limits and confidence intervals.

4.1.2.1 Beta Distribution Characteristics

The beta distribution is also called the rank distribution for its application to the cumulative distribution as described in Section 4.1.2. The beta distribution is based on the two-variable beta function $B(\alpha,\beta)$:

$$B(\alpha, \beta) = \int_0^1 x^{\alpha-1} \cdot (1-x)^{\beta-1} dx \tag{4.5}$$

This can also be written as:

$$B(\alpha, \beta) = \frac{\Gamma(\alpha) \cdot \Gamma(\beta)}{\Gamma(\alpha + \beta)} = B(\beta, \alpha) \tag{4.6}$$

where Γ is the gamma function (see Section 10.1.2 and Appendix I).

The beta distribution $B(x;\alpha,\beta)$ views the beta function from the perspective of variable X with α and β taken as parameters, and is normalized by dividing by $B(\alpha,\beta)$ in the form of Eq. (4.6):

$$B(x; \alpha, \beta) = \frac{\Gamma(\alpha + \beta)}{\Gamma(\alpha) \cdot \Gamma(\beta)} \int_0^x t^{\alpha-1} \cdot (1-t)^{\beta-1} dt \tag{4.7}$$

For $\alpha = \beta = 1$, the beta distribution $B(x;\alpha,\beta)$ reduces to the uniform distribution $U(x;0,1)$. Table 4.7 shows the characteristics of $B(x;\alpha,\beta)$.

The median $x_{50\%}$ (see Section 3.2) is the x value found by solving:

$$x_{50\%} = B^{inv}\left(\frac{1}{2}; \alpha, \beta\right) \tag{4.8}$$

Table 4.7 Characteristics of the beta distribution $B(x; \alpha, \beta)$.

Parameters	α, β with $\alpha > 0 \quad \wedge \quad \beta > 0$
Domain	$0 \leq x \leq 1$
Fundamental functions	
Cumulative distribution B	$B(x; \alpha, \beta) = \dfrac{\Gamma(\alpha + \beta)}{\Gamma(\alpha) \cdot \Gamma(\beta)} \displaystyle\int_0^x t^{\alpha-1} \cdot (1-t)^{\beta-1} dt$
Reliability R_B	$R_B(x) = 1 - B(x; \alpha, \beta)$
Distribution density b	$b(x; \alpha, \beta) = \dfrac{\Gamma(\alpha + \beta)}{\Gamma(\alpha) \cdot \Gamma(\beta)} x^{\alpha-1} \cdot (1-x)^{\beta-1}$
Hazard rate h	$h(x) = \dfrac{b(x; \alpha, \beta)}{1 - B(x; \alpha, \beta)}$
Cumulative hazard H	$H(x) = \displaystyle\int_0^x h(t) dt$

Measures			
Mean	Variance	Median	Mode
$\dfrac{\alpha}{\alpha + \beta}$	$\dfrac{\alpha \cdot \beta}{(\alpha + \beta)^2 \cdot (\alpha + \beta + 1)}$	$B^{inv}\left(\dfrac{1}{2}; \alpha, \beta\right)$ see Eq. (4.8)	$\dfrac{\alpha - 1}{\alpha + \beta - 2}$
Skewness		Excess kurtosis	
$\dfrac{2(\beta - \alpha)\sqrt{\alpha + \beta + 1}}{(\alpha + \beta + 2)\sqrt{\alpha\beta}}$		$6\dfrac{(\alpha - \beta)^2(\alpha + \beta + 1) - \alpha\beta(\alpha + \beta + 2)}{\alpha\beta(\alpha + \beta + 2)(\alpha + \beta + 3)}$	

where B^{inv} is the inverse function of B. Or, in other words, the median $x_{50\%}$ is the x value where:

$$B(x_{50\%}; \alpha, \beta) = \frac{\Gamma(\alpha + \beta)}{\Gamma(\alpha) \cdot \Gamma(\beta)} \int_0^{x_{50\%}} t^{\alpha-1} \cdot (1-t)^{\beta-1} dt = \frac{1}{2} \tag{4.9}$$

Applied to the case of the beta distribution of the ranked $F_{i,n}$ values (Section 4.1.2), the following is found. For the ith ranked asset it holds that out of n assets: 1 fails (namely the ith in ranking), $(i-1)$ failed before and $(n-i)$ survive. The respective probabilities are: dp, $F = p^{i-1}$ and $R = (1-p)^{n-i}$. The number of combinations in the group of n assets can be calculated with a multinomial. The probability $dF_{i,n}$ that the ith failure occurs during a given probability element dp is:

$$dF_{i,n} = f_{i,n}(p) \; dp = \frac{n!}{(i-1)! \cdot 1! \cdot (n-i)!} p^{i-1} \cdot (1-p)^{n-i} dp \tag{4.10}$$

In terms of the gamma function, Eq. (4.10) becomes:

$$dF_{i,n} = f_{i,n}(p) \; dp = \frac{\Gamma(n+1)}{\Gamma(i) \cdot \Gamma(n+1-i)} p^{i-1} \cdot (1-p)^{n-i} dp \tag{4.11}$$

This is the beta density function with $\alpha = i$ and $\beta = n + 1 - i$. The beta distribution that describes the cumulative distribution of the ith ranked failure out of n thus becomes:

$$B(F_{i,n}; i, n+1-i) = \frac{\Gamma(n+1)}{\Gamma(i) \cdot \Gamma(n+1-i)} \int_0^{F_{i,n}} p^{i-1} \cdot (1-p)^{n-i} dp \tag{4.12}$$

If $F_{i,n}$ is taken to be 100%, then Eq. (4.12) integrates over the full range and $B(100\%,i,n+1-i) = 1$. The beta function is a density distribution that can be used to determine the expected values of functions of ranked variables $F_{i,n}$. If a function $g(F)$ exists then its expected values $<g>$ can be determined with Eq. (3.3) as:

$$< g >= \int_0^1 g(F)dF \tag{4.13}$$

The expected value $<g_{i,n}> = <g_{i,n}(F)>$ can be found by using the beta function as weighing function:

$$
\begin{aligned}
< g_{i,n} > &= \frac{\Gamma(n+1)}{\Gamma(i) \cdot \Gamma(n+1-i)} \int_0^1 g_{i,n}(F) \cdot p^{i-1} \cdot (1-p)^{n-i} dp \\
&= \frac{\Gamma(n+1)}{\Gamma(i) \cdot \Gamma(n+1-i)} \int_0^1 g_{i,n}(p) \cdot p^{i-1} \cdot (1-p)^{n-i} dp
\end{aligned}
\tag{4.14}
$$

An important application of this is to determine the expected ranked probability $<F_{i,n}>$. From Eq. (4.14) the expected rank value $<F_{i,n}>$ follows as:

$$
\begin{aligned}
< F_{i,n} > &= \frac{\Gamma(n+1)}{\Gamma(i) \cdot \Gamma(n+1-i)} \int_0^1 F_{i,n}(p) \cdot p^{i-1} \cdot (1-p)^{n-i} dp \\
&= \frac{\Gamma(n+1)}{\Gamma(i) \cdot \Gamma(n+1-i)} \int_0^1 p \cdot p^{i-1} \cdot (1-p)^{n-i} dp \\
&= \frac{i}{n+1}
\end{aligned}
\tag{4.15}
$$

An interpretation of this expression is as follows. If, repeatedly, a random sample of n items is drawn from a large population and ranked by their ranking position in the complete population, then the mean ranking of the result of the ith item ($i = 1,\dots,n$) will correspond to $F = i/(n+1)$. That is, on average, the ith ranking item will correspond to ranking position $i/(n+1)$ in the total population. This important result is used for plotting graphs. The fundamental functions of the beta distribution applied to ranked cumulative distributions are shown in Table 4.8.

Again the median is found as a solution of $F_{i,n,50\%}$, that is a solution of:

$$B(F_{i,n,50\%}; i, n+1-i) = \frac{\Gamma(n+1)}{\Gamma(i) \cdot \Gamma(n+1-i)} \int_0^{F_{i,n,50\%}} p^{i-1} \cdot (1-p)^{n-i} dp = \frac{1}{2} \tag{4.16}$$

An often-used approximation of the median rank $F_{i,n,50\%}$ is:

$$F_{i,n,50\%} \approx \frac{i-0.3}{n+0.4} \tag{4.17}$$

Figure 4.3 shows the beta distribution of five rankings $i = 1,\dots,5$ for sample size $n = 5$.

Figure 4.4 shows the beta distribution of the second ($i = 2$) of $n = 15$ assets drawn from an asset population and tested. The figure shows the beta distribution density. The 0–5% and 95–100% extreme cumulative beta distribution regions are shaded. The region in between is the so-called 90% confidence interval, where 90% of the $F_{2,15}$ values can be expected. Also indicated are the mode, median and mean $F_{2,15}$ values.

Table 4.8 Characteristics of the beta distribution applied to ranked cumulative distribution values $F_{i,n}$. The parameters α and β are expressed in terms of the ranking parameters i and n. Instead of variable x, a variable p is used which otherwise has no impact on the expressions.

Parameters	$\alpha = i$ $\beta = n + 1 - i$		
Domain	$0 \leq p \leq 1$		
Fundamental functions			
Cumulative distribution B	$B(F_{i,n}; i, n+1-i) = \dfrac{\Gamma(n+1)}{\Gamma(i) \cdot \Gamma(n+1-i)} \displaystyle\int_0^{F_{i,n}} p^{i-1} \cdot (1-p)^{n-i} dp$		
Reliability R_B	$R_B(p) = 1 - B(p; i, n+1-i)$		
Distribution density b	$b(p; i, n+1-i) = \dfrac{\Gamma(n+1)}{\Gamma(i) \cdot \Gamma(n+1-i)} p^{i-1} \cdot (1-p)^{n-i}$		
Hazard rate h	$h(p) = \dfrac{b(p; i, n+1-i)}{1 - B(p; i, n+1-i)}$		
Cumulative hazard H	$H(p) = \displaystyle\int_0^p h(t)dt$		
Measures			
Mean	Variance	Median	Mode
$\dfrac{i}{n+1}$	$\dfrac{i \cdot (n+1-i)}{(n+1)^2 \cdot (n+2)}$	$B^{inv}\left(\dfrac{1}{2}; i, n+1-i\right)$ see Eq. (4.16)	$\dfrac{i-1}{n-1}$
Skewness		Excess kurtosis	
$\dfrac{2(n+1-2i)\sqrt{n+2}}{(n+3)\sqrt{i \cdot (n+1-i)}}$		$6\dfrac{(n+1-2i)^2(n+2) - i \cdot (n+1-i)(n+3)}{i \cdot (n+1-i)(n+3)(n+4)}$	

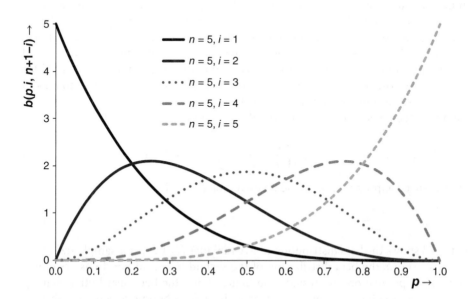

Figure 4.3 Example of the beta distributions of the rankings $i = 1, \ldots, 5$ for $n = 5$.

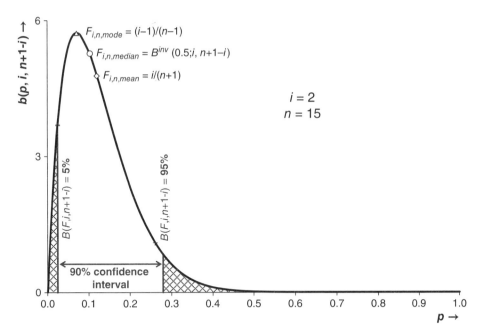

Figure 4.4 Beta distribution with the cumulative distribution F as variable. Here the distribution of F for $i = 2$ and $n = 15$ is shown, together with the mean, median and mode as well as the 90% confidence intervals where 90% of the $F_{2,15}$ values are shown.

In general, the $A\%$ confidence boundary is found as the solution $F_{i,n,A\%}$ of:

$$B(F_{i,n,A\%}; i, n + 1 - i) = \frac{\Gamma(n+1)}{\Gamma(i) \cdot \Gamma(n+1-i)} \int_0^{F_{i,n,A\%}} p^{i-1} \cdot (1-p)^{n-i} dp = A\% \quad (4.18)$$

Or in other words:

$$F_{i,n,A\%} = B^{inv}(A\%; i, n + 1 - i) \quad (4.19)$$

4.1.2.2 EXTRA: Moment Generating Function and Characteristic Function

In this section all indices are indicated by capitals I, J, M, N in order to avoid confusion with the imaginary number i that appears in the characteristic function φ.

The moment generating functions $G_X(t)$ (Section 3.1.4) for the beta distribution are shown in Table 4.9. The Kth moment can be found by taking the Kth derivative and setting $t = 0$ (cf. Eq. (4.2)).

Table 4.9 Moment generating function G of the beta distribution B.

Variable	Moment generating function	Distribution parameters
Continuous	$G_X(t) = 1 + \sum\limits_{J=1}^{\infty} \left(\prod\limits_{M=0}^{J-1} \frac{\alpha + M}{\alpha + \beta + M} \right) \frac{t^J}{J!}$	α, β with $\alpha > 0 \wedge \beta > 0$
Continuous	$G_P(t) = 1 + \sum\limits_{J=1}^{\infty} \left(\prod\limits_{M=0}^{J-1} \frac{I + M}{N + 1 + M} \right) \frac{t^J}{J!}$	$\alpha = I, \beta = N + 1 - I$ variable P instead of X

Table 4.10 Characteristic function φ of the beta distribution B. Note that i is the imaginary unity (i.e. $i^2 = -1$).

Variable	Characteristic function	Distribution parameters
Continuous	$\phi_X(t) = 1 + \sum_{J=1}^{\infty} \left(\prod_{M=0}^{J-1} \dfrac{\alpha + M}{\alpha + \beta + M} \right) \dfrac{(it)^J}{J!}$	α, β with $\alpha > 0 \wedge \beta > 0$
Continuous	$\phi_P(t) = 1 + \sum_{J=1}^{\infty} \left(\prod_{M=0}^{J-1} \dfrac{I + M}{N + 1 + M} \right) \dfrac{(it)^J}{J!}$	$\alpha = I, \beta = N + 1 - I$ variable P instead of X

The characteristic functions $\phi_X(t)$ (Section 3.1.5) for the beta distribution are shown in Table 4.10. Again, the Kth moment can be found with the Kth derivative of $\varphi_X(t)$ (cf. Eq. (4.4)).

4.2 Extreme Value Statistics

When assets are aged until failure in operational or test conditions, failure will occur at the weakest spot or (with redundant systems) at the last surviving part. The weakest spot is where the asset strength has the minimum or smallest extreme value; the last surviving part is where the asset strength has the maximum or largest extreme value. Of the many possible paths of strengths, naturally the extreme value appears.

The study of the minima and/or maxima is called extreme value statistics. Some relevant distributions are listed in Table 4.11.

If each asset by itself can be regarded as a large series of objects of which the weakest fails, then the asymptotic distributions for extreme values apply. For performance with a positive real value like breakdown times or breakdown strengths, the Weibull distribution is then applicable. If the performance is a real value in the range $(-\infty, \infty)$, the Gumbel or extreme value distribution applies (the latter name is somewhat misleading, because distributions of variables with other ranges than $(-\infty, \infty)$ can still be extreme value distributions). The Fréchet distribution is the third extreme value distribution.

Of course, the nature of the failure mechanism at each possible link in a chain may follow the Weibull distribution. However, even if the nature of all possible failing paths individually does not follow the Weibull distribution, it can be shown that the weakest link of a chain will still asymptotically follow the Weibull distribution [17, 18]. For this

Table 4.11 Extreme value distributions.

Domain	Distribution name	Parameters
$(-\infty, \infty)$	Gumbel or extreme value	
$[0, \infty)$	Weibull (two-parameter)	α, β
$[\delta, \infty)$	Weibull (three-parameter)	α, β, δ
$[0, \infty)$	Exponential	$\alpha = 0, \beta = 1$
$[0, \infty)$	Rayleigh	$\alpha, \beta = 2$

reason, the failure times with many ageing mechanisms follow a Weibull distribution or a combination of Weibull distributions when mixed distributions apply (cf. Section 2.5).

4.2.1 Weibull Distribution

The Weibull distribution is one of the most widely used distributions. The background as mentioned above is twofold. Firstly, many performance quantities are in the range $[0,\infty)$ – like breakdown voltage, breakdown strength, time to breakdown, and so on. Secondly, the Weibull distribution is the asymptotic distribution for variables where the smallest extreme value determines the outcome (i.e. situations where the model of the weakest link in a chain applies – for example, ageing in a cable will cause a breakdown at its weakest spot).

The Weibull distribution is usually applied in a two-parameter version, which implies that ageing starts at $t = 0$, but it may also be used in a three-parameter form (discussed in Section 4.2.2) or in a one-parameter form (discussed in Section 4.2.3).

4.2.1.1 Weibull-2 Distribution

The Weibull distribution is probably the most applied distribution for failures. The characteristics of the two-parameter distribution are summarized in Table 4.12. The

Table 4.12 Characteristics of the two-parameter Weibull distribution $(x;\alpha,\beta)$. For the Weibull-1 distribution the shape parameter β is substituted by a constant B.

Parameters	α, β with $\alpha > 0 \wedge \beta > 0$
Domain	$0 \leq x < \infty$
Fundamental functions	
Cumulative distribution F	$F(x; \alpha, \beta) = 1 - e^{-\left(\frac{x}{\alpha}\right)^{\beta}}$
Reliability R	$R(x; \alpha, \beta) = e^{-\left(\frac{x}{\alpha}\right)^{\beta}}$
Distribution density f	$f(x; \alpha, \beta) = \dfrac{\beta \cdot x^{\beta-1}}{\alpha^{\beta}} \cdot e^{-\left(\frac{x}{\alpha}\right)^{\beta}}$
Hazard rate h	$h(x; \alpha, \beta) = \dfrac{\beta \cdot x^{\beta-1}}{\alpha^{\beta}}$
Cumulative hazard H	$H(x) = \displaystyle\int_0^x h(t)dt = \left(\dfrac{x}{\alpha}\right)^{\beta}$
Measures	

Mean μ	Variance σ^2	Median	Mode
$\alpha \cdot \Gamma\left(1 + \dfrac{1}{\beta}\right)$	$\alpha^2 \cdot \left\{ \Gamma\left(1 + \dfrac{2}{\beta}\right) - \left[\Gamma\left(1 + \dfrac{1}{\beta}\right)\right]^2 \right\}$	$\alpha \cdot (\ln 2)^{\frac{1}{\beta}}$	$\alpha \cdot \left(1 - \dfrac{1}{\beta}\right)^{\frac{1}{\beta}}$ with $\beta > 1$ 0 with $\beta \leq 1$

Skewness γ_1	Excess kurtosis γ_2
$\dfrac{\alpha^3 \cdot \Gamma\left(1 + \dfrac{3}{\beta}\right) - 3\mu\sigma^2 - \mu^3}{\sigma^3}$	$\dfrac{\alpha^4 \cdot \Gamma\left(1 + \dfrac{4}{\beta}\right) - 4\gamma_1\sigma^3\mu - 6\mu^2\sigma^2 - \mu^4}{\sigma^4} - 3$

distribution has two parameters: the scale parameter α and the shape parameter β. Both parameters are positive numbers. The cumulative distribution F is:

$$F(t; \alpha, \beta) = 1 - e^{-\left(\frac{t}{\alpha}\right)^{\beta}}$$

(4.20)

The distribution density f is:

$$f(t; \alpha, \beta) = \frac{\beta}{\alpha^{\beta}} \cdot t^{\beta-1} \cdot e^{-\left(\frac{t}{\alpha}\right)^{\beta}}$$

(4.21)

Examples of the Weibull distribution are shown in Figure 4.5 for the cumulative distribution F and Figure 4.6 for the distribution density f.

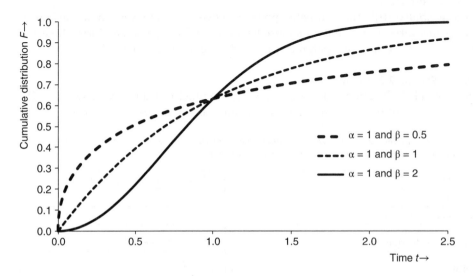

Figure 4.5 Examples of Weibull-2 cumulative distribution F.

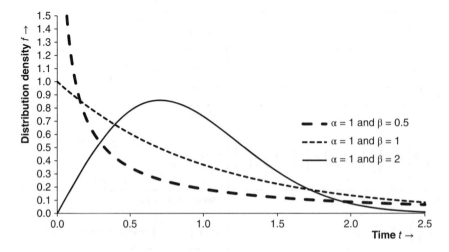

Figure 4.6 Examples of Weibull-2 distribution density f.

The hazard rate and the bath tub curve were introduced in Section 2.4.5. The Weibull hazard rate is:

$$h(t; \alpha, \beta) = \frac{\beta}{\alpha^\beta} \cdot t^{\beta-1} \tag{4.22}$$

The hazard rate decreases with time if the power of t is negative (i.e. if $\beta < 1$). A failure mechanism with declining hazard rate is associated with child mortality processes. A constant hazard rate has power zero (i.e. if $\beta = 1$). This is associated with a random failure mechanism. It increases with time if the power is positive (i.e. if $\beta > 1$). The Weibull shape parameter is therefore often elegantly taken as an indicator that points to a child mortality, random or wear-out mechanism. If the three processes occur in all samples simultaneously, then the resulting hazard rate curve takes the shape of the well-known bath tub model (Figure 4.7). This bath tub model was also discussed in Section 2.4.5.

Though a shape parameter $\beta < 1$ indicates a child mortality process, this statement may not be reversed. Not all child mortality processes have a Weibull distribution with $\beta < 1$. In practice, (hopefully) a minority of the samples in a batch may exhibit defects that make them wear out significantly faster than the regular samples. This subpopulation may have $\beta > 1$, but the most important feature is that the scale parameter is significantly lower than specified. The left side of the bath tub is no longer decreasing from a singularity at $t = 0$, but rather consists of a peak. An example of that situation is shown in Figure 4.8 (repeating Figure 2.10) and was discussed in Section 2.5.3.

4.2.1.2 Weibull-2 Distribution Moments and Mean

The moments of a distribution are defined as (Section 3.1.3):

$$\mu_k = <X^k> = \int_0^1 x^k dF = \int_{-\infty}^\infty x^k f(x) dx \tag{4.23}$$

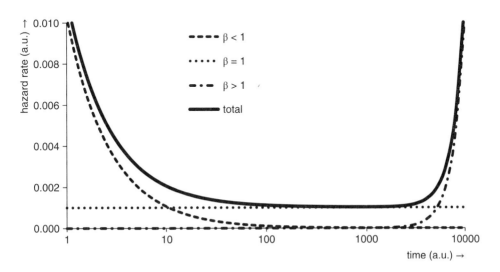

Figure 4.7 Example of the bath tub curve: three failure mechanisms with negative, constant and positive shape parameter β compete simultaneously in all samples. These are child mortality, random and wear-out mechanisms.

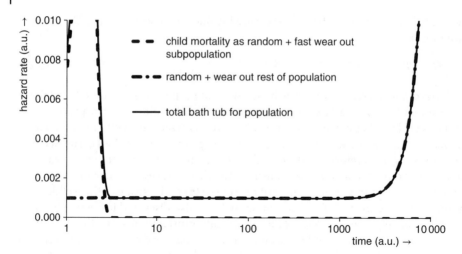

Figure 4.8 Example of the alternative bath tub curve that consists of a peak on the left due to the early extinction of a weak subpopulation and on the right a wear-out curve for the rest of the population. Both subpopulations additionally suffer from a random failure process.

For the Weibull-2 distribution the moments are given as:

$$\mu_k = \alpha^k \cdot \Gamma\left(1 + \frac{k}{\beta}\right) \tag{4.24}$$

Here Γ is the so-called gamma function (see Appendix I):

$$\Gamma(z) = \int_0^\infty t^{z-1}e^{-t}dt \tag{4.25}$$

The expression for μ_k can be achieved by substituting $t = (x/\alpha)^\beta$ and $dt = \beta(x^{\beta-1}/\alpha^\beta)dx$ in this expression for the gamma function, which yields:

$$\Gamma(z) = \int_0^\infty \beta\left(\frac{x^\beta}{\alpha^\beta}\right)^{z-1} \left(\frac{x^{\beta-1}}{\alpha^\beta}\right) e^{-\left(\frac{x}{\alpha}\right)^\beta} dx \tag{4.26}$$

For $z = 1 + k/\beta$ this expression becomes:

$$\Gamma\left(1 + \frac{k}{\beta}\right) = \int_0^\infty \beta\left(\frac{x^\beta}{\alpha^\beta}\right)^{1 + \frac{k}{\beta} - 1} \left(\frac{x^{\beta-1}}{\alpha^\beta}\right) e^{-\left(\frac{x}{\alpha}\right)^\beta} dx$$

$$= \int_0^\infty \beta \frac{x^k}{\alpha^k} \left(\frac{x^{\beta-1}}{\alpha^\beta}\right) e^{-\left(\frac{x}{\alpha}\right)^\beta} dx = \frac{<x^k>}{\alpha^k} \tag{4.27}$$

This is equivalent to Eqs (4.23) and (4.24). The expected or mean time to failure θ (i.e. μ_1 of the Weibull-2 distribution) thus becomes:

$$\theta = \alpha \cdot \Gamma\left(1 + \frac{1}{\beta}\right) \tag{4.28}$$

The ratio between θ and α is the gamma function $\Gamma(1 + 1/\beta)$. The relation between this gamma function and the shape parameter β is shown in Figure 4.9. The mean time θ equals α for $\beta = 1$, since $\Gamma(2) = 1$. For $\beta > 1$ the mean time θ and the scale parameter α

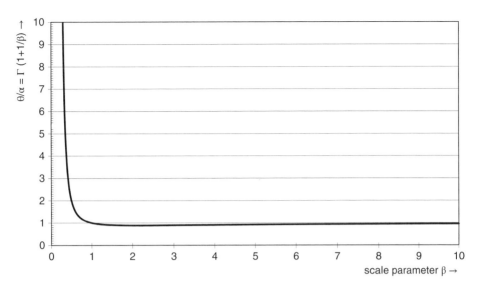

Figure 4.9 The ratio between the mean time θ to failure and the scale parameter α (i.e. the gamma function $\Gamma(1 + 1/\beta)$), as a function of shape parameter β.

resemble each other closely, and θ approaches α asymptotically with increasing β. In the region where $\beta > 1$, the greatest difference between θ and α is about 11.4%, which occurs at $\beta = 2$. For wear-out processes the scale parameter α is therefore a fair to very good approximation of the mean time to failure θ. However, for $\beta < 1$ the mean time to failure θ and the scale parameter α diverge rapidly with decreasing β, as shown in Figure 4.9.

4.2.1.3 Weibull-2 Distribution Characteristics
The characteristics of the Weibull-2 distribution are shown in Table 4.12.

4.2.1.4 EXTRA: Moment Generating Function
The moment generating function $G_X(t)$ (Section 3.1.4) for the Weibull distribution is:

$$G_X(t) = \int_0^\infty e^{tx} \frac{\beta}{\alpha^\beta} x^{\beta-1} e^{-\left(\frac{x}{\alpha}\right)^\beta} dx \tag{4.29}$$

Expanding e^{tx} as a Taylor series and using Eq. (4.24), this can be written as:

$$G_X(t) = \sum_{k=0}^\infty \frac{t^k \beta^k}{k!} \Gamma\left(1 + \frac{k}{\beta}\right) \tag{4.30}$$

The kth moment $<X^k>$ can be found by taking the kth derivative of $G_X(t)$ and setting $t = 0$:

$$\left. \frac{d^k}{dt^k} G_X(t) \right|_{t=0} = <X^k> \tag{4.31}$$

4.2.2 Weibull-3 Distribution

The Weibull three-parameter distribution is applicable if ageing by definition does not occur before a time threshold $t = \delta$ is passed, the so-called location or threshold parameter. For instance, assume that after 9 years a batch of installed cables suffers from an

Table 4.13 Characteristics of the three-parameter Weibull distribution $(x;\alpha,\beta,\delta)$.

Parameters	α,β,δ with $\alpha > 0 \wedge \beta > 0$
Domain	$\delta \leq x < \infty$
Fundamental functions	
Cumulative distribution F	$F(x;\alpha,\beta,\delta) = 1 - e^{-\left(\frac{x-\delta}{\alpha}\right)^{\beta}}$
Reliability R	$R(x;\alpha,\beta,\delta) = e^{-\left(\frac{x-\delta}{\alpha}\right)^{\beta}}$
Distribution density f	$f(x;\alpha,\beta,\delta) = \dfrac{\beta \cdot (x-\delta)^{\beta-1}}{\alpha^{\beta}} \cdot e^{-\left(\frac{x-\delta}{\alpha}\right)^{\beta}}$
Hazard rate h	$h(x;\alpha,\beta,\delta) = \dfrac{\beta \cdot (x-\delta)^{\beta-1}}{\alpha^{\beta}}$
Cumulative hazard H	$H(x) = \displaystyle\int_0^x h(t)dt = \left(\dfrac{x-\delta}{\alpha}\right)^{\beta}$

Measures

Mean μ	Variance σ^2	Median	Mode
$\delta + \alpha \cdot \Gamma\left(1+\dfrac{1}{\beta}\right)$	$\alpha^2 \cdot \left\{ \Gamma\left(1+\dfrac{2}{\beta}\right) - \left[\Gamma\left(1+\dfrac{1}{\beta}\right)\right]^2 \right\}$	$\delta + \alpha \cdot (\ln 2)^{\frac{1}{\beta}}$	$\delta + \alpha \cdot \left(1-\dfrac{1}{\beta}\right)^{\frac{1}{\beta}}$ with $\beta > 1$ 0 with $\beta \leq 1$

Skewness γ_1		Excess kurtosis γ_2	
$\dfrac{< (x+\delta)^3 >}{\sigma^3}$		$\dfrac{< (x+\delta)^4 >}{\sigma^4} - 3$	

extreme overvoltage that starts electrical treeing in the paper–oil insulation. The time between initiation of electrical trees (i.e. the 9 years) and their growth towards actual breakdown may be described by a Weibull distribution. The time before the 9 years, however, played no role in this ageing process and analysis of the breakdown data may prove that a Weibull three-parameter distribution is applicable with a threshold parameter δ of 9 years. The Weibull-3 distribution can be regarded as the Weibull-2 distribution shifted in time (see Table 4.13). However, a difference is that this shift is not fixed, but a parameter itself that comes with its own statistics when it is estimated from a data set.

4.2.3 Weibull-1 Distribution

The Weibull one-parameter distribution is used if the shape parameter β has a fixed value B and the threshold is taken as 0 generally (or the variable t is corrected for the threshold). Examples of the Weibull-1 distribution are the exponential distribution ($B = 1$) and the Rayleigh distribution ($B = 2$). Apart from theoretical reasons, there may also be practical reasons for assuming a fixed B, such as experience with the ageing mechanism under consideration. The Weibull-1 distribution thus becomes:

$$F(t;\alpha) = 1 - e^{-\left(\frac{t}{\alpha}\right)^{B}} \tag{4.32}$$

In the following the probably most important Weibull-1 distribution is discussed, namely the exponential distribution.

The characteristics of the Weibull-1 distribution are the same as those of the Weibull-2 distribution after substituting β by B in Table 4.12.

4.2.4 Exponential Distribution

The exponential distribution is used for processes that have no memory (i.e. the hazard rate is a constant in time). One example is radioactive decay. A second example is what are called cases of bad luck, where failure is related to external unpredictable causes.

The exponential distribution can be regarded as a Weibull distribution with threshold $\delta = 0$ and fixed shape parameter $\beta = 1$ (in terms of the Weibull-1 distribution: $B = 1$). The exponential distribution shows as:

$$F(t;\theta) = 1 - e^{-\frac{t}{\theta}} \tag{4.33}$$

The parameter θ is called the characteristic lifetime and equals the mean time to failure. In terms of a Weibull distribution, θ is the scale parameter. The hazard rate $h(t)$ of the exponential distribution is a constant:

$$h(t;\theta) = \frac{1}{\theta} \tag{4.34}$$

This shows that the hazard rate is not related to time (i.e. age), and for that feature the hazard rate of the exponential distribution is said to have 'no memory' of passed operational life.

4.2.4.1 Exponential Distribution and Average Hazard Rate

The exponential distribution is particularly important for maintained systems, where failure is prevented by timely maintenance. The underlying strategies are described in Sections 1.3.2–1.3.4 and concern, respectively, PBM, CBM and RBM. The result is that the hazard rate is theoretically repeatedly reset to its initial value at $t = 0$. Because of the periodic character, it is possible to represent the hazard rate by an average h_{ave}, as shown in Figure 4.10.

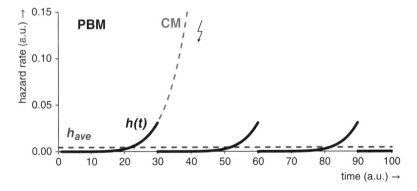

Figure 4.10 If the hazard rate $h(t)$ is repeatedly reset due to maintenance (here PBM), an average hazard rate h_{ave} can be determined as a first approximation of the original hazard rate $h(t)$. This constant h_{ave} defines an exponential distribution.

Interestingly, whenever the hazard rate of a distribution can be represented by an average hazard rate, the constant hazard rate means that the distribution can be approximated as an exponential distribution (cf. Eq. (4.34)). As discussed above, this applies to cases where the hazard rate is repeatedly reset, but the approach can also be applied when the timescale of a study is small compared to variations in the hazard rate. In the latter case the hazard rate is regarded as semi-constant and the hazard rate can be taken constant over a limited interval to approximate the distribution by an exponential distribution.

It is also noteworthy that so far no assumptions have been made about the original distribution with $h(t)$. The case of PBM (Figure 4.10) is further explored in order to find the relationship between the original distribution, the average hazard rate h_{ave} and the expected life θ.

Assume a period T as the time interval between servicing. As for time, only the operational time is considered (i.e. during servicing the clock is stopped as in Figure 4.10). Furthermore, assume that servicing resets ageing (i.e. the asset is as good as new). This means that the hazard rate is reset, but the reliability cannot be reset to 1 because in the previous periods T there was a non-zero hazard rate and thus non-zero probability of failure. After k periods T plus additional running time $t - kT$ (with $t < (k + 1)T$), the reliability $R_k(t)$ is the product of the reliabilities of all past periods:

$$R_k(t) = (R(T))^k \cdot R(t - kT) \tag{4.35}$$

The hazard rate is determined by the elapsed operational time after the last servicing and reset to the hazard rate at the beginning. The hazard rate after k operational periods is therefore:

$$h_k(t) = h(t - kT) \tag{4.36}$$

The average hazard rate h_{ave} can be calculated over the first period T as:

$$h_{ave}(T) = \frac{1}{T}\int_0^T h(t)dt = \frac{1}{T}\int_0^T \frac{-f(t)}{R(t)}dt = \frac{1}{T}\int_0^T \frac{-R'(t)}{R(t)}dt = \frac{-1}{T}\int_0^{R(T)} \frac{1}{R}dR \tag{4.37}$$

This yields (remember $R(0) = 1$ and $\ln(1) = 0$):

$$h_{ave}(T) = \frac{-1}{T} \cdot \ln(R(t))|_0^T = \frac{-\ln(R(T))}{T} \tag{4.38}$$

One might include the effect of the operational time $t - kT$ by adjusting Eqs (4.37) and (4.38). In that case the result is:

$$h_{ave}(t) = \frac{k \cdot [-\ln(R(T))] - \ln(R(t - k \cdot T))}{t} \tag{4.39}$$

Particularly with rarely serviced systems, the impact of the operational time after the last servicing may not be ignored and Eq. (4.39) is to be used, but for the sake of simplicity we proceed with Eq. (4.38) in the following.

Assuming the distribution can be approximated with an exponential distribution with hazard rate $h = h_{ave}$, the characteristic lifetime or expected lifetime θ_{PBM} of a periodically serviced asset follows from Eq. (4.34) as:

$$\theta_{PBM} = \frac{1}{h_{ave}} \tag{4.40}$$

In the following, three example cases of asset servicing are considered. Firstly, servicing of an asset that fails according to an exponential distribution; secondly, servicing of an asset that fails according to a Weibull wear-out distribution; and finally, servicing of assets that feature child mortality.

Example 4.1 *Servicing of an Asset with Exponential Distributed Ageing* Assume an asset is known to have a failure mechanism following an exponential distribution with characteristic lifetime θ as in Eq. (4.33). This is a random failure process. Assume the asset is also serviced after fixed periods T. For the sake of simplicity, the time of maintenance is not included in the asset age, so only operational time is counted. The average hazard rate h_{ave} is found with Eq. (4.38):

$$h_{ave}(T) = \frac{-\ln\left(e^{-\frac{T}{\theta}}\right)}{T} = \frac{T/\theta}{T} = \frac{1}{\theta} \tag{4.41}$$

The expected lifetime θ_{PBM} of the serviced asset following an exponential distribution is found with Eq. (4.40):

$$\theta_{PBM} = \frac{1}{h_{ave}} = \theta \tag{4.42}$$

The expected lifetime θ_{PBM} of a serviced asset that fails according to an exponential distribution equals the original expected lifetime θ. Servicing does not seem to influence the expected lifetime θ_{PBM}. This is not surprising. If an asset fails according to an exponential distribution, it means that there is no memory or ageing (i.e. the hazard rate is constant). Resetting the hazard rate to its value at $t = 0$ does not have any effect. Failure can be regarded as 'bad luck' and servicing is a waste of effort and may be avoided. The main purpose of inspection is to check the functionality of the asset.

As Example 4.1 shows, servicing does not improve the expected life of assets where failure is determined by purely random failure. In that case servicing is not effective. For assets with a wear-out failure mechanism the situation is different, as Example 4.2 shows.

Example 4.2 *Servicing of an Asset with a Weibull Wear-Out Failure Mechanism* Assume an asset is known to have a failure mechanism following a Weibull-2 distribution with a scale parameter α and furthermore a shape parameter $\beta > 1$ as in Eq. (4.20) (i.e. a wear-out mechanism). Again, assume the asset is serviced after fixed periods T and only the actual operational time is counted for the asset age. The average hazard rate h_{ave} is found with Eq. (4.38):

$$h_{ave}(T) = \frac{1}{T} \cdot -\ln\left[e^{-\left(\frac{T}{\alpha}\right)^{\beta}}\right] = \alpha^{-\beta} \cdot T^{\beta-1} \tag{4.43}$$

The hazard rate $h(T)$ for the Weibull-2 distribution is (cf. Eq. (4.22)):

$$h(T) = \beta \cdot \alpha^{-\beta} \cdot T^{\beta-1} \tag{4.44}$$

Therefore, the approximating exponential distribution has a constant hazard rate h_{ave}:

$$h_{ave}(T) = \frac{h(T)}{\beta} \tag{4.45}$$

Now, there is an effect of the period T on the average hazard rate h_{ave}. For the expected lifetime of a serviced asset, the result is:

$$\theta_{PBM} = \frac{1}{h_{ave}} = \frac{\beta}{h(T)} = \frac{\alpha^\beta}{T^{\beta-1}} \qquad (4.46)$$

The dependency of θ_{PBM} on T is investigated by differentiation:

$$\frac{d\theta_{PBM}}{dT} = (1-\beta) \cdot \frac{\alpha^\beta}{T^{\beta-2}} \qquad (4.47)$$

As $\beta > 1$ and α as well as T are positive, it follows that the gradient is negative and thus the expected life increases with decreasing T. In other words, the more servicing, the longer the expected life. In the limit of $T \to 0$, the expected lifetime θ_{PBM} becomes infinite. However, maintenance and servicing do come with costs and efforts. Decreasing the period T will increase the costs. Therefore, in practice, the period T will be optimized such that the cost of replacement and maintenance will be balanced. This is elaborated in Section 9.1.1. From Figure 4.10 it also follows that a shorter T lets $h(t)$ run up less, which results in a lower h_{ave} and thus a higher θ_{PBM}.

With wear-out mechanisms, servicing makes sense although costs must be taken into consideration in order to be efficient. Lastly, the situation with child mortality is described in the following Example 4.3.

Example 4.3 *Servicing of an Asset with a Weibull Child Mortality Process*
Assume an asset is known to have a failure mechanism following a Weibull-2 distribution with a shape parameter $\beta < 1$ as in Eq. (4.20) (i.e. a child mortality mechanism in the sense of a decreasing hazard rate). Again, assume the asset is serviced after fixed periods T. The time of maintenance is not included in the asset age. The expected lifetime of a serviced asset is again:

$$\theta_{PBM} = \frac{1}{h_{ave}} = \frac{\beta}{h(T)} = \frac{\alpha^\beta}{T^{\beta-1}} \qquad (4.48)$$

The dependency of θ_{PBM} on T is again given by:

$$\frac{d\theta_{PBM}}{dT} = (1-\beta) \cdot \frac{\alpha^\beta}{T^{\beta-2}} \qquad (4.49)$$

As $\beta < 1$ and α as well as T are positive, it follows that the gradient is positive and therefore the expected life increases with increasing T. This means the less servicing, the longer the expected life. The situation is quite different from Figure 4.10, because with child mortality the maximum $h(t)$ is at the start of each operational period (cf. Figure 4.7). A longer T drives $h(T)$ down, which results in a lower h_{ave} and thus in a higher θ_{PBM}. The optimum θ_{PBM} is reached with $T \to \infty$ or, in other words, no servicing at all. This is typical for assets where maintenance is likely to introduce child mortality (e.g. due to unbalance introduced with maintenance). Generators are known to be apt to fail after maintenance. In such assets it makes sense to use alternative strategies like partial discharge monitoring to assess the asset condition, but interfere as little as possible. If an asset population is suspected of having a subpopulation of assets suffering child mortality, then an often-applied method is to eradicate the subpopulation by accelerated ageing (see Section 2.6.2 and Example 2.10). This does not make sense if the complete population suffers a child mortality process. If all assets feature child mortality, then the best options are run to failure (i.e. corrective maintenance) and/or monitoring the condition by appropriate (non-interfering) sensors and adequate knowledge rules to plan maintenance if failure seems imminent. But servicing of the asset does not contribute to long life (servicing of wearing sensors may though). It should be noted that in practice a Weibull child mortality process does not necessarily follow a distribution with $\beta < 1$, but may as well have $\beta > 1$ in combination with an α that is much smaller than specified (see also Sections 2.5.3 and 4.2.1.1). The latter case is treated in Example 4.2.

The examples illustrate that periodic maintenance only makes sense for distributions with an increasing hazard rate, which results in an exponential distribution irrespective of the original distribution provided the maintenance makes the asset as good as new (i.e. resets the hazard rate periodically). For distributions with a constant or decreasing hazard rate the same principle holds, but maintenance is not enhancing the expected lifetime and therefore the maintenance costs and efforts are in vain. The exponential distribution is widely used to describe failures of maintained assets and its parameter, the characteristic lifetime, is also referred to as MTBF (mean time between failures).

In terms of the bath tub model (Section 2.4.5), the period where the exponential distribution applies is the base of the bath tub curve.

4.2.4.2 Exponential Distribution Characteristics
The characteristics of the exponential distribution are shown in Table 4.14.

Table 4.14 Characteristics of the exponential distribution $(x;\theta)$. The exponential distribution can be regarded as a Weibull-1 distribution with constant $B = 1$.

Parameter	θ with $\theta > 0$
Domain	$0 \leq x < \infty$
Fundamental functions	
Cumulative distribution F	$F(x; \theta) = 1 - e^{-\frac{x}{\theta}}$
Reliability R	$R(x; \theta) = e^{-\frac{x}{\theta}}$
Distribution density f	$f(x; \theta) = \frac{1}{\theta} \cdot e^{-\frac{x}{\theta}}$
Hazard rate h	$h(x; \theta) = \frac{1}{\theta}$
Cumulative hazard H	$H(x) = \int_0^x h(t)dt = \frac{x}{\theta}$
Measures	

Mean μ	Variance σ^2	Median	Mode
θ	θ^2	$\theta \cdot \ln 2$	0

Skewness γ_1		Excess kurtosis γ_2	
2		6	

4.3 Mean and Variance Statistics

When the mean or expected behaviour of a group is studied, distributions like the normal and the lognormal are used. They describe the mean value and the variance (i.e. scatter) around the mean. In the following, these two distributions are discussed.

4.3.1 Normal Distribution

The normal or Gaussian distribution is probably the most widely used statistical distribution. It is very useful to study the distribution of means by their mean μ and variance of variables. The variance is commonly replaced by its square root, namely the standard deviation σ. Where the Weibull distribution describes the smallest extreme or weakest link of a chain; the normal distribution describes the average. An example is the average mass of the links of a chain. The total mass of a chain may be measured and divided by the number of links. Studying various chains, the link mass averages form a distribution. The more links the chains consist of, the more this distribution resembles a normal distribution.

The reason why the normal distribution is widely used for describing the mean with its variance is that it is the asymptotic distribution for the means. This is a consequence of the central limit theorem, as will be discussed in Section 4.3.1.3. The shape of the standardized normal distribution is shown in Figure 4.11.

The density function $f(x)$ is:

$$f(x) = \frac{1}{\sigma\sqrt{2\pi}} \cdot e^{-\frac{1}{2}\left(\frac{x-\mu}{\sigma}\right)^2} \tag{4.50}$$

The two parameters are the mean μ and the standard deviation σ. The density function has two inflection points, namely at $x = \mu \pm \sigma$. To be more precise, the tangent of the

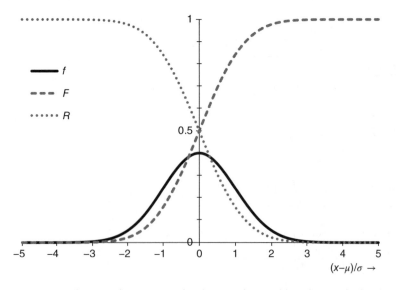

Figure 4.11 The normal or Gaussian distribution. The variable is the standardized variable $y = (x - \mu)/\sigma$.

density curve changes from increasing to decreasing (i.e. concave upward to concave downward) at $x = \mu - \sigma$. Vice versa, the tangent of the density curve changes from decreasing to increasing (i.e. from concave downward to concave upward) at $x = \mu + \sigma$.

The density function is symmetrical, that is:

$$f(-(x - \mu)) = f(x - \mu) \tag{4.51}$$

As a consequence, $F(x)$ and $R(x)$ are also mirrored in $x = \mu$:

$$F(x - \mu) = R(-(x - \mu)) = R(\mu - x) = 1 - F(\mu - x) \tag{4.52}$$

Unfortunately, there is no analytical expression for the cumulative distribution F (i.e. the integral of Eq. (4.50) does not yield an analytical expression). There are a number of approximations that can be helpful (see Section 10.6).

As there is no analytical expression for F and R, there is also no analytical expression for the hazard rate $h(x)$. In many software applications, like spreadsheets, numerical values for these functions can nevertheless be calculated conveniently. Section 10.6 discusses a series of approximation methods for the normal distribution.

The domain of the variable is $(-\infty, \infty)$. This fact is often ignored and can be the root of considerable inaccuracies when making forecasts. For instance, breakdown times in testing are in the range $[0, \infty)$. If the normal distribution is nevertheless applied to such breakdown times, then it is implied that a probability larger than zero exists that test objects fail even before they are tested or produced, and ultimately even before the universe came into existence. This is a fundamental problem, with many measures of performance with a lower finite boundary. However, if the mean is large compared to the standard deviation, then the error can become very small as will be shown below. If the mean μ is not large compared to the standard deviation σ and the variable has values in the range $[0, \infty)$, then other distributions may be more appropriate, like the lognormal or the Weibull distribution (which, however, studies the smallest extreme rather than a mean).

Table 4.15 shows which population portion falls within or outside intervals of an integer multiple of the standard deviation around the mean. The normal distribution is particularly useful for means and their accuracy. It should be noticed that the tails of the distribution may differ significantly from the actual distributions if the studied variable is not truly normally distributed.

4.3.1.1 Characteristics of the Normal Distribution

The characteristics of the normal distribution are given in Table 4.16. A special case is the normalized distribution with $\mu = 0$ and $\sigma = 1$ (see Table 4.17).

The normalized or standardized distribution can be translated into the normal distribution through the relation:

$$y(x) = \frac{x - \mu}{\sigma} \tag{4.53}$$

Table 4.15 Percentage of distribution falling inside (respectively outside) a range around the mean.

Range $[x_1, x_2] =$	$[\mu - 5\sigma, \mu + 5\sigma]$	$[\mu - 4\sigma, \mu + 4\sigma]$	$[\mu - 3\sigma, \mu + 3\sigma]$	$[\mu - 2\sigma, \mu + 2\sigma]$	$[\mu - \sigma, \mu + \sigma]$
$F(x_2) - F(x_1) =$	99.999943%	99.993666%	99.730020%	95.449974%	68.268949%
$1 - \{F(x_2) - F(x_1)\} =$	0.000057%	0.006334%	0.269980%	4.550026%	31.731050%

Table 4.16 Characteristics of the normal distribution $F(x;\mu,\sigma)$.

Parameters	μ : mean $\qquad\qquad -\infty < \mu < \infty$ σ : standard deviation $\quad \sigma > 0$					
Domain	$-\infty < x < \infty$					
Fundamental functions						
Cumulative distribution F	$F(x) \displaystyle\int_{-\infty}^{x} f(x)dx$ To be approximated (see Section 10.6)					
Reliability R	$R(x) = 1 - F(x)$					
Distribution density f	$f(x) = \dfrac{1}{\sigma\sqrt{2\pi}} \cdot e^{-\frac{1}{2}\left(\frac{x-\mu}{\sigma}\right)^2}$					
Hazard rate h	$h(x) = \dfrac{f(x)}{R(x)}$					
Cumulative hazard H	$H(x) = \displaystyle\int_{-\infty}^{x} \dfrac{f(x)}{R(x)}dx$					
Measures						
Mean	Variance	Skewness	Excess kurtosis	Median	Mode
μ	σ^2	0	0	μ	μ

Table 4.17 Characteristics of the normalized normal distribution $F(y;0,1)$.

Parameters	$\mu = 0$ $\sigma = 1$					
Domain	$-\infty < y < \infty$					
Fundamental functions						
Cumulative distribution F	$F(y) = \displaystyle\int_{-\infty}^{y} f(y)dy$ To be approximated (see Section 10.6)					
Reliability R	$R(y) = 1 - F(y) = F(-y)$					
Distribution density f	$f(y) = \dfrac{1}{\sqrt{2\pi}} \cdot e^{-\frac{1}{2}y^2}$					
Hazard rate h	$h(y) = \dfrac{f(y)}{R(y)}$					
Cumulative hazard H	$H(y) = \displaystyle\int_{-\infty}^{y} \dfrac{f(y)}{R(y)}dy$					
Measures						
Mean	Variance	Skewness	Excess kurtosis	Median	Mode
0	1	0	0	0	0

and:

$$dy(x) = \frac{\partial y(x)}{\partial x}dx = \frac{1}{\sigma}dx \qquad\qquad (4.54)$$

It follows that:

$$F(x) = \int_{-\infty}^{y(x)} \frac{1}{\sqrt{2\pi}} e^{-\frac{1}{2}y(x)^2} dy(x) = \int_{-\infty}^{x} \frac{1}{\sqrt{2\pi}} e^{-\frac{1}{2}\left(\frac{x-\mu}{\sigma}\right)^2} \frac{1}{\sigma} dx$$

$$= \int_{-\infty}^{x} \frac{1}{\sigma\sqrt{2\pi}} e^{-\frac{1}{2}\left(\frac{x-\mu}{\sigma}\right)^2} dx \qquad (4.55)$$

and:

$$f(x) = f(y(x))\frac{\partial y(x)}{\partial x} = \frac{1}{\sqrt{2\pi}} e^{-\frac{1}{2}\left(\frac{x-\mu}{\sigma}\right)^2} \cdot \frac{1}{\sigma} = \frac{1}{\sigma\sqrt{2\pi}} e^{-\frac{1}{2}\left(\frac{x-\mu}{\sigma}\right)^2} \qquad (4.56)$$

A table of values $F(y)$ is presented in Appendix G for $y \in [0,5]$. From this table the values for $y \in [-5,0]$ also follow, because $F(-y) = 1 - F(y)$ according to Eq. (4.52).

4.3.1.2 EXTRA: Moments, Moment Generating Function and Characteristic Function

The moments of the normal distribution can be calculated conveniently. Here, the moments of the standardized normal distribution $F(y;0,1)$ are discussed. Assume $n \in \mathbb{N}$, then:

$$< Y^n > = \int_{-\infty}^{\infty} y^n e^{-\frac{y^2}{2}} dy = \int_{-\infty}^{\infty} y^{n-1} y e^{-\frac{y^2}{2}} dy = -\int_{-\infty}^{\infty} y^{n-1} \frac{\partial e^{-\frac{y^2}{2}}}{\partial y} dy \qquad (4.57)$$

Integration in parts yields:

$$< Y^n > = -y^{n-1} e^{-\frac{y^2}{2}} \Big|_{-\infty}^{\infty} + \int_{-\infty}^{\infty} (n-1) y^{n-2} e^{-\frac{y^2}{2}} dy = 0 + (n-1) \cdot < Y^{n-2} > \qquad (4.58)$$

As $\mu = <Y>$ equals zero, all moments are zero for odd powers. This can also be recognized by taking notice of the symmetry of the density function (i.e. $f(y) = f(-y)$), in combination with the fact that -1 to an odd power is -1:

$$k \in \mathbb{N}: \quad (-y)^{2k+1} \cdot f(-y) = (-1)^{2k+1} \cdot y^{2k+1} \cdot f(y) = -y^{2k+1} \cdot f(y) \qquad (4.59)$$

Both tails of the integral will cancel out for that reason with odd powers. For even powers the moments are non-zero. In conclusion, for $k \in \mathbb{N}$:

$$< Y^{2k+1} > = 0 \qquad (4.60)$$

$$< Y^{2k} > = 1 \cdot 3 \cdot \ldots \cdot (2k-1) = \frac{(2k)!}{2^k \cdot k!} \qquad (4.61)$$

The moment generating function of the normal distribution is:

$$G_Y(t) = < e^{ty} > = \int_{-\infty}^{\infty} e^{ty} \frac{1}{\sqrt{2\pi}} e^{-\frac{y^2}{2}} dy = \frac{1}{\sqrt{2\pi}} \int_{-\infty}^{\infty} e^{-\frac{y^2}{2}+ty} dy = \frac{1}{\sqrt{2\pi}} \int_{-\infty}^{\infty} e^{-\frac{y^2}{2}+\frac{2ty}{2}} dy$$

$$(4.62)$$

$$G_Y(t) = \frac{1}{\sqrt{2\pi}} \int_{-\infty}^{\infty} e^{-\frac{(y-t)^2}{2}+\frac{t^2}{2}} dy = e^{\frac{t^2}{2}} \cdot \left(\frac{1}{\sqrt{2\pi}} \int_{-\infty}^{\infty} e^{-\frac{(y-t)^2}{2}} dy \right) = e^{\frac{t^2}{2}} \qquad (4.63)$$

The integral between brackets in Eq. (4.63) with any constant t is an integral of the normal density function over the full domain which equals 1.

The characteristic function of the normal distribution is:

$$\phi_Y(t) = <e^{it \cdot Y}> = e^{\frac{-t^2}{2}}$$

(4.64)

Interestingly, the characteristic function of the normal distribution appears to be a multiple of the density function $f_N(y)$ (the suffix N is added to stress that this is an exceptional property of the normal distribution):

$$\phi_{N,Y}(t) = \sqrt{2\pi} \cdot f_N(y)$$

(4.65)

Apart from the expressions in Eqs (4.94) and (4.95), the moments can be obtained from the moment generating function and the characteristic function through Eq. (4.2) (respectively Eq. (4.4)).

4.3.1.3 EXTRA: Central Limit Theorem

The foundation of the applicability of the normal distribution for the mean of large data sets is the central limit theorem. This theorem implies that for a large number of independent variables with an as yet arbitrary but identical distribution, their sum or average will follow a distribution that asymptotically approaches a normal distribution.

This feature is the reason for the importance of the normal distribution in statistics. That said, it may also be the reason for the abuse of the normal distribution, where other distributions should be applied. The normal distribution is typically suitable for describing the average behaviour of large populations, cancelling out the outliers. Its strength lies in describing the mean (or sum), not in situations at the extremities of the tail of the distribution, where the weakest and the strongest lie and extreme value statistics is more suitable.

This section first gives a simple proof that the distribution of means approaches the normal distribution for large sample sizes. Second, it is shown that the distribution of the sum of randomly drawn samples also approaches the normal distribution.

Assume a variable X with domain $(-\infty,\infty)$, an as yet undetermined density function $f(X)$, mean μ, variance σ^2 and standard deviation σ. Repeated sampling leads to a random sample of n outcomes X_j with ($j = 1,\ldots,n$). The mean $M_{X,n}$ of the outcome set is defined as:

$$M_{X,n} = \frac{1}{n}\sum_{j=1}^{n} X_j$$

(4.66)

If outcome sets with sample size n are retrieved repeatedly, means $M_{X,n,k}$ (with $k > 1$) are independent and distributed according to the same distribution. Using the fact that the variables X_j are independent, the mean, variance and standard deviation of $M_{X,n}$ are:

$$<M_{X,n}> = \frac{1}{n}\sum_{j=1}^{n} <X_j> = \frac{n}{n} \cdot \mu = \mu$$

(4.67)

$$\mathrm{var}(M_{X,n}) = <\sum_{j=1}^{n} \left(\frac{X_j - \mu}{n}\right)^2> = \frac{1}{n^2}\sum_{j=1}^{n} <(X_j - \mu)^2> = \frac{\sigma^2}{n}$$

(4.68)

$$\sigma_{M_{X,n}} = \frac{\sigma}{\sqrt{n}}$$

(4.69)

Now, a standard variable Y_X based on X is introduced:

$$Y_X = \frac{X - \mu}{\sigma} \tag{4.70}$$

The mean $<Y_X> = 0$ and the variance $\text{var}(Y_X) = 1$. The mean $M_{Y,n}$ of an outcome set $Y_{X,j}$ with $(j = 1,\ldots,n)$ is then defined as:

$$M_{Y,n} = \frac{1}{n} \sum_{j=1}^{n} Y_{X,j} \tag{4.71}$$

Again, using the fact that the variables $Y_{X,j}$ are independent, the mean, variance and standard deviation of $M_{Y,n}$ are:

$$< M_{Y,n} > = \frac{1}{n} \sum_{j=1}^{n} < Y_{X,j} > = \frac{n}{n} \cdot 0 = 0 \tag{4.72}$$

$$\text{var}(M_{Y,n}) = < \sum_{j=1}^{n} \left(\frac{Y_{X,j} - 0}{n} \right)^2 > = \frac{1}{n^2} \sum_{j=1}^{n} < Y_{X,j}^2 > = \frac{1}{n} \tag{4.73}$$

$$\sigma_{M_{Y,n}} = \frac{1}{\sqrt{n}} \tag{4.74}$$

In addition to the standard variable Y_X based on X, the standard variable $U_{MX,n}$ based on the mean $M_{X,n}$ is also defined:

$$U_{MX,n} = \frac{M_{X,n} - < M_{X,n} >}{\sqrt{\text{var}(M_{X,n})}} = \frac{M_{X,n} - \mu}{\sigma/\sqrt{n}} \tag{4.75}$$

This variable $U_{MX,n}$ can also be expressed in terms of Y:

$$U_{MX,n} = \frac{\frac{1}{n} \sum_{j=1}^{n} X_j - \mu}{\sigma/\sqrt{n}} = \frac{1}{\sqrt{n}} \sum_{j=1}^{n} \left(\frac{X_j - \mu}{\sigma} \right) = \frac{1}{\sqrt{n}} \sum_{j=1}^{n} Y_j \tag{4.76}$$

The mean and variance of $U_{MX,n}$ are:

$$< U_{MX,n} > = < \frac{M_{X,n} - \mu}{\sigma/\sqrt{n}} > = \frac{< M_{X,n} > - \mu}{\sigma/\sqrt{n}} = 0 \tag{4.77}$$

$$\text{var}(U_{MX,n}) = < U_{MX,n}^2 > = \frac{\text{var}(M_{X,n})}{\sigma^2/n} = 1 \tag{4.78}$$

As a next step, the characteristic functions for Y and $U_{MX,n}$ are related. Let φ_Y be the characteristic function of the distribution of Y:

$$\phi_Y(t) = < e^{itY} > \tag{4.79}$$

By Taylor expansion for small t and noting $\varphi_Y(0) = 1$, $\varphi_Y'(0) = <Y> = 0$ and $\varphi_Y''(0) = i^2<Y^2> = -1$ with $o(t^2)$ denoting a function of t that goes faster to zero than t^2:

$$\phi_Y(t) = \sum_{k=0}^{\infty} \frac{(it)^k}{k!} < Y^k > = 1 + 0 - \frac{t^2}{2!} + o(t^2) \tag{4.80}$$

Let φ_{Un} be the characteristic function of the distribution of $U_{MX,n}$:

$$\varphi_{Un}(t) = \; < e^{itU_{MX,n}} >$$

(4.81)

Because the Y_i are independent and distributed according to the same distribution, the relations in Eqs (3.40)–(3.42) can be applied to Eq. (4.76), which yields:

$$\phi_{Un}(t) = \left[\phi_Y \left(\frac{t}{\sqrt{n}} \right) \right]^n$$

(4.82)

By Taylor expansion for small t and large n and also noting that the terms t^k with $k > 2$ are divided by $n^{k/2}$, which approach 0 faster than $1/n$ for fixed t:

$$t \to 0: \qquad \phi_{Un}(t) = \left[1 + 0 - \frac{t^2}{2! \cdot n} + o(t^2) \right]^n \to \left[1 - \frac{t^2}{2n} \right]^n$$

(4.83)

Also noting that for $n \to \infty$:

$$\left(1 + \frac{A}{n} \right)^n = e^A$$

(4.84)

for large n the characteristic function φ_{Un} for the standard mean of X can therefore be written as:

$$\phi_{Un}(t) = e^{-\frac{t^2}{2}}$$

(4.85)

This is the characteristic function of the normal distribution (Eq. (4.64)) (i.e. the distribution of standard means approaches the standard normal distribution for large n). This result applies to the distribution of the mean of any variable X, since no assumption was made about the distribution $f(x)$ except for the domain $(-\infty,\infty)$.

The central limit theorem not only applies to the mean $M_{X,n}$ with large n, but also to the sum $Z_{X,n}$:

$$Z_{X,n} = \sum_{j=1}^{n} X_j$$

(4.86)

The mean, variance and standard deviation of $Z_{X,n}$ are:

$$< Z_{X,n} > = \sum_{j=1}^{n} < X_j > = n \cdot \mu$$

(4.87)

$$\text{var}(Z_{X,n}) = \; < \sum_{j=1}^{n} (X_j - \mu)^2 > = n \cdot \sigma^2$$

(4.88)

$$\sigma_{Z_{Y,n}} = \sqrt{n} \cdot \sigma$$

(4.89)

The standard sum $U_{ZX,n}$ can be rewritten as the standard mean $U_{MX,n}$:

$$U_{ZX,n} = \frac{Z_{X,n} - < Z_{X,n} >}{\sqrt{\text{var}(Z_{X,n})}} = \frac{n}{n} \frac{M_{X,n} - < M_{X,n} >}{\sqrt{\text{var}(M_{X,n})}} = U_{MX,n}$$

(4.90)

This leads to the same characteristic function and therefore the standard sum for $n \to \infty$ appears to approach the standard normal distribution as well.

Finally, the central limit theorem states that the distributions of the standard sum and of the standard mean both approach the standard normal distribution (see Eq. (4.85)).

The variance of the distribution of $M_{X,n}$ is finite and will approach zero when n approaches infinity. Moments with powers higher than 2 approach zero more rapidly than the variance (Eq. (4.80)), which is equivalent to the distribution of $M_{X,n}$ approaching the normal distribution for large n. As for sums $Z_{X,n}$, the variance increases with increasing n. The fact that the distribution of $Z_{X,n}$ approaches the normal distribution means that the variance dominates the higher moments.

4.3.2 Lognormal Distribution

The lognormal distribution is closely related to the normal distribution (see Section 4.3.1). If a variable X is distributed according to the normal distribution, then a variable T defined as:

$$T = e^X \iff X = \ln T \tag{4.91}$$

is distributed according to the lognormal distribution. As the domain of a normally distributed variable X is $(-\infty, \infty)$, the domain of a variable T distributed according to a lognormal distribution is $[0, \infty)$. The fundamental functions are shown in Figure 4.12.

Many observations, like elapsed time, weight, and so on, are in this range $[0, \infty)$. For the analysis of such cases, the lognormal distribution may therefore be more appropriate than the normal distribution. Other observations, like deviations from an average or logarithmic values of positive observations, may still have negative values in the range of the normal distribution. However, for observations in the range $[0, \infty)$, the lognormal distribution may be very useful.

Where the central limit theorem predicts the normal distribution for averages (or sums) of many random variables, the lognormal is expected to be the asymptotic distribution for products of many random variables based on Eq. (4.91).

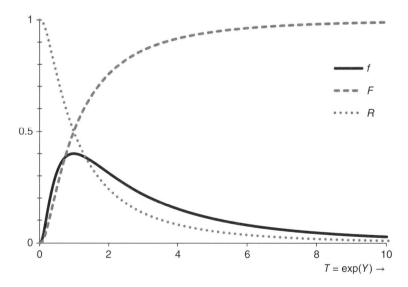

Figure 4.12 The lognormal distribution. The variable is the transformed normalized variable $T = \exp(Y) = \exp((X - \mu)/\sigma)$. The hazard rate h (not shown) lifts off fast. It becomes 1 where f and R intersect at about $T \approx 1.35$ and $h > 1$ as $f > R$ for $T \gtrsim 1.35$.

In practice, the lognormal distribution may be applied where the mean of a positive variable T is studied. A particular example is the mean battery life for equipment that requires frequent battery replacement. Another example concerns the repair times after failure that may show scatter.

If the standard deviation of T is very small compared to the mean of T, then the normal distribution is often assumed to apply based on the central limit theorem. However, if the standard deviation of T is not very small compared to the mean of T, then the natural logarithm of T may be studied in terms of the normal distribution. This is the normal distributed variable X in Eq. (4.91). Thus, T is assumed to behave according to the lognormal distribution. Confidence intervals of normal and lognormal distributions are compared in Table 4.18.

The lognormal distribution density $f_L(t)$ follows from the distribution density $f_N(x)$ of the normal distribution by substituting the right-hand side of Eq. (4.91) into Eq. (4.50):

$$f_L(t) = \frac{dF_L(t)}{dt} = \frac{\partial x}{\partial t}\frac{dF_N}{dx} = \frac{1}{t} \cdot f_N(x(t)) = \frac{1}{t\sigma\sqrt{2\pi}} \cdot e^{-\frac{1}{2}\left(\frac{\ln(t)-\mu}{\sigma}\right)^2} \tag{4.92}$$

Here, μ is a scale parameter and σ a shape parameter. The symbols are the same as in the normal distribution, but care must be taken with the definitions to avoid confusion:

$$\mu = <X> = <\ln T> \tag{4.93}$$

$$\sigma^2 = <(X-\mu)^2> = <(\ln T - \mu)^2> \tag{4.94}$$

So, μ is not the mean of T and neither is σ the standard deviation of T (see Table 4.19 for the mean of T and its variance, i.e. the square of the standard deviation).

An alternative expression replacing μ by $\ln(m)$ and rewriting is:

$$f_L(t) = \frac{1}{t\sigma\sqrt{2\pi}} \cdot e^{-\frac{1}{2}\left(\ln\left(\left(\frac{t}{m}\right)^{\frac{1}{\sigma}}\right)\right)^2} \tag{4.95}$$

This expression shows the character of the scale and shape parameter more explicitly.

The random variable T may be replaced by a variable Y with $Y > v$, which produces the three-parameter lognormal distribution where v is the location parameter:

$$f_L(y) = \frac{1}{(y-v)\sigma\sqrt{2\pi}} \cdot e^{\frac{1}{2}\left(\ln\left(\left(\frac{y-v}{m}\right)^{\frac{1}{\sigma}}\right)\right)^2} \tag{4.96}$$

As with the normal distribution, there is no analytical expression for the cumulative distribution.

Table 4.18 Percentage of distribution falling inside (respectively outside) a range around the mean for the standard normal distribution (first row) and the corresponding lognormal distribution. The variable Y is defined by Eq. (4.53) and T by Eq. (4.91) followed by Eq. (4.53).

Range $[y_1, y_2] =$	[−5,5]	[−4,4]	[−3,3]	[−2,2]	[−1,1]
Range $[t_1, t_2] =$	[0.0067,148]	[0.0183, 54.6]	[0.0498, 20.1]	[0.135, 7.39]	[0.368, 2.718]
$F(x_2) - F(x_1) =$	99.999943%	99.993666%	99.730020%	95.449974%	68.268949%
$1 - \{F(x_2) - F(x_1)\} =$	0.000057%	0.006334%	0.269980%	4.550026%	31.731050%

4.3.2.1 Characteristics of the Lognormal Distribution

The characteristics of the lognormal distribution are given in Table 4.19.

The mean of the lognormal distribution can be calculated as follows.

$$< T > = \int_0^\infty t \cdot f_L(t)\, dt = \int_0^\infty t \cdot \frac{1}{t\sigma\sqrt{2\pi}} \cdot e^{-\frac{1}{2}\left(\frac{\ln(t)-\mu}{\sigma}\right)^2}\, dt \tag{4.97}$$

This can be written in terms of the normal distribution using Eq. (4.91):

$$< T > = < e^X > = \int_{-\infty}^\infty \frac{1}{\sigma\sqrt{2\pi}} \cdot e^{-\frac{1}{2}\left(\frac{x-\mu}{\sigma}\right)^2} e^x dx = \int_{-\infty}^\infty \frac{1}{\sigma\sqrt{2\pi}} \cdot e^{x-\frac{1}{2}\left(\frac{x-\mu}{\sigma}\right)^2} dx \tag{4.98}$$

The expression can be rearranged by extracting a constant from the exponent:

$$< T > = < e^X > = \int_{-\infty}^\infty \frac{1}{\sigma\sqrt{2\pi}} \cdot e^{\frac{2\sigma^2 x}{2\sigma^2} - \frac{x^2 - 2\mu x + \mu^2}{2\sigma^2}} dx = e^{\mu+\frac{\sigma^2}{2}} \cdot \int_{-\infty}^\infty \frac{1}{\sigma\sqrt{2\pi}} e^{-\frac{(x-(\mu+\sigma^2))^2}{2\sigma^2}} dx \tag{4.99}$$

This means that $<T>$ is equal to the product of a constant and the integral of a normal density function over the full domain (i.e. the integral equals 1). This yields the mean in Table 4.19. In a similar fashion the variance can be found.

In general, the moments of the lognormal distribution are:

$$< T^k > = e^{\mu k + \frac{1}{2}\sigma^2 k^2} \tag{4.100}$$

This can be proven with the moment generating function in the next section.

Table 4.19 Characteristics of the two-parameter lognormal distribution $F_L(t;\mu,\sigma)$.

Parameters	μ: mean $\quad 0 < \mu < \infty$ σ: standard deviation $\quad \sigma > 0$				
Domain	$0 < t < \infty$				
Fundamental functions					
Cumulative distribution F	$F(t) = \int_0^t f(t) dt$ To be approximated as with the normal distribution				
Reliability R	$R(t) = 1 - F(t)$				
Distribution density f	$f(t) = \dfrac{1}{t\sigma\sqrt{2\pi}} \cdot e^{-\frac{1}{2}\left(\frac{\ln(t)-\mu}{\sigma}\right)^2}$				
Hazard rate h	$h(t) = \dfrac{f(t)}{R(t)}$				
Cumulative hazard H	$H(t) = \int_0^t \dfrac{f(t)}{R(t)} dt$				
Measures					
Mean	Variance	Skewness	Excess kurtosis	Median	Mode
$e^{\mu+\frac{\sigma^2}{2}}$	$(e^{\sigma^2} - 1) \cdot e^{2\mu+\sigma^2}$	$(e^{\sigma^2} + 2) \cdot \sqrt{e^{\sigma^2} - 1}$	$e^{4\sigma^2} + 2e^{3\sigma^2} + 3e^{2\sigma^2} - 6$	e^μ	$e^{\mu-\sigma^2}$

4.3.2.2   EXTRA: Moment Generating Function and Characteristic Function

In order to avoid confusion with T as defined in Eq. (4.91), the variable S will be used in the moment generating functions (i.e. $G(s)$). Let X have a normal distribution with mean μ and standard deviation σ. The moment generating function of this normal distribution is:

$$G_X(s) = < e^{s \cdot X} > \tag{4.101}$$

This becomes, in terms of the normalized variable Y:

$$G_X(s) = < e^{s \cdot (\mu + \sigma \cdot Y)} > = e^{s \cdot \mu} \cdot < e^{s \cdot \sigma \cdot Y} > = e^{s \cdot \mu} \cdot e^{\frac{s^2 \sigma^2}{2}} \tag{4.102}$$

The variable T has the corresponding lognormal as defined in Eq. (4.91). Its moments can be found by substituting Eq. (4.91) in Eq. (4.101) on the one hand and working out Eq. (4.102) on the other hand. This yields:

$$< e^{s \cdot \ln T} > = < T^s > = e^{\mu \cdot s + \frac{1}{2} \sigma^2 \cdot s^2} \tag{4.103}$$

This corresponds to Eq. (4.100), which defines the moments of the lognormal distributed variables.

4.3.2.3   Lognormal versus Weibull

The lognormal distribution and the Weibull distribution have the range $[0, \infty)$ in common. The appearance of the lognormal distribution bears a certain resemblance to the Weibull distribution for a wear-out process.

So, what is the difference? As mentioned in Section 4.2, the Weibull distribution is an asymptotic distribution for the smallest extreme value (i.e. for objects of which the strength is determined by the weakest link out of many links that make a chain). For instance, many breakdown paths through an XLPE insulation layer may be possible, but the weakest path will be the path for the breakdown and determines, for example, the time to failure.

The lognormal distribution is related to the normal distribution, which is an asymptotic distribution of the mean of many observations. The lognormal distribution is related to the exponential value of such observations. It is related to observations that can be regarded as a mean.

It is up to the experimenter how the observations are viewed as means or as extremes. From that perspective the distribution might be selected. The similarity index may be used to evaluate the (lack of) resemblance (cf. Section 3.6 and following).

4.4   Frequency and Hit Statistics

The last category of distributions in this chapter concerns series of observations and the statistics of hits and non-hits. This relates to repeated experiments where the number of elementary events per single experiment is finite, and usually binary (hit versus non-hit). In this context the focus is on the distribution of how many times a certain outcome occurs when an experiment is repeated.

In the following, most subsections discuss situations that can be regarded as so-called Bernoulli trials. These are experiments where each single experiment has two outcomes, like flipping a coin and studying how often heads turn up. These outcomes can be called either 'hit' versus 'non-hit' or 'success' versus 'non-success' or 'success' versus 'failure', and so on. Sequential experiments in Bernoulli trials are by definition independent (i.e. experiments do not influence the outcomes of other experiments). The probabilities of the outcomes 'hit' and 'non-hit' are constants p and q:

$$P(\text{hit}) = p \tag{4.104}$$

$$P(\text{non-hit}) = 1 - p = q \tag{4.105}$$

For instance, when throwing a dice, the set of outcomes Ω is $\{1,2,3,4,5,6\}$ (see Section 2.1). Assuming the dice is fair, each individual outcome has a probability of $1/6$. In the Bernoulli trials the hit or success may be defined as throwing a particular elementary outcome (e.g. 3) where $p = 1/6$ or an event that combines various outcomes, like throwing an odd number where $p = 1/2$. Of particular interest, then, are the statistics of such events occurring. For example, what is the probability that two odd numbers occur in 10 trials. An example related to grid assets is the situation where $n = 24$ post insulators are installed at a substation and experience (e.g. from other similar substations) shows that there is a probability p for each of the post insulators that they are defective. In order to estimate the seriousness of the problem at the substation, the asset manager may wish to evaluate the probabilities that certain numbers of defects exist.

In case of two complementary outcomes per single experiment, the binomial distribution can be used to describe the outcomes of series of experiments (Section 4.4.1). A variant of the Bernoulli trial experiments with larger outcome sets (e.g. throwing a dice with six possible outcomes; or insulators with multiple defect types) can be studied with the multinomial distribution (Section 4.4.5).

4.4.1 Binomial Distribution

The binomial distribution applies when a fixed number n of experiments is carried out and each single experiment has only two possible outcomes X. As mentioned above, these are usually indicated as 'hit' or 'success' on the one hand and 'non-hit', 'non-success' or 'failure' on the other hand. The probability of a hit is known, and therefore also the probability of a non-hit is known.

Following the convention of reliability and logic variables, 'hit' or 'success' is also indicated with the variable value $x = 1$ and 'non-hit' or 'non-success' with the variable value $x = 0$. In case of tossing a coin, event $x = 1$ may be associated with 'head' and $x = 0$ with 'tail'. Applied to the case of the post insulators event, $x = 1$ may be associated with detecting a defective item.

The binomial distribution has two parameters: the number of trials n and the probability p of success per trial. The binomial distribution has one variable, namely the number of successes Z in these n trials.

In more detail, let the variables X_j ($j = 1,\ldots,n$) represent outcomes of the individual trials. Following the convention above, the X_j can adopt values $x_j = 1$ (for hit) and $x_j = 0$ (for non-hit). A sequence of n trials produces a sample space Ω of which the vector $\vec{X} = (X_1,\ldots,X_n)$ is a random sample with size n. The sum of successes can be represented by variable Z:

$$Z = \sum_{j=1}^{n} X_j \tag{4.106}$$

The value z of Z ranges from zero to n (i.e. $z \in \{0,\ldots,n\}$). As Z is a discrete variable, the distribution has a probability mass function f (see Section 2.4.4). The meaning of the mass function is the probability that there are exactly z successes in n Bernoulli trials with given probability p of success per trial. The binomial mass function $f(z;n,p)$ is:

$$f(z; n, p) = \binom{n}{z} p^z (1 - p)^{n-z} \tag{4.107}$$

This is derived as follows. The probability that z trials have outcome 1 is p^z; likewise the probability that the complement of $n - z$ trials have outcome 0 is $(1 - p)^{n-z}$. The exact sequence of the outcomes is irrelevant, because only the number of successes is of interest. Therefore, the events with equal number of successes can be summed. The number of combinations of a subgroup of z trials out of a total of n trials equals the binomial coefficient 'n-over-z':

$$\binom{n}{z} = \frac{n!}{z! \cdot (n - z)!} = \frac{n \cdot (n - 1) \cdot \ldots \cdot (n - z + 1)}{z \cdot (z - 1) \cdot \ldots \cdot 1} \tag{4.108}$$

The product of the probabilities p^z, $(1 - p)^{n-z}$ and the binomial coefficient of Eq. (4.24) leads to the expression of the mass function $f(z;n,p)$ in Eq. (4.107).

This mass function has some noteworthy properties. A first relation is symmetry:

$$f(z; n, p) = \frac{n!}{z!(n - z)!} p^z (1 - p)^{n-z} = \frac{n!}{(n - z)!z!} (1 - p)^{n-z} p^z = f(n - z; n, 1 - p) \tag{4.109}$$

A second relation is recurrence:

$$\begin{aligned} f(z; n, p) &= \frac{n!}{z!(n - z)!} p^z (1 - p)^{n-z} \\ &= \frac{z + 1}{n - z} \cdot \frac{1 - p}{p} \cdot \frac{n!}{(z + 1)!(n - z - 1)!} p^{z+1} (1 - p)^{n-z-1} \\ &= \frac{z + 1}{n - z} \cdot \frac{1 - p}{p} \cdot f(z + 1; n, p) \end{aligned} \tag{4.110}$$

Reversely formulated:

$$f(z + 1; n, p) = \frac{n - z}{z + 1} \cdot \frac{p}{1 - p} \cdot f(z; n, p) \tag{4.111}$$

The cumulative distribution $F(z;n,p)$ describes the probability of up to z hits in n Bernoulli trials with given hit probability p per trial:

$$F(z; n, p) = \sum_{k=0}^{z} \binom{n}{k} p^k (1 - p)^{n-k} \tag{4.112}$$

This can also be expressed in terms of the gamma function:

$$F(z; n, p) = \Gamma(n + 1) \cdot \sum_{k=0}^{z} \frac{p^k (1 - p)^{n-k}}{\Gamma(k + 1) \cdot \Gamma(n - k + 1)} \tag{4.113}$$

An example of the binomial distribution is shown in Example 4.4.

Example 4.4 *Statistics of Defect Insulators with Given Probability p* As an example, the situation of 24 post insulators is studied. Assume these insulators are installed at a substation and the probability p of a defect is known to be 5%. As there are $n = 24$ insulators, the expected or average number of defective insulators is $n \cdot p = 1.2$, which means about 1 out of 24 would be defective. Due to redundancy of the circuits, a single defective insulator would not impose a critical situation, but more than one defective insulator might. Therefore, the utility wants to know the probability that two or more insulators are defective. Figure 4.13 shows the binomial mass function (see Eq. (4.107)) and the cumulative function (see Eq. (4.112)). Table 4.20 shows the corresponding values for $z = 0,\dots,8$. The probability f of having $z = 2$ or 3 defective insulators is 22.3% (respectively 8.6%). The probability that more than one insulator is defective can be calculated as:

$$P(Z > 1) = 1 - P(Z \le 1) = 1 - F(1; 24, 5\%) = 33.9\% \qquad (4.114)$$

At 33.9%, the probability of two or more defective insulators is significant and the utility may find it worthwhile to inspect all insulators.

A word of caution about the terms 'success' and 'failure' in the light of Bernoulli trials is in place here. In many descriptions of Bernoulli trials the outcome 'non-success' is called 'failure'. Above, the term 'failure' was avoided and replaced by the term 'non-hit' or 'non-success' for the following reason. Generally, the cumulative distribution F is

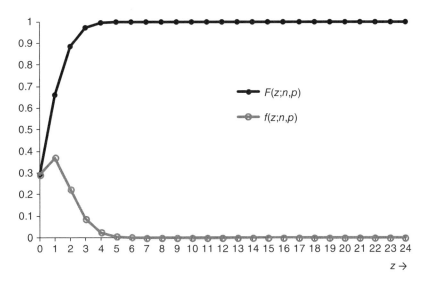

Figure 4.13 The mass function and the cumulative function of the binomial distribution with parameters $n = 24$ and $p = 5\%$.

Table 4.20 The values of the binomial mass function and cumulative function for $z = 0,\dots,8$ with parameters $n = 24$ and $p = 5\%$.

z	0	1	2	3	4	5	6	7	8
f	0.292	0.369	0.223	0.086	0.024	0.005	0.001	0.000	0.000
F	0.292	0.661	0.884	0.970	0.994	0.999	1.000	1.000	1.000

associated with failure, but actually is about the population fraction that 'successfully' underwent the (failure) event of which F is the distribution. The cumulative binomial distribution is then still consistent in that sense. However, generally, the reliability is the surviving fraction that did not undergo the event (i.e. the population fraction that did not yet fail). In case of the binomial distribution, the term $F(z;n,p)$ is the fraction of successes (or hits) and the reliability $R(z;n,p)$ is the fraction of failures (i.e. the fraction that 'failed to fail'). This may become confusing. Therefore, the term 'non-hit' or 'non-success' was used to avoid the situation that $R(z;n,p)$ would thus be called the fraction of failures.

4.4.1.1 Mean and Variance

The probability of a hit in a Bernoulli trial is p. This is also the expected value of the outcome X_j of a single trial j. Therefore:

$$< X_j > = p \tag{4.115}$$

For n (by definition) independent trials, the mean number of success outcomes $<Z>$ is:

$$< Z > = < \sum_{j=1}^{n} X_j > = \sum_{j=1}^{n} < X_j > = np \tag{4.116}$$

This can also be calculated from the mass function $f(z;n,p)$. The average number of hits when n Bernoulli trials are carried out with probability p is:

$$< Z > = \sum_{z=0}^{n} z \cdot f(z; n, p) = \sum_{z=0}^{n} z \cdot \binom{n}{z} p^z (1 - p)^{n-z} \tag{4.117}$$

This can be elaborated as:

$$< Z > = \sum_{z=0}^{n} z \cdot \frac{n!}{z!(n - z)!} p^z (1 - p)^{n-z} = np \cdot \sum_{z=1}^{n} \frac{(n - 1)!}{z - 1!(n - z)!} p^{z-1} (1 - p)^{n-z}$$

$$= np \cdot \sum_{z=1}^{n} \binom{n - 1}{z - 1} p^{z-1} (1 - p)^{n-z} = np \cdot \sum_{w=0}^{n-1} f(w; n - 1, p) = np \tag{4.118}$$

Similarly, the variance can be calculated:

$$\text{var}(Z) = np(1 - p) \tag{4.119}$$

4.4.1.2 Characteristics of the Binomial Distribution

The characteristics of the binomial distribution are given in Table 4.21.

The median and mode are both so-called floor or ceiling functions. With a floor function, the value $\lfloor z \rfloor$ equals the greatest preceding integer z and with a ceiling function, the value $\lceil z \rceil$ equals the least succeeding integer. For example, $\lfloor 2.1 \rfloor = 2$ and $\lfloor 3.9 \rfloor = 3$, while $\lceil 2.1 \rceil = 3$ and $\lceil 3.9 \rceil = 4$.

For various conditions the values of the mode or modes are shown in Table 4.22.

As for the median, the situation is more complicated (see Table 4.23).

Finally, it may be noted that the hazard rate h is not a function of time and is therefore a constant. In terms of ageing, the events thus happen randomly in time. This, of course, is due to the fact that p and q were taken to be constants. The population as such has an exponential distribution with respect to a binomial type of process. If the hazard rate

Table 4.21 Characteristics of the two-parameter binomial distribution $F(z;n,p)$.

Parameters	n: number parameter; $n \in \mathbb{N}$ p: probability parameter; $0 \leq p \leq 1$
Domain	$z \in [0, n]$
Fundamental functions	
Cumulative distribution F	$F(z) = \sum\limits_{j=0}^{z} \binom{n}{j} p^j (1 - p)^{n-j}$
Reliability R	$R(z) = 1 - F(z)$
Distribution density f	$f(z) = \binom{n}{z} p^z (1 - p)^{n-z}$
Hazard rate h	$h(z) = \dfrac{f(z)}{R(z)}$
Cumulative hazard H	$H(z) = \sum\limits_{j=0}^{z} \dfrac{f(j)}{R(j)}$

Measures					
Mean	Variance	Skewness	Excess kurtosis	Median	Mode
$n \cdot p$	$n \cdot p \cdot (1 - p)$	$\dfrac{1 - 2p}{\sqrt{n \cdot p \cdot (1 - p)}}$	$\dfrac{1 - 6p \cdot (1 - p)}{n \cdot p \cdot (1 - p)}$	$\lfloor np \rfloor$ or $\lceil np \rceil$	$\lfloor (n + 1)p \rfloor$ or $\lfloor (n + 1)p \rfloor - 1$

Table 4.22 Mean and modes for the binomial distribution $F(z;n,p)$.

Condition	Mean	Mode(s)
$p = 0$	$n \cdot p = 0$	0
$p = 1$	$n \cdot p = n$	n
$n \cdot p \in \{1, \ldots, n\}$	$n \cdot p$	$n \cdot p$
$(n + 1) \cdot p \in \{1, \ldots, n\}$	$n \cdot p$	$(n + 1) \cdot p$ and $(n + 1) \cdot p - 1$
Otherwise	$n \cdot p$	$\lfloor (n + 1)p \rfloor$

varies in time but can be assumed more or less constant over the period of interest, then a binomial study may be used as a good approximation.

4.4.1.3 EXTRA: Moment Generating Function
The moment generating function for the binomial distribution is:

$$G_Z(t) = \; <e^{t \cdot Z}> \; = \sum_{z=0}^{n} e^{t \cdot Z} \binom{n}{z} p^z (1 - p)^{n-z} \tag{4.120}$$

This can be rewritten as:

$$G_Z(t) = \sum_{z=0}^{n} \binom{n}{z} \cdot (e^t \cdot p)^z \cdot (1 - p)^{n-z} = (e^t \cdot p + (1 - p))^n \tag{4.121}$$

Table 4.23 Mean and median(s) for the binomial distribution $F(z;n,p)$.

Condition	Mean	Median
$p = 0$	$n \cdot p = 0$	0
$p = 1$	$n \cdot p = n$	n
$n \cdot p \in \{1, \dots, n\}$	$n \cdot p$	$n \cdot p$
$p = 0.5$ and n odd	$n \cdot p$	non-unique: $[(n-1)p, (n+1)p]$
$p \leq 1 - \ln 2 \quad \vee$		
$p \geq \ln 2 \quad \vee$	$n \cdot p$	Round(np)
$\lvert \text{Round}(np) - np \rvert \leq \min(p, \ 1-p)$		

The moments are found by differentiation:

$$\frac{d^k}{dt^k} G_Z(t) \bigg|_{t=0} = \langle Z^k \rangle \tag{4.122}$$

4.4.2 Poisson Distribution

The Poisson distribution is used when the expected number of hits μ during a period is known, but the total number of trials n may remain unknown. So, while the binomial distribution applies to a number of trials n, the Poisson distribution applies to a period (which may or may not be mentioned explicitly as a parameter). Unlike the situation with the binomial distribution and due to an unknown number of trials during the observation period, the probabilities p of a hit and q of a non-hit may also remain unknown with the Poisson distribution. However, the product of n and p equals μ. Furthermore, the events are supposed to be independent and the hits are countable. The variable of the Poisson distribution is the number of hits z during the observation period.

An example can be the number of lightning strikes per square kilometre per year. If the average is 2 strikes/km^2·yr, what is the probability that 1, 2, 3 or 4 strikes will occur in a given year?

The Poisson mass function $f(z;\mu)$ is defined as:

$$f(z; \mu) = \frac{\mu^z}{z!} e^{-\mu} \tag{4.123}$$

The cumulative function is then:

$$F(z; \mu) = \sum_{k=0}^{z} \frac{\mu^k e^{-\mu}}{k!} \tag{4.124}$$

4.4.2.1 Characteristics of the Poisson Distribution

The characteristics of the Poisson distribution are given in Table 4.24.

Finally, it may be noted that the hazard rate h is not a function of time and therefore a constant. In terms of ageing, the events thus happen randomly in time. This, of course, is due to the fact that the frequency of hits μ was taken independent of time. The population as such has an exponential distribution with respect to a Poisson type of process. If the hazard rate varies in time but can be assumed more or less constant over the period of interest, then a Poisson study may be used as a good approximation.

Table 4.24 Characteristics of the Poisson distribution $F(z;\mu)$.

Parameter	μ: frequency of hits; $\mu \in (0,\infty)$
Domain	$z \in \mathbb{N}$
Fundamental functions	
Cumulative distribution F	$F(z) = \sum_{k=0}^{z} \dfrac{e^{-\mu}\mu^k}{k!}$
Reliability R	$R(z) = 1 - F(z) = 1 - \sum_{k=0}^{z} \dfrac{e^{-\mu}\mu^k}{k!}$
Distribution density f	$f(z) = \dfrac{\mu^z}{z!}e^{-\mu}$
Hazard rate h	$h(z) = \dfrac{f(z)}{R(z)} = \dfrac{\mu^z}{z!\left(e^{\mu} - \sum_{k=0}^{z} \dfrac{\mu^k}{k!}\right)}$
Cumulative hazard H	$H(z) = \sum_{j=0}^{z} \dfrac{f(j)}{R(j)}$
Measures	

Mean	Variance	Skewness	Excess kurtosis	Median [19]	Mode
μ	μ	$\dfrac{1}{\sqrt{\mu}}$	$\dfrac{1}{\mu}$	$\in \left[\mu - \log 2, \mu + \dfrac{1}{3}\right)$	$\lfloor \mu \rfloor$

4.4.2.2 Derivation of the Poisson Distribution

The Poisson distribution describes the statistics of the event frequency around a known mean frequency μ. The Poisson distribution can be derived from the Taylor expansion of the function e^{μ} in powers of μ:

$$e^{\mu} = \sum_{k=0}^{\infty} \frac{\mu^k}{k!} \tag{4.125}$$

Dividing both sides by e^{μ} yields:

$$1 = \sum_{k=0}^{\infty} \frac{\mu^k e^{-\mu}}{k!} \tag{4.126}$$

This equation can be viewed as a cumulative function $F(\infty;\mu) = 1$ and the sum of density functions $f(k;\mu)$ with $k = \{0,\ldots,\infty\}$. From this, Eqs (4.123) and (4.124) follow.

4.4.2.3 Homogeneous Poisson Process

In the previous subsections the events were not specified in terms of whether a population was to be viewed as a draw with or without replacement. If the population consists of a single repairable system, failure can occur over and over again. Depending on the mechanism of failure, the Poisson distribution might apply.

Assume such a repairable system (i.e. a system that is repaired after each failure and put back into operation). Examples are a connection or a cable. A function $N(t)$ is defined that counts the cumulative number of events (failures) at operational lifetime t.

If many similar systems are observed, then at each operational lifetime t an average cumulative number of events is counted. This average increases with time and is the expected number of failures at time t. It is defined as the cumulative function $M(t)$:

$$M(t) = <N(t)> \tag{4.127}$$

The derivative $m(t)$ of $M(t)$ is the so-called rate of change of frequency (ROCOF):

$$m(t) = \frac{dM(t)}{dt} = \frac{d<N(t)>}{dt} \tag{4.128}$$

The functions $m(t)$ and $M(t)$ look respectively like the hazard rate $h(t)$ (Section 2.4.5) and the cumulative hazard rate $H(t)$ (Section 2.4.6), but there is a difference. The (cumulative) hazard rates $h(t)$ and $H(t)$ apply to objects that survived up to t and may fail in a coming period. The average counting functions $m(t)$ and $M(t)$ apply to objects that survived only since the last repair, but the time is the total (or cumulative) operational lifetime of the repairable systems.

The so-called homogeneous Poisson process (HPP) is defined as a process where the intensity function $m(t)$ is a constant λ. As a consequence, $M(t)$ is linear with time t (with the assumption $M(0) = 0$):

$$m_{HPP}(t) = \lambda \tag{4.129}$$

$$M_{HPP}(t) = \lambda \cdot t \tag{4.130}$$

This is applicable when the intervals between failures (times to failure) are independent and distributed according to an exponential distribution (Section 4.2.4). Now the value of $m(t)$ and $h(t)$ is the same (and a constant). Although a function of time t that started at $t = 0$, the cumulative function $\Phi(\tau)$ concerns the probability of failure at time τ since the last failure:

$$\Phi(\tau) = 1 - e^{-\lambda \tau} \tag{4.131}$$

The MTBF is $1/\lambda$ (cf. Eq. (4.34) with $\lambda = 1/\theta$).

The cumulative number of failures is $N(t)$. The average number of failures during a period t is $M_{HPP}(t) = \lambda t$. The Poisson distribution describes the statistics of $N(t)$ being z according to Eq. (4.123):

$$f(z; \mu = \lambda t) = \frac{(\lambda t)^z}{z!} e^{-\lambda t} \tag{4.132}$$

In Section 4.2.4.1 it was shown how a maintained (or ultimately repeatedly repaired) system can be approximated as a process with a constant hazard rate, that is featuring an exponential distribution. The present description uses the Poisson distribution and yields the same result. Both can be used to describe the period which forms the base of the bath tub curve (Section 2.4.5) and represents the bulk of the useful life of assets.

4.4.2.4 Non-Homogeneous Poisson Process

Following the HPP in the previous section (i.e. Section 4.4.2.3), the non-homogeneous Poisson process (NHPP) is defined such that the average number of failures $M(t)$ is a power function of time t with constant $a > 0$ and power $b > 0$:

$$M_{NHPP}(t) = a \cdot t^b \tag{4.133}$$

The derivative is the power law intensity function $m_{NHHP}(t)$:

$$m_{NHPP}(t) = ab \cdot t^{b-1} \tag{4.134}$$

For $0 < b < 1$ this is a decreasing function, which means with time and repairing after failures the system becomes more reliable (i.e. the periods between failures are growing). For $b > 1$ this is an increasing function, which means with time and repairing after failures the system becomes less reliable (i.e. the intervals between failures shorten). For $b = 1$ this reduces to the HPP case in Section 4.4.2.3 and the intervals between failures scatter about their mean $1/\lambda$.

The intensity function $m_{NHHP}(t)$ can be regarded as an average hazard rate and with $b = \beta$ and $a = \alpha^{-\beta}$ this is the hazard rate of a Weibull distribution.

This does not mean that the underlying distribution(s) of the individual assets need to be of Weibull shape. The requirement is that the average hazard rate obeys Eq. (4.134).

4.4.2.5 Poisson versus Binomial Distribution

The binomial distribution and the Poisson distribution resemble each other for large n and constant μ. The probabilities p of hits and q of non-hits can be written as:

$$p = \frac{\mu}{n} \tag{4.135}$$

$$q = 1 - p = 1 - \frac{\mu}{n} \tag{4.136}$$

The constant μ implies that the probability p of a hit becomes small and the probability q of a non-hit approaches 1. Substitution of Eqs (4.135) and (4.136) in the binomial mass function $f(z;n,p)$ of Eq. (4.107) yields $f(z;n,\mu)$:

$$f(z; n, \mu) = \binom{n}{z} \cdot \left(\frac{\mu}{n}\right)^z \cdot \left(1 - \left(\frac{\mu}{n}\right)\right)^{n-z} \tag{4.137}$$

This can be rewritten as:

$$f(z; n, \mu) = \frac{n \cdot \ldots \cdot (n - z + 1)}{n^z} \cdot \frac{\mu^z}{z!} \cdot \left(1 - \left(\frac{\mu}{n}\right)\right)^n \cdot \left(1 - \left(\frac{\mu}{n}\right)\right)^{-z} \tag{4.138}$$

For $n \to \infty$ the first term approaches 1:

$$\lim_{n \to \infty} \frac{n \cdot \ldots \cdot (n - z + 1)}{n^z} = \lim_{n \to \infty} \left(\frac{n}{n} \cdot \ldots \cdot \frac{n - z + 1}{n}\right)$$

$$= \lim_{n \to \infty} \left(1 \cdot \ldots \cdot \left(1 - \frac{z - 1}{n}\right)\right) = 1^z = 1 \tag{4.139}$$

The same applies to the fourth term:

$$\lim_{n \to \infty} \left(1 - \left(\frac{\mu}{n}\right)\right)^{-z} = 1^{-z} = 1 \tag{4.140}$$

The third term is a well-known limit and approaches $e^{-\mu}$:

$$\lim_{n \to \infty} \left(1 - \left(\frac{\mu}{n}\right)\right)^n = e^{-\mu} \tag{4.141}$$

In conclusion, for large n, constant $\mu = np$ and thus inherently small p, the binomial distribution is found to resemble the Poisson distribution (cf. Eq. (4.123)):

$$\lim_{n \to \infty} f(z; n, \mu) \equiv \lim_{n \to \infty} \binom{n}{z} \cdot \left(\frac{\mu}{n}\right)^z \cdot \left(1 - \left(\frac{\mu}{n}\right)\right)^{n-z} = \frac{\mu^z}{z!} e^{-\mu} \tag{4.142}$$

4.4.3 Hypergeometric Distribution

The binomial distribution requires subsequent trials to be independent. This means that the outcomes of the various trials are not related, which is called sampling with replacement. However, in case the number of hits is limited and one or more trials are carried out, the probabilities of consecutive trial outcomes are influenced.

For instance, assume there are N items of which M are defective and consequently $N - M$ are non-defective. For this group the probability of a hit is $p = M/N$ and of a non-hit $q = 1 - M/N$. For testing, assume n times a test item is randomly selected from the group of N items and after each trial the tested item returns to the group and may or may not be subject to the next trial. This sampling with replacement implies that the probabilities p and q do not change and the probability of z hits in n trials is the binomial distribution density function:

$$f(z; n, N, M) = \binom{n}{z} \cdot \left(\frac{M}{N}\right)^z \left(1 - \frac{M}{N}\right)^{n-z} \tag{4.143}$$

Here, $z \in \{0,\ldots,n\}$ again and p is replaced by $p = M/N$.

This density function changes if the tested items do not return to the group. This is called sampling without replacement. Assume that n test items are randomly selected and removed from the group and subsequently tested. The tested items don't return to the subgroup of n items. The probability of z hits in these n trials with $z \in \{0,\ldots,n\}$ is the so-called hypergeometric distribution:

$$f(z; n, N, M) = \frac{\binom{M}{z} \cdot \binom{N - M}{n - z}}{\binom{N}{n}} \tag{4.144}$$

This mass function contains three binomial coefficients (cf. Eq. (4.108)), which in clockwise order are referred to as M-over-z, $(N - M)$-over-$(n - z)$ and N-over-n. There are M-over-z combinations of z defective items out of the total group of M defective items and there are $(N - M)$-over-$(n - z)$ combinations of $n - z$ non-defective items out of the total group of $N - M$ non-defective items. The product M-over-z and $(N - M)$-over-$(n - z)$ is the number of ways to combine z defective and $n - z$ non-defective items. There are N-over-n combinations of n items out of the total group of N items, regardless of the number of defective items (i.e. the total number of combinations for all $z \in \{0,\ldots,n\}$). The ratio of the number of combinations with z defective items and the total number of combinations is the probability of z hits in n trials from a total group of N items with M defective items (i.e. this is the hypergeometric distribution density function of Eq. (4.144)).

The cumulative hypergeometric distribution is defined as:

$$F(z; n, N, M) = \frac{1}{\binom{N}{n}} \sum_{k=0}^{z} \binom{M}{k} \cdot \binom{N - M}{n - k} \tag{4.145}$$

A case of working with the hypergeometric distribution is elaborated in Example 4.5 following the description of the mean and variance.

4.4.3.1 Mean and Variance of the Hypergeometric Distribution

The mean and variance of the hypergeometric distribution are:

$$<Z> = \frac{nM}{N} \tag{4.146}$$

$$\text{var}(Z) = \frac{nM}{N}\left(1 - \frac{M}{N}\right) \cdot \frac{N - n}{N - 1} \tag{4.147}$$

Noting that $p = M/N$, the means of the binomial and the hypergeometric distribution are the same (cf. Eqs (4.116) and (4.146)). However, the variances of the binomial and the hypergeometric distribution differ (cf. Eqs (4.119) and (4.147)) by a factor of $(N - n)/(N - 1)$. This factor corrects for the finite population in case of the hypergeometric distribution and reduces the variance compared to the variance of the binomial distribution.

In practical cases a batch of defective items may blend in with the regular non-defective items and a customer like a utility may not have access to the values of N and M. From own and other experience though, $p = M/N$ may be known as the ratio of all hits and all trials. Instead of the hypergeometric distribution, the binomial distribution may then be used because that only requires the probability p and the number of trials n, which might be the number of assets managed by the utility.

As an example, the case of the 24 post insulators in Example 4.4 is treated with the hypergeometric distribution and compared to the binomial distribution in Example 4.5 below.

Example 4.5 *Statistics of Defect Insulators with Given Probability p*   With reference to Example 4.4 where a case of $n = 24$ post insulators with probability $p = 5\%$ was studied, a comparison of the binomial and hypergeometric distributions is carried out in the present example. Assume that the total number of items N and the total number of defective items $M = Np$ are known. Then it is possible to determine both the binomial and the hypergeometric distribution. The distribution density functions are indicated as $fB(z;n,p)$ and $fH(z;n,N,M)$, respectively, and defined in Eqs (4.107) and (4.144). The cumulative distribution functions are indicated as $FB(z;n,p)$ and $FH(z;n,N,M)$, respectively, and defined in Eqs (4.112) and (4.145).

For the sake of comparison, two group sizes $N = 300$ and $N = 1200$ are considered. Consequently, with $p = 5\%$, the group sizes of the defective items are $M = 15$ and $M = 60$, respectively. In both cases the utility has $n = 24$ post insulators at the station and has knowledge of N and M. In total, three distributions are obtained to describe the probability of z hits in n trials: the binomial distribution with $p = 5\%$ and two hypergeometric distributions with $(N,M) = (300,15)$, respectively $(N,M) = (1200,60)$. The results are shown in Figure 4.14 and Table 4.25.

As for the graphical representation, the distributions resemble each other very well. The table shows more details of the differences. The variance Z of the hypergeometric distribution differs most from that of the binomial distribution when the population size N is the smallest (i.e. when $N = 300$). This is in agreement with the correction for the finite population size. The distribution density of that hypergeometric distribution is also the most peaked, albeit only slightly more than the other two distributions. This difference is hardly noticeable in Figure 4.14 and is better observable in Table 4.25.

In this particular case, the differences between the distributions are very small and the use of the binomial distribution will probably not lead to other inferences and

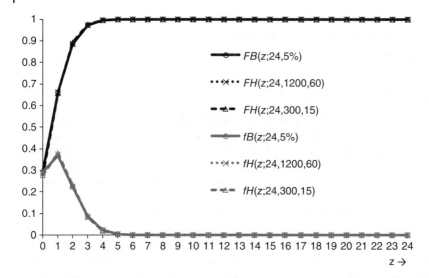

Figure 4.14 The mass function and the cumulative function of the binomial distribution with parameters $n = 24$ and $p = 5\%$ in comparison to the hypergeometric distributions with $(N,M) = (300,15)$, respectively $(N,M) = (1200,60)$. The difference among the cumulative functions and the mass functions is quite small. Note in all cases, $p = M/N = 5\%$.

Table 4.25 The mass function and the cumulative function of the binomial and the two given hypergeometric mass functions for $z = 0,...,8$ (see text and Figure 4.14).

$n = 24$ z	$p = 5\%$ $<Z> = 1.2$ var$(Z) = 1.140$		$N = 1200, M = 60$ $<Z> = 1.2$ var$(Z) = 1.118$		$N = 300, M = 15$ $<Z> = 1.2$ var$(Z) = 1.052$	
	fB	FB	fH	FH	fH	FH
0	0.2920	0.2920	0.2884	0.2884	0.2774	0.2774
1	0.3688	0.6608	0.3718	0.6603	0.3812	0.6586
2	0.2232	0.8841	0.2257	0.8859	0.2334	0.8920
3	0.0862	0.9702	0.0858	0.9717	0.0843	0.9762
4	0.0238	0.9940	0.0229	0.9946	0.0200	0.9963
5	0.0050	0.9990	0.0046	0.9992	0.0033	0.9996
6	0.0008	0.9999	0.0007	0.9999	0.0004	1.0000
7	0.0001	1.0000	0.0001	1.0000	0.0001	1.0000
8	0.0000	1.0000	0.0000	1.0000	0.0000	1.0000

conclusions than when the hypergeometric distribution is used. A population size effect is noticeable though, and in case of doubt with such analyses it may be checked whether the hypergeometric distribution is more appropriate than the binomial distribution.

4.4.3.2 Characteristics of the Hypergeometric Distribution

The characteristics of the hypergeometric distribution are given in Table 4.26.

Finally, it may be noted that the hazard rate h is not a function of time and therefore a constant. In terms of ageing, the events thus happen randomly in time. This is, of course,

Table 4.26 Characteristics of the hypergeometric distribution $F(z;n,N,M)$.

Parameters	n: number of trials; $n \in [0,N]$
	N: total populations size; $N \in \mathbb{N}$
	M: total number of hits in population; $M \in [0,N]$
Domain	$z \in [0,n]$

Fundamental functions	
Cumulative distribution F	$F(z) = \dfrac{1}{\binom{N}{n}} \sum\limits_{k=0}^{z} \binom{M}{k} \cdot \binom{N-M}{n-k}$
Reliability R	$R(z) = 1 - F(z) = 1 - \dfrac{1}{\binom{N}{n}} \sum\limits_{k=0}^{z} \binom{M}{k} \cdot \binom{N-M}{n-k}$
Distribution density f	$f(z) = \dfrac{\binom{M}{z} \cdot \binom{N-M}{n-z}}{\binom{N}{n}}$
Hazard rate h	$h(z) = \dfrac{f(z)}{R(z)} = \dfrac{\binom{M}{z} \cdot \binom{N-M}{n-z}}{\binom{N}{n} - \sum\limits_{k=0}^{z} \binom{M}{k} \cdot \binom{N-M}{n-k}}$
Cumulative hazard H	$H(z) = \sum\limits_{j=0}^{z} \dfrac{f(j)}{R(j)}$

Measures		
Mean	Variance	Skewness
$\dfrac{nM}{N}$	$\dfrac{nM}{N}\left(1 - \dfrac{M}{N}\right) \cdot \dfrac{N-n}{N-1}$	$\dfrac{(N-2M) \cdot \sqrt{(N-1)} \cdot (N-2n)}{\sqrt{nM \cdot (N-M) \cdot (N-n)} \cdot (N-2)}$
Excess kurtosis		Mode
$\dfrac{\begin{array}{c} N^2 \cdot (N-1) \cdot [N(N+1) - 6n(N-n) - 6M(N-M)] \\ + 6nM(N-M) \cdot (N-n) \cdot (5N-6) \end{array}}{nM \cdot (N-n) \cdot (N-M) \cdot (N-2) \cdot (N-3)}$		$\left\lfloor \dfrac{(n+1)(M+1)}{N+2} \right\rfloor$

due to the fact that N and M, thus p, were taken constants. The population as such has an exponential distribution with respect to a hypergeometric type of process. If the hazard rate varies in time but can be assumed more or less constant over the period of interest, then a hypergeometric study may be used as a good approximation.

4.4.4 Normal Distribution Approximation of the Binomial Distribution

The normal distribution can be used to approximate the binomial distribution when the number of trials n is large. The binomial distribution was defined in Eq. (4.107) and is repeated here:

$$f(z; n, p) = \binom{n}{z} p^z (1 - p)^{n-z} \tag{4.148}$$

The number of hits in n trials is z and the probability of a hit in a single trial is p. The outcomes of the trials are supposedly independent.

The mean number of hits $<Z>$ with the binomial distribution is (see Eq. (4.116)):

$$< Z > = np \tag{4.149}$$

The variance of Z with the binomial distribution can be calculated as:

$$\text{var}(Z) = np(1 - p) \tag{4.150}$$

The normal distribution is discussed in Section 4.3.1 and defined as (see Eq. (4.50)):

$$f_N(x) = \frac{1}{\sigma \sqrt{2\pi}} \cdot e^{-\frac{1}{2}\left(\frac{x-\mu}{\sigma}\right)^2} \tag{4.151}$$

with mean μ and variance σ^2. The suffix N is added to explicitly indicate the normal distribution.

For very large n, the binomial distribution can be approximated with a normal distribution with the parameters:

$$\mu = < Z > = np \tag{4.152}$$

$$\sigma^2 = \text{var}(Z) = np(1 - p) \tag{4.153}$$

which implies that the ratio of the standard deviation and the mean is:

$$\frac{\sigma}{\mu} = \frac{\sqrt{np(1 - p)}}{np} = \sqrt{\frac{(1 - p)}{np}} \tag{4.154}$$

This yields for large n:

$$f(z) \approx f_N(z) \frac{1}{\sqrt{np(1 - p)} \sqrt{2\pi}} \cdot e^{-\frac{1}{2}\left(\frac{z-np}{\sqrt{np(1-p)}}\right)^2} \tag{4.155}$$

As for determining boundaries, the limit theorem of de Moivre [20] and de Laplace [21] applies. It states that the probability that Z can be found in the range $[a,b]$ can be calculated as:

$$P(a \leq Z \leq b) = \sum_{z=a}^{b} \binom{n}{z} p^z (1 - p)^{n-z} \approx F_N(\beta) - F_N(\alpha) \tag{4.156}$$

Here, F_N is the cumulative standardized normal distribution (see Section 4.3.1, and for approximations see Section 10.6). The values α and β are defined as:

$$\alpha = \frac{a - np - 0.5}{np(1 - p)} \tag{4.157}$$

$$\beta = \frac{b - np + 0.5}{np(1 - p)} \tag{4.158}$$

The 0.5 terms are related to the change from a discrete to a continuous distribution.

With $z \in [0,\infty)$, the binomial distribution has a different domain than the normal distribution where $x \in (-\infty,\infty)$. Particularly if the binomial density function at small z is not negligible, the approximation with the normal distribution falls short. An indicator of this can be the ratio σ/μ that should be sufficiently small. The approximation of the binomial distribution by the normal distribution is deemed fairly good firstly if $\mu = np > 5$ and $p \leq 0.5$ or secondly if $n(1-p) > 5$ and $(1-p) \leq 0.5$.

The implication of these inequalities can be elaborated in terms of the ratio σ/μ as a measure of the error for small z. In the first case $0.5 \leq 1-p \leq 1$ and $1/np < 0.2$, which means that the ratio $\sigma/\mu < 0.45$ and therefore μ is at least about 2σ. According to Table 4.15 about 4.55% of the variable X lies outside the range $[\mu - 2\sigma, \mu + 2\sigma]$, which indicates that the error on the left side of the interval at $z = 0$ is typically on the order of a few percent. Because of the symmetry of the binomial distribution and the binomial coefficient (i.e. n-over-z is the same as n-over-$(n-z)$), the second case leads to the same considerations and result.

The lognormal distribution having a similar domain as the binomial distribution might therefore also be a good option as an approximation. However, even then the ratio σ/μ as in Eq. (4.154) should be sufficiently small, because a significant binomial $f(0)$ would impose a problem for the lognormal distribution for which $f_L(0) = 0$ by definition.

In conclusion, the normal (and lognormal) distribution offer good approximations for samples with a large sample size n. This is convenient because the binomial coefficient and the powers in calculation of the binomial distribution (see Eqs (4.107) and (4.148)) may cause accuracy problems for large n. The ratio σ/μ must be sufficiently small, so the binomial $f(0)$ should be negligibly small.

4.4.5 Multinomial Distribution

A generalization of the binomial distribution (Section 4.4.1) is the multinomial distribution. Consider again n trials of which the outcomes V_j ($j = 1,\dots,n$) are mutually exclusive and independent. Instead of only two possible outcomes per trial as with the binomial distribution, it is now assumed that there are k possible outcomes E_m ($m = 1,\dots,k$) for each trial. The probability of each outcome E_m is p_m ($m = 1,\dots,k$). As the outcomes are mutually exclusive, the sum of the probabilities must equal 1:

$$\sum_{m=1}^{k} p_m = 1 \tag{4.159}$$

Generalizing the case of the binomial distribution, let the number of outcomes E_m in n trials be the variable Z_m. All values z_m are natural numbers. As there are n trials, the sum of the number of hits per outcome must equal the number of trials, so:

$$\sum_{j=1}^{k} Z_j = n \tag{4.160}$$

An example might be rolling a six-sided dice which has six elementary outcomes E_m. The set of elementary outcomes $\{E_1, E_2, E_3, E_4, E_5, E_6\} = \{1,2,3,4,5,6\}$ and $k = 6$. Rolling the dice n times produces a sequence of n outcomes and each trial outcome $V_j \in \{1,\dots,6\}$ with $j = 1,\dots,n$. For a fair dice the probability of each outcome E_m is $p_m = 1/6$.

An example related to asset management concerns the situation where the group of surge arrestors has been investigated and the condition of each surge arrestor was categorized with a health index (see Section 1.3.3.4). The number of health index categories is the number of possible outcomes k per trial and is not (yet) standardized. Therefore, k varies per company asset management system. In Section 1.3.3.4, Figure 1.4 an example with $k = 4$ was shown where the categories follow the traffic-light convention (green, yellow, red) plus an additional colour (purple). The meaning of these categories in that example is:

- green = good;
- orange/yellow = fair, but action required and possible to return to green;
- red = suspect, plan for replacement or refurbishment;
- purple = urgent attention required.

Inspections, interpretations and experience indicate the probabilities p_m $(m = 1,...,k)$ and define the distribution of probabilities.

The probability mass function $f(z_1,...,z_k;n,p_1,...,p_k)$ of the multinomial distribution is:

$$f(z_1, \ldots, z_k; n; p_1, \ldots, p_k) = \frac{n!}{z_1! \cdot \ldots \cdot z_k!} p_1^{z_1} \cdot \ldots \cdot p_k^{z_k} \tag{4.161}$$

The conditions in Eqs (4.159) and (4.160) must be met, otherwise the mass function $f(z_1,...,z_k;n,p_1,...,p_k)$ equals 0. In some notations in the literature n is left out as a parameter because this parameter can be replaced by the sum of the z_k with reference to the condition in Eq. (4.160). Here it is included to explicitly indicate the number of trials. Instead, the last variable z_k (or any other) could be left out, because due to the condition in Eq. (4.160) the variables are not independent. For instance, z_k is given by:

$$z_k = n - \sum_{m=1}^{k-1} z_m \tag{4.162}$$

This was implicitly stated for the binomial distribution where the probability of a non-hit q was written as $(1 - p)$ in Eq. (4.107).

The probability mass function $f(z_1,...,z_k;n,p_1,...,p_k)$ can also be expressed in terms of the gamma function:

$$f(z_1, \ldots, z_k; n; p_1, \ldots, p_k) = \Gamma(n + 1) \cdot \frac{\prod_{m=1}^{k} p_m^{z_m}}{\prod_{m=1}^{k} \Gamma(z_m + 1)} \tag{4.163}$$

Again, either one of the variables or the parameter n might be left out on the left-hand side of the equation due to the boundary conditions (Eqs (4.159) and (4.160)).

It can be checked conveniently that in case the number of possible outcomes k equals 2, Eqs (4.161) and (4.163) indeed reduce to the expressions of the binomial distribution.

4.4.5.1 Mean, Variances and Moment Generating Function

The expressions for the mean and variance are basically the same as for the binomial distribution (see Section 4.4.1.1). The probability of outcome E_m in a single trial is p_m

$(m = 1,\ldots,k)$ and the expected number $<Z_m>$ of times that E_m occurs in n trials is:

$$< Z_m >= np_m \tag{4.164}$$

The variance is:

$$\text{var}(Z_m) = np_m(1 - p_m) \tag{4.165}$$

The covariance of the numbers Z_i and Z_j of two distinct elementary outcomes E_i and E_j is:

$$i \neq j : \quad \text{cov}(Z_i, Z_j) = -np_i p_j \tag{4.166}$$

The negative sign reflects the fact that with a fixed number of trials n an increase of one number consequently reduces the number of another outcome.

The moment generating function G is:

$$G(t) = \left(\sum_{m=1}^{k} p_m e^{t_m} \right)^n \tag{4.167}$$

and can be used to derive the moments of the multinomial distribution.

4.4.6 Multivariate Hypergeometric Distribution

As with the binomial distribution (Section 4.4.1), the multinomial distribution (Section 4.4.5) concerns sampling with replacement. For the case of sampling without replacement and two possible outcomes (e.g. hit versus non-hit), the hypergeometric distribution should be turned to as explained in Section 4.4.3. Generalizing from binary outcomes to $k > 2$ outcomes, the corresponding distribution is the multivariate hypergeometric distribution.

As with the multinomial distribution, assume there are k elementary outcomes E_m $(m = 1,\ldots,k)$ of a single trial and the number of times that outcome E_m occurs in n trials is Z_m. As with the hypergeometric distribution, assume that items are tested in the trials to assess one of the k outcomes to that item. The items do not return to the group of untested items (i.e. the tests are sampling without replacement). Assume there is a total group of N items with $N \in \mathbb{N}$. In this group the number of items with test outcome E_m is M_m. The probability of each outcome E_m in the first trial is therefore $p_m = M_m/N$.

Assume that n test items are randomly selected, removed from the group and subsequently tested. The tested items also don't return to the subgroup of n. The probability of z_m outcomes E_m in these n trials is described by the so-called multivariate hypergeometric distribution:

$$f(z_1, \ldots, z_k; n, N, M_1, \ldots, M_k) = \frac{1}{\binom{N}{n}} \prod_{m=1}^{k} \binom{M_m}{z_m} \tag{4.168}$$

Two boundary conditions in this case are:

$$\sum_{m=1}^{k} M_m = N \tag{4.169}$$

With substitution $p_m = M_m/N$ this is equivalent to the sum of probabilities being 1 as in Eq. (4.159). The second condition is:

$$\sum_{j=1}^{k} Z_j = n \qquad (4.170)$$

The mean is the same as with the multinomial distribution:

$$<Z_m> = \frac{n \cdot M_m}{N} = n \cdot p_m \qquad (4.171)$$

The variance differs from that of the multinomial distribution. The variance of the multivariate hypergeometric distribution is:

$$\text{var}(Z_m) = \frac{nM_m}{N}\left(1 - \frac{M_m}{N}\right) \cdot \frac{N - n}{N - 1} \qquad (4.172)$$

For $k = 2$ the suffices can be omitted and Eqs (4.171) and (4.172) are the same as for Eqs (4.146) and (4.147) for the hypergeometric distribution.

For the multivariate hypergeometric distribution the covariance of the numbers Z_i and Z_j of two distinct elementary outcomes E_i and E_j is:

$$i \neq j : \quad \text{cov}(Z_i, Z_j) = -\frac{nM_iM_j}{N^2} \cdot \frac{N - n}{N - 1} \qquad (4.173)$$

Again, the negative sign reflects the fact that with fixed number of trials n an increase of one number consequently reduces the number of another outcome.

4.5 Summary

This chapter discussed a range of specific distributions with their characteristics. These distributions were grouped into four categories of statistics: fractions and ranking (Section 4.1); extreme values (Section 4.2); mean and variance (Section 4.3); frequency and hit statistics (Section 4.4).

As for fractions and ranking statistics (Section 4.1), the uniform distribution (Section 4.1.1) and the beta or rank distribution (Section 4.1.2) were discussed. These are particularly useful for describing the ranking of observed data and operations on the cumulative distribution values F.

As for extreme value statistics (Section 4.2), we have the two-parameter Weibull or Weibull-2 distribution (Section 4.2.1), the Weibull-3 distribution (Section 4.2.2), the Weibull-1 distribution (Section 4.2.3) and the exponential distribution (Section 4.2.4). Particular attention is paid to child mortality processes (Section 4.2.1.1). In the literature, these are usually associated with Weibull shape parameters $\beta < 1$, but it is discussed that in practice child mortality appears to be associated with wear-out processes in (hopefully) limited subpopulations on the one hand and that are characterized by a very small-scale parameter on the other hand. Another subject is the matter of maintained systems that are repeatedly renewed and then become characterized by random failure. As a result, such systems feature an exponential distribution on average (Section 4.2.4.1).

As for mean and variance statistics (Section 4.3), the normal or Gaussian distribution (Section 4.3.1) and the lognormal distribution (Section 4.3.2) are discussed. Particular attention is paid to various ways to approximate the normal distribution (Section 10.6) with Taylor series, polynomial expressions and a power function that is introduced for approximating asymptotic behaviour (see also Section 10.2.3). The central limit theorem is the background for the extensive use of the normal distribution for describing the means and variance statistics. The extreme statistics refers to the tails of the distributions, which is not the area where the normal distribution is the optimum solution. The lognormal distribution is closely related to the normal distribution. If a variable X is normally distributed, then the variable $T = \exp(X)$ is lognormally distributed. Lognormal is more suitable than the normal distribution for variables T that by definition are larger than a finite minimum (e.g. breakdown times $T \in [0,\infty)$). But again, the lognormal and normal distributions are asymptotic distributions for the mean, while other distributions such as the Weibull are asymptotic distributions for extreme values.

As for frequency and hit statistics (Section 4.4), various distributions are discussed that describe the probability that an event will take place within a given number of trials or periods. Most concern so-called Bernoulli trials, which are experiments with binary outcomes: yes versus no, hit versus non-hit, and so on, with probabilities p, respectively $q = 1 - p$. The binomial distribution describes the probabilities of each combination of hits and non-hits in a fixed number n of experiments (Section 4.4.1). For example, if it is known that an overall percentage 15% of assets produce audible noise, what is the probability that at a certain substation 30% produce audible noise? It is assumed that the noise does not require replacement, so this is sampling without replacement.

The Poisson distribution describes the probabilities of the number of hits, without knowledge of n and the number of non-hits (Section 4.4.2). For example, if two lightning strikes occur per square kilometre per year on average, what is the probability that 1, 2, 3 or 4 strikes occur in a given year?

The hypergeometric distribution (Section 4.4.3) is similar to the binomial distribution, but concerns sampling with replacement. For example, inspections are carried out and defective assets are replaced, which reduces the fraction of defective assets. For large n the normal distribution can be used to approximate the binomial distribution (Section 4.4.4). The multinomial distribution also resembles the binomial distribution by sampling without replacement, but now the outcomes are not binary, there are multiple elementary events instead (Section 4.4.5). For example, throwing a dice with its six possible outcomes is a well-known example. The multivariate hypergeometric distribution describes sampling with replacement and multiple outcomes (Section 4.4.6). For example, inspections on assets and replacement in case one of multiple possible defects is found.

Questions

4.1 With reference to fractions and ranking statistics (Section 4.1):
The uniform distribution can be applied to discrete and continuous distributions. Throwing a dice and tossing a coin are two very old random variable generators.
    a. Assume a fair dice (i.e. the probabilities for each elementary event are the same). What is the probability of throwing an even number once, twice and three times?

b. Assume a game where two fair dice are thrown. After one throw, one dice with the highest score is set apart. What is the probability that throwing the second dice will give a higher score than the first?

4.2 With reference to fractions and ranking statistics (Section 4.1): Figure 4.3 shows the case of five ranked probabilities (i.e. $n = 5$ and $i = 1,...,5$).

a. For each (i,n) combination: find the mode (if feasible), median and mean probability.

b. Determine the 5% and the 95% confidence limit.

4.3 With reference to extreme value statistics (Section 4.2):

A given generator is known to have a failure behaviour according to a Weibull distribution with parameters α and β. Assume the generator costs €250k. Without maintenance, the generator has a mean operational life of 7 years. With maintenance at a cost of €30k this generator can be brought back into a new state.

a. Firstly, assume that no maintenance is carried out. What is the failure rate expressed in terms of the Weibull parameters?

b. Still assume no maintenance is carried out and ignore interest effects: what is the cost rate (depreciation) expressed in €/yr?

c. Secondly, assume maintenance is carried out periodically with time intervals T. What is the failure rate per interval?

d. Still assume maintenance is carried out and ignore interest effects: what is the cost rate based on depreciation and maintenance costs?

e. Find the optimum period T to minimize the total cost rate.

4.4 With reference to mean and variance statistics (Section 4.3):

The distribution densities of the Weibull-2 and the normal distribution may resemble each other. Assume a normal distribution $f_N(x)$ with $\mu = 1$ and $\sigma = 1$.

a. Find a Weibull distribution $f_W(x;\alpha,\beta)$ that has the same cumulative probability at $\mu - \sigma$ and $\mu + \sigma$ as the given normal distribution (i.e. find Weibull parameters α and β such that $F_W(\mu - \sigma;\alpha,\beta) = F_N(\mu - \sigma;\mu,\sigma)$ and $F_W(\mu + \sigma;\alpha,\beta) = F_N(\mu + \sigma;\mu,\sigma)$).

b. Also, find a lognormal distribution that has the same cumulative probability at these two variable values (i.e. find lognormal parameters m_L and σ_L such that $F_W(\mu - \sigma;m_L,\sigma_L) = F_N(\mu - \sigma;\mu,\sigma)$ and $F_W(\mu + \sigma;m_L,\sigma_L) = F_N(\mu + \sigma;\mu,\sigma)$).

c. What is the probability for $x < 0$ in case of this normal distribution?

d. What is the probability for $x < 0$ in case of this Weibull distribution?

e. What is the probability for $x < 0$ in case of this lognormal distribution?

f. What is the mean $<x>$ of this Weibull distribution?

g. What is the mean $<x>$ of this lognormal distribution?

h. Assume the choice for μ is changed to $\mu = 2$. What would change with respect to the probability of $x < 0$ for the normal and Weibull distributions? And what with respect to the mean $<x>$ of the Weibull distribution?

i. Explain whether a Weibull-2 shape parameter exists that makes the Weibull and the normal distribution resemble each other by the criteria in subquestion (a) above.

4.5 With reference to the frequency and hit statistics (Section 4.4):

The number of lightning strikes per square kilometre per year is about 2 in NW Europe, about 30 in Florida and about 50 in Central Africa.

a. What distribution is applicable to estimate a given number of lightning strikes per square kilometre for a given year?

b. What is the probability for a year with the following frequencies of lightning strikes: 1, 3, 10, 30, 100 in NW Europe?

c. Similarly in Central Africa?

d. Similarly in Florida?

5

Graphical Data Analysis

With descriptive statistics many characteristics such as parameters and central moments as explained in Chapter 3 are possible. Although such a set of numerical measures can be very informative, humans are generally capable of recognizing patterns and shapes in the graphical way in which situations can be evaluated. The phrase 'a picture is worth a thousand words' seems not to be as old as often claimed, being traced back to the quote 'Use a picture. It's worth a thousand words' in 1911 [22]. Writers have implied this observation earlier and the vast collection of paintings, cartoons and sculptures dating back to ancient times underlines the impact that visualizations can have.

In statistics, graphs or plots provide a very powerful means to visualize the meaning of data. Graphs are known from the fourteenth century, as described by Oresme [23]. The use of graphs serves various purposes:

- To observe and recognize trends in data.
- To identify outliers (i.e. data that appear erroneous or irrelevant).
- To evaluate whether the data should be split up into groups of data.
- To test the adequacy of distributions.
- To estimate distribution parameters and confidence limits.
- And more, like forecasting under various scenarios (inferential statistics).

It is essential to understand and master how to make a plot and how to interpret it. This chapter discusses the plotting of data and particularly techniques for making probability plots in order to address the purposes listed above.

It starts with collecting the data to be used in plots. Usually, these data come from practical experience and/or from testing. The data are measured performances like lifetime, mechanical strength, electrical strength, and so on. In discussions, often the performance is expressed as time to failure, but one should be aware that any measure other than time could be used as well.

As for the data and unless stated otherwise, it is assumed that the total (manufactured or existing) population of objects consists of N units of which a limited number n is involved in the observational data. This subgroup of n tested units is a sample from the population and the number n is called the sample size. In some cases not all n will have failed (i.e. only a limited number r objects may have been observed to fail). This situation is called a case of censored data and the set of r known failure times is called 'uncensored' observations, whereas the remaining unknown $n - r$ data are called 'censored' observations.

Reliability Analysis for Asset Management of Electric Power Grids, First Edition. Robert Ross.
© 2019 John Wiley & Sons Ltd. Published 2019 by John Wiley & Sons Ltd.
Companion website: www.wiley.com/go/ross/reliabilityanalysis

For instance, $N = 100\,000$ vehicles may have been (or will be) produced, of which a fleet of $n = 95$ were bought by a certain company. If 23 out of 95 did break down, there are $r = 23$ uncensored observations and $95 - 23 = 72$ censored observations. So, unless stated otherwise, the following holds:

- The size of the total population of units is N.
- The test population (i.e. sample size) is n.
- The number of uncensored observations (i.e. known failure times) is r.
- The observations may be ranked in increasing order and labelled with an index i.

The impact of censoring will be discussed, and how to deal with such cases.

5.1 Data Quality

The analysed data can be of various quality levels. For instance, if it is known that the ageing of an asset like a cable depends directly on the time of operation and the load, then collecting such data would be excellent for determining the distribution that describes the ageing process. However, that data may not be available.

With old cable systems that often date back to times before the use of computers, the commissioning date may be available allowing determination of the age of the cable, but not the actual time of operation. The load may not be available as a time function, but there may be an estimate or impression of the average load. If only age and impression of load exist rather than actual time of operation and actual load as a function of time, then the data is of lower quality. However, it may still serve to perform a less accurate, but nevertheless sufficiently useful, analysis. An analysis that supports decision-making.

An important lesson is that quality with AM not only concerns the statistical analysis, but also the gathering of good-quality data. Care should be taken to use: the right quantity, like time of operation instead of calendar time; proper measures of operational conditions, which may allow the application of the power law (Section 2.6.1); and clear failure definitions.

The last may seem trivial, but in practice there are many incidents where an asset is temporarily out of order but starts to function again without repair. If an engine shows this behaviour and ceases to run after several incidents, then when did this engine fail? The first time that it tripped but could be restarted? The last time after which it could not be started anymore? Another example is mass-impregnated paper lead cable that can have a failure after which the circuit trips. The oil heated by the breakdown energy can get thinner and can migrate, which may interrupt the breakdown channel. If that happens the cable may be energized again without immediate tripping. There is probably still some damage that weakens the cable locally and after some months the cable may break down permanently. What must be regarded as time of failure depends on the definition of failure: the moment of the initial interruption or the final permanent failure. The choice of definition may depend on the purpose of the analysis.

5.2 Parameter-Free Graphical Analysis

Graphs form a very powerful method to represent data. The display of data in graphs may allow the comparison of distributions and the identification of data that likely do

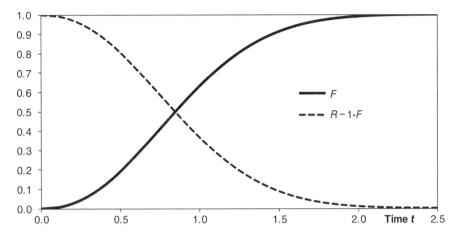

Figure 5.1 Graph of cumulative distribution *F* and the reliability function *R* against time *t*.

or do not belong in the distribution under consideration. Section 2.5 presented cases where data from different phenomena are mixed and graphs can be an excellent way to detect such situations.

5.2.1 Basic Graph of a Population Sample

A basic graph shows the distribution of observed or simulated data against a performance variable such as time (see Figure 5.1). The failure distribution *F* is along the vertical axis and the performance variable (here time *t*) is along the horizontal axis. The other functions that represent the distribution (i.e. the reliability *R*, the distribution density *f* and/or the hazard rate *h*) can of course also be drawn (see Section 2.4).

In practice, the distribution is not known or has to be checked. Instead of the known distribution a set of observations may be available, which is then regarded as a representative sample of the total set of possible observations (see Chapter 2). A graph that best fits the data is determined according to an optimization procedure, for which various methods exist.

An example of observations is shown in the top two rows of Table 5.1. These observations are values of the variable T_i. In this example the 15 ranked observations are completely known, but in other cases several observations may (yet) be unknown, because the assets did not fail as yet or may have been removed for reasons other than the failure mechanism under investigation. Such unknown observations are called 'censored' or 'suspended' (see Section 5.2.2). For now, the case is continued with all assets failed and their failure times all known.

A graph requires a data set of *n* variable values (t_i) and function values (e.g. $F(t_i)$) in order to produce the coordinates ($t_i, F(t_i)$) in a two-dimensional plot with $i = 1, \dots, n$. The observations are along the horizontal axis, but at this stage the vertical values are yet to be determined. These values are called the plotting positions which must be assigned to each observation. The expected value of $F(t_i)$ is an appropriate choice. This expected value was discussed in Section 4.1.2 and specified in Eq. (4.15):

$$< F_{i,n} > = \frac{i}{n+1} \tag{5.1}$$

Table 5.1 Example of observations t_i and expected cumulative distribution $<F_i>$ and $F(<Z_i>)$.

Index	1	2	3	4	5	6	7	8	9	10	11	12	13	14	15
Observation t_i (hr)	0.22	0.38	0.47	0.52	0.61	0.71	0.77	0.82	0.92	1.01	1.10	1.14	1.30	1.45	1.61
$<F_{i,n}>$ (%)		6.3	12.5	18.8	25.0	31.3	37.5	43.8	50.0	56.3	62.5	68.8	75.0	81.3	87.8	93.8

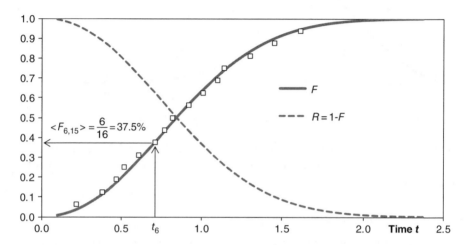

Figure 5.2 The observations of Table 5.1 plotted with the expected plotting positions $<F_{i,n}>$. The construction of the sixth point is explicitly shown (i.e. the point $(t_6, <F_{6,15}>)$). F (and R) is the distribution of the complete population from which the sample of 15 was drawn.

These values are shown in the third row of Table 5.1. The graph of the observations t_i along the horizontal axis and the expected plotting positions $<F_{i,n}>$ are shown in Figure 5.2.

Another choice for the plotting position is the median $F_{i,n,50\%}$. This position can be calculated with Eq. (4.16). An approximation was given in Eq. (4.17):

$$F_{i,n,50\%} \approx \frac{i - 0.3}{n + 0.4} \tag{5.2}$$

The merits of expected versus median values were discussed in Section 3.4.

It may be noted that for the construction of the graph, no knowledge of the type of distribution was necessary, nor were any distribution parameters used. Therefore, this representation is called parameter-free.

5.2.2 Censored Data

As mentioned in the previous sections, observations may not always form a complete set. If values are completely unknown, such data may be discarded as being not part of the sample. In fact, if the 15 observations in Table 5.1 apply to a sample of 15 tested objects taken from a larger set of, for example, 1000 objects, then the 985 that are not involved in the test are discarded in that way. However, situations also occur where data remain partly known. For instance, if the test had been ended after 1 hour, then

Table 5.2 Types of censoring.

Left censoring	Unknown value < a maximum value
Right censoring	Unknown value > a minimum value
Interval censoring	Minimum < unknown value < maximum
Type I censoring	Experiment stops after predetermined time. The number of censored data is a variable. The data of the surviving test objects are right-censored
Type II censoring	Experiment stops after a predetermined number of observations are made. The closing time of the test is a variable. The data of the surviving test objects are right-censored
Random censoring	Test objects are removed from a test without relation to the failure time. This may happen when some test objects fail due to another unrelated phenomenon. Such data are also right-censored

the breakdown time of items $i = 10$ up to 15 would have remained unknown, except that their value would be larger than 1 hour. The fact that their value exceeds 1 hour is very relevant information, even though the exact value remains unknown (i.e. censored). Data may remain unknown for various reasons (see Table 5.2).

If the situation in Table 5.1 was to be evaluated at 0.75 hour (i.e. after the sixth failure occurred), then the remaining nine failure times would remain right-censored and the censoring might also be regarded as Type I at the moment the evaluation was set. It would be Type II if the choice had been made in advance to evaluate after six failures. The situation in Table 5.3 would arise.

What implications would that have for plotting? The choice of the plotting position would remain the same as in Eq. (5.2) and is applied to $i = 1, …, 6$ since these are the uncensored observations and since it is also known that the test involves $n = 15$ objects. The result is shown in Figure 5.3.

In practice, it may be difficult to assign a certain censoring type. For instance, suppose 10 assets are put into service and 3 out of 10 fail surprisingly soon after commissioning. As a consequence, a statistical analysis may be carried out driven by the high number of failures in a short time. It is the combination of number and time that urges the analysis, which is neither Type I nor Type II.

In a grid, the situation can be even more complex. Assets may have been installed and replaced after failure. Since assets may have been commissioned at different dates, it is possible that some assets have been in service for a shorter time than already failed assets. It may even be the case that some assets which are still in service will eventually appear to have shorter failure times than already failed assets. The various ranking

Table 5.3 Observations of Table 5.1 if the results are evaluated after the sixth failure.

Index	1	2	3	4	5	6	7	8	9	10	11	12	13	14	15
Observation t_i (hr)	0.22	0.38	0.47	0.52	0.61	0.71	x	x	x	x	x	x	x	x	x
$<F_{i,n}>$ (%)		6.3	12.5	18.8	25.0	31.3	37.5	43.8	50.0	56.3	62.5	68.8	75.0	81.3	87.8	93.8

Note: The remaining test objects have right-censored failure times (> 0.71) and are therefore known to have an index $i > 6$. They are indicated with 'x'.

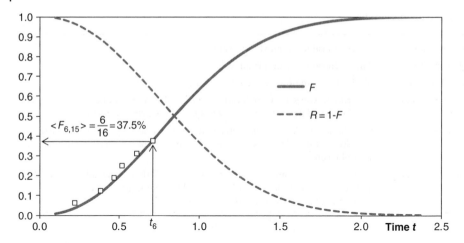

Figure 5.3 A graph of the observations in Table 5.1 if only the first six observations are known and the other remain right-censored. It is known that nine observations are to follow with times $> t_6$.

possibilities can be averaged to obtain a plot of the censored data. Example 5.1 explains how this can be done by averaging.

Example 5.1 *Plotting with Random Censored Data* A series of 10 components was commissioned at various dates. At the moment of evaluation, the situation is as seen in Table 5.4. Six components failed with known times to failure and four are still in service. The question is how to plot this.

At the moment of evaluation there are six uncensored and four censored lifetimes. For plotting it is necessary to determine what ranking positions are possible for the uncensored observations, but the ranking of the lifetimes can only be provisional at this stage, because the censored lifetimes may surpass some or even all the presently observed lifetimes. For example, the component that failed at 6.5 years may still rank as $i = 2$, 3 or 4, depending on whether it is survived or not by the components with a present operational life of 4.9 and 5.7 years. This is a matter of possible permutations. Table 5.5 shows how many ranking possibilities exist for each lifetime. Each ranking is also associated with the mean (i.e. expected) plotting position that follows from Eq. (5.1). The table also shows the mean ranking and the mean plotting position. The latter can be used to plot this set of data (see Figure 5.4).

Table 5.4 Evaluation of component lifetimes.

Operational t (yr)	3.9	4.9	5.7	6.5	7.8	8.2	9.0	9.4	10.2	12.2
Failed?	Yes	No	No	Yes	Yes	No	No	Yes	Yes	Yes
Preliminary index i	1	—	—	2	3	—	—	4	5	6

Note: Part of the components failed and part of the lifetimes are each right-censored. The total data set is partly randomly censored due to the different service times. In this case the first known failure will always be ranked as the smallest.

Table 5.5 For each uncensored (i.e. observed) lifetime the number of possible rankings is shown.

		Number of ranking possibilities for the six observations					
Time (yr)		3.9	6.5	7.8	9.4	10.2	12.2
Ranking	$<F_{i,n}>$						
1	0.091	1440					
2	0.182		1120				
3	0.273		280	840			
4	0.364		40	480	360		
5	0.455			120	480	120	
6	0.545				360	288	24
7	0.636				180	396	96
8	0.727				60	384	228
9	0.818					252	420
10	0.909						672
Number of permutations		1440	1440	1440	1440	1440	1440
Mean ranking I		1	2.25	3.5	5.375	7.25	9.125
Mean plotting position		0.091	0.205	0.318	0.489	0.659	0.830

Note: With each ranking the plotting position according to Eq. (5.1) is shown. From the number of permutations follows the average ranking position and the average plotting position.

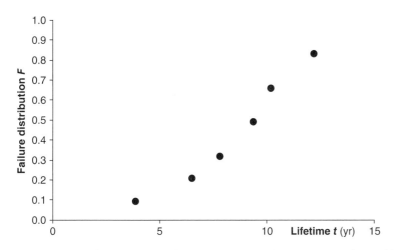

Figure 5.4 Plot of censored data of Table 5.4 using the plotting positions from Table 5.5.

The number of permutations is calculated by multiplying the possible rankings that the censored data may achieve in the end. The right-censored lifetime of 9 years may obtain four positions, namely before 9.4 years, between 9.4 and 10.2 years, between 10.2 and 12.2 years and finally after 12.2 years. Now, having a set of four lifetimes in the end, the right-censored lifetime of 8.2 years can end up in five positions, namely before, between or after this set. Now, having in total seven lifetimes after the right-censored lifetime of 5.7 years, that component may rank eight positions. Likewise, the right-censored lifetime of 4.9 years can achieve nine positions. The total number of permutations is therefore $9 \times 8 \times 5 \times 4$ (i.e. 1440). Although Table 5.5 shows the results for all 1440 permutations, it is not necessary to consider all censored data to find the average ranking position of, for example, 6.5 years. Only the permutations of the censored data preceding 6.5 years appear relevant.

The mean plotting positions are subsequently calculated by averaging $<F_{i,n}>$ in the second column weighed by the number of permutations. So, the mean plotting position for observation $t = 6.5$ years is calculated as $(1120 \cdot 0.182 + 280 \cdot 0.273 + 40 \cdot 0.364)/1440$, which yields 2.25 years. Figure 5.4 shows the points $t(i)$, $<F_{I(i),n}>$.

The procedure in Example 5.1 can be replaced by the adjusted ranking method [24]. This method calculates an adjusted index $I(i,n)$ based on i, n and the sum of the number of censored and uncensored data up to i. The equation to determine the adjusted rank $I(i,n)$ reads:

$$I(i) = I(i-1) + \frac{n+1-I(i-1)}{n+2-C_i} \tag{5.3}$$

C_i is the sum of the number of censored and uncensored observations, including the failure i. $I(0)$ is zero by definition. A censored observation before i means that it is right-censored. Below, the adjusted rank is calculated for $i = 1$ and $i = 2$:

$$I(1) = I(0) + \frac{n+1-I(0)}{n+2-C_1} = 0 + \frac{10+1-0}{10+2-1} = 1 \tag{5.4}$$

$$I(2) = I(1) + \frac{n+1-I(1)}{n+2-C_2} = 1 + \frac{10+1-1}{10+2-4} = \frac{18}{8} = 2.25 \tag{5.5}$$

Table 5.6 shows the results for all the data of Table 5.4. The results are the same as the averaging in Example 5.1.

The mean plotting position is calculated using the mean ranking $I(i)$ in Eq. (5.1):

$$< F_{I,n} > = \frac{I(i)}{n+1} \tag{5.6}$$

If the ageing is continued, all components will fail. As an example, for the group of components of Table 5.4 the results in Table 5.7 may be obtained.

For comparison, the censored data set with adjusted ranking (Table 5.6) and the uncensored data set with final ranking (Table 5.7) are plotted in Figure 5.5. Some

Table 5.6 Determining the adjusted rank and plotting position through the adjusted rank method using Eq. (5.3).

Operational t (yr)	3.9	4.9	5.7	6.5	7.8	8.2	9.0	9.4	10.2	12.2
Failed?	Yes	No	No	Yes	Yes	No	No	Yes	Yes	Yes
Preliminary index i	1	—	—	2	3	—	—	4	5	6
Sum C_i	1			4	5			8	9	10
Adjusted rank $I(i)$	1	—	—	2.25	3.5	—	—	5.38	7.25	9.13
$<F_{I,n}>$	0.091			0.205	0.318			0.489	0.659	0.830

Table 5.7 A possible outcome after all components failed. The adjusted rank is not needed.

Operational t (yr)	3.9	5.5	6.5	7.8	8.8	9.4	10.2	10.9	12.2	14.1
Failed?	Yes	Yes	Yes	Yes	Yes	Yes	Yes	Yes	Yes	Yes
Index i	1	2	3	4	5	6	7	8	9	10
$<F_{i,n}>$	0.091	0.182	0.273	0.364	0.454	0.545	0.636	0.727	0.818	0.909

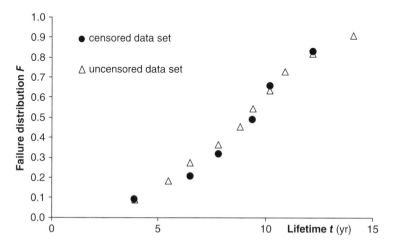

Figure 5.5 Combined plot of censored data at an earlier evaluation moment and the uncensored data after complete exhaust of the component group. Although the censored data set lacks information, the plots bear a reasonable resemblance.

change in the ranking occurred. For instance, with the censored data set the longest time to failure was 12.2 years, but after all components failed the longest time appears to be 14.1 years.

5.2.3 Kaplan–Meier Plot

Another approach to a parameter-free plot is the plot of the so-called Kaplan–Meier estimator [25], which is an estimator of the surviving fraction (i.e. reflecting the reliability function; see Section 2.4.2). The surviving fraction is stepwise reduced at every moment of failure or period that includes one or more failures. The Kaplan–Meier approach explicitly takes into account how many components are under test at the moment a failure occurs and explicitly involves (right)-censored data in the analysis.

Different types of inspection can be used. The failure data as well as right-censored data are used to determine the Kaplan–Meier estimator. However, inspections at discrete times that indicate the number of failed and interval-censored data are also used. An example of the first type is a test environment where times to failure and censoring (removal from the test) are recorded for each individual test object. This has been applied in the medical sector to patients who may pass away or leave the hospital alive. An example of the second is an inventory where the response of patients to treatment is inventoried daily. Example 5.2 illustrates a case of the first kind.

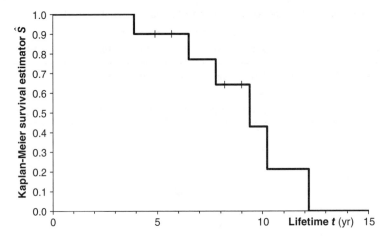

Figure 5.6 Kaplan–Meier plot of failure and censored data in Tables 5.4 and 5.8. The censored data are indicated by a '+' mark, but do not affect the value of the Kaplan–Meier estimator $\hat{S}$. At every failure event the Kaplan–Meier estimator takes a step that is calculated with Eq. (5.7).

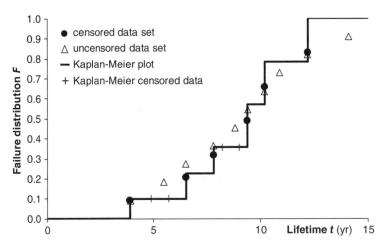

Figure 5.7 The plotting methods of adjusted ranking and Kaplan–Meier and the uncensored data set in one graph.

The Kaplan–Meier estimator is plotted against time usually as in Figures 5.6 and 5.7 (with data from Tables 5.4 and 5.8). This estimator basically divides the number of surviving objects by the number of objects at risk. Objects that failed or are removed for any other reason are excluded from the number of objects at risk. There are times related to censoring and times related to failure. Only the times to failure will be counted for the estimator. The Kaplan–Meier estimator $\hat{S}$ at moment t is defined as:

$$\hat{S}(t) = \prod_{t_i < t} \frac{n_i - d_i}{n_i} = \prod_{t_i < t} \left(1 - \frac{d_i}{n_i}\right) \tag{5.7}$$

here, n_i is the number of objects at risk prior to the failure at t_i; d_i is the number of failing objects at t_i. The interpretation of the estimator is that at each t_i the survival

Table 5.8 Data of Table 5.4 prepared for the Kaplan–Meier plot.

Operational t (yr)	3.9	4.9	5.7	6.5	7.8	8.2	9.0	9.4	10.2	12.2
Failed?	Yes	No	No	Yes	Yes	No	No	Yes	Yes	Yes
Index i	1	—	—	2	3	—	—	4	5	6
Failure time t_i	t_1	—	—	t_2	t_3	—	—	t_4	t_5	t_6
Number n_i prior to failure	10	—	—	7	6	—	—	3	2	1
Failing number d_i	1	—	—	1	1	—	—	1	1	1
$(n_i - d_i)/n_i$	9/10	—	—	6/7	5/6	—	—	2/3	1/2	0
$\hat{S}(t_i)$	0.900	—	—	0.771	0.643	—	—	0.429	0.214	0
$F(t_i) = 1 - \hat{S}(t_i)$	0.100	—	—	0.229	0.357	—	—	0.571	0.786	1

fraction or probability is calculated by multiplying the probabilities of surviving all periods up to t.

If it can be assumed that individual actual failure times are each recorded and two failure events always differ in time (however small the interval), then d_i is always equal to 1. If the number of combined failed and withdrawn test objects at t_i is denoted as j_i (note: including the breakdown at t_i), then the estimator can be written as:

$$\hat{S}(t) = \prod_{t_i < t} \frac{n - j_i}{n - j_i + 1} = \prod_{t_i < t} \left(1 - \frac{1}{n - j_i + 1}\right) \tag{5.8}$$

The procedure for the Kaplan–Meier plot is:

- *Step 1.* Rank all data in increasing order (failure and censored data alike).
- *Step 2.* Index the actual failure times only.
- *Step 3.* Assign to each failure time the number n_i of objects at risk just prior to the failure event.
- *Step 4.* Assign to each failure time the number d_i of objects at risk that fail during the event. If the breakdown time of each object is separately recorded, then generally d_i equals 1, but if the data are collected per period of time (e.g. day or year), then d_i can be larger than 1.
- *Step 5.* Calculate the Kaplan–Meier estimator $\hat{S}$ according to Eq. (5.7) or (5.8).
- *Step 6.* Plot $\hat{S}$ against time t as in Figure 5.6.
- *Step 7.* Censored data can be indicated '+' markers.
- *Step 8.* The failure distribution F can be found by $F = 1 - R$ or $F = 1 - \hat{S}$ as in Figure 5.7.

Example 5.2   *Data of Table 5.4 in Kaplan–Meier Approach*   Assume the history of a series of components is known, as represented in Table 5.4. The Kaplan–Meier method is then applied as in Table 5.8.

For comparison, the cumulative failure distribution F is plotted in Figure 5.7 according to the adjusted rank method and the Kaplan–Meier method. The Kaplan–Meier method particularly deviates at the extremes of the population, which is partly related to the fact that $\hat{S}$ remains 1 (and so F remains 0) until the first failure and $\hat{S}$ runs to 0 (and so F remains 1) with the last failure, while the expected plotting position never starts at 0 or ends at 1 (except in case of the limit $n \to \infty$). Various adjustments to the

Kaplan–Meier estimator have been proposed, which bring the two methods more in line. The Kaplan–Meier estimator as an estimator for reliability tends to be too optimistic at the start and too pessimistic at the end.

The Kaplan–Meier approach generally leads to a finite time interval where the cumulative distribution F varies from 0 to 1 and consequently the distribution density f is non-zero. Modified Kaplan–Meier estimators are also proposed, for instance based on median ranking (e.g. [26]).

An advantage of the Kaplan–Meier approach is the relative simplicity of handling censored data compared to the adjusted ranking method in Section 5.2.2. A disadvantage is the appearance of a step function and the finite time interval of the non-zero distribution density function (unless modified).

5.2.4 Confidence Intervals Around a Known Distribution

If the failure distribution $F(t)$ as a function of t is known (or assumed) and a limited random selection of n assets is taken in the operation, then these are regarded as a limited random sample of n assets taken from the total population of assets. The ranked failure time of each asset corresponds to a value of F of the total population. Since the n assets are randomly selected, it is unknown to which exact value of $F_{i,n}$ their failure times t_i correspond ($i = 1,...,n$) and in fact for each i and n there is a distribution describing what these values $F_{i,n}$ can be. This is the beta distribution or rank distribution, as discussed in Section 4.1.2. The average or expected plotting position $<F_{i,n}>$ follows from Section 5.2.1, Eq. (5.1):

$$< F_{i,n} > = \frac{i}{n+1} \tag{5.9}$$

The median plotting position $F_{i,n,50\%}$ is the F below which 50% of the $F_{i,n}$ lie with given i and n. A good approximation of the median $F_{i,n,50\%}$ is given in Eq. (5.2):

$$F_{i,n,50\%} \approx \frac{i - 0.3}{n + 0.4} \tag{5.10}$$

The exact median of a uniformly distributed variable x (here the variable is F) follows from the beta distribution (see Section 4.1.2.1) and in particular Eq. (4.8):

$$x_{50\%} = B^{inv}\left(\frac{1}{2}; \alpha, \beta\right) \tag{5.11}$$

Applied to $F_{i,n}$, the expression is the inverse of Eq. (4.16):

$$B(F_{i,n,50\%}; i, n + 1 - i) = \frac{\Gamma(n+1)}{\Gamma(i) \cdot \Gamma(n+1-i)} \int_0^{F_{i,n,50\%}} p^{i-1} \cdot (1-p)^{n-i} dp = \frac{1}{2} \tag{5.12}$$

which becomes:

$$F_{i,n,50\%} = B^{inv}\left(\frac{1}{2}; i, n + 1 - i\right) \tag{5.13}$$

Likewise, not only for ½ (i.e. 50%) but also for any percentage $A\%$ the $F_{i,n,A\%}$ can be found as:

$$F_{i,n,A\%} = B^{inv}(A\%; i, n + 1 - i) \tag{5.14}$$

This function can also be found in popular applications such as spreadsheets. The method provides a possibility to not only calculate the expected $<F_{i,n}>$, but also indicate an interval $[A\%, 1 - A\%]$ in which $F_{i,n}$ with $1 - 2A\%$ confidence will fall. For instance, the observation of the batch of n samples that ranks $i = 3$ will on average correspond to $<F_{i,n}> = 3/(n + 1)$ of the total population. Based on Eq. (5.14) it can be calculated that in 95% of all cases a sample with $i = 3$ will correspond to $F_3 > F_{3,n,5\%} = B^{inv}(5\%; 3, n + 1 - 3)$ and also $F < F_{3,n,95\%} = B^{inv}(95\%; 3, n + 1 - 3)$. This means that in 90% of all cases $F_{3,n,5\%} = B^{inv}(5\%; 3, n + 1 - 3) < F_3 < F_{3,n,95\%} = B^{inv}(95\%; 3, n + 1 - 3)$. The 5% and 95% confidence limits together form the 90% confidence interval $[F_{3,n,5\%}, F_{3,n,95\%}]$. This is further explained in Example 5.3.

Example 5.3 *Confidence Intervals of a Limited Sample* As an example of confidence intervals, a sample of $n = 10$ assets is studied. In Table 5.7 a possible result of a test on a set of 10 components was shown, but in the present example the focus is not on plotting a single set of outcomes. The focus is rather on the intervals surrounding the underlying (i.e. true) distribution F of the total population from which limited sets of $n = 10$ – like in Table 5.7 – are drawn randomly. The confidence intervals are shown in Table 5.9.

In the present case Table 5.9 shows, for instance, that there is a probability of:

- 1% that the component with $F_{1,10}$ belongs to the subpopulation with $F \leq 0.001$.
- 99% that the component with $F_{1,10}$ belongs to the subpopulation with $F \leq 0.369$.
- 98% that the component with $F_{1,10}$ belongs to the subpopulation with $0.001 \leq F \leq 0.369$.
- 90% that the component with $F_{4,10}$ belongs to the subpopulation with $0.150 \leq F \leq 0.607$.

Figure 5.8 shows the $F(t)$ curve of the total population. On the curve the expected $<F_{i,n}>$ for the case $n = 10$ are marked. The 90% confidence intervals of each $F_{i,n}$ are

Table 5.9 The expected and median $F_{i,n}$ are calculated with Eqs (5.9) and (5.13), respectively.

Ranking i	Expected $F_{i,n}$: $<F_{i,n}>$	Median $F_{i,n}$: $F_{i,n,50\%}$	98% (i.e. [1%,99%]) confidence interval $F_{i,n,1\%}$	$F_{i,n,99\%}$	90% (i.e. [5%,95%]) confidence interval $F_{i,n,5\%}$	$F_{i,n,95\%}$
1	0.091	0.067	0.001	0.369	0.005	0.259
2	0.182	0.162	0.016	0.504	0.037	0.394
3	0.273	0.259	0.048	0.612	0.087	0.507
4	0.364	0.355	0.093	0.703	0.150	0.607
5	0.455	0.452	0.150	0.782	0.222	0.696
6	0.544	0.548	0.218	0.850	0.304	0.778
7	0.636	0.645	0.297	0.907	0.393	0.850
8	0.727	0.741	0.388	0.952	0.493	0.913
9	0.818	0.838	0.496	0.984	0.606	0.963
10	0.909	0.933	0.631	0.999	0.741	0.995

Note: Next the confidence intervals are calculated using Eq. (5.14); $n = 10$ here.

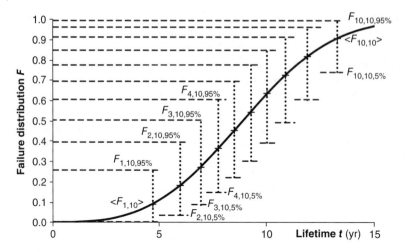

Figure 5.8 Graph of $F(t)$ of the total population, with marked expected plotting positions $<F_{i,n}>$ together with the 90% confidence interval for each $F_{i,n}$ ($n = 10$).

shown around these expected values $<F_{i,n}>$ as vertical dotted lines. Note that the same confidence intervals can be drawn around the median $F_{i,n,50\%}$. The interval boundaries are shown as horizontal lines and the confidence intervals could be considered horizontal regions.

Under the assumption that a certain distribution $F(t)$ (defined by graph or specified parameters) applies, the confidence intervals for t can be derived as well. Since $F(t)$ is known (or at least assumed to be true), the corresponding time follows from the inverse distribution, including the boundaries of the confidence intervals. Therefore, the confidence interval of the failure times can be constructed as time regions.

This can be done graphically by following a dotted horizontal confidence boundary $F_{i,n,A\%}$ until it crosses the curve (see Figure 5.9 for the case $i = 4, n = 10$). Then, drop a line from that point to the t-axis. This vertical line is the corresponding confidence boundary $t_{i,n,A\%}$. For example, the 90% confidence interval around $<F_{4,10}>$ can be converted into a 90% confidence interval around the corresponding time $t_{4,10} = t(<F_{4,10}>)$. Remember that $<F_{4,10}>$ follows from Eq. (5.9) as 4/11.

If the cumulative distribution is expressed in an equation, then the time interval can be calculated (but usually parameters are involved, so then the method is no longer parameter-free, strictly speaking):

$$t_{i,n,A\%} = F^{inv}(F_{i,n,A\%}(t)) \tag{5.15}$$

At this stage it is possible to construct confidence intervals in terms of both cumulative distribution and failure time. The cumulative distribution is assumed to be known in this approach. The next section will discuss the situation with censored data.

Often, the confidence intervals are shown as continuous lines rather than the horizontal bars in Figure 5.10. In Section 5.2.2 the adjusted rank method yielded rankings i that were no longer integers, but real numbers. This provides a means to draw continuous confidence intervals. The line that represents the $A\%$ confidence boundary consists of

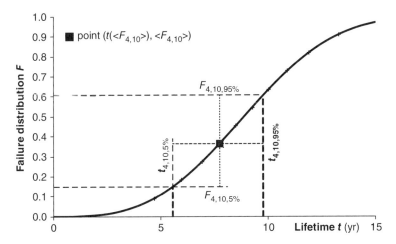

Figure 5.9 Construction of the t confidence interval $[t_{4,10,5\%}, t_{4,10,95\%}]$ around $t(<F_{4,10}>)$ from the F confidence interval $[F_{4,10,5\%}, F_{4,10,95\%}]$ around $<F_{4,10}>$. The known or assumed distribution $F(t)$ translates $F_{4,10,5\%}$ into $t_{4,10,5\%}$ and translates $F_{4,10,95\%}$ into $t_{4,10,95\%}$ by taking the inverse F^{inv}.

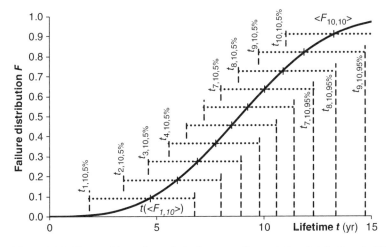

Figure 5.10 The confidence boundaries in terms of cumulative distribution F of Figure 5.8 are used to construct the confidence boundaries in terms of failure time t under the assumption that $F(t)$ (i.e. the solid curve) applies. $t_{10,10,95\%}$ is 17.1 years, which is beyond the scale maximum.

the points $(t_{i,n,A\%}, <F_{i,n}>)$, where $0 < i \leq n$. Again, $<F_{i,n}>$ follows from Eq. (5.9) and $t_{i,n,A\%}$ is calculated through Eqs (5.15) and (5.14). The result is shown in Figure 5.11.

It should be noted that $F(t)$ (so also $F_{i,n,A\%}(t)$), if exactly known, does not depend on the plotting position, but the graphical representations of the confidence intervals do. With the expected plotting position the points of the $A\%$ boundary are $(t_{i,n,A\%}, <F_{i,n}>)$ whereas with the median plotting position they become $(t_{i,n,A\%}, F_{i,n,50\%})$. It is therefore important to know the type of plotting position if inferences are made.

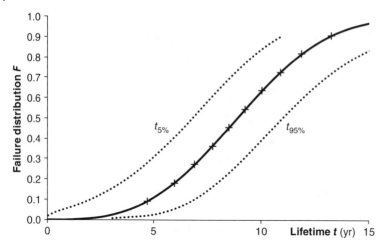

Figure 5.11 The 90% *t* confidence intervals represented with dotted curves around the *F(t)* curve. The points consisting of the expected *F* values $<F_{i,n}>$ and corresponding $t(<F_{i,n}>)$ are indicated with a '+' mark. The fact that the confidence intervals are horizontal is lost in this representation.

5.2.5   Confidence Intervals with Data

If observations have been made, a complete or censored set of data is available. Sections 5.2.1 and 5.2.2 showed how the data can be represented in a graph. As a next step, these data can be compared with a reference function $F(t)$. This function may be the true function (if known) to which the observations should belong, but it could also be a best fit to the data. The confidence intervals mark a region where the data can be expected to fall with a certain probability.

The combination of the observations and the reference function with its confidence intervals requires a choice of the plotting position. Here the expected plotting position is demonstrated, but the procedure is basically the same for the median plotting position.

As an example, first the complete data set of Table 5.7 is elaborated (see Table 5.10 and Figure 5.12) and next the censored set of Table 5.6 is demonstrated (see Table 5.11 and Figure 5.13). Both sets have the underlying function that was used in Section 5.2.4. The expected plotting position (Eq. (5.9)) was used, but the approach for the median plotting position (Eq. (5.10)) would follow a very similar approach.

Table 5.10 The data of the completed test with 5% and 95% boundaries of time and 90% confidence intervals.

Operational t (yr)	3.9	5.5	6.5	7.8	8.8	9.4	10.2	10.9	12.2	14.1
Failed?	Yes	Yes	Yes	Yes	Yes	Yes	Yes	Yes	Yes	Yes
Index i	1	2	3	4	5	6	7	8	9	10
$<F_{i,10}>$	0.091	0.182	0.273	0.364	0.454	0.545	0.636	0.727	0.818	0.909
$t_{i,10,5\%}$	1.83	3.47	4.62	5.57	6.41	7.20	8.00	8.83	9.77	11.02
$t_{i,10,95\%}$	6.78	8.00	8.94	9.78	10.58	11.40	12.29	13.33	14.70	17.10

Note: The data are represented in Figure 5.12.

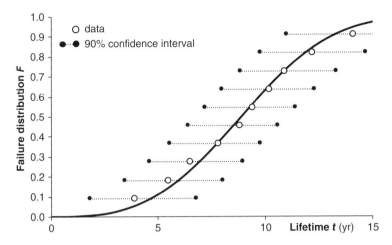

Figure 5.12 Data of Table 5.10 combined with the underlying function $F(t)$ and the confidence intervals. The scatter of the data is small compared to the width of the confidence intervals. The data are plotted at the expected plotting positions $<F_{i,10}>$.

Table 5.11 The set of Table 5.6 contains censored data.

Operational t (yr)	3.9	4.9	5.7	6.5	7.8	8.2	9.0	9.4	10.2	12.2
Failed?	Yes	No	No	Yes	Yes	No	No	Yes	Yes	Yes
Preliminary index i	1	—	—	2	3	—	—	4	5	6
Sum C_i	1			4	5			8	9	10
Adjusted rank $I(i)$	1	—	—	2.25	3.5	—	—	5.38	7.25	9.13
$<F_{I,n}>$	0.091			0.205	0.318			0.489	0.659	0.830
$t_{i,10,5\%}$	1.83			3.78	5.11			6.71	8.20	9.90
$t_{i,10,95\%}$	6.78			8.25	9.37			10.89	12.54	14.92

Note: The adjusted plotting positions are determined with the procedures in Section 5.2.2. The data are represented in Figure 5.13.

Here the representation with horizontal bars representing the t confidence intervals (see Figure 5.10) is used in order to explicitly show the meaning of the interval, but of course the continuous curves for the confidence boundaries could be drawn as well (see Figure 5.11).

What would happen if the set contained censored data? The approach remains the same, but the censored set must be plotted at the plotting position of choice. A censored set was shown in Table 5.6. The adjusted ranking can be determined following the procedure in Section 5.2.2. The confidence boundaries (here 5% and 95%) follow again from Eqs (5.14) and (5.15). If F^{inv} is not known as an expression, then the confidence boundaries can be constructed from the $F_{I,n,A\%}$ around $<F_{I,n}>$ as in Figure 5.9. Note that instead of i the index is I, indicating the use of an adjusted ranking. Also note that although only six observations are uncensored, the total number n remains 10 (i.e. the total of censored and uncensored data).

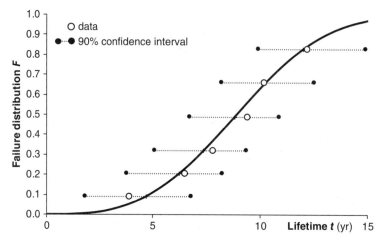

Figure 5.13 Data of Table 5.11 combined with the underlying function $F(t)$ and the confidence intervals. Considering the width of the confidence intervals, the scatter of the data could have been stronger. The data are plotted at the expected plotting positions $<F_{i,10}>$.

5.2.6 Alternative Confidence Intervals

The background of the confidence intervals discussed in Sections 5.2.4 and 5.2.5 is based on the assumption that the scatter in the variable t is due to random selection. There is a (real or virtual) large distribution F ranking all possible outcomes, from which a limited sample is drawn with sample size n. Each outcome is assumed a random selection. The ith ranking time in the sample n has F value expected to be $<F_{i,n}>$ with a bandwidth that can be calculated by the beta distribution.

Is it possible that the model does not apply or that other effects may be stronger? Yes, it is. The assumption that the sample is drawn from a large population may not be true. The random selection may not be true either. Further, F may not be the distribution of a homogeneous population, but rather a combined distribution or an inhomogeneous distribution (cf. Section 2.5.2.2). Finally, the beta confidence boundaries ignore lower ranking observations. These four remarks are discussed here briefly.

As for the sample being drawn from a large population, the other extreme is that the population only consists of the sample itself. Since there is just one possible choice for the sample, the corresponding confidence interval is tightly surrounding the data points and much smaller than the beta distribution. The regression-based confidence limits about the best fit (see Section 10.3) remain, but are also much narrower than those based on the beta distribution usually (see also Sections 10.4, 10.4.1 and 10.4.3). The greatest reason why regression-based confidence intervals are narrower is that the regression error is not determined against the true (as yet unknown) distribution, but against the fit. The flexibility of the best fit by adjusting the distribution parameters largely reduces the scatter. This is illustrated in Section 10.4.3.

As for random selection, certain parts of the population may be eradicated before operational life. This happens with quality control by destructive testing to take out child mortality. As shown in Section 10.4.3.1 and particularly in Figure 10.19, stress tests reduce certain parts of the population more, making certain samples less likely to occur in practice (which may be a good thing). That also changes the statistics of parameter estimators and confidence intervals. Another cause of non-random selection is a likely dependency of the failure distribution on production parameters like extruder setting, material properties, and so on. If a batch of assets (e.g. a length of cable reeled on a number of drums) is produced on one day, the production parameters may be more constant than if the batch is produced over many days. Breakdown data of cable produced on one day probably correlate more than that of cables produced on various days reducing the likelihood of certain combinations.

As for F being an inhomogeneous distribution, this happens with varying production parameters which is more likely with a longer production run, multiple runs or incidents with temporary changes (production flaws). Within the population different subpopulations may occur (which tends to broaden intervals rather than narrowing them). A batch that is produced in one day after tuning the production line may have a fairly constant set of parameters. Data combinations mixing the breakdown values of different days may simply not occur if batches are delivered on order. The beta distribution may therefore not apply. With the cable short-term homogeneous production process a higher correlation of breakdown data is quite likely. This can be different if, for instance, products like switching gear are manufactured piece by piece (i.e. each batch would consist of one product). If the variations in the production process do vary randomly and faster than the completion of a batch, then the products may be regarded as independent and the rank distribution might apply.

Finally, when a graph with observations is drawn, the data are ranked, implying that the observed variable t is increasing. Since the data are ranked the next failure comes later than the previous. With this condition the confidence intervals are limited on the lower side. However, the beta distribution just indicates where an ith ranked observation might fall in a similar experiment and does not take into account the previous actual observation. This is addressed in Section 9.4.1.3 and 9.4.2.2.

In summary, there are reasons why a sample may not be fully random and why the confidence intervals might have to be adjusted from the beta-based confidence limits. Some of these reasons are: products produced in the same batch may have a strong correlation; products may undergo testing that selectively reduces parts of the population. However, the main reasons why regression-based confidence intervals are narrow is that part of the scatter is absorbed in the deviation of the best-fit distribution parameters from the true distribution parameters, and secondly, that the beta distribution does not take the ranked observations into account. These two types of confidence interval have different meanings.

- Beta distribution-based confidence intervals are a measure of the scatter of data about the true distribution.
- Regression-based confidence intervals are a measure of the accuracy of the best fit given a sample drawn from the full population. The next sample may, however, barely overlap in the confidence interval. But both samples would probably largely fall in the beta distribution-based confidence interval as they should, being samples from the same population.

Background information can be found in Section 10.4 and following, while an extensive example of using both methods can be found in Section 9.4.1.2.

5.3 Model-Based or Parametric Graphs

In the previous section graphs were discussed that represent the cumulative distribution F or the reliability function R versus time t. Such graphs do not require prior knowledge of the statistical distribution of the data. A disadvantage of such graphs is that curve fitting, identification of outliers, extrapolation and interpolation can be quite inaccurate.

For instance, Figure 5.14 shows the data of Figure 5.3 together with the (in this case known) underlying distribution F up to the sixth observation.

If there is no knowledge of the failure data to come and no knowledge of the underlying distribution, then it would be barely possible to correctly draw the dotted part of the graph by visual evaluation.

This can be overcome by manipulating the axis and processing the data such that the graph yields a straight line. In order to achieve such a linearized graph, some sort of linear relationship is required between the processed F and the processed t:

$$Y(F) = y_0 + y' \cdot X(t) \tag{5.16}$$

where y_0 is the Y intercept (i.e. $Y(X(t) = 0)$) and y' is the slope (i.e. derivative dY/dX). A linear graph is aimed for that easily enables extrapolation and interpolation. It may be noted that the linear relationship can also be stated as:

$$X(t) = x_0 + x' \cdot Y(F) \tag{5.17}$$

It is customary to construct graphs where the variable is along the horizontal axis and the covariable along the vertical axis. This means that to each variable value, a covariable is assigned. However, failure distribution graphs can be viewed differently: the plotting positions are the chosen variable values and the observed times are the covariables assigned to the chosen plotting position. In that approach the plotting positions are without error, but the time observations do feature scatter. Compared to other graphs, it would make sense to swap the axes, but this is very uncommon.

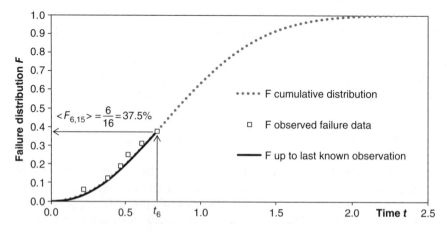

Figure 5.14 The data up to the sixth observed failure time (a.u.). The underlying distribution F is drawn with a solid line up to the last observation and beyond with a dotted line. If this dotted line is as yet unrevealed, the extrapolation of the line from the sixth point up is barely possible by visual evaluation. Actually, a best fit might not produce the solid line either.

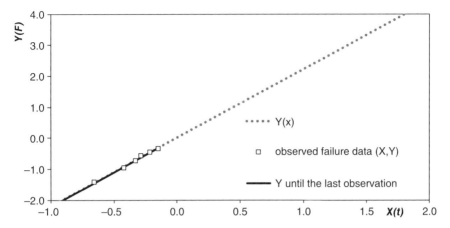

Figure 5.15 The data up to the sixth observation. Here F and t have been processed into Y and X, which have a linear relationship. The axes (a.u.) here are in terms of Y and X but may be replaced by (usually non-linear) scales in terms of F and t, as explained in the following sections.

In another approach it is recognized that both Y and X feature scatter. The background is that the confidence intervals are essentially due to not knowing the exact ranking of the object with respect to the grant population from which the tested objects are just a sample. In Figure 5.8 the plotting positions are $<F_{i,n}>$ for which $i/(n+1)$ is taken, but that is an approach. The confidence interval in that graph indicates in what range the true ranking lies. The observed times can then be regarded to scatter about the best fit due to inaccurate time measurements. This inaccuracy scatter is another type of scatter than the one related to the confidence interval due to random sampling. This subject is revisited in Section 6.3.

Whatever the choice of variable and covariable, the challenge is to find a linear relationship. For various distributions discussed in Chapter 4, an analytical expression exists either theoretically or in good approximation. If such a relationship is reformulated in the shape of Eq. (5.16) then a graph with a straight line can be obtained (Figure 5.15).

If a series of $t_1, \ldots, t_n$ is measured, this results in a series of $x(t_1), \ldots, x(t_n)$ or for short $x_1, \ldots, x_n$. As in Section 5.2.1, a choice must be made for the corresponding so-called plotting positions $y_1, \ldots, y_n$. Again, the main choices are the expected values $<y_{i,n}>$ and the median $y_{i,n,50\%}$. The expected values are taken when x_i is regarded as an estimate of $<x_i>$ and the median is taken if x_i is interpreted as an estimate of the median $x_{i,n,50\%}$.

The expected values $<y_{i,n}>$ have an advantage when the plot is also used in combination with distribution parameter estimation using linear regression analysis (see Section 6.3). Having consistency with plotting and parameter estimation is a strong argument to use the expected plotting position.

However, the median value $y_{i,n,50\%}$ is also used quite often, which follows conveniently once $F_{i,n,50\%}$ and the function $Y(F)$ are known. The expected plotting position $<y_{i,n}>$ does not follow from $<F_{i,n}>$ generally. This is because the mean is not a conserved property as the function $Y(F)$ is usually not a linear operation. Hence, the $<y_{i,n}>$ values will have to be determined otherwise (as will be shown in the next chapters).

The analysis of a graph can be used to evaluate:

- Whether the data do support the assumption of the underlying distribution.

- Whether the data may belong to mixed distributions (see Section 2.5).
- Whether some data should be omitted as outliers (i.e. data that seem to belong to a separate group).
- The estimates of parameters.

Furthermore, linear graphs enable extrapolation and interpolation conveniently.

If the type of distribution is known or assumed (e.g. Weibull-2), then such distributions generally have parameters (such as the scale and the shape parameter). Working based on the selected underlying distribution is called a parametric approach.

The following discusses various parametric plots together with their confidence intervals.

5.4 Weibull Plot

The Weibull distribution was introduced in Section 4.2.1. The cumulative distribution F is defined in Eq. (4.20) as:

$$F_{wb}(x; \alpha, \beta) = 1 - e^{-\left(\frac{x}{\alpha}\right)^{\beta}} \tag{5.18}$$

A relationship in the shape of Eq. (5.16) using t as variable is obtained by:

$$F_{wb} = 1 - e^{-(t/\alpha)^{\beta}} \quad \Leftrightarrow \quad \left(\frac{t}{\alpha}\right)^{\beta} = -\ln(1 - F_{wb})$$

$$\Leftrightarrow \quad \beta \cdot \log t = \beta \cdot \log \alpha + \log(-\ln(1 - F_{wb}))$$

$$\Leftrightarrow \quad \log t = \log \alpha + \frac{1}{\beta} \cdot \log(-\ln(1 - F_{wb})) = \log \alpha + \frac{1}{\beta} \cdot Z \tag{5.19}$$

Because of the extensive expression, a variable Z is introduced here:

$$Z = \log(-\ln(1 - F_{wb})) \tag{5.20}$$

Therefore, the linear relationship is now found as:

$$Z(F_{wb}) = \log(-\ln(1 - F_{wb})) = \beta[\log t - \log \alpha] \tag{5.21}$$

The Weibull plot shows Z_i along the vertical axis, though the Z marks are usually replaced by the corresponding $F(Z)$ values to be calculated through the inverse of Eq. (5.20).

Parameter estimation is subject of Chapter 6, but also follows from the graph. For $t = \alpha$ the value of β does not matter, because F will always be $1 - e^{-1}$ (i.e. about 0.63):

$$F(t = \alpha) = 1 - e^{-\left(\frac{t}{\alpha}\right)^{\beta}} = 1 - e^{-1} \approx 63\% \tag{5.22}$$

Consequently, $F = 0.63$ can be used to identify the scale parameter α and is often shown by a dotted line (cf. Appendix A). The shape parameter β is found as the slope of the graph. It can be determined from a given plot by comparing the point $(t = \alpha, Z = 0)$ with an arbitrary other point $(t, Z(t))$ or if the vertical scale is expressed in terms of F: $(t = \alpha, F = 1 - e^{-1})$ with $(t, F(Z(t)))$ and subsequently determining β as:

$$\beta = \frac{Z(t)}{\log t - \log \alpha} = \frac{\log[-\ln(1 - F(t))]}{\log t - \log \alpha} \tag{5.23}$$

5.4.1 Weibull Plot with Expected Plotting Position

As mentioned above, the expected plotting position is preferred if graphic representation is combined with parameter estimation (which is the subject of Chapter 6) using linear regression. The resulting consistency between plotting and parameter estimation is very recommendable.

The expected values of the plotting position $<Z_{i,n}>$ can be calculated analytically as [27]:

$$< Z_{i,n} >= \left[-\gamma + i \binom{n}{i} \sum_{j=0}^{i-1} \binom{i-1}{j} \frac{(-1)^{i-j} \cdot \ln(n-j)}{n-j} \right] \cdot \log e \tag{5.24}$$

In this expression γ is the Euler constant: $\gamma \approx 0.577215665$. From Eq. (5.24) the cumulative distribution can be calculated exactly through the inverse of Eq. (5.20):

$$F(< Z_i >) = 1 - \exp(-10^{<Z_i>}) \tag{5.25}$$

These expressions require sufficient precision due to differences of large numbers with larger i and n, but with the increasing calculation force of computers this problem is decreasing. At the end of the previous century various approximations were developed for $<Z_{i,n}>$ [28]. Though Eqs (5.24) and (5.25) may be preferred if sufficient calculation precision is available, approximations remain useful to forecast or recognize trends and are therefore discussed here.

One observation is that the plot of $F(<Z_{i,n}>)$ appears to vary almost (but not exactly) linearly with i (see Figure 5.16). Like the case of the median, this opens the door to approximations of the form:

$$F_{i,n} \approx \frac{i - A}{n + B} \tag{5.26}$$

For the median and the expected cumulative distribution B equals $1 - 2A$, which is a consequence of the fact that the sum of F_i has to equal $n/2$. However, this does not apply to $F(<Z_{i,n}>)$, because this is derived through a non-linear relationship with Z_i.

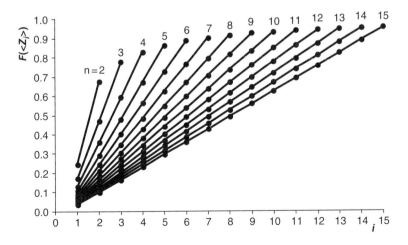

Figure 5.16 With given sample size n, the cumulative distribution F derived by calculation from $<Z_i>$ appears to have a practically linear relationship with ranking i.

A first approximation which is also recommended by IEEE 930 [24] and IEC 62539 [29] is defined as:

$$F(<Z_{i,n}>) \approx \frac{i - 0.44}{n + 0.25} \tag{5.27}$$

This means:

$$<Z_{i,n}> \approx \log\left(-\ln\left(1 - \frac{i - 0.44}{n + 0.25}\right)\right) = \log\left(-\ln\left(\frac{n + 0.69 - i}{n + 0.25}\right)\right) \tag{5.28}$$

This expression was derived by comparison of approximations for $n = 5(1)20(10)50$, meaning sample sizes 5, 6, …, 19, 20, 30, 40 and 50. This result worked slightly better than an older result in [30] with $A = 0.5$ and $B = 0.25$:

$$F_{Blom}(<Z_{i,n}>) \approx \frac{i - 0.5}{n + 0.25} \tag{5.29}$$

A third approximation is marked in Figures 5.17 and 5.18 as 'F1 interpolate Fn' ($n = 5$, 10, 15) and is defined as:

$$F(<Z_{i,n}>) \approx F(<Z_{1,n}>) + (i - 1) \cdot \frac{F(<Z_{n,n}>) - F(<Z_{1,n}>)}{n - 1} \tag{5.30}$$

A fourth approximation interpolates between $i = 2$ and $i = n$ and approximates $F(<Z_{n,n}>)$ by $1 - 1/1.5n$ [28]. It is marked in Figures 5.17 and 5.18 as 'F1, F2 interpolate Fn(=1/1.5n)' ($n = 5$, 10, 15) and is defined as:

$$i \leq 2: \qquad F(<Z_{i,n}>)$$

$$2 < i < n: \quad F(<Z_{i,n}>) \approx F(<Z_{2,n}>) + (i - 2) \cdot \frac{\left(1 - \frac{1}{1.5n}\right) - F(<Z_{2,n}>)}{n - 2}$$

$$i = n: \qquad F(<Z_{i,n}>) \approx 1 - \frac{1}{1.5n} \tag{5.31}$$

Figures 5.17 and 5.18 compare the plotting positions according to their deviation from the analytically calculated plotting position.

As a conclusion, calculating the exact plotting positions $F(<Z_i>)$ with integer i can be done, but in some cases approximations are convenient or required for non-integers that appear in adjusted ranking. All approximations give very workable values. The approximation provided by Blom [30] is most accurate in the midrange of i, whereas the

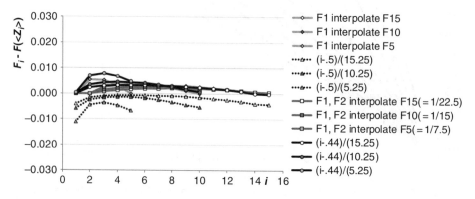

Figure 5.17 A comparison of plotting positions in terms of F_i. The graph shows the difference with the analytically calculated $F(<Z_i>)$ from Eq. (5.33).

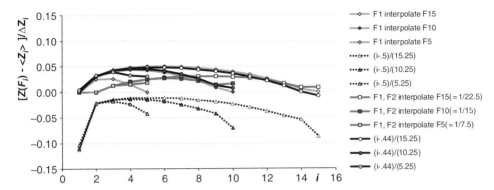

Figure 5.18 A comparison of plotting positions in terms of Z_i. The graph shows the difference with the analytically calculated $<Z_i>$ divided by the difference between two successive plotting positions. Therefore, the figure shows the relative deviations.

Table 5.12 Comparison of plotting position approximations.

Method	Eq. (5.29)	Eq. (5.27)	Eq. (5.30)	Eq. (5.31)
References	[30]	[24]	[28]	—
at n = 5				
$F_i - F(<Z_i>)$: min; max	−0.004; −0.011	0.000; 0.008	0; 0.005	0; 0.003
$Z_i - <Z_i>/\Delta Z_i$: min; max	−0.103; −0.018	0.004; 0.039	0; 0.026	0; 0.018
at n = 10				
$F_i - F(<Z_i>)$: min; max	−0.001; −0.006	0.000; 0.005	0; 0.004	0; 0.003
$Z_i - <Z_i>/\Delta Z_i$: min; max	−0.110; −0.013	0.001; 0.045	0; 0.043	0; 0.027
at n = 15				
$F_i - F(<Z_i>)$: min; max	−0.001; −0.004	0.000; 0.003	0; 0.003	0; 0.002
$Z_i - <Z_i>/\Delta Z_i$: min; max	−0.112; −0.011	0.000; 0.048	0; 0.050	0; 0.031

Note: ΔZ_i is defined in the text as the difference between two successive plotting positions: for $i = 1$ it is $<Z_2> - <Z_1>$; for $i = n$ it is $<Z_n> - <Z_{n-1}>$; for $1 < i < n$ it is $(<Z_{i+1}> - <Z_{i-1}>)/2$.

other approximations are more accurate at the start and the end. The low i is interesting in a plot if early failures have to be plotted most accurately.

Approximations in Eqs (5.27) and (5.29) are most convenient if a single formula is desired for an approximation of $F(<Z_i>)$. The deviations of approximations from the true values are compared for various sample sizes n in Table 5.12.

Figure 5.19 shows the data of Table 5.7 plotted according to the exact plotting position $F(<Z_i>)$ and the four approximations discussed above. The graph takes the shape of a Weibull plot (i.e. the vertical axis is linear in Z, but the marks are in F and the horizontal axis is a logarithm of t). Particularly the method of Eq. (5.29) stands out, with primarily the first point that is noticeably plotted lower than the exact plotting position, but the other plotting positions barely make a difference with the exact plotting position $<Z_i>$. For ease of use, Eq. (5.27) may be preferred.

In the following subsections the Weibull plots will use plotting positions according to Eq. (5.27) unless stated otherwise.

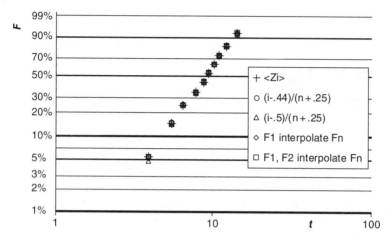

Figure 5.19 Comparison of plotting positions applied to the data of Table 5.7.

An advantage of Weibull plots with the expected plotting position is consistency with linear regression parameter estimation. This means that the parameters of best fit in a Weibull plot are exactly the same as the estimated parameters, which means that graphical representation and parameter estimation are fully consistent (see also Sections 6.3.2 and particularly 6.3.2.7).

5.4.2 Weibull Plot with Median Plotting Position

The median values $Z_{i,n,50\%}$ are calculated as:

$$Z_{i,n,50\%} = \log(-\ln(1 - F_{i,n,50\%})) \tag{5.32}$$

A good approximation using Eq. (5.2) is:

$$Z_{i,n,50\%} \approx \log\left(-\ln\left(\frac{n-i+0.7}{n+0.4}\right)\right) \tag{5.33}$$

The data of Table 5.7 are plotted in Figure 5.20 according to the expected, exact median and approximated median plotting positions in, respectively, Eqs (5.24), (5.32) and (5.33).

A first observation is that the exact median and approximated median plotting position agree very well. Compared to the expected plotting position, the median plotting position is a bit higher for low i and a bit lower for high i. The median plotting position therefore produces graphs with a somewhat smaller slope and suggests a somewhat lower shape parameter.

5.4.3 Weibull Plot with Expected Probability Plotting Position

Though it is not recommended, it is possible to produce a Weibull plot and choose a plotting position based on the expected probability as in Eq. (5.1):

$$< F_{i,n} >= \frac{i}{n+1} \tag{5.34}$$

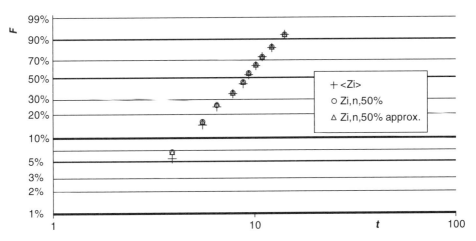

Figure 5.20 Comparison of exact expected, exact median and approximated median with the data of Table 5.7.

The plotting position $Z(<F_{i,n}>)$ thus becomes:

$$Z(< F_{i,n} >) = \log(-\ln(1- < F_{i,n} >)) \tag{5.35}$$

Therefore, the plotting positions $Z_{i,n}$ based on the expected probability are:

$$Z_{i,n} = \log\left(-\ln\left(1 - \frac{i}{n+1}\right)\right) = \log\left(-\ln\left(\frac{n+1-i}{n+1}\right)\right) \tag{5.36}$$

If a sample is drawn from a Weibull-distributed population with size n, then the expected variables along the horizontal axis are $<\log t_i>$ with $i = 1, \ldots, n$. The expected plotting position yields a straight line (naturally). The median plotting position is close to the expected plotting position but slightly curved. The plotting position based on the expected probability is most off, more or less S-shaped and tends to indicate a lower shape parameter β. The three plotting methods are compared in Figure 5.21.

5.4.4 Weibull Plot with Censored Data

Censored data can also be processed in Weibull plots. The adjusted ranking procedure was introduced in Section 5.2.2 and is based on the number of ranking position combinations of the actual failure data.

There are two ways to apply this method to Weibull plots. The first is the application of adjusted ranking to the cumulative failure distribution [24]. The second acknowledges the non-linear operation from i to $<Z_{i,n}>$ and determines the adjusted plotting positions as the average possible plotting positions.

The first method determines the adjusted ranking I according to Eq. (5.3). From the adjusted ranking, the corresponding Z_I can then be calculated through one of the methods in Section 5.4.1. For example, Eq. (5.27) is recommended in [24] and:

$$F(I, n) \approx \frac{I - 0.44}{n + 0.25} \tag{5.37}$$

Subsequently:

$$< Z_{I,n} > \approx \log\left(-\ln\left(\frac{n+I+0.69}{n+0.25}\right)\right) \tag{5.38}$$

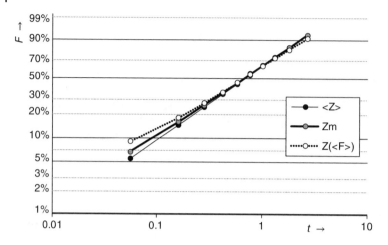

Figure 5.21 The expected variables $\langle \log t_i \rangle$ with $i = 1, ..., n$ plotted with various plotting positions: line $\langle Z \rangle$ with expected plotting positions $\langle Z_{i,n} \rangle$ (Eq. (5.24)); line Zm with median plotting positions $Z_{i,n,50\%}$ (Eqs (5.32) and (4.16)); line $Z(\langle F \rangle)$ with plotting positions based on expected probabilities (Eq. (5.36)).

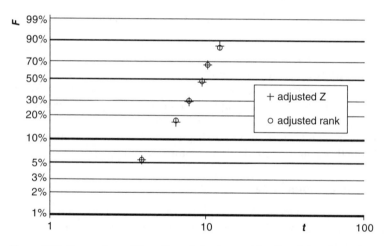

Figure 5.22 Comparison of the adjusted plotting position Z and the adjusted rank method.

Note that the adjusted rank I does not need to be an integer. However, Eqs (5.37) and (5.38) do not require I to be an integer either, unlike Eq. (5.24). A fundamental problem with this method is that Eq. (5.38) is not a linear operation and the method may therefore be useful as an approximation, but is not exact.

The second method adjusts not the ranking position I, but the plotting position Z. It follows the procedure in Example 5.1 but applies the adjustment to $\langle Z_{i,n} \rangle$ rather than to $\langle F_{i,n} \rangle$. The method is worked out in Example 5.4 and is recommended for plotting censored data. However, as shown in Figure 5.22, the difference between the adjusted ranking method in [24] and [29] does not seem to differ much from the adjusted plotting position in Example 5.4.

Table 5.13 Evaluation of component lifetimes.

Operational t (yr)	3.9	4.9	5.7	6.5	7.8	8.2	9.0	9.4	10.2	12.2
Failed?	Yes	No	No	Yes	Yes	No	No	Yes	Yes	Yes
Preliminary index i	1	—	—	2	3	—	—	4	5	6
Ranking time j	1	2	3	4	5	6	7	8	9	10

Note: Repeated Table 5.4.

Example 5.4   *Weibull Plot with Random Censored Data*   Consider the same series of 10 components as in Table 5.4. Six components failed with known times to failure and four are still in service (Table 5.13). A Weibull plot is prepared.

Again, the permutations are investigated. The total number of permutations can be calculated by starting from the last censored observation, which can take $n + 1 - j$ positions ultimately. Since $n = 10$ and the last censored observation is at $j = 7$, this results in four possibilities. Next, the censored observation at $j = 6$ can take five positions. The third last censored observation can take eight positions and the one at $j = 2$ can take nine positions. The total is therefore $4 \times 5 \times 8 \times 9 = 1440$ permutations.

Table 5.14 shows how many ranking possibilities exist for each lifetime. For example, for the uncensored time 12.2 years at present $j = 10$ can end up as sixth ranking time, namely if all four presently censored times will ultimately fall beyond 12.2 years. There are 4! permutations of the four. The uncensored time 12.2 years can also end up as seventh ranking time, which means that three out of four presently censored times may pass by ultimately. These three can be ordered in $3! = 6$ permutations. One out of three presently censored times will occur before 12.2 years; the ones at $j = 7$ and $j = 6$ have three possible positions and the ones at $j = 3$ and $j = 2$ have five possible positions (i.e. 16 in total). So, there are $16 \times 6 = 96$ permutations with 12.2 years at the seventh position. And so on.

The table also shows the average plotting position as the adjusted plotting position $Z_{i,n,adj}$, from which a cumulative failure function $F(Z_{i,n,adj})$ and a ranking can be derived. The adjusted plotting position and the adjusted ranking method are compared in Figure 5.22. Though the results of the two methods do differ, the difference is relatively small and it may be tempting to select the adjusted ranking method for its convenience.

5.4.5   Confidence Intervals in Weibull Plots

Confidence boundaries can be drawn with Weibull plots following the same approach as in Section 5.2.4.

The data of Table 5.10 lead to the graph in Figure 5.23. The complete procedure for producing this plot is:

1. Rank the observed data in increasing order and assign a plotting position $<Z_i>$ to each observation. In a Weibull plot the axes are processed in such a way that $F(<Z_i>)$ is read from the vertical scale (and t_i from the horizontal scale). $F(<Z_i>)$ follows from Eq. (5.25). With censored data use the adjusted rank or the adjusted plotting position method (Section 5.4.4).

Table 5.14 For each uncensored (i.e. observed) lifetime the number of possible plotting positions is shown.

Time (yr)		3.9	6.5	7.8	9.4	10.2	12.2
Ranking	$<Z_{i,n}>$						
1	−1.2507	1440					
2	−0.7931		1120				
3	−0.5503		280	840			
4	−0.3770		40	480	360		
5	−0.2361			120	480	120	
6	−0.1118				360	288	24
7	0.0052				180	396	96
8	0.1232				60	384	228
9	0.2539					252	420
10	0.4299						672
Number of permutations		1440	1440	1440	1440	1440	1440
Result – adjusted plotting position method							
Average $Z_{i,n}$: $Z_{i,n,adj}$		−1.251	−0.734	−0.466	−0.195	0.037	0.293
$F(Z_{i,n,adj})$		0.055	0.168	0.289	0.472	0.663	0.859
$I(F(Z_{i,n,adj}))$		1.000	2.166	3.407	5.275	7.237	9.249
Result – adjusted rank method							
Adjusted rank: I		1.000	2.250	3.500	5.375	7.250	9.125
$F(I,n)$		0.055	0.177	0.299	0.482	0.664	0.847
$Z(F(I,n))$		−1.250	−0.712	−0.450	−0.183	0.038	0.274

Note: From the number of permutations follows the average plotting position, which is the adjusted plotting position. This is compared to the previously obtained results in Table 5.5.

2. Determine $A\%$ confidence limits $F_{i,n,A\%}$. The $A\%$ confidence limits in Figure 5.23 are the 5% and 95% limits. The $F_{i,n,A\%}$ are found with Eq. (5.14) and B^{inv} the inverse beta function:

$$F_{i,n,A\%} = B^{inv}(A\%; i, n + 1 - i) \tag{5.39}$$

3. Using Weibull parameters a and b translates these limits into $A\%$ confidence limits $t_{i,n,A\%}$ with Eq. (5.15) and F^{inv} the inverse Weibull function:

$$t_{i,n,A\%} = F^{inv}(F_{i,n,A\%}(t)) = a \cdot [-\ln(1 - F_{i,n,A\%})]^{\frac{1}{b}} \tag{5.40}$$

4. Plot these $A\%$ confidence limits $t_{i,n,A\%}$ at the same plotting positions as the observed data. The limits are indicated by '+' markers in Figure 5.23.
5. The rest of the confidence limit is calculated here by plotting the $A\%$ confidence limits $t_{j,n,A\%}$ for any j other than the observation indices i at the corresponding plotting

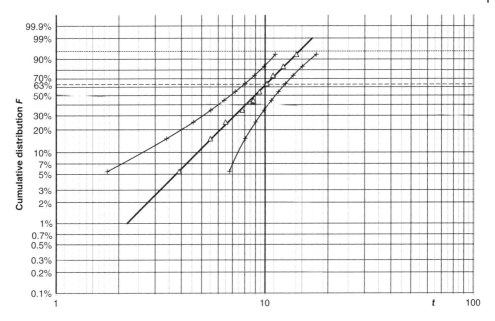

Figure 5.23 Weibull plot of the data of Table 5.15 with the expected plotting position. These observations are indicated by triangle markers. The 5% and 95% confidence limits $t_{i,n,A\%}$ are drawn about the best fit.

Table 5.15 The data of the completed test with 5% and 95% boundaries of time and 90% confidence intervals.

Operational t (yr)	3.9	5.5	6.5	7.8	8.8	9.4	10.2	10.9	12.2	14.1
Failed?	Yes	Yes	Yes	Yes	Yes	Yes	Yes	Yes	Yes	Yes
Index i	1	2	3	4	5	6	7	8	9	10
$<F_{i,10}>$	0.091	0.182	0.273	0.364	0.454	0.545	0.636	0.727	0.818	0.909
$t_{i,10,5\%}$	1.83	3.47	4.62	5.57	6.41	7.20	8.00	8.83	9.77	11.02
$t_{i,10,95\%}$	6.78	8.00	8.94	9.78	10.58	11.40	12.29	13.33	14.70	17.10

Note: Repeated Table 5.10.

positions. This is very convenient if the plotting positions are approximated with expressions like those of Eqs (5.27) and (5.37).

Generally, the observations tend to scatter less than might be expected based on the confidence limits. One explanation is that the confidence limits belong to the best-fit line through the observations. If the true distribution were known, the corresponding line would most likely differ from the best fit due to random scatter. This difference between true distribution and best fit is, however, unknown and the limits are seemingly too wide. If multiple experiments were carried out, the confidence limits would prove adequate if the ruling mechanism is random selection. For other possibilities, see Section 5.2.6.

5.5 Exponential Plot

For the exponential distribution the cumulative distribution F is:

$$F_{exp} = 1 - e^{-t/\theta} \tag{5.41}$$

which leads to a linear relationship between t and $\ln(1 - F)$:

$$-\ln(1 - F_{exp}) = \frac{1}{\theta} \cdot t \tag{5.42}$$

This linear relationship between $\ln(1 - F_{exp})$ and t can be used to produce a linear graph.

Alternatively, the exponential distribution can be regarded as a special case of the Weibull distribution with shape parameter β equal to 1 (cf. Eq. (5.21)):

$$\log(-\ln(1 - F_{exp})) = \log t - \log \theta \tag{5.43}$$

Therefore, the exponential distributed data can also be plotted in a Weibull plot and it can be evaluated whether the data are exponentially distributed indeed by checking whether the shape parameter β is equal to 1. The Weibull plot was treated in Sections 5.4 through 5.4.5. In the following sections the exponential plot following from Eq. (5.42) will be discussed.

5.5.1 Exponential Plot with Expected Plotting Position

As mentioned above, the expected plotting position is preferred if graphic representation is combined with parameter estimation (which is the subject of Chapter 6) using linear regression. The consistency between plotting and parameter estimation is very recommendable.

An estimate of the (inverted) characteristic time θ follows from the slope. The expected plotting position $<Z_i>$ becomes:

$$< Z_i >= - \ln(1 - F_{exp,i}) = - \int_0^1 \ln(1 - p) \cdot i \binom{n}{i} \cdot p^{i-1} \cdot (1 - p)^{n-i} dp \tag{5.44}$$

The solution is:

$$< Z_i > = - \int_0^1 \ln(q) \cdot i \binom{n}{i} \cdot (1 - q)^{i-1} \cdot q^{n-i} dq$$

$$= -i \binom{n}{i} \sum_{j=0}^{i-1} \binom{i-1}{j} (-1)^j \int_0^1 q^{n-i+j} \ln q \; dq$$

$$= -i \binom{n}{i} \sum_{j=0}^{i-1} \binom{i-1}{j} (-1)^j \left[\frac{q^{n-i+j+1}}{n-i+j+1} \ln q - \frac{q^{n-i+j+1}}{(n-i+j+1)^2} \right]\Bigg|_0^1$$

$$= i \binom{n}{i} \sum_{j=0}^{i-1} \binom{i-1}{j} \frac{(-1)^j}{(n-i+j+1)^2} \tag{5.45}$$

The corresponding exact $F(<Z_i>)$ is:

$$F(< Z_i >) = 1 - \exp(- < Z_i >) \tag{5.46}$$

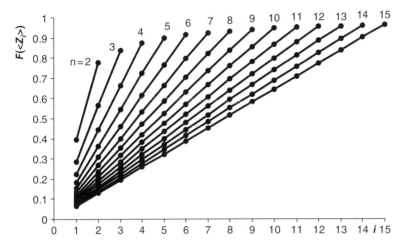

Figure 5.24 With given sample size n, the cumulative distribution F derived by calculation from $<Z_i>$ appears to have a linear relationship with ranking i.

As with the Weibull expected plotting position, the exponential expected plotting position also appears practically linear with i (see Figure 5.24).

The first plotting position for a distribution of n observations can be calculated as:

$$F(< Z_{1,n} >) = 1 - \exp(-(1/n)) \tag{5.47}$$

For the nth plotting position, empirically a good approximation appears to be:

$$F(< Z_{n,n} >) \approx 1 - \frac{0.562047488}{n + 0.518922085} \tag{5.48}$$

An approximation to calculate $F(<Z_i>)$ within typically better than 0.7% is achieved with the following linear formula:

$$F(< Z_{i,n} >) \approx F(< Z_{1,n} >) + \frac{i-1}{n-1} \cdot (F(< Z_{n,n} >) - F(< Z_{1,n} >)) \tag{5.49}$$

For $F(<Z_{1,n}>)$ and $F(<Z_{n,n}>)$, Eqs (5.47) and (5.48) can be used. This is a similar method as the approximated expected plotting position for Weibull as defined with Eq. (5.30).

5.5.2 Exponential Plot with Median Plotting Position

As with the Weibull plot the median plotting position is an alternative to the expected plotting position. The median values $Z_{i,n,50\%}$ are calculated as:

$$Z_{i,n,50\%} = -\ln(1 - F_{i,n,50\%}) \tag{5.50}$$

A good approximation using Eq. (5.2) is:

$$Z_{i,n,50\%} \approx -\ln\left(\frac{n-i+0.7}{n+0.4}\right) \tag{5.51}$$

From this, the $F(Z_{i,n,50\%})$ can be calculated as:

$$F(Z_{i,n,50\%}) = 1 - \exp(-Z_{i,n,50\%}) \tag{5.52}$$

5.5.3 Exponential Plot with Censored Data

Just as with the Weibull plot, censored data can be processed in exponential plots. In Section 5.2.2 the adjusted ranking was presented. Again, the procedure was based on the number of ranking position combinations of the actual failure data.

There are two ways to apply this method to exponential plots. The first is the application of adjusted ranking to the cumulative failure distribution [24]. The second, more correct, way is to determine adjusted plotting positions.

The first method determines the adjusted ranking I according to Eq. (5.3). From the adjusted ranking the corresponding Z_I can then be calculated through one of the methods in Section 5.5.1. For example, Eq. (5.49) yields:

$$F(I, n) \approx F(< Z_{1,n} >) + \frac{I - 1}{n - 1} \cdot (F(< Z_{n,n} >) - F(< Z_{1,n} >)) \tag{5.53}$$

And subsequently:

$$< Z_{I,n} > \approx - \ln(1 - F(I, n)) \tag{5.54}$$

Note that the adjusted rank I does not need to be an integer. However, Eqs (5.41) and (5.42) do not require I to be an integer, unlike Eq. (5.24). A principle problem with this method is that Eq. (5.42) is not a linear operation. Even though the method is not exact, it may still be very useful as an approximation.

The second method adjusts not the ranking position, but the plotting position. It follows the procedure in Example 5.1 but applies it to $<Z_{i,n}>$ rather than to $<F_{i,n}>$. The method is worked out in Example 5.4 and is considered the correct way to plot censored data. However, as shown in Figure 5.22, the adjusted ranking methods in [24] and in Example 5.4 do not seem to differ much.

5.5.4 Exponential Plot with Confidence Intervals

Confidence intervals for the exponential distribution can be plotted in the same way as for the Weibull distribution following Section 5.2.4.

The complete procedure for producing this plot is:

1. Consider making a graph as an exponential plot or as a Weibull plot (where the shape parameter should show as close to 1).
2. Rank the observed data in increasing order and assign a plotting position $<Z_i>$ to each observation. With uncensored data use the adjusted rank or the adjusted plotting position method (Section 5.4.4).
3. Determine $A\%$ confidence limits $F_{i,n,A\%}$ with Eq. (5.14). The limits $A\%$ (5% and 95%) are calculated similarly as for the Weibull (Figure 5.25) and exponential (Figure 5.23) distributions.
4. Translate these limits into $A\%$ confidence limits $t_{i,n,A\%}$ with Eq. (5.15).
5. Plot these $A\%$ confidence limits $t_{i,n,A\%}$ at the same plotting positions as the observed data. The limits are indicated by '+' markers in Figures 5.25 and 5.23.
6. The rest of the confidence limit can be calculated by plotting the $A\%$ confidence limits $t_{j,n,A\%}$ for any j other than the observation indices i at the corresponding plotting positions or by drawing a smooth line through the calculated confidence limits.

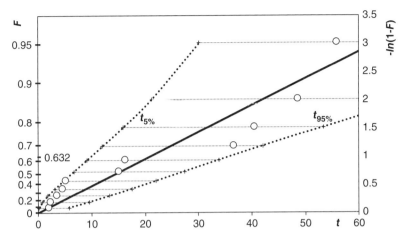

Figure 5.25 Exponential plot of the data in Table 5.16. The best fit seems a distribution with $\theta = 21$ years. Under the assumption that this distribution applies, the confidence intervals between $t_{5\%}$ and $t_{95\%}$ are determined, which show as horizontal dotted lines.

Table 5.16 Supposedly exponentially distributed data.

Operational t (yr)	2.00	2.27	3.47	4.52	5.19	15.1	16.3	36.7	40.6	48.7	55.9
Failed?	Yes	Yes	Yes	Yes	Yes	Yes	Yes	Yes	Yes	Yes	Yes
Index i	1	2	3	4	5	6	7	8	9	10	11
$F_{5\%}$ (%)	0.5	3.3	7.9	14	20	27	35	44	53	64	76
$F_{95\%}$ (%)	24	36	47	56	65	73	80	86	92	97	99.5
$t_{i,5\%}$ (yr)	0.098	0.711	1.72	3.05	4.67	6.64	9.04	12.0	15.8	21.2	30.1
$t_{i,95\%}$ (yr)	5.72	9.51	13.3	17.4	22.0	27.4	33.8	42.0	53.3	71.4	113

To illustrate the plotting of supposedly exponentially distributed data, two examples are given (an exponential plot and a Weibull plot) in, respectively, Examples 5.5 and 5.6.

The graphs have quite a different appearance. The best fits do differ slightly, and the time confidence intervals consequently differ as well. The exponential graph has a linear timescale (see Eq. (5.42)) and spreads out considerably with increasing time. The Weibull plot has a log–double log character (see Eq. (5.43)), which optically compresses the higher values both along the vertical as well as along the horizontal axis. The inferences will be practically the same, as Tables 5.16 and 5.17 show.

Example 5.5 *Exponential Graph* Consider the data in Table 5.16. These assets are supposed to have an exponential distribution with a characteristic lifetime $\theta = 20$ years. The data are shown in Table 5.16.

In case some data are unknown, the methods for censored data in Sections 5.2.2 and 5.4.4 can be used to construct adjusted plotting positions for the exponential plots of censored data with confidence intervals. See also Section 5.4.5 for the comparable case

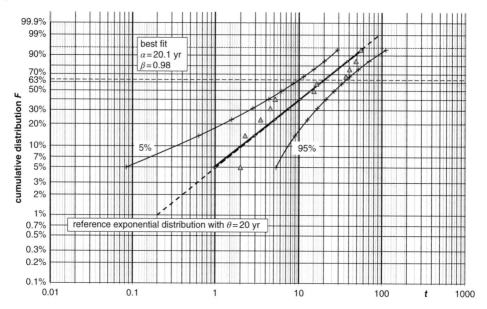

Figure 5.26 Weibull plot of the exponential data of Table 5.16 with the expected plotting position. These observations are indicated by triangle markers and the solid line is the best fit. Around the best fit, the 5% and 95% confidence limits $t_{i,n,A\%}$ are drawn. The sloped dotted line is the reference line with $\theta = 20$ years.

Table 5.17 Supposedly exponentially distributed data analysed as a Weibull distribution.

Operational t (yr)	2.00	2.67	3.47	4.26	5.19	15.1	16.3	36.7	40.6	48.7	55.9
Failed?	Yes	Yes	Yes	Yes	Yes	Yes	Yes	Yes	Yes	Yes	Yes
Index i	1	2	3	4	5	6	7	8	9	10	11
$F_{5\%}$ (%)	0.5	3.3	7.9	14	20	27	35	44	53	64	76
$F_{95\%}$ (%)	24	36	47	56	65	73	80	86	92	97	99.5
$t_{i,5\%}$ (yr)	0.084	0.635	1.57	2.80	4.34	6.21	8.50	11.4	15.1	20.3	29.0
$t_{i,95\%}$ (yr)	5.33	8.96	12.6	16.6	21.1	26.4	32.7	40.8	52.0	70.1	112

of Weibull-distributed data with confidence intervals where, of course, the Weibull plotting positions are used instead of the exponential plotting positions (Section 5.5.1).

Example 5.6 *Weibull Plot for Exponentially Distributed Data* Consider again the data in Table 5.16 in the previous example. As the exponential distribution can be regarded as a Weibull distribution with shape parameter $\beta = 1$, the data can also be plotted in a Weibull plot.

The best fit has a shape parameter β very close to 1 (which may deviate more in practice). Estimating the best-fit parameters will be discussed in Chapter 6. There are also methods to determine the Weibull distribution with a fixed shape parameter. So, for exponentially distributed data the shape parameter could be set to 1.

The confidence intervals are slightly different than in Table 5.16, because the distribution used in the Weibull plot in Figure 5.26 is the best-fit Weibull distribution with scale parameter $\alpha = 20.1$ years and shape parameter $\beta = 0.98$. In contrast, the best fit with an exponential distribution in Figure 5.25 has a characteristic lifetime of $\theta = 21$ years and assumed shape parameter $B = 1$.

5.6 Normal Distribution

The normal distribution was introduced in Section 4.3.1. The distribution density function is defined in Eq. (4.50) as:

$$f(x) = \frac{1}{\sigma\sqrt{2\pi}} \cdot e^{-\frac{1}{2}\left(\frac{x-\mu}{\sigma}\right)^2} \tag{5.55}$$

The cumulative distribution F cannot be derived analytically, but there are numerical tables (cf. Appendix G) and approximations (Section 10.6) that yield values for F. Generally software spreadsheets generate these values also. Once an $F(x)$ can be assigned to each x, an inverse function F^{-1} can be given that assigns an x value to each F:

$$F^{-1}(F(x)) = x \tag{5.56}$$

When plotting the data, a straight graph can be achieved by plotting the expected values of some kind of $<F^{-1}(F_i)>$ against the observed x_i. In order to find this $<F^{-1}(F_i)>$, the normalized distribution is helpful. This was introduced with the normalized variable Y in Section 4.3.1.1. The value of the normalized variable $y(x)$ is defined as:

$$y(x) = \frac{x - \mu}{\sigma} \tag{5.57}$$

and:

$$dy(x) = \frac{\partial y(x)}{\partial x}dx = \frac{1}{\sigma}dx \tag{5.58}$$

If a normal-distributed population consists of n random samples, then the observations (ranked in increasing order of magnitude) are $X_1, \ldots, X_n$. The corresponding normalized observations $Y_1, \ldots, Y_n$ follow from Eq. (5.57). With repeated experiments we obtain the mean normalized variables $<Y_1>, \ldots, <Y_n>$. For each $<Y_i>$ it holds that:

$$< Y_i >= \int_{-\infty}^{\infty} y \cdot f(y) \cdot (F(y))^{i-1} \cdot (R(y))^{n-i}dy \tag{5.59}$$

Unfortunately, there is no analytical expression for F and R, but the $<y_i>$ values can be calculated numerically. Where necessary, the sample n can be shown explicitly by extending the suffix of y as $y_{i,n}$. An $F_{i,n}$ can be associated with each $<y_{i,n}>$ by (numerically) calculating the cumulative normal distribution value:

$$F_{i,n} = F(< y_{i,n} >) \tag{5.60}$$

A linearized normal probability plot can be derived by actually plotting the variables $<Y>$ along the vertical axis against X along the horizontal axis. By expressing Y in terms of $F_{i,n}$, a normal probability plot is achieved that yields a straight line, as in Figure 5.27.

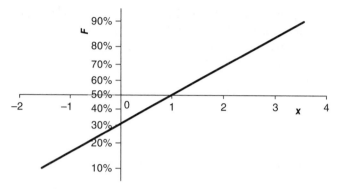

Figure 5.27 Normal probability plot with $x_{i,n}$ values along the horizontal axis and $<y_{i,n}>$ values along the vertical axis. Though the actual plot is $<y_{i,n}>$ against $x_{i,n}$, the $<y_{i,n}>$-axis is expressed in units of $F_{i,n}$ (Eq. (5.60)), which makes this a probability plot.

This approach is similar to Weibull plots, where the graph of Z and $\log t$ forms the actual linear graph, but the units along the axes are expressed as F and t (cf. Figure 5.23).

The intercept of the graph with the horizontal axis at $y = 0$ (i.e. $F = 0.5$) yields the mean μ and the slope of the graph is $1/\sigma$. In Figure 5.27 the mean appears as $\mu = 1$. If the plot is dimensioned such that the units of the x and $<y>$ axes are the same, then the slope can be determined directly. In this example the standard deviation was taken to be $\sigma = 2$, so the slope would then be $1/\sigma = 0.5$.

5.6.1 Normal Plot with Expected Plotting Position

As mentioned above, the expected plotting position is preferred if graphic representation is combined with parameter estimation (which is the subject of Chapter 6) using linear regression. Consistency between plotting and parameter estimation is highly recommended.

The plotting position in a probability plot such as Figure 5.27 is the expected plotting position $<Y_{i,n}>$. The scaling is in terms of $F_{i,n}$ through Eq. (5.60). These cumulative probabilities cannot be analytically determined.

For $n = 2,...,10$ the cumulative probabilities $F_{i,n}$ are shown in Figure 5.28. The cumulative probabilities appear quite linear with i, though again this is not exactly true.

This practical linearity has led to quite a number of models of the shape:

$$F_{i,n} \approx \frac{i - A}{n + B} \tag{5.61}$$

Because of the symmetry of the normal distribution, the following relationship holds:

$$< y_{i,n} >= - < y_{i,n+1-i} > \tag{5.62}$$

As a consequence (cf. Eq. (4.52)):

$$F_{i,n} = 1 - F_{i,n+1-i} \tag{5.63}$$

If the model of Eq. (5.61) holds, then:

$$F_{i,n} \approx \tilde{F}_{i,n} = \frac{i - A}{n + 1 - 2A} \tag{5.64}$$

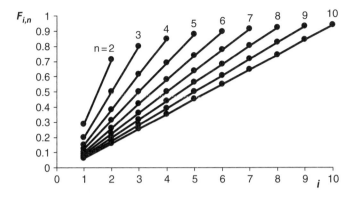

Figure 5.28 Plotting positions $F_{i,n} = F(<y_{i,n}>)$ as a function of i. As with previous plotting positions.

A closer study, however, shows that firstly A and B are not constants, but do at least depend on n. Secondly, similar to the case of the plotting positions for the exponential and the Weibull distributions particularly, the plotting positions according to such formulas deviate the most from the exact values at the extremes (i.e. towards $i = 1$ and $i = n$; cf. Section 5.4.1).

An alternative approach is to determine the slope and intercept of the graphs in Figure 5.28 and approach these parameters. In this approach the formula for $F_{i,n}$ becomes:

$$F_{i,n} \approx \tilde{F}_{i,n} = C_n \cdot i + D_n \tag{5.65}$$

C_n and D_n are constants with respect to i, but vary with n. Slope C_n was fit with a three-parameter power function of n. This method is discussed further in Section 10.2.3. Subsequently, D_n was derived as the intercept at $i = 0$ by subtracting $<i> = (n+1)/2$ times the slope C_n from $<F_{i,n}> = 1/2$. For the plots in Figure 5.28, the following approximations were found:

$$C_n \approx \frac{1.01}{n + 0.36} \tag{5.66}$$

$$D_n \approx \frac{1}{2} - C_n \cdot \frac{n+1}{2} = \frac{1}{2} \cdot (1 - C_n \cdot (n+1)) \tag{5.67}$$

The formula for $F_{i,n}$ thus becomes:

$$F_{i,n} \approx \tilde{F}_{i,n} = \frac{1.01}{n + 0.36} \cdot \left(i - \frac{n+1}{2} \right) + \frac{1}{2} = \frac{1}{2} \cdot \left[1 - 1.01 \cdot \frac{n + 1 - 2i}{n + 0.36} \right] \tag{5.68}$$

In the form of Eq. (5.65):

$$F_{i,n} \approx \tilde{F}_{i,n} = \frac{1.01}{n + 0.36} \cdot i - \frac{1}{200} \cdot \frac{n + 65}{n + 0.36} \tag{5.69}$$

Table 5.18 provides an overview of options for plotting positions and the sum of the squared error for each n. This squared error is a measure for the accuracy of the approximated $F_{i,n}$ and is defined as:

$$S_n^2 = \sum_{i=1}^{n} \left(F_{i,n} - \tilde{F}_{i,n} \right)^2 \tag{5.70}$$

Table 5.18 Comparison of plotting positions for normal probability plots.

Approach	Blom	Expected F	Median	Slope and intercept
Parameters	$A = 0.375$	$A = 0$	$A = 0.3$	C_n (see Eq. (5.66))
	$B = 0.25$	$B = 1$	$B = 0.4$	D_n (see Eq. (5.67))
$F_{i,n}$	Eq. (5.61)	Eq. (5.61)	Eq. (5.61)	Eq. (5.68)
n	S_n^2	S_n^2	S_n^2	S_n^2
2	1.5×10^{-4}	0.004	6×10^{-5}	2×10^{-7}
3	8×10^{-5}	0.005	1.0×10^{-4}	9×10^{-7}
4	4×10^{-5}	0.005	1.3×10^{-4}	6×10^{-6}
5	2×10^{-5}	0.005	1.4×10^{-4}	4×10^{-6}
6	1.4×10^{-5}	0.005	1.5×10^{-4}	4×10^{-6}
7	8×10^{-6}	0.004	1.5×10^{-4}	3×10^{-6}
8	5×10^{-6}	0.004	1.5×10^{-4}	3×10^{-6}
9	3×10^{-6}	0.004	1.5×10^{-4}	2×10^{-6}
10	2×10^{-6}	0.004	1.5×10^{-4}	2×10^{-6}

Note: The parameters A and B to be used in Eq. (5.61) (respectively the parameters C and D to be used in Eq. (5.65)) are shown. The latter yields Eq. (5.68).

The approach of Blom [30] is widely used and appears to perform very well according to Table 5.18, only to be surpassed by the slope and intercept approach for $n < 10$.

5.6.2 Normal Probability Plot with Confidence Intervals

For confidence intervals the principle is the same as for the parameter-free plots and the other previously discussed plots. Confidence intervals can be plotted in the same way as for the Weibull distribution following Section 5.2.4. For censored data the same approaches can be used as in Sections 5.2.2 and 5.4.4.

The complete procedure for producing this plot is:

1. Rank the observed data in increasing order and assign a plotting position $<Y_i>$ to each observation. With uncensored data use the adjusted rank or the adjusted plotting position method (Section 5.4.4).
2. Determine $A\%$ confidence limits $F_{i,n,A\%}$ with Eq. (5.14). The limits $A\%$ (5% and 95%) are calculated for Figures 5.25 and 5.23.
3. Translate these limits into $A\%$ confidence limits $t_{i,n,A\%}$ with Eq. (5.15) (i.e. for the normal distribution, Eq. (5.56)).
4. Plot these $A\%$ confidence limits $t_{i,n,A\%}$ at the same plotting positions as the observed data. The limits are indicated by '+' markers in Figures 5.25 and 5.23.
5. The rest of the confidence limits can be calculated by plotting the $A\%$ confidence limits $t_{j,n,A\%}$ for any j other than the observation indices i at the corresponding plotting positions or by drawing a smooth line through the calculated upper and lower confidence limits.

5.6.3 Normal Plot and Lognormal Data

The lognormal distribution is strongly related to the normal distribution as discussed in Section 4.3.2. If a variable T is distributed according to a lognormal distribution, the reverse of Eq. (4.91) can be used to obtain variable X that is distributed according to the normal distribution:

$$X = \ln T \tag{5.71}$$

In this way a variable T with domain $[0,\infty)$ can be converted into a variable X with domain (∞,∞), which solves the domain problem. In this way variables like time to failure, voltage and other non-negative variables can still be analysed with a normal plot.

5.7 Power Law Reliability Growth

The Duane and Crow AMSAA plots aim to represent the growth of reliability (i.e. tracking whether improvement programmes are successful or not). Originally developed to monitor learning curves, in particular progress of improved manufacturing of airplanes, these plots can also be applied to grids, power plants, machinery and organizations, provided failure can be defined and quantified. The concept behind the plots is that a power law rules the improvement.

This may apply to the performance of asset management systems as well. For instance, if a new maintenance strategy is introduced, aimed at reducing the number of failures per year, then these plots may be used not only to check, but also to monitor the improvement and forecast performance.

5.7.1 Duane and Crow AMSAA Plots and Models

As an empirical finding, Duane noted that reliability improvement programmes on many large systems lead to a linear learning curve when the cumulative MTBF θ_c is plotted against the cumulative time t on a log–log scale as in Figure 5.29 [31, 32]. Similarly, the learning curve becomes straight when the cumulative failure rate is plotted against the cumulative time t as in Figure 5.30. These Duane plots aim to show this log–log relationship.

Let $N(t)$ be the number of failure events that occur over a period $[0,t]$. The MTBF used by Duane plots is the cumulative operating time t divided by the number of encountered

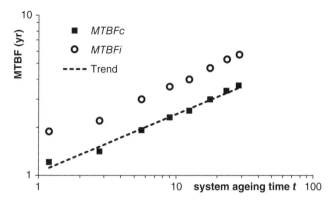

Figure 5.29 Duane MTBF plot of the system with failure data in Table 5.20. In addition to the cumulative $MTBF_c$, the instantaneous $MTBF_i$ is also shown. The latter is not standard in Duane plots.

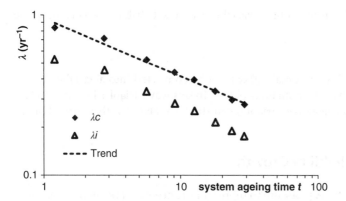

Figure 5.30 Duane $N(t)/t$ or λ plot of the system with failure data in Table 5.21. In addition to the cumulative λ_c, the instantaneous λ_i is also shown. The latter is not standard in Duane plots.

failures $N(t)$. This is called the cumulative MTBF, $MTBF_c$ or θ_c:

$$MTBF_c = \theta_c = \frac{t}{N(t)} \tag{5.72}$$

The linear relationship on a log–log scale means:

$$\log(\theta_c) = B \cdot \log(t) - \log(A^B) \quad \Leftrightarrow \quad \theta_c = \left(\frac{t}{A}\right)^B \tag{5.73}$$

A and B are parameters yet to be determined. Parameter A must be larger than 0. For systems where the reliability improves, the failure rate decreases (i.e. $B > 0$). If the reliability is not improving but deteriorating, then $B < 0$. If indeed the model applies, and $B = 0$, then θ_c is time-independent.

In [18] the derivation of the Duane model starts with $N(t)$ and leads to the same relations. The linear relationship between $N(t)/t$ and t on a log–log scale means:

$$\log\left(\frac{N(t)}{t}\right) = \log(A) - B \cdot \log(t) \quad \Leftrightarrow \quad \frac{N(t)}{t} = \left(\frac{t}{A}\right)^{-B} \tag{5.74}$$

Both Eqs (5.73) and (5.74) lead to the same power law relationship for the number of failures:

$$N(t) = \left(\frac{t}{A}\right)^{1-B} \tag{5.75}$$

The instantaneous failure rate λ_i is the time derivative of $N(t)$:

$$\lambda_i(t) = \frac{dN(t)}{dt} = (1 - B) \cdot \left(\frac{t}{A}\right)^{-B} \tag{5.76}$$

The instantaneous MTBF, $MTBF_i$, is defined as the inverse of the instantaneous failure rate:

$$MTBF_i = \frac{1}{\lambda_i(t)} = \frac{1}{(1 - B)} \cdot \left(\frac{t}{A}\right)^{B} \tag{5.77}$$

The parameter B is called the reliability growth parameter by Duane. With actual improvement, $B > 0$. In addition to the instantaneous failure rate λ_i and $MTBF_i$, the

Table 5.19 Collecting data for a Duane MTBF plot.

Failure number $N(t)$	System age at last failure t	$MTBF_c$
1	t_1	$t_1/1$
i	t_i	t_i/i
n	t_n	t_n/n

cumulative failure rate λ_c and $MTBF_c$ are defined:

$$\lambda_c = \frac{N(t)}{t} = \left(\frac{t}{A}\right)^{-B} \tag{5.78}$$

$$MTBF_c = \frac{1}{\lambda_c} = \frac{t}{N(t)} = \left(\frac{t}{A}\right)^{B} \tag{5.79}$$

The instantaneous and cumulative failure rates differ by a factor of $1 - B$:

$$\lambda_i = (1 - B) \cdot \lambda_c \tag{5.80}$$

$$MTBF_i = \frac{1}{1 - B} \cdot MTBF_c \tag{5.81}$$

On a log–log scale this factor is a constant shift of the graph.

A Duane plot is easily made. In time, subsequent failures are collected as in Table 5.19. Subsequently, the data of the third and fourth column are plotted against those of the second column. An example is given in Example 5.7. The slope of the graph B is the reliability growth or improvement rate and shows how the MTBF is growing during the lifetime of the system. The increase can be due to measures taken in an ongoing improvement programme.

Example 5.7 *Duane Plot of MTBF* Consider a system that is put into operation and improved over the years. This might be a circuit consisting of various components where the asset management methodology is changed. After each failure the system is repaired and put back into service. The system fails at the system ages (i.e. operating times) in Table 5.20. The data are subsequently plotted in the log–log graph of Figure 5.29. The reliability growth (i.e. the slope of the graph B) is 0.36 in this case.

The Duane plot shows that the system is still improving. If the slope started to increase in time (i.e. gradients turning larger), then the MTBF would apparently grow faster, which would indicate an increasingly effective improvement programme. If the slope levelled off (i.e. gradient becoming 0), then no improvements would occur apparently.

The Duane plot can also be made as a cumulative failure rate plot, as shown in Example 5.8.

Example 5.8 *Duane Plot of Cumulative Failure Rate* Using the same data as in Example 5.7, Table 5.21 is produced in preparation of the cumulative failure rate plot. The data are subsequently plotted in the log–log graph of Figure 5.29. The reliability growth (i.e. the negative slope of the graph B) is 0.36 in this case.

Table 5.20 Example of system failure data for a Duane plot based on Eqs (5.77) and (5.79).

Failure number N(t)	System age at failure (t (yr))	$MTBF_c$ (t/N(t) (yr))	$MTBF_i$ (yr)
1	1.2	1.20	1.88
2	2.8	1.40	2.20
3	5.7	1.90	2.98
4	9.1	2.28	3.57
5	12.6	2.52	3.95
6	17.8	2.97	4.65
7	23.6	3.37	5.29
8	29	3.63	5.69

Table 5.21 Example of system failure data for a Duane plot based on Eqs (5.76) and (5.78).

Failure number N(t)	System age at failure (t (yr))	Cumulative failure rate (N(t)/t (yr^{-1}))	Instantaneous failure rate (N(t)/t (yr^{-1}))
1	1.2	0.83	0.53
2	2.8	0.71	0.46
3	5.7	0.53	0.34
4	9.1	0.44	0.28
5	12.6	0.40	0.25
6	17.8	0.34	0.21
7	23.6	0.30	0.19
8	29	0.28	0.18

The Duane plot shows that the system is still improving. If the slope starts to decrease in time (gets steeper), then the failure intensity apparently slows down, which would indicate an increasingly effective improvement programme. If the slope levelled off and the gradient became 0, then no improvements would occur apparently.

5.7.2 NHPP Model in Duane and Crow AMSAA Plots

Though the Duane model was close, Crow [33, 34] connected the Duane model to the non-homogeneous Poisson process (NHPP, Section 4.4.2.4). With this process the failure intensity function $m_{NHPP}(t)$ was defined as ($\alpha > 0$ and $\beta > 0$):

$$m_{NHPP}(t) = \alpha\beta \cdot t^{\beta-1} \tag{5.82}$$

The average or expected number of failures $M_{NHPP}(t)$ is:

$$M_{NHPP}(t) = \alpha \cdot t^{\beta} \tag{5.83}$$

The parameter α is referred to as the scale parameter and β as the shape parameter. For $\beta < 1$ ($\beta > 1$) there is positive (respectively negative) reliability growth. This means that for improving systems, $0 < \beta < 1$. For $\beta = 1$ there is no reliability growth, since the failure intensity function is a constant which corresponds to the homogeneous Poisson process (HPP, Section 4.4.2.3).

Plotting the cumulative number of failures N (as an estimate of M_{NHPP}) against the cumulative time t yields a straight line:

$$\log(N(t)) \approx M_{NHPP}(t) = \log(\alpha) - \beta \cdot \log(t) \tag{5.84}$$

With reliability growth the slope appears as $\beta < 1$. This is shown in Example 5.9.

In order to match the NHPP equations with the equations in the previous Section 5.7.1, note that m_{NHHP} as the derivative of M_{NHPP} corresponds to the instantaneous failure rate λ_i and therefore:

$$\beta = 1 - B \tag{5.85}$$

$$\alpha = A^B \tag{5.86}$$

The cumulative failure rate λ_c equals M_{NHPP}/t:

$$\lambda_c = \frac{M_{NHPP}(t)}{t} = \alpha \cdot t^{\beta-1} \tag{5.87}$$

In terms of the NHPP model, the instantaneous and the cumulative MTBF become:

$$MTBF_i = \frac{1}{\lambda_i(t)} = \frac{1}{m_{NHPP}(t)} = \frac{1}{\alpha\beta} \cdot t^{1-\beta} \tag{5.88}$$

$$MTBF_c = \frac{1}{\lambda_c(t)} = \frac{1}{\alpha} \cdot t^{1-\beta} \tag{5.89}$$

From Eq. (5.83), the times of failure can be predicted:

$$t = \left(\frac{M_{NHPP}}{\alpha}\right)^{-\beta} \tag{5.90}$$

Example 5.9 *Plot of Cumulative Failures*   Using the same data as in Example 5.7, Table 5.22 follows directly from the gathered failure data. The data are subsequently plotted in the log–log graph of Figure 5.31. The slope of the graph β is 0.64 in this case, which corresponds well to the fact that in Examples 5.7 and 5.8 a reliability growth of $B = 0.36$ was found, and $\beta = 1 - B$ Eq. (5.85).

The plot with $\beta = 0.64 < 1$ shows that the system is still improving. If the slope decreased further in time, then the failure intensity would apparently slow down more, which would indicate an increasingly effective improvement programme. If the slope increased towards 1, then improvements would become less effective.

With respect to starting time $t = 0$, the question is what to take. If these plots are used to assess the instantaneous failure rate as a measure of reliability growth, basically any period and starting point can be taken. If the annual performance is of interest, then plots can be made over one or more calendar years and the slope (i.e. parameters β and B) can be determined for each year. For periods where $\beta < 1$, the reliability growth was positive. If β_X of a certain year is smaller than β_{X-1} of the year before, then reliability growth is accelerating.

Table 5.22 Example of system failure data for a Duane plot based on Eq. (5.75) which corresponds to Eq. (5.83).

Failure number (N(t))	System age at failure (t (yr))
1	1.2
2	2.8
3	5.7
4	9.1
5	12.6
6	17.8
7	23.6
8	29

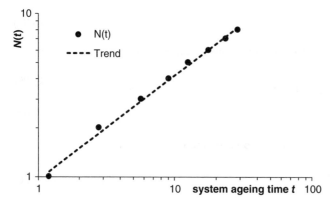

Figure 5.31 $N(t)$ plot of the system with failure data in Table 5.22. For the slope of the graph, $\beta = 0.64$ is found.

Another approach can be to mark beginnings and endings of certain improvement programmes and evaluate from the graph whether or not the shape parameter $\beta < 1$. Remember the Duane reliability growth parameter $B = 1 - \beta$ and positive reliability growth means $B > 0$.

For further reading, [35] is recommended.

5.8 Summary

This chapter discusses graphical data analysis: Data quality (Section 5.1); parameter-free graphs (Section 5.2); parametric graphical analysis (Section 5.3); Weibull plots (Section 5.4); exponential plots (Section 5.5); normal plots (Section 5.6); power law reliability growth plots (Section 5.7).

As for data quality (Section 5.1), it is self-explanatory that keeping good records of age, load levels, operational time and failure times greatly facilitates adequate analysis and decision-making. Less obvious is that failure events require good definitions. The reason is that before ultimate breakdown performance interruptions may sometimes have occurred already. It is a matter of definition which is considered as the failure: the ultimate breakdown or the earlier temporary interruptions. These will have different values, and the exact choice has an impact on the analysis results and conclusions.

As for parameter-free graphs, attention is paid to the expected plotting position $<F_{i,n}> = i/(n+1)$ with i the ranking and n the sample size (Section 5.2.1) and how to deal with censored data (i.e. missing data) through the adjusted ranking method (Section 5.2.2). Another type of parameter-free graph is the Kaplan Meier plot, which can also handle censored data (Section 5.2.3). Finally, confidence intervals around a known distribution plot based on the beta distribution are discussed (Sections 5.2.4 and 5.2.5). These confidence intervals are related to the scatter in the ranking. Alternative confidence intervals are discussed in qualitative terms (Section 5.2.6).

As for parametric graphical analysis (Section 5.3), the data are assumed to follow a given distribution. Because distributions are characterized by their parameters, this is called a parametric model. Of particular interest is the ability to represent data in a linear graph, because it facilitates both interpolation and extrapolation of the best fit. Subsequently, the plots for various distributions are discussed.

As for distributions, particularly the Weibull-2 distribution (Section 5.4), the exponential distribution (Section 5.5), which is also an example of Weibull-1, and the normal distribution (Section 5.6) are discussed. For each, the expected plotting position Z, the matter of censored data and confidence intervals based on the beta distribution are discussed. With censored data there are two adjustment techniques: the adjusted ranking method as recommended in IEEE 930 and IEC 62539 and the adjusted plotting position technique. Both are elaborated on in the Weibull discussion.

Finally, as for power law reliability growth (Section 5.7), the Duane model (Section 5.7.1) and the Crow AMSAA model (Section 5.7.2) are discussed. These are connected to the Poisson distribution and are very useful for monitoring improvement programmes and checking the overall trend in reliability development.

Questions

5.1 With reference to censored data (Section 5.2.2), the Weibull plot (Section 5.4), Weibull plots with censored data (Section 5.4.4), accelerated ageing (Section 2.6.2) and lognormal data (Section 5.6.3):

A set of 10 insulators is tested. The test voltage was $3\,kV$ and the times to breakdown $t_1, \ldots, t_{10}$ were recorded. However, the test facility was only available for 10 days, after which the test was stopped. A second problem occurred with the recorder that missed the fifth time to failure, but its ranking was certain. Therefore, t_5 is indicated as X_5. The test data (in days) are: 0.7, 1.2, 2.8, 3.2, X_5, 6.3, 8.3. The remaining three test objects did not fail in the 10-day test period. It is assumed that the data are Weibull-2 distributed with parameters α and β.

a. Produce a table of the failure times with corresponding failure times if these data are to be plotted in a Weibull plot.

b. Put the data in a Weibull plot and draw a best fit (either by eye or by calculation).

    c. Estimate the scale parameter α and the shape parameter β from the plot.

    d. Mark on the best fit the expected positions of the four censored data and estimate their values (in days).

    e. Based on the best fit, estimate the required test period in which 99% of the tested population would fail.

    f. Assume that there is a power law relationship between the time to breakdown and the test voltage level. Also assume that the power $p = 8$. What test voltage would be required to have 99% of the tested population fail within 10 days?

    g. Assume that the data do not follow a Weibull, but a lognormal distribution. Plot the data in a normal plot.

    h. Based on the best fit in the normal plot, compare the required test period in which 99% of the tested population would fail with the required 99% test period obtained for the Weibull distribution.

5.2 With reference to the Weibull plot (Section 5.4) and the exponential plot (Sections 5.5 and 4.2.4.1):

Wall bushings are insulators that enable high-voltage conductors to pass through a wall. Assume these are planned to be used near a polluted area. Some bushings failed in the past in that area and it was noticed that a Weibull distribution applies. It is decided to periodically clean the bushings, after which they are as good as new. An accelerated test at $U_{test} = 2U_0$ (so at twice the operating voltage) is planned in order to find the ruling Weibull distribution. From previous experiments it is known that a power of $p = 10$ applies. Table 5.23 shows the test results after 11.6 hours, after which the results are evaluated.

    a. Complete Table 5.23 with:

       i. the times to failure $t_{0,i}$ at nominal U_0 (mind the units);

       ii. the plotting positions for a Weibull plot F_i.

    b. Put the $t_{0,i}$ data in a Weibull plot and draw a best fit by eye or by calculation.

    c. Estimate the Weibull parameters α and β from the best fit.

    d. Assume the maximum allowable hazard rate is 0.0025/day. Give an expression for the hazard rate of the Weibull distribution.

    e. At what time $t = T$ is the maximum allowable hazard rate of 0.0025/day reached with the Weibull parameters as estimated in subquestion (c)?

    f. This period T is now adopted as the maintenance interval. Estimate the average hazard rate h_{ave} when this period T is used.

Table 5.23 Test results at $U_{test} = 2U_0$ for 10 insulators.

Index i	1	2	3	4	5	6	7	8	9	10
Times to failure $t_{t,i}$ (hr) at $U_{test} = 2U_0$	3.2	5.5	6.6	7.1	8.1	9.9	10.3	11.6	> 11.6	> 11.6
Times to failure $t_{0,i}$ (days) at U_0										
Plotting position F_i										

Table 5.24 Failure events in the system requiring repair.

No.	Time (t)	No.	Time (t)	No.	Time (t)	No.	Time (t)
1	0.434	13	3.443	25	6.154	37	11.829
2	0.816	14	3.573	26	6.570	38	12.366
3	1.180	15	3.700	27	6.997	39	12.913
4	1.533	16	3.820	28	7.434	40	13.470
5	1.878	17	3.938	29	7.881	41	14.036
6	2.216	18	4.052	30	8.340	42	14.611
7	2.526	19	4.166	31	8.808	43	15.196
8	2.702	20	4.281	32	9.286	44	15.789
9	2.865	21	4.603	33	9.775	45	16.392
10	3.019	22	4.974	34	10.274	46	17.003
11	3.167	23	5.356	35	10.783	47	17.624
12	3.309	24	5.750	36	11.301	48	18.253

g. Draw the line for the exponential distribution in the Weibull plot of subquestion (b).

h. Calculate the time t at which 63% of the bushings failed if:
   i. no maintenance is carried out (Weibull distribution applies);
   ii. maintenance is carried out (exponential distribution with h_{ave} applies).

5.3 With reference to power reliability growth (Section 5.7):
In a system, the reliability seems to decrease slowly. Failure events occur that are repaired, but the rate of failures is believed to increase. At two stages, measures are taken to improve the reliability. Measure A is found to successfully improve the system, whereas measure B is unfortunate and rather decreases the reliability of the system. For now, the moments of these measures are not disclosed.

a. Plot the data of Table 5.24 in an appropriate graph.

b. Identify the moment when each measure became effective.

c. Estimate the reliability growth for the three distinguishable stages.

6

Parameter Estimation

Chapter 5 discussed data analysis by graphical techniques. Such techniques can give a good insight into the suitability of distribution models, trends, the presence of various subpopulations and possible outliers (i.e. data that might be ignored). From the graphs some parameters can be determined, like the Weibull scale parameter α at $F = 1 - e^{-1} \approx 0.63$ and probabilities versus times. However, for estimating the distribution parameters and confidence intervals, quantitative techniques are necessary.

The present chapter aims to discuss techniques to estimate parameters and their accuracies. Graphical analysis and parameter estimation can be treated as distinct fields, but consistency between the two fields is desired. The best fit in a graph has parameters and these should preferably be the same as parameter estimators. It will be shown that linear regression (LR) and parametric plotting using the expected plotting position can be fully consistent.

Once the distribution parameters are established, inferences are possible. For example, it will be possible to make prognoses about component and system behaviour, such as expected lifetime, failure rates and required stock of spare parts. Parameter estimation is therefore often crucial for decision-making about investments, preventive measures, and so on.

This chapter studies two types of parameter estimation: maximum likelihood (ML) (Section 6.2) and linear regression (Section 6.3). Their accuracy and how they relate to graphical analysis are included. This will be elaborated for the Weibull distribution, the exponential distribution and the normal distribution.

6.1 General Aspects with Parameter Estimation

Chapter 4 presented various statistical distributions like the Weibull, exponential and normal distributions. Distributions have a set of m parameters $(u_1, \ldots, u_m)$, such as the normal distribution ($m = 2$: parameters μ and σ), Weibull distribution ($m = 2$: parameters α and β; $m = 3$: parameters α, β and δ; $m = 1$: parameter α) or the exponential distribution ($m = 1$: parameter θ). Furthermore, distributions have one or more variables. Assuming one variable, the distribution density function f can be written as:

$$f(x) = f(x \mid u_1, \ldots, u_m) \tag{6.1}$$

Reliability Analysis for Asset Management of Electric Power Grids, First Edition. Robert Ross.
© 2019 John Wiley & Sons Ltd. Published 2019 by John Wiley & Sons Ltd.
Companion website: www.wiley.com/go/ross/reliabilityanalysis

This notation has the format 'function(variable|parameters)', meaning the variable x with a distribution parameter set $(u_1, \ldots, u_m)$. For instance, with the exponential distribution (see Table 4.14):

$$f_{Exp}(x) = f(x \mid \theta) = \frac{1}{\theta}e^{-\frac{x}{\theta}} \tag{6.2}$$

If the parameters are known, then $f(x)$ is completely defined. However, often, the distribution parameters are not known, but a limited sample of observations $(x_1, \ldots, x_n)$ may be available. It is also possible that n objects are put to a test, but that only the failure data of $r \leq n$ objects are known (i.e. uncensored), leaving the other $n - r$ as censored data. The subject of censored data was discussed in Section 5.2.2.

At this point it may still be a subject to explore whether a single distribution actually applies or whether there are multiple failure modes and subpopulations (cf. Section 2.5) at work. The graphical analysis described in Chapter 5 may provide adequate insight with respect to what data might be regarded as one population (even if only hypothetically). Furthermore, the similarity index (Section 3.6) may evaluate whether it is useful to apply a more complicated distribution or stick to a simpler basic distribution.

Parameter estimation methods use (sometimes only partly) known failure data $x_1, \ldots, x_n$ to find best estimates of the parameters $u_1, \ldots, u_m$. Different data $x_1, \ldots, x_n$ will generally lead to different best estimates of $u_1, \ldots, u_m$. Now the roles of variables and parameters are swapped. There is a distribution $v_n(u|x)$ of parameter estimator values which are now the variables and there are data which are now the parameters of $v_n(u|x)$:

$$v_n(u_1, \ldots, u_m \mid x_1, \ldots, x_n) \tag{6.3}$$

For instance, assume a population of N variable values t that are exponentially distributed with parameter θ (Eq. (6.2)). According to Section 4.2.4.2, the exponential distribution mean $<t>$ equals θ. If a first set of observations $(t_1, \ldots, t_n)$ is sampled with $n < N$, then an estimator $\hat{u}_{n,1}(t_1, \ldots, t_n)$ for the parameter θ can be the average of the observations $\sum t_i/n$. Example 5.5 showed a case of $n = 11$ observations, as shown in Table 6.1. Taking the mean of the observations in the table yields $\hat{u}_{n,1} = 20.9$ years. If a second sample of 11 observations is drawn from the distribution, there will be other values $t'_1, \ldots, t'_n$ and the estimate $\hat{u}_{n,2}(t'_1, \ldots, t'_n)$ most likely differs from $\hat{u}_{n,1}(t_1, \ldots, t_n)$. Repeated samples of size n will yield a distribution $v_n(u|T_1, \ldots, T_n)$ for estimator $\hat{u}$, where $T_1, \ldots, T_n$ have the role of parameters. This distribution is characterized by a mean $<\hat{u}>$ and higher moments similar to the specific distributions described in Chapter 4.

In the following sections some characteristics of estimators are discussed in greater depth, with additional focus on the asymptotic behaviour of estimated parameters and their moments.

Table 6.1 Supposedly exponentially distributed data. The average observation $\hat{u}$ is 20.9 years, which is used as estimator of the exponential distribution parameter θ.

Operational t (yr)	2.00	2.27	3.47	4.52	5.19	15.1	16.3	36.7	40.6	48.7	55.9
Failed?	Yes	Yes	Yes	Yes	Yes	Yes	Yes	Yes	Yes	Yes	Yes
Index i	1	2	3	4	5	6	7	8	9	10	11

6.1.1 Fundamental Properties of Estimators

Before introducing two families of parameter estimators in Sections 6.2 and 6.3, three properties of parameter estimators are discussed. These are: bias, efficiency and consistency.

In brief, these concern the systematic error in estimators, the variance of estimators and finally the convergence of the estimator to the true value with increasing sample size n. In the following, the three properties are discussed in more detail.

6.1.1.1 Bias

Bias is the difference between the expected parameter estimator value $<\hat{u}>$ which depends on the estimation method and the value of the true parameter v (Greek lowercase character upsilon). The mean is obtained by averaging over all possible observation samples $(x_1, \ldots, x_n)$. Another name for bias is systematic error. The bias $B(\hat{u}, n; v)$ is defined as:

$$B(\hat{u}, n; v) = <\hat{u}_n> - v \tag{6.4}$$

The suffix n denotes the number of observations on which the estimated parameter $\hat{u}_n$ is based. An estimator is unbiased if the difference between the expected value $<\hat{u}>$ and the true value v is zero.

If a certain estimator is biased, but the bias $B(\hat{u}, n; v)$ is known, then an unbiased estimator $\breve{u}$ is found by simply subtracting the bias from the estimator:

$$\breve{u}_n = \hat{u}_n - B(\hat{u}, n; v) \tag{6.5}$$

With Eq. (6.4) it follows directly that the expected unbiased estimator equals the true value:

$$<\breve{u}_n> = v \tag{6.6}$$

Even for the same distribution, various estimators for the distribution parameters may be possible. In Section 6.2 the so-called maximum likelihood and in Section 6.3 the linear regression estimators are discussed. Each features its own bias and other characteristics such as efficiency and consistency.

Correction of an estimator for its bias is called 'unbiasing'. This step can have a very significant impact on the analysis of small data sets particularly.

6.1.1.2 Efficiency

Efficiency concerns the variance of the expected parameter estimator $\hat{u}_n$ with respect to the true parameter value v. Knowledge of the efficiency gives an indication of the accuracy of estimated parameters and of how firm inferences actually are. In the following it is discussed that the variance of estimators features a lower bound (i.e. the variance of estimators is equal to or larger than a theoretical minimum that depends on the distribution).

The squared error of an estimator $\hat{u}_n$ is:

$$<(\hat{u}_n - v)^2> = <(\breve{u}_n + B(\hat{u}, n; v) - v)^2> = <[(\breve{u}_n - v) + B(\hat{u}, n; v)]^2> \tag{6.7}$$

$B(\hat{u}, n; v)$ is the bias as defined in Eq. (6.4) and $\breve{u}$ the unbiased estimator of Eq. (6.5). As the bias itself is a constant, since it is already the result of an integration over all samples, the squared error can be split into the variance of the unbiased estimator plus the square of the bias:

$$<(\hat{u}_n - v)^2> = <(\breve{u}_n - v)^2> + B(\hat{u}, n; v)^2 = \text{var}(\breve{u}_n) + B(\hat{u}, n; v)^2 \tag{6.8}$$

The Cramér–Rao inequality [36, 37] states a relationship between the Fisher information and the unbiased parameter estimator $\breve{u}$:

$$\text{var}(\breve{u}_n) \geq \frac{1}{I(v)} \tag{6.9}$$

Here, $I(v)$ is the so-called Fisher information, defined as:

$$I(v) = <\left(\frac{\partial}{\partial v} \ln f(X_1, \ldots, X_n, v)\right)^2> \tag{6.10}$$

where f is the distribution density function of $(X_1, \ldots, X_n)$. If the X_i are independent, then the product of the individual densities $f(X_i)$ can be taken. This product is also called the likelihood function of f:

$$I(v) = <\left(\frac{\partial}{\partial v} \ln\left(\prod_{i=1}^{n} f(X_i)\right)\right)^2> \tag{6.11}$$

If $\ln(f(X|v)$ can be differentiated twice with respect to v and the function, then $I(v)$ can also be written as:

$$I(v) = -<\frac{\partial^2}{\partial v^2} \ln\left(\prod_{i=1}^{n} f(X_i)\right)> \tag{6.12}$$

Equation (6.9) means that $1/I(v)$ appears to be the lower bound of the variance that an unbiased estimator $\breve{u}$ for a given distribution $f(X)$, parameter v and sample size n can have.

A biased estimator $\hat{u}_n$ has a higher lower bound of its variance:

$$\text{var}(\hat{u}_n) \geq \frac{\left(1 + \frac{\partial}{\partial v} B(\hat{u}, n; v)\right)^2}{I(v)} \tag{6.13}$$

In practice, the variance of estimators may be calculated analytically or may have to be determined numerically. This variance will then be equal to or larger than the lower bound in Eq. (6.13), which reduces for an unbiased estimator to the lower bound in Eq. (6.9).

The efficiency $e(\hat{u}_n)$ quantifies how well the variance of an estimator approaches the lower bound of the variance [38]:

$$e(\hat{u}_n) = \frac{[I(v)]^{-1}}{\text{var}(\hat{u}_n)} \leq 1 \tag{6.14}$$

In other words, the efficiency is the ratio between the lowest possible variance of the unbiased estimator and the actual variance of the parameter estimator.

6.1.1.3 Consistency

Consistency concerns the accuracy of a parameter estimator and its dependence on the sample size n of the observations $x_1, \ldots, x_n$ on which the estimator is based. If a parameter estimator $\hat{u}_n$ is consistent, then its distribution converges towards the true parameter v for $n \to \infty$. Mathematically formulated:

$$\lim_{n \to \infty} P(|\hat{u}_n - v| < \varepsilon) = 1 - \lim_{n \to \infty} P(|\hat{u}_n - v| \geq \varepsilon) = 1 \qquad (6.15)$$

For each, however small, $\varepsilon > 0$, the probability that the difference between the estimator $\hat{u}_n$ and the true parameter v will differ by less than ε approaches 1 as the sample size n approaches infinity. In short, if an estimator is consistent, then the true parameter value can be obtained with any desired accuracy by gathering a sufficiently large data set $x_1, \ldots, x_n$.

This concept of consistency was introduced by Fisher [38, 39]. Since it only concerns the limit $n \to \infty$, it underestimates the importance of the low end, namely small sample sizes. For asset management of strategic infrastructures, extending the concept of consistency towards small sample sizes n is very much worthwhile.

In addition to the Fisher meaning of consistency, therefore, a stronger consistency concept is preferable for small n. There is a minimum n_{min} for n; the sample size n must always be larger than the number n_{min} of independent distribution parameters in order to allow an estimation. From this n_{min} on, the concept of consistency may be sharpened. In case the estimator $\hat{u}_n$ is not unbiased, for all $n > n_{min}$ the bias of an $n + 1$-sized sample must be equal to or less than the n-sized sample:

$$\text{for } B(\hat{u}, n; v) \neq 0 : \qquad \left| \frac{B(\hat{u}, n + 1; v)}{B(\hat{u}, n; v)} \right| < 1 \qquad (6.16)$$

The absolute bias is then strictly decreasing with increasing n. If the '<' sign is replaced by '$\leq$', the requirement would be that the bias is monotonically decreasing with increasing n. In that case, however, the as yet non-zero bias might not improve at certain sample size n with more observations. For that reason, estimators with a strictly decreasing absolute bias are required.

Likewise, the estimator variance should decrease strictly with increasing sample size n:

$$\left| \frac{\text{var}(\hat{u}_{n+1})}{\text{var}(\hat{u}_n)} \right| < 1 \qquad (6.17)$$

Estimators that meet these stricter consistency requirements make collecting additional data for analysis always rewarding in terms of greater accuracy.

6.1.2 Why Work with Small Data Sets?

The impact of bias and scatter can be very large on statistical inferences. Systematic and random errors, together with consistency aspects, make one wonder: why not just collect a sufficient amount of data to perform accurate statistical analyses?

For accurate statistics it is indeed advisable to work with as much data as possible. However, in asset management of strategic infrastructures, decisions regularly have to be made on scarce data (i.e. small sample sizes n are often unavoidable).

The background is at least twofold: testing and unexpected (possibly high-impact) failures. As for testing, the costs of test objects and the costs of testing in high-voltage and high-power labs can be considerable. As for the first, assets like large generators, transformers and parts like generator bars can be very expensive. As a consequence, it is not economically feasible to use large test series. In practice, test series can be as small as $n = 5$ in quality control (and often concern only parts of a single asset).

As for unexpected (possibly high-impact) failures, there is a need for timely decision-making. Due to the urgency of preventing interruption of services for such infrastructures, as soon as some failures alert the organization, a status update is required, possible scenarios have to be evaluated and choices have to be made. As a consequence, the number of observations on which a decision must be based can be very small.

The principle of improving accuracy, as in the previous sections, by collecting and analysing a sufficient amount of data is highly recommended, but that may not be feasible in crisis situations or with expensive testing. On the other hand, if many decisions have to be made, it may be worthwhile having a high mean success rate (i.e. on average, the decisions appear right and efficient). The challenge is therefore to also perform well with scarce data.

6.1.3   Asymptotic Behaviour of Estimators

Consistency implies that parameter estimators asymptotically approach the true parameters with increasing observation sample size n (and n larger than the number of independent distribution parameters).

If the parameter estimators can be analytically expressed as a function of n, then the asymptotic behaviour follows from that expression and it may be checked for consistency (Section 6.1.1.3). However, an analytical expression is not always possible, and then simulations and numerical analysis are required to evaluate the asymptotic behaviour. Based on simulations with various sample sizes n, the bias and variance may be estimated.

Even for parameter estimators where the bias and variance as a function of n cannot be derived analytically, it may still be possible to find an empirical expression based on the numerical analysis. As a model, a three-parameter power function was introduced that appeared analytically correct in various cases, and otherwise could serve as an approximation [27]. Assume E_n and E_∞ are expected values of an estimator based on finite and infinite sample sizes, respectively. The difference D_n between them is expressed in this three-parameter power function model as:

$$D_n = E_n - E_\infty = Q \cdot (n - R)^P \qquad (6.18)$$

The parameters P, Q and R of this power function are to be determined analytically or numerically. The parameters should not be confused with quantities like probability, reliability, and so on. Particularly the parameter R has a large influence in describing the asymptotic behaviour at small sample sizes n.

This expression can be applied to the bias, but also to the variance (or standard deviation) and has been used successfully to describe other asymptotic behaviour as well. For unbiased estimators, the power function parameter $Q = 0$. For strictly decreasing absolute biases and other estimators, $P < 0$. The power function D_n is discussed further in Section 10.2.3.

6.2   Maximum Likelihood Estimators

There are various methods to estimate the distribution parameters $u_1, \ldots, u_m$, of which ML is discussed in the present section. First the method, as such, will be explained for the uncensored case in Section 6.2.1 and the censored case in Section 6.2.2. Subsequently, the method is applied to the Weibull-2 distribution in Section 6.2.3, the exponential distribution in Section 6.2.4 and the normal distribution in Section 6.2.5.

6.2.1 ML with Uncensored Data

The concept of the ML method is to estimate distribution parameters $u_1, \ldots, u_m$ such that the so-called likelihood function L is maximized. The procedure is as follows. Suppose a set of n uncensored observations $(x_1, \ldots, x_n)$ is obtained. The observations are assumed to be independent and identically distributed. The distribution is supposed to have m parameters $(u_1, \ldots, u_m)$, as in Section 6.1.

As a next step, the roles of the variables and parameters are swapped. The likelihood function L is a function with variable set $(u_1, \ldots, u_m)$ and the set observations $(x_1, \ldots, x_n)$ are the parameters. The likelihood function is now defined as:

$$L(u_1, \ldots, u_m; x_1, \ldots, x_n) = \prod_{i=1}^{n} f(x_i | u_1, \ldots, u_m) \tag{6.19}$$

A semicolon ';' separates the variables and the parameter input (the parameters of f).

It is convenient to work with the logarithm of the likelihood function, which turns the product into a sum:

$$\ln[L(u_1, \ldots, u_m; x_1, \ldots, x_n)] = \sum_{i=1}^{n} \ln f(x_i | u_1, \ldots, u_m) \tag{6.20}$$

The parameters $u_1, \ldots, u_m$ are estimated by finding the set of values $\hat{u}_1, \ldots, \hat{u}_m$ that maximize $\ln[L]$. The set $\hat{u}_1, \ldots, \hat{u}_m$ are the so-called maximum likelihood estimators (MLEs) of the parameters $u_1, \ldots, u_m$. For the maximum:

$$d \ln[L(u_1, \ldots, u_m; x_1, \ldots, x_n)] = \sum_{i=1}^{n} \frac{\partial \ln[L(u_1, \ldots, u_m | x_1, \ldots, x_n)]}{\partial u_i} du_i = 0 \tag{6.21}$$

Each partial derivative $\partial \ln L / \partial u_i$ in Eq. (6.21) must be zero. This produces a set of m equations from which the MLEs $\hat{u}_1, \ldots, \hat{u}_m$ are solved. These equations cannot always be solved analytically (e.g. in the case of the Weibull distribution) and then a numerical method like the Newton–Raphson method can be employed to maximize $\ln[L]$. Since the logarithmic function is monotonically increasing, the likelihood function L will have its maximum at the same set $\hat{u}_1, \ldots, \hat{u}_m$.

The condition of independent variables $x_1, \ldots, x_n$ may not be met. An adaptation of the likelihood function is possible that acknowledges the correlations between the variables [40]. For the sake of completeness, the joint probability function for non-independent variables is given as [40]:

$$f(x_1, \ldots, x_n) = \frac{1}{(2\pi)^{n/2} \sqrt{\det(COV)}}$$
$$\cdot \exp\left(-\frac{1}{2}[x_1 - \mu_1, \ldots, x_n - \mu_n]COV^{-1}[x_1 - \mu_1, \ldots, x_n - \mu_n]^T\right) \tag{6.22}$$

with COV the covariance matrix and $\mu_1, \ldots, \mu_n$ the means of the variables $x_1, \ldots, x_n$. Theoretically wrong or not, the correlation is often ignored because the covariance between the variables may be unknown and/or experimenters may be unaware of covariance implications. Therefore, Eq. (6.20) is rather used in practice without notice.

6.2.2 ML for Sets Including Censored Data

If a test stops, after which r out of n test objects have failed and $k = n - r$ have not failed as yet, the set contains right-censored data (i.e. failure lies beyond a known time, see Section 5.2.2 and Table 5.2). The likelihood function L can be adapted to include the corresponding survival probabilities of the censored data. With Type I censoring there is a predetermined maximum of the results, say ξ. Most often this maximum concerns the test period θ, regardless of the number of failed objects and therefore all observed failure times $t_i \leq \theta$. In a step voltage test it could also be the voltage, because the maximum feasible test voltage may be reached while not all test objects have failed. With Type I censoring the largest observed value $x_r \leq \xi$. It should be noted that r is not fixed. The likelihood function then becomes:

$$L(u_1, \ldots, u_m; n, r, \xi; x_1, \ldots, x_r) = (1 - F(\xi))^{n-r} \cdot \prod_{i=1}^{r} f(x_i | u_1, \ldots, u_m) \tag{6.23}$$

With Type II censoring there is a predetermined number r of the results (e.g. the test is stopped if r failures are observed regardless of the time it takes). With Type II censoring, usually the largest observed value $x_r = \xi$:

$$L(u_1, \ldots, u_m; n, r, \xi; x_1, \ldots, x_r) = (1 - F(\xi))^{n-r} \cdot \prod_{i=1}^{r} f(x_i | u_1, \ldots, u_m) \tag{6.24}$$

If $x_r \neq \xi$ then the equation is the same as Eq. (6.23), but it should be noted that with Type II the number of observed failures r is fixed and ξ is not fixed.

More generally, the likelihood function can apply to the case of r uncensored data x_i $(i = 1, \ldots, r)$ and censored data ξ_j $(j = 1, \ldots, n - r)$. This situation often applies to a grid in which components are replaced and the times of censoring may be shorter than the already observed breakdown times. The likelihood function then becomes:

$$L(u_1, \ldots, u_m; \xi_1, \ldots, \xi_{n-r}; x_1, \ldots, x_r) = \prod_{j=1}^{n-r} (1 - F(\xi_j)) \cdot \prod_{i=1}^{r} f(x_i | u_1, \ldots, u_m) \tag{6.25}$$

Whether r or ξ is fixed becomes relevant in Monte Carlo experiments, where repeatedly test series are generated. Type I would censor the data of each series beyond ξ, which results in comparing series with different numbers of observations r. Type II would censor the data of each series after the fixed number of r smallest observations. All test series would then have the same number of uncensored data (namely r). The situation of Eq. (6.25) can be part of various investigations, including random censoring by time or number of observations.

A further extension of the method could involve items that failed at an unknown moment before a censoring time ξ. This would introduce a product of terms with $F(\xi_k)$ in the likelihood function L (analogous to the terms $1 - F(\xi_j)$ for the surviving items in Eq. (6.25)). This is not elaborated further here.

6.2.3 ML for the Weibull Distribution

As mentioned in Section 4.2.1, the Weibull distribution can be used with one, two or three parameters. The most widely used form is the two-parameter Weibull distribution, for which ML is elaborated in the following. First the case of uncensored data is discussed, followed by censored data in Section 6.2.3.1.

6.2.3.1 ML Estimators for Weibull-2 Uncensored Data

The Weibull-2 distribution is (see Section 4.2.1.1):

$$F(x; \alpha, \beta) = 1 - e^{-\left(\frac{t}{a}\right)^{\beta}} \tag{6.26}$$

The distribution density f is:

$$f(x; \alpha, \beta) = \frac{\beta}{\alpha^{\beta}} \cdot t^{\beta-1} \cdot e^{-\left(\frac{t}{a}\right)^{\beta}} \tag{6.27}$$

The ML parameter estimators a and b for the Weibull-2 distribution parameters α and β based on a set of n uncensored observations $x_1, \ldots, x_n$ are discussed in the following. The logarithmic likelihood function thus becomes:

$$\ln L(a, b; x_i) = \sum_{i=1}^{n} \ln f(x_i \mid a, b)$$

$$= n \cdot \ln b - n \cdot b \ln a + (b-1) \cdot \sum_{i=1}^{n} \ln x_i - \sum_{i=1}^{n} \left(\frac{x_i}{a}\right)^{b} \tag{6.28}$$

Maximizing $\ln(L)$ yields:

$$\frac{\partial \ln L(a, b; x_i)}{\partial a} = 0 = -\frac{n \cdot b}{a} + \frac{b}{a^{b+1}} \cdot \sum_{i=1}^{n} x_i^{b} \tag{6.29}$$

$$\frac{\partial \ln L(a, b; x_i)}{\partial b} = 0 = \frac{n}{b} - n \ln a + \sum_{i=1}^{n} \ln x_i - \sum_{i=1}^{n} \left(\frac{x_i}{a}\right)^{b} \ln \frac{x_i}{a} \tag{6.30}$$

Rewriting Eq. (6.29) yields:

$$a^{b} = \frac{1}{n} \cdot \sum_{i=1}^{n} x_i^{b} \tag{6.31}$$

With this expression, a can be eliminated from Eq. (6.30). This yields an expression which contains estimator b alone:

$$\frac{\sum_{i=1}^{n} x_i^{b} \ln x_i^{b}}{\sum_{i=1}^{n} x_i^{b}} - \frac{1}{n} \cdot \sum_{i=1}^{n} \ln x_i = \frac{1}{b} \tag{6.32}$$

Equation (6.32) cannot be solved analytically and therefore numerical methods like the Newton–Raphson method are applied to estimate b. Estimator a is found by substituting the numerically determined b in Eq. (6.31).

6.2.3.2 ML Estimators for Weibull-2 Censored Data

Assume again a sample with size n under test. In case of censored data a subgroup with size r failed, yielding known observations $x_1, \ldots, x_r$ $(r < n)$, while the data of a complementary subgroup with size $n - r$ remains unknown and censored at ξ_j $(j = 1, \ldots, n - r)$. As mentioned in Section 6.2.2, there is an expression censoring in general (Eq. (6.25)) and there are also expressions for two specific types of censoring, namely Type I (i.e. by value, for example breakdown time) and Type II (i.e. by number r). In the general case the logarithmic likelihood function becomes:

$$
\ln L(u_1, \ldots, u_m; \xi_1, \ldots, \xi_{n-r}; x_1, \ldots, x_r)
$$

$$
= \sum_{j=1}^{n-r} \ln(1 - F(\xi_j)) + \sum_{i=1}^{r} \ln f(x_i \mid u_1, \ldots, u_m)
$$

$$
= -\sum_{j=1}^{n-r} \left(\frac{\xi_j}{a}\right)^b + r \cdot \ln b - r \cdot b \ln a + (b-1) \cdot \sum_{i=1}^{r} \ln x_i - \sum_{i=1}^{r} \left(\frac{x_i}{a}\right)^b \qquad (6.33)
$$

In principle, each surviving object j can have its own censored value. In practice, this happens when assets are in service and replaced by new assets after failure, which is quite normal in electricity grids. For Type I (censoring by value) the ξ_j are defined and usually by a single fixed value ξ. In practice, this may be because the test period is limited by availability and the program is aborted, or it can be because a test is still carrying on, but intermediate results are wanted. For Type II the number r is defined and censoring takes place at the last observed value x_r, and therefore for each j: $\xi_j = \xi = x_r$.

Maximizing $\ln(L)$ yields:

$$
\frac{\partial \ln L(a,b)}{\partial a} = 0 = -\frac{r \cdot b}{a} + \frac{b}{a^{b+1}} \cdot \left(\sum_{i=1}^{r} x_i^b + \sum_{j=1}^{n-r} \xi_j^b\right) \qquad (6.34)
$$

$$
\frac{\partial \ln L(a,b)}{\partial b} = 0 = \frac{r}{b} - r \cdot \ln a + \sum_{i=1}^{r} \ln x_i - \sum_{i=1}^{r} \left(\frac{x_i}{a}\right)^b \ln \frac{x_i}{a} - \sum_{j=1}^{n-r} \left(\frac{\xi_j}{a}\right)^b \ln \frac{\xi_j}{a} \qquad (6.35)
$$

From Eq. (6.34) follows:

$$
a^b = \frac{1}{r} \cdot \left(\sum_{i=1}^{r} x_i^b + \sum_{j=1}^{n-r} \xi_j^b\right) \qquad (6.36)
$$

And from Eq. (6.35) with substitution of the expression for a:

$$
\frac{\sum_{i=1}^{n} x_i^b \ln x_i^b + \sum_{j=1}^{n-r} \xi_j^b \ln \xi_j^b}{\sum_{i=1}^{n} x_i^b + \sum_{j=1}^{n-r} \xi_j^b} - \frac{1}{r} \cdot \sum_{i=1}^{r} \ln x_i = \frac{1}{b} \qquad (6.37)
$$

The censoring values ξ_j depend on the type of censoring, as mentioned above. In a single experiment or test run the difference may not seem significant, but when expected values are determined by repeating experiments (e.g. Monte Carlo experiments) the choice for the type of censoring becomes relevant indeed. Expected values are discussed in the following sections.

6.2.3.3 Expected ML Estimators for the Weibull-2 Distribution

The ML method fulfils the requirements of an efficient estimator. As discussed in Section 6.1.1.3, the consistency property is related to the behaviour with large sample sizes n. With asset management and particularly incident management (see Section 1.4), working with small data sizes is practically standard. Therefore, the matter of bias and scatter is of great interest, especially for small sample sizes r and/or n, even down to $3, \ldots, 10$. Naturally, the scatter will be large, which leads to inaccurate information for decision-making. However, for some decisions, estimates in terms of orders of magnitude can be sufficient.

As the bias for the shape parameter cannot be determined analytically, the bias and scatter must be investigated numerically. Fortunately, not all combinations need to be simulated. Pivotal quantities and functions can be investigated, after which the results can be translated to combinations in general. Let y be distributed with the normalized Weibull distribution (i.e. $\alpha = \beta = 1$), which corresponds to the exponential distribution:

$$F(y; \alpha = 1, \beta = 1) = 1 - e^{-\left(\frac{y}{1}\right)^{1}} = 1 - e^{-y} \tag{6.38}$$

The ML estimators of the scale parameter $\alpha = 1$ and the shape parameter $\beta = 1$ are indicated as $a_{1,1}$ and $b_{1,1}$. Any Weibull-2 distribution can be produced by substituting:

$$y = \left(\frac{x}{\alpha}\right)^{\beta} \tag{6.39}$$

The variable y is a pivotal quantity. The model does not change with the choice of the α and β parameters. These ML estimators $a_{1,1}$ and $b_{1,1}$ are the solutions of:

$$\frac{\sum_{i=1}^{n} y_i^{b_{1,1}} \ln y_i^{b_{1,1}}}{\sum_{i=1}^{n} y_i^{b_{1,1}}} - \frac{1}{n} \cdot \sum_{i=1}^{n} \ln y_i = \frac{1}{b_{1,1}} \tag{6.40}$$

and:

$$a_{1,1}^{b_{1,1}} = \frac{1}{n} \cdot \sum_{i=1}^{n} y_i^{b_{1,1}} \tag{6.41}$$

The ML estimators of the general scale parameter α and shape parameter β are indicated as $a_{\alpha,\beta}$ and $b_{\alpha,\beta}$. Substitution of Eq. (6.39) in Eqs (6.40) and (6.41) with subsequent comparison with Eqs (6.32) and (6.31) shows that:

$$b_{\alpha,\beta} = \beta \cdot b_{1,1} \quad \Leftrightarrow \quad \frac{1}{b_{\alpha,\beta}} = \frac{1}{\beta \cdot b_{1,1}} \tag{6.42}$$

and:

$$\left(\frac{a_{\alpha,\beta}}{\alpha}\right)^{b_{\alpha,\beta}} = a_{1,1}^{b_{1,1}} \quad \Leftrightarrow \quad b_{\alpha,\beta}(\ln a_{\alpha,\beta} - \ln \alpha) = b_{1,1} \ln a_{1,1} \tag{6.43}$$

These expressions, like $b_{\alpha,\beta}/\beta = b_{1,1}$, are pivotal (i.e. independent of actual α and β). The relationships in Eqs (6.42) and (6.43) can be used to describe the expected ML estimators for Weibull-2 in general. Several relationships for expected estimators are:

$$<b_{\alpha,\beta}> = \beta \cdot <b_{1,1}> \tag{6.44}$$

$$<\frac{1}{b_{\alpha,\beta}}> = \frac{1}{\beta} \cdot <\frac{1}{b_{1,1}}> \tag{6.45}$$

$$\left< \left(\frac{a_{\alpha,\beta}}{\alpha} \right)^{b_{\alpha,\beta}} \right> = <a_{1,1}{}^{b_{1,1}}> \tag{6.46}$$

$$<b_{\alpha,\beta} \ln a_{\alpha,\beta}> = <b_{1,1} \ln a_{1,1}> + <b_{1,1}> \cdot \beta \ln \alpha \tag{6.47}$$

The expected parameter estimators show which entities have to be averaged.

6.2.3.4 Formulas for Bias and Scatter

It should be noted that the ML estimators are functions of the number of censored and uncensored observations. As consistent estimators, the bias in the expected estimators approaches zero with increasing number of observations (see Section 6.1.1.3). Generally, the bias in the parameter estimators cannot be ignored for small data sets that often have to be dealt with in incident management. However, the above relations can be used for unbiasing. The unbiased parameters are indicated with suffix U: b_U, $(1/b)_U$ and $(b \ln a)_U$:

$$<b_U> = \beta \quad \rightarrow \quad b_U = \frac{b_{\alpha,\beta}}{<b_{1,1}>} \tag{6.48}$$

$$\left< \left(\frac{1}{b} \right)_U \right> = \frac{1}{\beta} \quad \rightarrow \quad \left(\frac{1}{b} \right)_U = \frac{1}{b_{\alpha,\beta}} \cdot \left(<\frac{1}{b_{1,1}}> \right)^{-1} \tag{6.49}$$

$$<(b \ln a)_U> = \beta \ln \alpha \quad \rightarrow \quad (b \ln a)_U = \frac{b_{\alpha,\beta} \ln a_{\alpha,\beta} - <b_{1,1} \ln a_{1,1}>}{<b_{1,1}>} \tag{6.50}$$

For purely uncensored data and also for a limited combination of censored data, the pivotal functions have been reported in various studies [27, 41, 42]. In such studies, Monte Carlo simulations were carried out for a range of sample sizes and the resulting bias was reported in tables. Particularly in [27] the random number generator was tested to produce expected values for a range of test functions (see Section 10.4.4).

Such tables have been used in [27] to derive a power expression for the bias as well as for the standard deviation in the shape of the power function approximation in Eq. (6.18):

$$D_n = E_n - E_\infty = Q \cdot (n - R)^P \tag{6.51}$$

Here, E_n and E_∞ are expected and asymptotic values of an estimator based on finite and infinite sample sizes and D_n is the difference between them; P, Q and R are the parameters of this power function determined from the tables. As P is negative, $D_n \rightarrow 0$ with $n \rightarrow \infty$. For the shape parameter it is found that:

$$<b_{1,1}>_n - <b_{1,1}>_\infty = <b_{1,1}>_n - 1 \approx \frac{1.32}{n - 2} \tag{6.52}$$

So, $P = -1$, $Q = 1.32$ and $R = 2$. In [27] the question was raised whether R should be an integer. The background is that Eq. (6.52) follows from solving Eqs (6.40) and (6.41), where two parameters require two degrees of freedom.

The unbiased shape parameter estimator b_U expressed in terms of the empirical function of sample size n becomes [27]:

$$b_U = b_{\alpha,\beta} \cdot \frac{1}{<b_{1,1}>} \approx b_{\alpha,\beta} \cdot \frac{n - 2}{n - 0.68} \tag{6.53}$$

In statistics, the degrees of freedom play a role and $R = 2$ could be linked to two degrees of freedom taken by the two parameters. From the analysis this cannot be excluded, but in other cases R is quite certainly not an integer and so far there is no theoretical need for P, Q and R to be integers. In fact, in [27] other values are also mentioned which give practically the same results, like $P = -0.9995$, $Q = 1.367$ and $R = 1.92$, which unbias with an accuracy better than 0.3%. The choice in Eq. (6.52) is simple and effective in the sense that it reduces the bias to below 0.6% for $n > 3$ and to about 2% for $n = 3$.

For the standard deviation of the unbiased b_U the asymptotic value $E_\infty = 0$, therefore D_n equals E_n. The ML estimated standard deviation of b_U is found to be [27]:

$$\sigma_{b_U} \approx \frac{0.8}{\sqrt{n-3}} \cdot \beta \tag{6.54}$$

This power function relation offers the possibility to determine the sample size n that would be needed to reach a specified precision in b with experiments and subsequent ML analysis. As for the mean b_U from M trials with sample size n, the accuracy is $\sigma_{bU}/\sqrt{M}$ (see Eq. (4.69)). The number of trials in [27] was 20 000 and therefore the standard deviation of $<b_U>/\beta$ equalled $0.8/\sqrt{(20\,000(n-3))}$.

If the shape parameter estimator is not unbiased, then the standard deviation of b is:

$$\sigma_b = <b_{1,1}> \cdot \sigma_{b_U} \approx \frac{n-0.68}{n-2} \cdot \frac{0.8}{\sqrt{n-3}} \cdot \beta \tag{6.55}$$

The inverse shape parameter $(1/b)$ can also be approached with the asymptotic power expression of Eq. (6.51):

$$<\frac{1}{b_{1,1}}>_n - <\frac{1}{b_{1,1}}>_\infty = <\frac{1}{b_{1,1}}>_n - 1 \approx \frac{-0.76}{n-0.18} \tag{6.56}$$

So, $P = -1$, $Q = -0.76$ and $R = 0.18$. The unbiased inverse shape parameter estimator $(1/b)_U$ expressed in terms of the empirical function of sample size n becomes [27]:

$$\left(\frac{1}{b}\right)_U = \frac{1}{b_{\alpha,\beta}} \cdot \left(<\frac{1}{b_{1,1}}>\right)^{-1} \approx \frac{1}{b_{\alpha,\beta}} \cdot \frac{n-0.18}{n-0.94} \tag{6.57}$$

The ML estimated standard deviation of $1/b_U$ is found to be [27]:

$$\sigma_{1/b_U} \approx \frac{0.8}{\sqrt{n-1.2}} \cdot \frac{1}{\beta} \tag{6.58}$$

Finally, the product $(b \ln a)$ can be approached with the asymptotic power expression of Eq. (6.51) too [27]:

$$<b_{1,1} \ln a_{1,1}>_n - <b_{1,1} \ln a_{1,1}>_\infty = <b_{1,1} \ln a_{1,1}>_n - 0 \approx -0.26 \cdot (n-1.9)^{-1.23} \tag{6.59}$$

Here, $P = -1.23$, $Q = -0.26$ and $R = 1.9$. The unbiased product $b \ln a$ follows from Eqs (6.50), (6.52) or (6.53) and the empirical formula (6.59) as:

$$(b \ln a)_U = \frac{b_{\alpha,\beta} \ln a_{\alpha,\beta} - <b_{1,1} \ln a_{1,1}>}{<b_{1,1}>}$$

$$\approx \left(b_{\alpha,\beta} \ln a_{\alpha,\beta} + 0.26 \cdot (n-1.9)^{-1.23}\right) \cdot \frac{n-2}{n-0.68} \tag{6.60}$$

The standard deviation of $(b \ln a)_U$ was not reported in [27].

So far the uncensored case was studied. Censoring (see Section 5.2.2) can be of Type I (suspension after predetermined time) or Type II (suspension after predetermined number of failures r out of the total n). For Type II censoring a limited study has been carried out on the shape parameter [27]. It was found that:

$$<b_{1,1}>_{n,r} - <b_{1,1}>_{\infty} = <b_{1,1}>_{n,r} - 1 \approx \frac{1.37}{n-1.92} \cdot \sqrt{\frac{n}{r}} \tag{6.61}$$

This expression was consistent with the findings in [43], where $P = -1$, $Q = 1.37$ and $R = 1.92$ for the uncensored case. In order to make this consistent with Eq. (6.52), the parameters of that equation can be used, yielding:

$$<b_{1,1}>_{n,r} - <b_{1,1}>_{\infty} = <b_{1,1}>_{n,r} - 1 \approx \frac{1.32}{n-2} \cdot \sqrt{\frac{n}{r}} \tag{6.62}$$

6.2.3.5 Effect of the ML Estimation Bias in Case of Weibull-2

If $\beta > 1$, then the bias of α is generally on the order of a few percent (typically $< 10\%$) and is often ignored. The bias in the estimated shape parameter b affects the slope of the Weibull plot. With the assumption of a negligible bias in the scale parameter a, the effect of bias is primarily that the Weibull plot rotates about its point of gravity at $t = \alpha$ (see Figure 6.1).

As shown in Figure 6.1, the slope of the graph increases due to the bias in the shape parameter with the rotation point roughly at $(t \approx \alpha, F \approx 1 - e^{-1})$. The steeper graph makes the times, such as $t_{1\%}$ and $t_{10\%}$ at low failure probabilities $F = 1\%$ and $F = 10\%$, seem

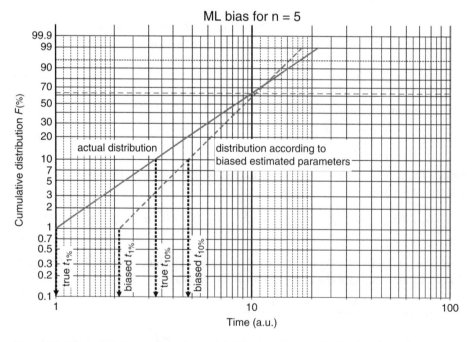

Figure 6.1 Effect of ML bias on the estimated distribution. The true distribution shown has parameters $\alpha = 10$ and $\beta = 2$. The ML parameter estimation produces a bias in both parameters. The strongest effect is a rotation of the line about the point $(t \approx \alpha, F \approx 1 - e^{-1})$.

to appear at later times compared to the true distribution. In other words, early failures will occur sooner in reality than the biased graph predicts. The unbiased ML parameter estimation therefore suggests a higher quality in that region than is actually true. Particularly for tests with small sample sizes, the overestimation of quality can be considerable. In the shown case of $n = 5$, the biased estimated time of 1% failure probability $t_{1\%}$ is more than twice the actual (i.e. true) time $t_{1\%}$. For the time of 10% failure probability this effect is smaller, but the predicted biased time $t_{10\%}$ is still about 50% larger than the true time $t_{10\%}$. The lower the demanded probability of failure F, the more significant is the quality overestimation effect due to the bias (i.e. the larger the increase of t_F).

With expensive equipment, the number of objects for destructive testing is kept to a minimum while the required failure probability tends to be lower at given times. Test sample sizes of $n = 5$ are quite common when testing generator bars for instance. In light of the discussion above, it is important to be aware of the effect of bias caused by ML parameter estimation.

6.2.4 ML for the Exponential Distribution

The exponential distribution shows as (see Section 4.2.4):

$$F(t; \theta) = 1 - e^{-\frac{t}{\theta}} \tag{6.63}$$

There is only one parameter, namely the characteristic lifetime θ. The exponential distribution density f is:

$$f(t; \theta) = \frac{1}{\theta} \cdot e^{-\frac{t}{\theta}} \tag{6.64}$$

The ML estimation of characteristic life parameter based on a set of n (uncensored) observations $t_1, \ldots, t_n$ is shown in the following. With θ_{ML} the ML parameter estimator of the characteristic life θ, the logarithmic likelihood function becomes:

$$\ln L(\theta_{ML}; t_i) = \sum_{i=1}^{n} \ln f(t_i \mid \theta_{ML}) = -n \cdot \ln \theta_{ML} - \sum_{i=1}^{n} \frac{t_i}{\theta_{ML}} \tag{6.65}$$

Maximizing $\ln(L)$ yields:

$$\frac{\partial \ln L(\theta_{ML}; t_i)}{\partial \theta_{ML}} = 0 = -\frac{n}{\theta_{ML}} + \sum_{i=1}^{n} \frac{t_i}{\theta_{ML}^2} \quad \Rightarrow \quad \theta_{ML} = \frac{1}{n} \sum_{i=1}^{n} t_i \tag{6.66}$$

The ML estimator θ_{ML} of the characteristic lifetime is therefore the mean of the lifetimes and can be calculated analytically. Given an uncensored data set this is an unbiased estimator and coincides with the definition of the characteristic life in Section 4.2.4.

As for the censored case with r uncensored t_i values ($i = 1, \ldots, r$) and $n - r$ censoring times ϑ_j ($j = 1, \ldots, n - r$), the logarithmic likelihood $\ln L$ is adapted according to Section 6.2.2 (Eq. (6.25)) for the general case:

$$\ln L(\theta_{ML}; t_i) = \sum_{j=1}^{n-r} \ln(1 - F(\vartheta_j \mid \theta_{ML})) + \sum_{i=1}^{r} \ln f(t_i \mid \theta_{ML})$$

$$= -\sum_{j=1}^{n-r} \frac{\vartheta_j}{\theta_{ML}} - r \cdot \ln \theta_{ML} - \sum_{i=1}^{r} \frac{t_i}{\theta_{ML}} \tag{6.67}$$

Maximizing $\ln L$ for the censored case yields:

$$\frac{\partial \ln L(\theta_{ML}; t_i)}{\partial \theta_{ML}} = 0 = \sum_{j=1}^{n-r} \frac{\vartheta_j}{\theta_{ML}^2} - \frac{r}{\theta_{ML}} + \sum_{i=1}^{r} \frac{t_i}{\theta_{ML}^2}$$

$$\Rightarrow \theta_{ML} = \frac{1}{r} \left(\sum_{i=1}^{r} t_i + \sum_{j=1}^{n-r} \vartheta_j \right) \tag{6.68}$$

This estimator is not necessarily unbiased.

6.2.5 ML for the Normal Distribution

The density function $f(x)$ of the normal distribution is (see Section 4.3.1):

$$f(x) = \frac{1}{\sigma\sqrt{2\pi}} \cdot e^{-\frac{1}{2}\left(\frac{x-\mu}{\sigma}\right)^2} \tag{6.69}$$

The two parameters are the mean μ and the standard deviation σ. The ML estimation based on a set of n (uncensored) observations $x_1, \ldots, x_n$ is shown in the following. With m and s the ML parameter estimators of the mean and the standard deviation, the logarithmic likelihood function becomes:

$$\ln L(m, s; x_i) = \sum_{i=1}^{n} \ln f(x_i \mid m, s) = -n \cdot \ln(s\sqrt{2\pi}) - \frac{1}{2} \sum_{i=1}^{n} \frac{(x_i - m)^2}{s^2} \tag{6.70}$$

Maximizing $\ln L$ yields:

$$\frac{\partial \ln L(m, s; x_i)}{\partial m} = 0 = -\sum_{i=1}^{n} \frac{x_i}{s^2} + \frac{n \cdot m}{s^2} \quad \Rightarrow \quad m = \frac{1}{n} \sum_{i=1}^{n} x_i \tag{6.71}$$

and:

$$\frac{\partial \ln L(m, s; x_i)}{\partial s} = 0 = -\frac{n}{s} + \sum_{i=1}^{n} \frac{(x_i - m)^2}{s^3} \quad \Rightarrow \quad s^2 = \frac{1}{n} \sum_{i=1}^{n} (x_i - m)^2 \tag{6.72}$$

The ML estimators m and s of the mean and standard deviations coincide with the definitions of mean, variance and standard deviation such as in Eqs (4.67) and (4.68). These ML estimators can be calculated analytically. Given the data set is uncensored, these are unbiased estimators.

As for the censored case with r uncensored x_i values ($i = 1, \ldots, r$) and $n - r$ censoring times ξ_j ($j = 1, \ldots, n-r$), the logarithmic likelihood $\ln L$ is adapted according to Section 6.2.2 (Eq. (6.25)) for the general case:

$$\ln L(m, s; x_i) = \sum_{j=1}^{n-r} \ln(1 - F(\xi_j \mid m, s)) + \sum_{i=1}^{r} \ln f(x_i \mid m, s)$$

$$= \sum_{j=1}^{n-r} \ln(1 - F(\xi_j \mid m, s)) + -r \cdot \ln(s\sqrt{2\pi}) - \frac{1}{2} \sum_{i=1}^{r} \frac{(x_i - m)^2}{s^2} \tag{6.73}$$

There is no analytical expression for $F(\xi_j)$. The approximations in Section 10.6 might be used instead. These are defined in terms of the normalized variable y. The values x

(and ξ) are obtained from the normalized variable y, the standard deviation σ and the mean μ as:

$$x = \sigma \cdot y + \mu \tag{6.74}$$

Most normal distribution approximations contain a summation which is not convenient when taking the logarithm $\ln[1 - F(x)]$. A second complication is that the approximations usually differ for positive versus negative y (i.e. x being larger or smaller than the mean μ). The ML estimators can be expected biased for the censored case.

6.3 Linear Regression

Linear regression is an alternative technique for the estimation of distribution parameters $u_1, \ldots, u_m$ based on a data set $t_1, \ldots, t_n$. First the technique is discussed, after which it is applied to the actual variables and parameters.

6.3.1 The LR Method

Linear regression employs an assumed linear relationship between Y and X. This approach is similar to the linear graphs based on distribution models (see Section 5.3) and can be regarded as a member of the family of linear regression methods for polynomials (discussed in Section 10.2.2.2).

The linear relationship is:

$$Y = A_0 + A_1 \cdot X \tag{6.75}$$

where X is the variable, Y the covariable, A_0 the location parameter or Y intercept (i.e. $Y(X = 0)$) and A_1 the regression coefficient or slope (i.e. derivative dY/dX). Assume there is a series of variable values x_i ($i = 1, \ldots, n$) for which a covariable y_i ($i = 1, \ldots, n$) is observed. The observed points (x_i, y_i) usually do not fit the model of Eq. (6.75) exactly, but deviate from it with an error ε_i. These errors are deviations of the observations from the model, of which an example is shown in Figure 6.2:

$$y_i = A_0 + A_1 \cdot x_i + \varepsilon_i \tag{6.76}$$

The error ε_i is also called residual (then often indicated as r_i), expressing the difference between the observed data and the model. With k observations, the linear regression parameter estimators a_0 and a_1 for A_0 and A_1 are found by minimizing the n residuals ε_i:

$$\varepsilon_i = y_i - a_0 - a_1 \cdot x_i \tag{6.77}$$

Minimizing the n residuals can be done according to various criteria. The choice of a criterion can have an impact on the bias, efficiency and consistency of estimators.

The (unweighted) least-squares (LS) method is probably the most popular method. The criterion for a best fit of the LS method is to minimize the sum of the squared errors ε_i. This method is discussed in Section 6.3.1.1. A more efficient method is the weighted least-squares (WLS) method that minimizes a weighted sum of the squared errors ε_i. The background is that the confidence intervals may vary between the different y_i. The more accurate observations may weigh heavier than the less accurate ones. Acknowledging a possible covariance between a_0 and a_1 can further increase the efficiency of the estimators. The WLS method is discussed in Section 6.3.1.2.

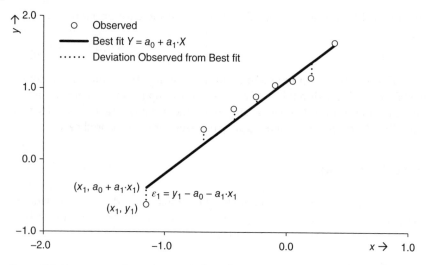

Figure 6.2 Linear regression estimates the best-fit parameters a_0 and a_1 by minimizing the squared errors or deviations of the observations (x_i, y_i) from the best fit.

Linear regression, with its linear relationship, is closely related to graphical analysis with a linear plot (cf. Section 5.3). Using graphical analysis with linear plots and parameter estimation by linear regression in parallel has the advantage that estimated parameters and plots are consistent (i.e. the best fit in the graphical analysis coincides with the best fit with the (W)LS estimated parameters).

6.3.1.1 LR by Unweighted Least Squares

Linear regression using the unweighted LS method estimates parameters by minimizing the sum of the n (unweighted) squared residuals that are the differences between the observations y_i and a model $Y(x_i; \beta)$, where β represents the parameters of the model. In the present case the model is the linear relationship defined in Eq. (6.75) with parameters a_0 and a_1. The squared sum S of the errors or residuals ε_i is:

$$S = \sum_{i=1}^{n} \varepsilon_i^2 = \sum_{i=1}^{n} (y_i - a_0 - a_1 \cdot x_i)^2 \tag{6.78}$$

The least-squares estimators (LSEs) are obtained by minimizing S, which is done by setting the derivative of S with respect to each parameter equal to zero (and checking that this is indeed a minimum). The present case yields:

$$\frac{\partial S}{\partial a_0} = -2 \cdot \sum_{i=1}^{n} (y_i - a_0 - a_1 \cdot x_i) = -2 \cdot \left(\sum_{i=1}^{n} y_i - n \cdot a_0 - a_1 \cdot \sum_{i=1}^{n} x_i \right) = 0 \tag{6.79}$$

$$\frac{\partial S}{\partial a_1} = -2 \cdot \sum_{i=1}^{n} x_i \cdot (y_i - a_0 - a_1 \cdot x_i) = -2 \cdot \left(\sum_{i=1}^{n} x_i y_i - a_0 \cdot \sum_{i=1}^{n} x_i - a_1 \cdot \sum_{i=1}^{n} x_i^2 \right) = 0 \tag{6.80}$$

The average value $\bar{u}$ of a variable set u_i $(i = 1, \dots, n)$ can be written as:

$$\bar{u} = \frac{1}{n} \sum_{i=1}^{n} u_i \tag{6.81}$$

In the following, u is replaced by either x, y, x^2 or $x \cdot y$. Dividing Eqs (6.79) and (6.80) by $-2n$ yields expressions in terms of averages (cf. Section 3.1):

$$\bar{y} - a_0 - a_1 \cdot \bar{x} = 0 \tag{6.82}$$

$$\overline{x \cdot y} - a_0 \cdot \bar{x} - a_1 \cdot \overline{x^2} = 0 \tag{6.83}$$

From these equations follow expressions for the parameter estimators a_0 and a_1, where the three expressions for a_1 are equivalent alternatives:

$$a_0 = \bar{y} - a_1 \cdot \bar{x} \tag{6.84}$$

$$a_1 = \frac{\overline{x \cdot y} - \bar{y} \cdot \bar{x}}{\overline{x^2} - \bar{x}^2} = \frac{\overline{(x - \bar{x}) \cdot y}}{\overline{(x - \bar{x})^2}} = \frac{\overline{(x - \bar{x}) \cdot (y - \bar{y})}}{\overline{(x - \bar{x}) \cdot (x - \bar{x})}} \tag{6.85}$$

Equation (6.85) can also be expressed in terms of summations in various ways:

$$a_1 = \frac{\sum\limits_{i=1}^{n}(x_i \cdot y_i) - \sum\limits_{i=1}^{n} y_i \cdot \frac{1}{n}\sum\limits_{i=1}^{n} x_i}{\sum\limits_{i=1}^{n}(x_i^2) - \frac{1}{n}\left(\sum\limits_{i=1}^{n} x_i\right)^2} = \frac{\sum\limits_{i=1}^{n}(x_i \cdot y_i) - n \cdot \bar{y} \cdot \bar{x}}{\sum\limits_{i=1}^{n}(x_i^2) - n \cdot \bar{x}^2}$$

$$= \frac{\sum\limits_{i=1}^{n}((x_i - \bar{x}) \cdot (y_i - \bar{y}))}{\sum\limits_{i=1}^{n}((x_i - \bar{x}) \cdot (x_i - \bar{x}))} = \frac{\sum\limits_{i=1}^{n}((x_i - \bar{x}) \cdot y_i)}{\sum\limits_{i=1}^{n}\left((x_i - \bar{x})^2\right)} \tag{6.86}$$

Example 6.1 **_Linear Regression on Weibull-2 Model_** As an example, the variables in case of the Weibull distribution could be the expected plotting positions $x_i = <Z_i> = <\log[-\ln(1 - F_i)]>$ and the covariables the observed log times $y_i = \log[t_i]$ (cf. Section 5.4.1). Note that this case happens to deviate from the intuitive convention of the variables along the horizontal and the covariables along the vertical axis. The linear relationship (Eq. (5.21)) can be written as:

$$Y = \log(t) = \log(\alpha) + \frac{1}{\beta}Z = y_0 + y' \cdot X \tag{6.87}$$

So, in that case $A_0 = \log(\alpha)$ and $A_1 = 1/\beta$.

Some notations are useful to compact expressions (see also Section 10.3, Eqs (10.64)–(10.66) for alternative equivalents):

$$S_{XX} = \sum\limits_{i=1}^{n}(x_i - \bar{x})^2 \tag{6.88}$$

$$S_{YY} = \sum\limits_{i=1}^{n}(y_i - \bar{y})^2 \tag{6.89}$$

$$S_{XY} = \sum\limits_{i=1}^{n}(x_i - \bar{x})(y_i - \bar{y}) \tag{6.90}$$

The solutions of the LS estimators method are analytical. Furthermore, it is possible to calculate various measures of accuracy analytically. The variances of the estimators a_0, a_1 and of y are derived and discussed further in Section 10.3.

Let σ^2_{int,a_0} and σ^2_{int,a_1} be the variances due to the internal error of a_1 and a_0. These are the variances $(\Delta a_1)^2$ and $(\Delta a_0)^2$ in Section 10.3. As before, variances are functions of expected values, which are generally unknown. Approximations are achieved by summations and averages of the observations. The variance σ^2_{int,a_1} of the regression coefficient or slope a_1 is approximated by s^2_{int,a_1}:

$$\sigma^2_{int,a_1} \approx s^2_{int,a_1} = \frac{1}{\sum\limits_{i=1}^{n}(x_i - \bar{x})^2} \cdot \frac{\sum\limits_{i=1}^{n}(y_i - a_0 - a_1 \cdot x_i)^2}{n-2} = \frac{1}{S_{XX}} \cdot \frac{S}{n-2} \tag{6.91}$$

The short notation on the right uses Eq. (6.78) for S and Eq. (10.64) for S_{XX}. In Section 10.3 the squared sum of residuals S is also indicated as $n \cdot s^2$ (cf. Eq. (10.92)), with s the observed standard deviation of the sample. The variance σ^2_{int,a_0} of the location parameter or Y intercept a_0 is approximated by s^2_{int,a_0}:

$$\sigma^2_{int,a_0} \approx s^2_{int,a_0} = \left(\frac{1}{n} + \frac{\bar{x}^2}{\sum\limits_{i=1}^{n}(x_i - \bar{x})^2}\right) \cdot \frac{\sum\limits_{i=1}^{n}(y_i - a_0 - a_1 \cdot x_i)^2}{n-2}$$

$$= \left(\frac{1}{n} + \frac{\bar{x}^2}{S_{XX}}\right) \cdot \frac{S}{n-2} \tag{6.92}$$

Again, the short notation on the right uses Eqs (6.78) and (10.64).

The correlation coefficient ρ_{XY} is a measure of the validity of the regression model. It is expressed in terms of expected values, which are generally unknown but can be estimated. The correlation coefficient was discussed in Section 3.5.2 and defined in Eq. (3.85) as:

$$\rho_{XY} = \frac{\sigma_{XY}}{\sigma_X \sigma_Y} \tag{6.93}$$

The correlation coefficient ρ_{XY} for a sample can be approximated with averages rather than expected values through the empirical correlation coefficient r_{XY} as:

$$\rho_{XY} \approx r_{XY} = \frac{\sum\limits_{i=1}^{n}((x_i - \bar{x}) \cdot (y_i - \bar{y}))}{\sqrt{\sum\limits_{i=1}^{n}(x_i - \bar{x})^2} \cdot \sqrt{\sum\limits_{i=1}^{n}(y_i - \bar{y})^2}} = \frac{S_{XY}}{\sqrt{S_{XX}} \cdot \sqrt{S_{YY}}} \tag{6.94}$$

The approximation of σ_{XY} by s_{XY} was discussed in Section 3.5.1 with Eq. (3.84) and the approximation of σ_X by s_X in Section 3.1.8 with Eq. (3.70). There the term $(n-1)$ appears. The subtraction of 1 is one degree of freedom that was required for the mean in Section 3.5.1. The present model requires two parameters rather than one parameter for the mean in Section 3.5.1 and therefore $(n-2)$ appears. If it seems tricky to come to the right number of degrees of freedom, using the sum as in Eqs (6.88), (6.89) and (6.90) prevents such considerations.

As discussed at the end of Section 3.5.2, the correlation coefficient is defined as a non-linear function of expected values and care must be taken when calculating the correlation coefficient of repeated experiments: when combining the experimental results by averaging the results and then determining the combined correlation coefficient, first the combined σ_{XY}, σ_X^2 and σ_Y^2 should be averaged, followed by determining the combined correlation coefficient r_{XY} as an approximation of ρ_{XY}.

It may be noted that from Eqs (6.86) and (6.94), a_1 can be written as:

$$a_1 = \frac{S_{XY}}{S_{XX}} = r_{XY} \cdot \frac{s_Y}{s_X} \tag{6.95}$$

The accuracy of the estimated parameters a_0 and a_1 can also be estimated. In this respect there are two error types of interest: the internal and the external error. The internal error is due to the scatter in the process of random sampling and the LS method. The external error relates to other errors like measuring inaccuracies. Both internal and external errors in the parameter estimators a_1 and a_0 are discussed below.

The external error may differ from asset to asset and even per individual y_i. An example is recording the failure time of equipment like cables. Particularly for decades-old cables, only the year of installation may be known. The accuracy of the time to failure may then be estimated on the order of ± 0.5 years. But for other assets, the month of installation or even the day of commissioning is known, including a good record of operational time. In that case the time to failure is much more accurate. Each y_i may have its specific error Δy_i. If the errors Δy_i are independent, a function $A(y_1,..,y_n)$ is found to have the error ΔA:

$$(\Delta A)^2 = \sum_{i=1}^{n} \left(\frac{\partial A}{\partial y_i}\right)^2 (\Delta y_i)^2 \tag{6.96}$$

If the Δy_i are not independent, then their covariances must be taken into account:

$$(\Delta A)^2 = \sum_{i=1}^{n} \left(\frac{\partial A}{\partial y_i}\right)^2 (\Delta y_i)^2 + \sum_{i=1}^{n}\sum_{i=1}^{n} \left(\frac{\partial A}{\partial y_i}\right)\left(\frac{\partial A}{\partial y_j}\right) \Delta y_i \Delta y_j \tag{6.97}$$
$$i \neq j$$

Assuming that the external errors Δy_i are indeed independent, the squared external error s_{ext,a_1}^2 for a_1 and s_{ext,a_0}^2 for a_0 are found as:

$$s_{ext,a_1}^2 = \sum_{i=1}^{n} \left(\frac{(x_i - \bar{x})}{\sum_{i=1}^{n}\left((x_i - \bar{x})^2\right)} \right)^2 \cdot (\Delta y_i)^2 \tag{6.98}$$

$$s_{ext,a_0}^2 = \left(\frac{1}{n}\right)^2 \sum_{i=1}^{n} (\Delta y_i)^2 \tag{6.99}$$

The internal and external errors should be about the same in experiments, as the largest determines the error of the parameters a_1 and a_0. If the internal error is larger, then collecting more data will improve accuracy (although that may hinder timely decision-making). If the external error is larger, it is worthwhile exploring possibilities to improve the accuracy of the y_i (e.g. in case of lifetimes by diving into the archives to better assess the operational life of the assets under study).

A final remark about the LS method is in order. The two coefficients a_0 and a_1 are not independent. Two truly independent coefficients are $\bar{y}$ and a_1, and the linear regression model may be perceived as finding the regression coefficient a_1 with its error Δa_1 and the point of rotation $(\bar{x}, \bar{y})$ with error $\Delta \bar{y}$. It might therefore be preferred to write the best fit for the regression model as:

$$y = (\bar{y} \pm \Delta \bar{y}) + (a_1 \pm \Delta a_1) \cdot (x - \bar{x}) \tag{6.100}$$

and the error in y would then be written as:

$$(\Delta y)^2 = (\Delta \bar{y})^2 + (x - \bar{x})^2 \cdot (\Delta a_1)^2. \tag{6.101}$$

In Section 10.3, Eq. (10.75) a parameter b was introduced, defined as:

$$b = a_0 + a_1 \cdot \bar{x} = \bar{y} = \frac{1}{n} \sum_{i=1}^{n} y_i \tag{6.102}$$

The variance $(\Delta y)^2$ of y is derived in Section 10.3 as (see Eq. (10.96)):

$$\begin{aligned}(\Delta y)^2 &= (\Delta b)^2 + (x - \bar{x})^2 \cdot (\Delta a_1)^2 \\ &\approx \frac{1}{n-2}\left(1 + \frac{n \cdot (x - \bar{x})^2}{S_{XX}}\right) \cdot s^2 \end{aligned} \tag{6.103}$$

The smallest error in y can be expected at $x = \bar{x}$.

With reference to Section 5.2.6, there are options for the confidence intervals. The first option is through the beta distribution (Sections 5.2.4 and 5.2.5); the second is based on the standard deviation Δy.

If the distribution of y about the true line can be assumed to follow the normal distribution, then confidence intervals based on the scatter of data can also be determined using the standardized normal distribution, of which a table is included in Appendix G. For instance, the $F = 95\%$ confidence limit corresponds to 1.645σ. Then, the $F = 5\%$ confidence limit corresponds to -1.645σ. As Δy is proportional to σ, the 90% confidence interval would be $[-1.645\Delta y, 1.645\Delta y]$.

6.3.1.2 LR by Weighted Least Squares

With unweighted linear regression (Section 6.3.1.1) all observation pairs (x_i, y_i) have the same weight in the various averaging procedures like in Eq. (6.78). However, there can be good reasons to apply different weights to the pairs. As mentioned before, certain observations y_i may have a larger standard deviation or scatter than other observations y_j. Linear regression can be more efficient and consistent if the more accurate y_i observations have a greater weight in averaging.

Linear regression by WLS assigns weights w_i ($i = 1, \ldots, n$) to the individual observations. In general, this method applies to any choice of weights, but most common is to choose the weights $w_i = w_{\text{var}(y),i}$ as the inverse variance of each y_i:

$$w_{\text{var}(y),i} = \frac{1}{\langle y_i^2 \rangle - \langle y_i \rangle^2} = \frac{1}{\sigma_{y_i}^2} \tag{6.104}$$

The weight may be normalized by dividing $w_{\text{var}(y),i}$ by the sum $\sum w_{\text{var}(y),i}$, which makes it dimensionless and the sum of all weights equal to 1. But this may be chosen at will.

It may also be noted that choosing all $w_i = 1$ reduces the WLS method to the LS method.

Since this choice can be made arbitrarily, a more general notation w_i is used in the equations below. Linear regression to adopt the WLS method is then adapted as follows. The squared sum S of the errors or residuals ε_i becomes (cf. Eq. (6.78)):

$$S = \sum_{i=1}^{n} w_i \varepsilon_i^2 = \sum_{i=1}^{n} w_i (y_i - a_0 - a_1 \cdot x_i)^2 \tag{6.105}$$

The derivatives in Eqs (6.79) and (6.80) become:

$$\frac{\partial S}{\partial a_0} = -2 \cdot \left(\sum_{i=1}^{n} w_i y_i - \sum_{i=1}^{n} w_i \cdot a_0 - a_1 \cdot \sum_{i=1}^{n} w_i x_i \right) = 0 \tag{6.106}$$

$$\frac{\partial S}{\partial a_1} = -2 \cdot \left(\sum_{i=1}^{n} w_i x_i y_i - a_0 \cdot \sum_{i=1}^{n} w_i x_i - a_1 \cdot \sum_{i=1}^{n} w_i x_i^2 \right) = 0 \tag{6.107}$$

The weighted average value $\bar{u}_w$ of a variable set u_i $(i = 1, \ldots, n)$ becomes (cf. Eq. (6.81)):

$$\bar{u}_w = \frac{\displaystyle\sum_{i=1}^{n} w_i u_i}{\displaystyle\sum_{i=1}^{n} w_i} \tag{6.108}$$

The solutions for the parameter estimators a_0 and a_1 are similar to the expressions in Eqs (6.84) and (6.85), but be very aware that all averages $\bar{u}$ are now defined by Eq. (6.108):

$$a_0 = \bar{y}_w - a_1 \cdot \bar{x}_w \tag{6.109}$$

$$a_1 = \frac{\overline{(x \cdot y)}_w - \bar{y}_w \cdot \bar{x}_w}{\overline{(x^2)}_w - (\bar{x}_w)^2} = \frac{\overline{((x - \bar{x}_w) \cdot y)}_w}{\overline{(x - \bar{x}_w)^2}} = \frac{\overline{((x - \bar{x}_w) \cdot (y - \bar{y}_w))}_w}{\overline{((x - \bar{x}_w) \cdot (x - \bar{x}_w))}_w} \tag{6.110}$$

In terms of the (weighted) summations, the expressions are (with all averages $\bar{u}_w$ again according to Eq. (6.108)):

$$a_1 = \frac{\displaystyle\sum_{i=1}^{n} w_i (x_i \cdot y_i) - \sum_{i=1}^{n} w_i y_i \cdot \frac{1}{\sum_{i=1}^{n} w_i} \sum_{i=1}^{n} w_i x_i}{\displaystyle\sum_{i=1}^{n} w_i (x_i^2) - \frac{1}{\sum_{i=1}^{n} w_i} \left(\sum_{i=1}^{n} w_i x_i \right)^2}$$

$$= \frac{\displaystyle\sum_{i=1}^{n} w_i (x_i \cdot y_i) - \sum_{i=1}^{n} w_i \cdot \bar{y}_w \cdot \bar{x}_w}{\displaystyle\sum_{i=1}^{n} w_i (x_i^2) - \sum_{i=1}^{n} w_i \cdot (\bar{x}_w)^2}$$

$$= \frac{\displaystyle\sum_{i=1}^{n} w_i ((x_i - \bar{x}_w) \cdot (y_i - \bar{y}_w))}{\displaystyle\sum_{i=1}^{n} w_i ((x_i - \bar{x}_w) \cdot (x_i - \bar{x}_w))} = \frac{\displaystyle\sum_{i=1}^{n} w_i ((x_i - \bar{x}_w) \cdot y_i)}{\displaystyle\sum_{i=1}^{n} w_i ((x_i - \bar{x}_w)^2)} \tag{6.111}$$

$$a_0 = \frac{\sum\limits_{i=1}^{n} w_i y_i}{\sum\limits_{i=1}^{n} w_i} - a_1 \cdot \frac{\sum\limits_{i=1}^{n} w_i x_i}{\sum\limits_{i=1}^{n} w_i} \tag{6.112}$$

The estimated variance s_{w,int,a_1}^2 for the variance $\sigma_{w,\text{int},a_1}^2$ (i.e. $(\Delta a_1)^2$ in Section 10.3) related to the internal error of the regression coefficient or slope a_1 now becomes (with all averages $\bar{u}$ again according to Eq. (6.108)):

$$\sigma_{w,\text{int},a_1}^2 \approx s_{w,\text{int},a_1}^2 = \frac{1}{n-2} \cdot \frac{\sum\limits_{i=1}^{n} w_i (y_i - a_0 - a_1 \cdot x_i)^2}{\sum\limits_{i=1}^{n} w_i (x_i - \bar{x}_w)^2} \tag{6.113}$$

The estimated variance s_{w,int,a_0}^2 for the variance $\sigma_{w,\text{int},a_0}^2$ (i.e. $(\Delta a_0)^2$ in Section 10.3) becomes:

$$\sigma_{w,\text{int},a_0}^2 \approx s_{w,\text{int},a_0}^2 = \left(\frac{1}{\sum\limits_{i=1}^{n} w_i} + \frac{(\bar{x}_w)^2}{\sum\limits_{i=1}^{n} w_i (x_i - \bar{x}_w)^2} \right) \cdot \frac{\sum\limits_{i=1}^{n} w_i (y_i - a_0 - a_1 \cdot x_i)^2}{n-2} \tag{6.114}$$

The empirical correlation coefficient r_{wXY} (cf. Eq. (6.94)) becomes:

$$\rho_{XY} \approx r_{wXY} = \frac{\sum\limits_{i=1}^{n} w_i((x_i - \bar{x}_w) \cdot (y_i - \bar{y}_w))}{\sqrt{\sum\limits_{i=1}^{n} w_i(x_i - \bar{x}_w)^2} \cdot \sqrt{\sum\limits_{i=1}^{n} w_i(y_i - \bar{y}_w)^2}} = \frac{s_{wXY}}{s_{wX} \cdot s_{wY}} \tag{6.115}$$

The external error calculations yield:

$$s_{w,\text{ext},a_1}^2 = \sum\limits_{i=1}^{n} w_i \left(\frac{(x_i - \bar{x}_w)}{\sum\limits_{i=1}^{n} w_i(x_i - \bar{x}_w)^2} \right)^2 \cdot (\Delta y_i)^2 \tag{6.116}$$

$$s_{w,\text{ext},a_0}^2 = \left(\frac{1}{n} \right)^2 \sum\limits_{i=1}^{n} w_i (\Delta y_i)^2 \tag{6.117}$$

The weights for the internal errors and the external errors may be chosen differently (i.e. to let them depend on the internal, respectively external, error; cf. Eq. (6.104)).

Similar to the unweighted LS, the variance $(\Delta y)^2$ of y can be estimated (see also Section 10.3) as:

$$(\Delta y)^2 = (\Delta b)^2 + (x - \bar{x}_w)^2 \cdot (\Delta a_1)^2$$

$$\approx \left(\frac{1}{\sum\limits_{i=1}^{n} w_i} + \frac{(x - \bar{x}_w)^2}{\sum\limits_{i=1}^{n} w_i(x_i - \bar{x}_w)^2} \right) \cdot \frac{\sum\limits_{i=1}^{n} w_i(y_i - a_1 \cdot x_i - a_0)^2}{n-2} \tag{6.118}$$

The smallest error in y can be expected at $x = \bar{x}_w$. Once $(\Delta y)^2$ is determined and if the distribution of y about the true line can be assumed to follow the normal distribution, then confidence intervals based on the scatter of data can also be determined using the standardized normal distribution, of which a table is included in Appendix G. For instance, the $F = 95\%$ confidence limit corresponds to 1.645σ. Then the $F = 5\%$ confidence limit corresponds to -1.645σ. As Δy is proportional to σ, the 90% confidence interval would be $[-1.645\Delta y, 1.645\Delta y]$.

6.3.1.3 LR with Censored Data

Censored or suspended data occur when objects under test have not failed at the moment of evaluation. Examples are: objects that survive the test because their lifetime is apparently longer than the test duration period; objects that fail to another mechanism; objects that were withdrawn from the test for reasons other than failure. It also applies to the situation where the failure statistics of a group of assets are investigated, while only a part of the population failed and the remaining part survives, including the ones that were installed later. These examples show that censored data are very common in daily practice.

The treatment of censored data follows the line of graphical analysis of censored data with the adjusted rank method (see Section 5.2.2) or the adjusted plotting position method (see Section 5.4.4) by combinatorics. Assume that the population under investigation has size n, of which r objects failed at known times t_i $(i = 1, ..., r)$. This means that the failed fraction has sample size r and the surviving fraction has sample size $n - r$.

In case objects entered the test at different moments (for instance due to the common practice of replacing failed assets in the grid by new assets), it is possible that some presently surviving objects can be in operation for a shorter time than some of the recorded breakdown times of already failed objects. An example was given in Table 5.4 (Example 5.1 in Section 5.2.2), and is repeated here in Table 6.2.

The component with operational life of 4.9 years may very well end up as the longest-living component, but it could also fail before 6.5 years and rank as the second out of 10.

With the adjusted rank method all combinations are explored and the ranking index i is adjusted (see Section 5.2.2). The procedure of the adjusted ranking method [24] calculates an adjusted index $I(i,n)$ based on i, n and the sum of the number of censored and uncensored data up to i. The equation to determine the adjusted rank $I(i,n)$ reads (see Eq. (5.3) in Section 5.2.2):

$$I(i) = I(i - 1) + \frac{n + 1 - I(i - 1)}{n + 2 - C_i} \tag{6.119}$$

Table 6.2 Evaluation of component lifetimes. Part of the components failed; part of the lifetimes right-censored. The total data set is partly randomly censored due to the different service times. Here, the first known failure will always be ranked as the smallest.

Operational t (yr)	3.9	4.9	5.7	6.5	7.8	8.2	9.0	9.4	10.2	12.2
Failed?	Yes	No	No	Yes	Yes	No	No	Yes	Yes	Yes
Preliminary index i	1	—	—	2	3	—	—	4	5	6

C_i is the sum of the number of censored and uncensored observations, including the failure i. $I(0)$ is zero by definition. A censored observation before i means that it is right-censored. The treatment of the Table 6.2 case is elaborated in Section 5.2.2.

From this $I(i)$ an adjusted cumulative probability $F_{I(i)}$ follows which is input to (usually) variable x_I, but might alternatively be used as input to covariable y_I in the linear regression method.

The adjusted plotting position method follows the same idea but recognizes that there are non-linear operations to calculate either x_i or y_i from ranking index i. Averaging is therefore done in the respective X or Y domain. The adjusted plotting position method applied to the Table 6.2 case is elaborated in Example 5.4 (Section 5.4.4) and compared with the adjusted rank method.

The methods for censored data will be detailed with each of the specific distributions in the following subsections.

6.3.1.4 LR with Fixed Origin

The previous section concerned the LS method with two parameters a_0 and a_1. The location parameter is the value of Y at $x = 0$ (i.e. $a_0 = y(x = 0)$). In case a_0 has a fixed value by definition, in particular (after possible rescaling of Y at $a_0 = 0$) only one parameter has to be found, namely a_1. The model becomes simplified to:

$$Y = a_1 \cdot X \tag{6.120}$$

If $a_0 = 0$ then Eq. (6.78) becomes

$$S = \sum_{i=1}^{n} \epsilon_i^2 = \sum_{i=1}^{n} (y_i - a_1 \cdot x_i)^2 \tag{6.121}$$

Minimizing S with respect to yields:

$$\frac{\partial S}{\partial a_1} = -2 \cdot \sum_{i=1}^{n} x_i \cdot (y_i - a_1 \cdot x_i) = -2 \cdot \left(\sum_{i=1}^{n} x_i y_i - a_1 \cdot \sum_{i=1}^{n} x_i^2 \right) = 0 \tag{6.122}$$

The solution for m follows as:

$$a_1 = \frac{\sum_{i=1}^{n} (x_i \cdot y_i)}{\sum_{i=1}^{n} (x_i^2)} \tag{6.123}$$

In terms of averages:

$$a_1 = \frac{\overline{x \cdot y}}{\overline{x^2}} \tag{6.124}$$

The variance $\sigma_{\text{int},a_1}^2$ of the regression coefficient or slope a_1 is approximated by s_{int,a_1}^2:

$$\sigma_{\text{int},a_1}^2 \approx s_{\text{int},a_1}^2 = \frac{1}{\sum_{i=1}^{n} x_i^2} \cdot \frac{\sum_{i=1}^{n} (y_i - a_1 \cdot x_i)^2}{n - 1} \tag{6.125}$$

With weighted linear regression the expressions become:

$$a_1 = \frac{\sum\limits_{i=1}^{n} w_i(x_i \cdot y_i)}{\sum\limits_{i=1}^{n} w_i x_i^2} \qquad (6.126)$$

In terms of weighted averages as defined in Eq. (6.108):

$$a_1 = \frac{\overline{(x \cdot y)}_w}{\overline{(x^2)}_w} \qquad (6.127)$$

The variance σ_{int,a_1}^2 of the regression coefficient or slope a_1 is approximated by s_{int,a_1}^2:

$$\sigma_{int,a_1}^2 \approx s_{int,a_1}^2 = \frac{1}{\sum\limits_{i=1}^{n} w_i x_i^2} \cdot \frac{\sum\limits_{i=1}^{n} w_i(y_i - a_1 \cdot x_i)^2}{n-1} \qquad (6.128)$$

The correlation coefficients can be calculated as before.

6.3.1.5 Which is the (Co)variable?

Section 5.3 discussed the linear graph representation of distributions. There are various ways in which the linear graphs can be perceived and the same applies to parameter estimation. These are:

- The variables are based on ranking (the plotting positions) and assumed exact. Under the assumption of the applied distribution, the observed time-related covariables scatter. The confidence intervals are related to the beta distribution (see Section 4.1.2). This approach is the background for the graphs in Section 5.3 and following sections. Additionally, the observations may suffer an inaccuracy (the external error, see Section 6.3.1.1) that should not be neglected and the scatter of the covariables thus becomes a composition of two uncertainties. The plotting positions are usually regarded as the variables $Y(F)$ and the measured observations are the covariables $X(t)$.
- The variables are based on the observed times and assumed exact. The covariables are related to the ranking and scatter according to the beta distribution (reflecting that it is not known which test objects are exactly in the population sample). In this case the variables $X(t_i)$ are based on the observed times t_i and the covariables $Y(F_i)$ are the estimated plotting positions based on F_i that actually scatter.
- There can also be a combination of measuring times with an external error while the plotting positions have an internal error. In this case both Y and X scatter, but whether the plotting position is related to the variable or to the covariable is debatable.

The subsequent choices have an impact on the way linear regression is performed, as elaborated below. IEEE 930 [24] and IEC 62539 [29] follow the line of relating the rank to the variable and the measured entity (often time to failure) to the covariable. Unless stated otherwise, that line is also followed here.

Common practice with plotting is to use the horizontal axis for the variable and the vertical axis for the covariable. Often, the horizontal axis is even referred to as the X-axis and the vertical axis as the Y-axis. However, it is worth noticing that probability plots as in [24, 29] are an exception to that practice. The expected plotting positions are regarded as the variables and are along the *vertical* axis, whereas the Eq. (6.86) observations are regarded as the covariables and scaled along the *horizontal* axis.

6.3.2 LR for the Weibull Distribution

The Weibull distribution was introduced in Section 4.2.1. It describes the statistics of the weakest link in a chain (or weakest path through insulation, etc.) and particularly the Weibull-2 distribution is probably the most widely used to describe failure data:

$$F_{wb}(t; \alpha, \beta) = 1 - e^{-\left(\frac{t}{\alpha}\right)^{\beta}} \tag{6.129}$$

Here α and β are the scale, respectively the shape parameter. Applying linear regression to the Weibull-2 distribution calls for a linear relationship as in Eq. (6.75):

$$Y = a_0 + a_1 \cdot X \tag{6.130}$$

In Section 5.4 a relationship was also needed to produce a linear Weibull plot and formulated in Eq. (5.21) as:

$$Z(F_{wb}) = \beta[\log t - \log \alpha] \tag{6.131}$$

The plotting position variable Z was introduced for compactness of equations, which will be pursued in the following as well. Z was introduced in Eq. (5.20) as:

$$Z(F_{wb}) = \log(-\ln(1 - F_{wb})) \tag{6.132}$$

Here 'log' is the logarithm to base 10 and 'ln' the natural logarithm. In the literature the 'log' term is also expressed in terms of 'ln', which is not a fundamental difference, though the inverse operation must be based on e rather than on 10 of course. In line with [24, 29] the variable Z is chosen of which the ranked expected values $<Z_{i,n}>$ can be calculated analytically as [27]:

$$<Z_{i,n}> = \left[-\gamma + i \binom{n}{i} \sum_{j=0}^{i-1} \binom{i-1}{j} \frac{(-1)^{i-j} \cdot \ln(n-j)}{n-j}\right] \cdot \log e \tag{6.133}$$

In this expression γ is the Euler constant: $\gamma \approx 0.577215665$. See also Section 5.4.1, where various approximations of $<Z_{i,n}>$ with their accuracy are discussed. The approximation for $F(<Z_{i,n}>)$ in Eq. (5.27) is adopted in [24, 29]:

$$F(<Z_{i,n}>) \approx \frac{i - 0.44}{n + 0.25} \tag{6.134}$$

With Eq. (6.132) this leads to the approximation for $<Z_{i,n}>$:

$$<Z_{i,n}> \approx \log\left(-\ln\left(1 - \frac{i - 0.44}{n + 0.25}\right)\right) = \log\left(-\ln\left(\frac{n - i + 0.69}{n + 0.25}\right)\right) \tag{6.135}$$

The covariable is $\log(t)$. The linear relationship in the shape of Eq. (6.130) thus becomes:

$$\log t = \log \alpha + \frac{1}{\beta} \cdot Z(F_{wb}) \tag{6.136}$$

with $A_0 = \log(\alpha)$ and $A_1 = 1/\beta$.

6.3.2.1 LR by Unweighted LS for the Weibull Distribution

Assume n observation pairs $(<Z_i>, \log(t_i))$ are available. The averages are defined as (cf. Eq. (6.81)):

$$\overline{<Z>} = \frac{1}{n} \sum_{i=1}^{n} <Z_i> \tag{6.137}$$

$$\overline{\log t} = \frac{1}{n} \sum_{i=1}^{n} \log t_i \tag{6.138}$$

The unweighted LS estimators a_{LS} and b_{LS} according to Section 6.3.1.1 are now found as:

$$\log a_{LS} = \overline{\log t} - \frac{1}{b_{LS}} \cdot \overline{<Z>} \quad \Leftrightarrow \quad a_{LS} = 10^{\overline{\log t} - \frac{1}{b_{LS}} \cdot \overline{<Z>}} \tag{6.139}$$

$$\frac{1}{b_{LS}} = \frac{\sum_{i=1}^{n} \left(<Z_i> - \overline{<Z>} \right) \cdot \left(\log t_i - \overline{\log t} \right)}{\sum_{i=1}^{n} (<Z_i> - \overline{<Z>})^2}$$

$$\Leftrightarrow b_{LS} = \frac{\sum_{i=1}^{n} (<Z_i> - \overline{<Z>})^2}{\sum_{i=1}^{n} (<Z_i> - \overline{<Z>}) \cdot (\log t_i - \overline{\log t})} \tag{6.140}$$

The correlation coefficient $r_{Z,\log t}$ can be analytically calculated as in Section 6.3.1.1 (cf. Eq. (6.94)):

$$r_{Z,\log t} = \frac{\sum_{i=1}^{n} ((<Z_i> - \overline{<Z>}) \cdot (\log t_i - \overline{\log t}))}{\sqrt{\sum_{i=1}^{n} (<Z_i> - \overline{<Z>})^2} \cdot \sqrt{\sum_{i=1}^{n} (\log t_i - \overline{\log t})^2}} = \frac{s_{Z,\log t}}{s_Z \cdot s_{\log t}} \tag{6.141}$$

The internal errors $s_{int,1/b}^2$ and $s_{int,\log a}^2$ as in Eqs (6.91) and (6.92):

$$s_{int,1/b}^2 = \frac{1}{\sum_{i=1}^{n} (<Z_i> - \overline{<Z>})^2} \cdot \frac{\sum_{i=1}^{n} \left(\log t_i - \log a_{LS} - \frac{1}{b_{LS}} \cdot Z_i \right)^2}{n-2} \tag{6.142}$$

$$s^2_{int,\log a} = \left(\frac{1}{n} + \frac{\overline{<Z>}^2}{\sum\limits_{i=1}^{n}(<Z_i> - \overline{<Z>})^2}\right) \cdot \frac{\sum\limits_{i=1}^{n}\left(\log t_i - \log a_{LS} - \frac{1}{b_{LS}} \cdot Z_i\right)^2}{n-2} \tag{6.143}$$

The external errors $s^2_{ext,1/b}$ and $s^2_{ext,\log a}$ as in Eqs (6.98) and (6.99):

$$s^2_{ext,1/b} = \sum\limits_{i=1}^{n}\left(\frac{(<Z_i> - \overline{<Z>})}{\sum\limits_{i=1}^{n}(<Z_i> - \overline{<Z>})^2}\right)^2 \cdot (\Delta \log t_i)^2 \tag{6.144}$$

$$s^2_{ext,\log a} = \left(\frac{1}{n}\right)^2 \sum\limits_{i=1}^{n}(\Delta \log t_i)^2 \tag{6.145}$$

These errors can be processed with Eqs (6.96) or (6.97) to find estimators of the corresponding errors in a_{LS} and b_{LS}.

The confidence intervals can be determined with the beta distribution (Sections 5.2.4, 5.2.5 and 5.4.5) if the data are considered to be randomly drawn from a large population (cf. Section 5.2.6) or based on the variance if the scatter can be considered normally distributed (cf. Sections 6.3.1.1 and 6.3.1.2).

6.3.2.2 LR by Weighted LS for the Weibull Distribution

Linear regression by WLS follows the approach of Section 6.3.1.2. The weights w_i are defined as (n is omitted as suffix below):

$$w_i = \frac{1}{<Z_i^2> - <Z_i>^2} \tag{6.146}$$

The expected plotting position $<Z_i>$ was defined in Eq. (6.133). The expected value $<Z_i^2>$ is found as:

$$<Z_i^2> = i\binom{n}{i}\sum\limits_{m=0}^{2}\binom{2}{m}(-1)^{2-m}\frac{\partial^m}{\partial s^m}\Gamma(s+1)|_{s=0}$$

$$\times \sum\limits_{j=0}^{i-1}\binom{i-1}{j}(-1)^{i-1-j}\frac{(\ln[n-j])^{2-m}}{n-j} \tag{6.147}$$

with:

$$\Gamma^{(0)}(1) = 1$$

$$\Gamma^{(1)}(1) = -\gamma \approx -0.577215665$$

$$\Gamma^{(2)}(1) = \gamma^2 + \frac{\pi^2}{6} \approx 1.978111991 \tag{6.148}$$

The weighted averages of the variable and covariable are defined as (cf. Eq. (6.108)):

$$\overline{<Z>}_w = \frac{\sum\limits_{i=1}^{n} w_i <Z_i>}{\sum\limits_{i=1}^{n} w_i} \tag{6.149}$$

$$\overline{\log t}_w = \frac{\sum\limits_{i=1}^{n} w_i \log t_i}{\sum\limits_{i=1}^{n} w_i} \tag{6.150}$$

The weighted LS estimators a_{WLS} and b_{WLS} according to Section 6.3.1.2 are now found as:

$$\frac{1}{b_{WLS}} = \frac{\sum\limits_{i=1}^{n} w_i(<Z_i> - \overline{<Z>}_w) \cdot (\log t_i - \overline{\log t}_w)}{\sum\limits_{i=1}^{n} w_i(<Z_i> - \overline{<Z>}_w)^2}$$

$$\Leftrightarrow b_{WLS} = \frac{\sum\limits_{i=1}^{n} w_i(<Z_i> - \overline{<Z>}_w)^2}{\sum\limits_{i=1}^{n} w_i(<Z_i> - \overline{<Z>}_w) \cdot (\log t_i - \overline{\log t}_w)} \tag{6.151}$$

$$\log a_{WLS} = \overline{\log t}_w - \frac{\overline{<Z>}_w}{b_{WLS}} \quad \Leftrightarrow \quad a_{WLS} = 10^{\overline{\log t}_w - \frac{\overline{<z>}_w}{b_{WLS}}} \tag{6.152}$$

The internal errors $s^2_{w,\text{int},1/b}$ and $s^2_{w,\text{int},\log a}$ as in Eqs (6.113) and (6.114):

$$s^2_{w,\text{int},1/b} = \frac{1}{\sum\limits_{i=1}^{n} w_i(<Z_i> - \overline{<Z>}_w)^2} \cdot \frac{\sum\limits_{i=1}^{n} w_i\left(\log t_i - \log a_{WLS} - \frac{1}{b_{WLS}} \cdot Z_i\right)^2}{n - 2} \tag{6.153}$$

$$s^2_{w,\text{int},\log a} = \left(\frac{1}{n} + \frac{(\overline{<Z>}_w)^2}{\sum\limits_{i=1}^{n} w_i(<Z_i> - \overline{<Z>}_w)^2}\right) \cdot \frac{\sum\limits_{i=1}^{n} w_i\left(\log t_i - \log a_{WLS} - \frac{1}{b_{WLS}} \cdot Z_i\right)^2}{n - 2} \tag{6.154}$$

The correlation coefficient $r_{Z,\log t,w}$ can be analytically calculated as discussed in Section 6.3.1.2 (cf. Eq. (6.115)):

$$r_{w,Z,\log t} = \frac{\sum\limits_{i=1}^{n} w_i \cdot (<Z_i> - \overline{<Z>}_w) \cdot (\log t_i - \overline{\log t}_w)}{\sqrt{\sum\limits_{i=1}^{n} w_i(<Z_i> - \overline{<Z>}_w)^2} \cdot \sqrt{\sum\limits_{i=1}^{n} w_i(\log t_i - \overline{\log t}_w)^2}} = \frac{s_{w,Z,\log t}}{s_{w,Z} \cdot s_{w,\log t}} \tag{6.155}$$

The external errors $s^2_{w,\text{ext},1/b}$ and $s^2_{w,\text{ext},\log a}$: as in Eqs (6.116) and (6.117):

$$s^2_{w,\text{ext},1/b} = \sum\limits_{i=1}^{n} w_i\left(\frac{(<Z_i> - \overline{<Z>}_w)}{\sum\limits_{i=1}^{n} w_i(<Z_i> - \overline{<Z>}_w)^2}\right)^2 \cdot (\Delta \log t_i)^2 \tag{6.156}$$

$$s^2_{w,\text{ext,log}\,a} = \left(\frac{1}{n}\right)^2 \sum_{i=1}^{n} w_i (\Delta \log t_i)^2 \tag{6.157}$$

These errors can be processed with Eqs (6.96) or (6.97) to find estimators of the corresponding errors in a_{WLS} and b_{WLS}.

The confidence intervals can be determined with the beta distribution (Sections 5.2.4, 5.2.5 and 5.4.5) if the data are considered to be randomly drawn from a large population (cf. Section 5.2.6) or based on the variance if the scatter can be considered normally distributed (cf. Sections 6.3.1.1 and 6.3.1.2).

6.3.2.3 Processing Censored Data with the Adjusted Rank Method

An introduction to censored data and the application of linear regression was given in Section 6.3.1.3. It is assumed that the population under investigation has size n, of which r objects failed at known times t_i $(i = 1, ..., r)$. This means that the failed fraction has sample size r and the surviving fraction has sample size $n - r$. It was explained that two methods are in use to find the ranking possibilities for the r observations: the adjusted rank method and the adjusted plotting position method. The approach to processing censored data is the same as with the graphical analysis in Section 5.4.4.

The variables for the linear regression are the $<Z_{i,n}>$:

$$<Z_{i,n}> = \left[-\gamma + i \binom{n}{i} \sum_{j=0}^{i-1} \binom{i-1}{j} \frac{(-1)^{i-j} \cdot \ln(n-j)}{n-j}\right] \cdot \log e \tag{6.158}$$

but in the above the index i is defined within the subgroup of failed objects which has sample size r and not in relation to the entire population. Within the total population with sample size n the ranking index is yet to be determined, since the ranking is likely subject to change until all objects have ultimately failed. This is particularly true if some objects have a shorter operational life than the $r < n$ already observed times to failure at the evaluation moment, as explained in Section 6.3.1.3. A way to solve this is to inventory all possible rankings and evaluate for each failed sample with index i what its average ranking index $I(i)$ or plotting position $Z_{I,n}$ is. The latter is discussed in the next section.

With the adjusted rank method, first all uncensored data t_i $(i = 1, ..., r)$ and censored data τ_j $(j = 1, ..., n - r)$ are collected as in Table 6.2. The uncensored data are attributed to rank index i and subsequently an adjusted rank index $I(i)$ is calculated through the equation (see Eq. (5.3) in Section 5.2.2):

$$I(i) = I(i-1) + \frac{n + 1 - I(i-1)}{n + 2 - C_i} \tag{6.159}$$

C_i is the sum of the number of censored and uncensored observations, including the failure i. $I(0)$ is zero by definition. In case all censored observations are greater than the uncensored observations (which happens if all objects were simultaneously put to test from the beginning), then $C_i - 1$ equals $I(i-1)$ and all $I(i)$ are simply equal to i $(i = 1, ..., r)$.

Generally, the $I(i)$ are neither necessarily integers themselves nor are their differences necessarily integers. The summation in Eq. (6.158) is therefore hard to carry out. However, since $Z_{i,n}$ is a function of $F_{i,n}$ and the cumulative function $F(<Z_{i,n}>)$ for the Weibull distribution can be approximated with Eq. (6.134), the cumulative function $F(<Z_{I,n}>)$ can conveniently be approximated as:

$$F(<Z_{I,n}>) \approx \frac{I - 0.44}{n + 0.25} \qquad (6.160)$$

It is noteworthy that this expression does not require I to be an integer. From this the expected plotting position with the adjusted ranking I follows as:

$$<Z_{I,n}> \approx \log\left(-\ln\left(1 - \frac{I - 0.44}{n + 0.25}\right)\right) = \log\left(-\ln\left(\frac{n - I + 0.69}{n + 0.25}\right)\right) \qquad (6.161)$$

As for $F(<Z_{I,n}>)$, it is not required for the approximation of $<Z_{I,n}>$ that I is an integer. The observed failure data t_i are now attributed to the adjusted rank, that is:

$$t_I = t_{I(i)} = t_i \qquad (6.162)$$

With the calculated $<Z_{I(i),n}>$ and the observed failure data $t_{I(i)}$, unweighted linear regression is fairly straightforward for a set of r pairs of variables and covariables $(x_{I(i)}, y_{I(i)})$ with:

$$x_I = x_{I(i)} = <Z_{I(i),n}> \qquad (6.163)$$

$$y_I = y_{I(i)} = \log t_{I(i)} \qquad (6.164)$$

It should be noted that the analysis is based on r pairs, and therefore n must be replaced by r when summations over $I(i)$ (basically over i) are carried out in the equations of Section 6.3.1.1. From that perspective, the averages are defined as (cf. Eq. (6.81)):

$$\overline{<Z>} = \frac{1}{r}\sum_{i=1}^{r} <Z_{I(i)}> \qquad (6.165)$$

$$\overline{\log t} = \frac{1}{r}\sum_{i=1}^{r} \log t_{I(i)} \qquad (6.166)$$

For censored data with r uncensored observations, the unweighted LS estimators a_{LS} and b_{LS} according to Section 6.3.1.1 are now found as (cf. Eqs (6.139) and (6.140)):

$$\log a_{LS} = \overline{\log t} - \frac{1}{b_{LS}} \cdot \overline{<Z>} \quad \Leftrightarrow \quad a_{LS} = 10^{\overline{\log t} - \frac{1}{b_{LS}} \cdot \overline{<Z>}} \qquad (6.167)$$

$$\frac{1}{b_{LS}} = \frac{\sum_{i=1}^{r}\left(<Z_{I(i)}> - \overline{<Z>}\right) \cdot \left(\log t_{I(i)} - \overline{\log t}\right)}{\sum_{i=1}^{r}\left(<Z_{I(i)}> - \overline{<Z>}\right)^2}$$

$$\Leftrightarrow b_{LS} = \frac{\sum_{i=1}^{r}\left(<Z_{I(i)}> - \overline{<Z>}\right)^2}{\sum_{i=1}^{r}\left(<Z_{I(i)}> - \overline{<Z>}\right) \cdot \left(\log t_{I(i)} - \overline{\log t}\right)} \qquad (6.168)$$

with the averages as in Eqs (6.165) and (6.166), $I(i)$ from Eq. (6.159), $<Z_{I(i),n}>$ from Eq. (6.161) and $t_{I(i)}$ from Eq. (6.162). The correlation coefficient and the errors are found in the same way through the equations in Section 6.3.2.1 by replacing n with r.

Table 6.3 Evaluation of component lifetimes.

Operational t (yr)	3.9	4.9	5.7	6.5	7.8	8.2	9.0	9.4	10.2	12.2
Failed?	Yes	No	No	Yes	Yes	No	No	Yes	Yes	Yes
Preliminary index i	1	—	—	2	3	—	—	4	5	6

Note: Repeated Tables 5.4 and 5.13.

Unfortunately, a convenient approximation for $<Z_I^2>$ is not yet available, which makes this method not suitable for weighted linear regression as yet. But if a suitable expression is found, the estimator can be worked out with Eqs (6.151) and (6.152), where n is replaced by r and i in the suffices is replaced by $I(i)$. For unweighted linear regression the adjusted rank is recommended in [29].

6.3.2.4 EXTRA: Processing Censored Data with the Adjusted Plotting Position Method

The censored data and the application of linear regression were introduced in Section 6.3.1.3. Again it is assumed that the population under investigation has size n, of which r objects failed at known times t_i ($i = 1, ..., r$). This means that the failed fraction has sample size r and the surviving fraction has sample size $n - r$. It was explained that two methods are in use to find the ranking possibilities for the r observations: the adjusted rank method and the adjusted plotting position method. The latter is discussed in the present section. The approach to process censored data is the same as with the graphical analysis in Section 5.4.4.

Again, the permutations are investigated and from that the adjusted plotting position $Z_{i,n,adj}$ is determined as the average of all possible Z_i to which the observed $\log(t_i)$ can be assigned. Section 5.4.4 discussed an example based on the data in Table 6.3 (copied from Table 5.13). Table 6.4 (a copy of Table 5.14) shows how many ranking possibilities exist for each lifetime. The table also shows the average plotting position as the adjusted plotting position $Z_{i,n,adj}$. These $Z_{i,n,adj}$ are the variables to be used in the linear regression, comparable to $<Z_{I,n}>$ in the previous section. The covariables remain the $\log(t_i)$.

The adjusted plotting position and the adjusted ranking method are compared in Figure 5.22. Though the results of the two methods do differ, the difference is relatively small and it may be tempting to select the adjusted ranking method for its convenience.

Again, it should be noted that the analysis is based on r pairs and therefore n must be replaced by r when summations over i are carried out in the equations of Section 6.3.1.1. In that respect, the averages are defined as (cf. Eq. (6.81)):

$$\overline{<Z>} = \frac{1}{r} \sum_{i=1}^{r} Z_{i,n,adj} \tag{6.169}$$

$$\overline{\log t} = \frac{1}{r} \sum_{i=1}^{r} \log t_i \tag{6.170}$$

For censored data with r uncensored observations, the unweighted LS estimators a_{LS} and b_{LS} according to Section 6.3.1.1 are now found as (cf. Eqs (6.139) and (6.140)):

$$\log a_{LS} = \overline{\log t} - \frac{1}{b_{LS}} \cdot \overline{<Z>} \quad \Leftrightarrow \quad a_{LS} = 10^{\overline{\log t} - \frac{1}{b_{LS}} \cdot \overline{<Z>}} \tag{6.171}$$

Table 6.4 For each uncensored (i.e. observed) lifetime the number of possible plotting positions is shown. The adjusted plotting position $Z_{i,n,adj}$ is the average plotting position. It follows from the number of permutations.

Ranking	$<Z_{i,n}>$	Number of ranking possibilities for the six observations					
Time t_i (yr)		3.9	6.5	7.8	9.4	10.2	12.2
Log t_i (log-yr)		0.591	0.813	0.892	0.973	1.009	1.086
1	−1.2507	1440					
2	−0.7931		1120				
3	−0.5503		280	840			
4	−0.3770		40	480	360		
5	−0.2361			120	480	120	
6	−0.1118				360	288	24
7	0.0052				180	396	96
8	0.1232				60	384	228
9	0.2539					252	420
10	0.4299						672
Number of permutations		1440	1440	1440	1440	1440	1440
Result – adjusted plotting position method							
Average $Z_{i,n}$: $Z_{i,n,adj}$		−1.251	−0.734	−0.466	−0.195	0.037	0.293
Result – adjusted rank method							
$<Z_{1,n}>$		−1.250	−0.712	−0.450	−0.183	0.038	0.274

Note: Repeated Table 5.14.

$$\frac{1}{b_{LS}} = \frac{\sum_{i=1}^{r} (Z_{i,n,adj} - \overline{<Z>}) \cdot (\log t_i - \overline{\log t})}{\sum_{i=1}^{r} (Z_{i,n,adj} - \overline{<Z>})^2}$$

$$\Leftrightarrow b_{LS} = \frac{\sum_{i=1}^{r} (Z_{i,n,adj} - \overline{<Z>})^2}{\sum_{i=1}^{r} (Z_{i,n,adj} - \overline{<Z>}) \cdot (\log t_i - \overline{\log t})} \tag{6.172}$$

with the averages as in Eqs (6.169) and (6.170) and $Z_{i,n,adj}$ as determined from the permutations. The correlation coefficient and the errors are found in the same way through the equations in Section 6.3.2.1, by replacing n with r.

This method can be adapted for the WLS. In that case all permutations must be treated separately. For each permutation there is a set of rankings $I(i)$, a set of variables $Z_{I(i),n}$ and a set of covariables $\log(t_i)$, all with $i = 1, \ldots, r$. Unlike the situation in the previous section, each $I(i)$ is now an integer and the weighs w_i can be determined as in Eq. (6.146) for the $I(i)$:

$$w_i = w_{I(i)} = \frac{1}{<Z_{I(i)}^2> - <Z_{I(i)}>^2} \tag{6.173}$$

For each permutation the WLS estimators $(\log(a))_{WLS,k}$ and $(1/b)_{WLS,k}$ are determined. The index k is used to tell the permutations apart. So, in the case of Table 6.4 there are 210 possible combinations (i.e. $k = 1, ..., 210$). For example, $\{I(1), I(2), I(3), I(4), I(5), I(6)\} = \{1,2,3,4,5,6\}, \{1,2,3,4,5,7\}, \{1,2,3,4,5,8\}, \{1,2,3,4,5,9\}, \{1,2,3,4,5,10\}, \{1,2,3,4,6,7\}$, and so on. However, not all are feasible, since there are only two censored data items preceding t_2 and t_3 and these observed times can ultimately (when all 10 objects fail) not rank higher than 4 and 5, respectively.

The combinations consist of six rankings, because only six observations are uncensored. The non-appearing rankings will be occupied by the presently censored data once all objects fail. As shown and explained in Example 5.4, the total number of possible permutations is 1440.

Subsequently, these permutations can be used to obtain a weighted average of the WLS estimators $(\log(a))_{WLS,k}$ and $(1/b)_{WLS,k}$ in order to obtain the ultimate WLS estimators $(\log(a))_{WLS}$ and $(1/b)_{WLS}$.

6.3.2.5 Expected LS and WLS Estimators for the Weibull-2 Distribution

The LS method fulfils the requirements of an efficient estimator. The text below largely follows Section 6.2.3.3, which discussed the expected ML estimators, but is adjusted to LS estimators. As discussed in Section 6.1.1.3, the consistency property is related to the behaviour with large sample sizes n. With asset management and particularly incident management (see Section 1.4), working with small sample sizes is practically standard. Therefore, the matter of bias and scatter is of great interest, especially for small sample sizes r and/or n, even down to 3, ..., 10. Naturally, the scatter will be large, which leads to inaccurate information for decision-making. However, for some decisions, orders of magnitude can be sufficient and a high accuracy may not be necessary.

In case of LS, the bias for the shape parameter can be determined analytically, but it can also be investigated numerically, as discussed below. Pivotal quantities and functions can be investigated, after which the results can be translated to combinations in general. Let u be distributed to the normalized Weibull distribution (i.e. $\alpha = \beta = 1$), which corresponds to the exponential distribution:

$$F(u; \alpha = 1, \beta = 1) = 1 - e^{-\left(\frac{u}{1}\right)^1} = 1 - e^{-u} \tag{6.174}$$

The LS estimators of the scale parameter $\alpha = 1$ and the shape parameter $\beta = 1$ are indicated as $a_{1,1}$ and $b_{1,1}$. Any Weibull-2 distribution can be produced by substituting:

$$u = \left(\frac{t}{\alpha}\right)^\beta \tag{6.175}$$

The variable u is a pivotal quantity. The model does not change with the choice of the α and β parameters. The LS estimators $\log(a_{1,1})$, $a_{1,1}$, $1/b_{1,1}$ and $b_{1,1}$ for the uncensored case are (cf. Eqs (6.139) and (6.140)):

$$\log a_{1,1} = \overline{\log u} - \frac{1}{b_{1,1}} \cdot \overline{<Z>} \Leftrightarrow a_{1,1} = 10^{\overline{\log u} - \frac{1}{b_{1,1}} \cdot \overline{<Z>}} \tag{6.176}$$

$$\frac{1}{b_{1,1}} = \frac{\sum_{i=1}^{n}(<Z_i> - \overline{<Z>}) \cdot (\log u_i - \overline{\log u})}{\sum_{i=1}^{n}(<Z_i> - \overline{<Z>})^2}$$

$$\Leftrightarrow b_{1,1} = \frac{\sum_{i=1}^{n}(<Z_i> - \overline{<Z>})^2}{\sum_{i=1}^{n}(<Z_i> - \overline{<Z>}) \cdot (\log u_i - \overline{\log u})} \tag{6.177}$$

with:

$$\overline{<Z>} = \frac{1}{n}\sum_{i=1}^{n}<Z_i> \tag{6.178}$$

$$\overline{\log u} = \frac{1}{n}\sum_{i=1}^{n}\log u_i \tag{6.179}$$

The LS estimators of the general scale parameter α and shape parameter β are indicated as $a_{\alpha,\beta}$ and $b_{\alpha,\beta}$. Substitution of Eq. (6.175) in Eqs (6.176) and (6.177) with subsequent comparison with Eqs (6.139) and (6.140) shows that:

$$\frac{1}{b_{\alpha,\beta}} = \frac{1}{\beta \cdot b_{1,1}} \quad \Leftrightarrow \quad b_{\alpha,\beta} = \beta \cdot b_{1,1} \tag{6.180}$$

and:

$$\log a_{\alpha,\beta} = \frac{1}{\beta} \cdot \log a_{1,1} + \log \alpha \quad \Leftrightarrow \quad a_{\alpha,\beta} = \alpha \cdot a_{1,1}^{\frac{1}{\beta}} \tag{6.181}$$

These expressions are pivotal (i.e. independent of actual α and β). The relationships in Eqs (6.180) and (6.181) can be used to describe the expected LS estimators for Weibull-2 in general, such as:

$$<\frac{1}{b_{\alpha,\beta}}> = \frac{1}{\beta} \cdot <\frac{1}{b_{1,1}}> \tag{6.182}$$

$$<b_{\alpha,\beta}> = \beta \cdot <b_{1,1}> \tag{6.183}$$

$$<\log a_{\alpha,\beta}> = \frac{1}{\beta}<\log a_{1,1}> + \log \alpha \tag{6.184}$$

The expected parameter estimators show which entities have to be averaged in order to obtain unbiasing factors.

6.3.2.6 Formulas for Bias and Scatter for LS and WLS

As for the ML estimators also for the LS estimators, the bias and scatter are investigated and the results are fit with the asymptotic power function described in Section 10.2.3:

$$D_n = E_n - E_\infty = Q \cdot (n - R)^P \tag{6.185}$$

Again, E_n and E_∞ are respectively the expected and asymptotic values of an estimator based on finite and infinite sample sizes and D_n is the difference between them; P, Q and R are the parameters of this power function determined from the tables. As P is negative, $D_n \to 0$ as $n \to \infty$.

As consistent estimators, the bias in the expected LS estimators approaches zero with increasing number of observations (see Section 6.1.1.3). Generally, the bias in the parameter estimators cannot be ignored for the small data sets that often have to be dealt with in incident management. However, the above relations in Eqs (6.182)–(6.184) can be used for unbiasing. The unbiased parameters are indicated with suffix U (i.e. b_U, $(1/b)_U$, $(\ln a)_U$ and a_U.

The LS estimator for $1/b$ is unbiased. From Eq. (6.174) it follows that $u = -\ln(1 - F)$ and substitution in Eq. (6.177) yields:

$$\left\langle \frac{1}{b_{1,1}} \right\rangle = \left\langle \frac{\sum_{i=1}^{n}(<Z_i> - \overline{<Z>}) \cdot (\log u_i - \overline{\log u})}{\sum_{i=1}^{n}(<Z_i> - \overline{<Z>})^2} \right\rangle$$

$$= \frac{\sum_{i=1}^{n}(<Z_i> - \overline{<Z>}) \cdot <(\log u_i - \overline{\log u})>}{\sum_{i=1}^{n}(<Z_i> - \overline{<Z>})^2}$$

$$= \frac{\sum_{i=1}^{n}(<Z_i> - \overline{<Z>}) \cdot (<Z_i> - \overline{<Z>})}{\sum_{i=1}^{n}(<Z_i> - \overline{<Z>})^2} = 1 \qquad (6.186)$$

The same holds for the weighted LS estimator. Therefore, for both LS and WLS with the unbiased estimator $(1/b)_U$ it follows from Eq. (6.182) that:

$$\left\langle \left(\frac{1}{b}\right)_U \right\rangle = \frac{1}{\beta} \quad \rightarrow \quad \left(\frac{1}{b}\right)_U = \frac{1}{b_{\alpha,\beta}} \cdot \left(<\frac{1}{b_{1,1}}>\right)^{-1} = \frac{1}{b_{\alpha,\beta}} \qquad (6.187)$$

For the unbiased estimator b_U it follows from Eq. (6.183) that:

$$<b_U> = \beta \quad \rightarrow \quad b_U = \frac{b_{\alpha,\beta}}{<b_{1,1}>} \qquad (6.188)$$

For the unbiased estimator $(\log a)_U$ it follows from Eq. (6.184) that:

$$<(\log a)_U> = \log \alpha \quad \rightarrow$$

$$(\log a)_U = \log a_{\alpha,\beta} - \frac{1}{\beta} \cdot <\log a_{1,1}> = \log a_{\alpha,\beta} - \frac{1}{b_{\alpha,\beta}} \cdot \frac{<\log a_{1,1}>}{<\left(\frac{1}{b_{1,1}}\right)>} \qquad (6.189)$$

For purely uncensored data and also for limited combinations of censored data, the pivotal functions have been reported in various studies (see [27, 41, 42]). In such studies Monte Carlo simulations were carried out for a range of sample sizes and the resulting bias was reported in tables. Particularly in [27], the random number generator was tested to produce expected values for a range of test functions (see Section 10.4.4).

Such tables have been used in [27] to derive a power expression for the bias as well as for the standard deviation in the shape of Eq. (6.185). For the expected LS estimator $<1/b>$ the power function is trivial: the estimator is unbiased and therefore $D_n = 0$ and $Q = 0$. The same holds for the WLS estimator.

The standard deviation of the expected estimator $<(1/b)_U>$ has limit $E_\infty = 0$. For the LS estimator with $P = -0.47$, $Q = 0.9$ and $R = 0.9$:

$$<\sigma_{(1/b_{1,1})_U}>_n \approx \frac{0.9}{(n - 0.9)^{0.47}} \tag{6.190}$$

For the WLS estimator with $P = -0.5$, $Q = 0.8$ and $R = 1.2$:

$$<\sigma_{(1/b_{1,1})_U}>_n \approx \frac{0.8}{(n - 1.2)^{0.5}} \tag{6.191}$$

The more negative power value P of the WLS estimator ($P = -0.5$) than of the LS estimator ($P = -0.47$) means that the WLS estimator approaches the asymptote faster than the LS estimator (i.e. the scatter is reduced faster). The power P provides a means to quantify the efficiency of an estimator.

The unbiased LS (respectively WLS) estimators $(1/b)_{U,LS}$ and $(1/b)_{U,WLS}$ with their standard deviations are:

$$\left(\frac{1}{b}\right)_{U,LS} \pm \sigma_{(1/b)_{U,LS}} \approx \left(\frac{1}{b}\right)_{LS} \pm \frac{0.9}{(n - 0.9)^{0.47}} \cdot \frac{1}{\beta} \tag{6.192}$$

$$\left(\frac{1}{b}\right)_{U,WLS} \pm \sigma_{(1/b)_{U,WLS}} \approx \left(\frac{1}{b}\right)_{WLS} \pm \frac{0.8}{\sqrt{n - 1.2}} \cdot \frac{1}{\beta} \tag{6.193}$$

For the LS shape parameter estimator $<b_{1,1}>$ with $P = -0.9$, $Q = 0.68$ and $R = 2$:

$$<b_{1,1}>_n - <b_{1,1}>_\infty = <b_{1,1}>_n - 1 \approx \frac{0.68}{(n - 2)^{0.9}} \tag{6.194}$$

For the WLS shape parameter estimator with $P = -1$, $Q = 0.66$ and $R = 2$:

$$<b_{1,1}>_n - <b_{1,1}>_\infty = <b_{1,1}>_n - 1 \approx \frac{0.66}{n - 2} \tag{6.195}$$

The more negative power value P of the WLS estimator ($P = -1$) than of the LS estimator ($P = -0.9$) means that the WLS estimator approaches the asymptote faster than the LS estimator (i.e. the bias reduces faster). Again, the power P provides a means to quantify the efficiency of an estimator.

The standard deviation of the expected estimator $<b_U>$ also has limit $E_\infty = 0$. For the LS estimator with $P = -0.45$, $Q = 0.8$ and $R = 2.9$:

$$<\sigma_{(b_{1,1})_U}>_n \approx \frac{0.8}{(n - 2.9)^{0.47}} \tag{6.196}$$

For the WLS estimator with $P = -0.5$, $Q = 0.8$ and $R = 2.9$:

$$<\sigma_{(b_{1,1})_U}>_n \approx \frac{0.8}{\sqrt{n - 2.9}} \tag{6.197}$$

The unbiased LS (respectively WLS) estimators $b_{U,LS}$ and $b_{U,WLS}$ with their standard deviations are:

$$b_{U,LS} \pm \sigma_{b_{U,LS}} \approx \frac{(n - 2)^{0.9}}{(n - 2)^{0.9} + 0.68} \cdot b_{LS} \pm \frac{0.8}{(n - 2.9)^{0.45}} \cdot \beta \tag{6.198}$$

$$b_{U,WLS} \pm \sigma_{b_{U,WLS}} \approx b_{WLS} \pm \frac{0.8}{\sqrt{n - 1.2}} \cdot \beta \tag{6.199}$$

For the expected LS and WLS estimators we found $Q = 0$, which makes $\log(a)$ an unbiased estimator of $\log(\alpha)$ despite the corrective term in Eqs (6.184) and (6.189).

The standard deviations of the LS and WLS estimators appear practically the same with $P = -0.5$, $Q = 1.06$ and $R = 0.2$:

$$<\sigma_{(\log a_{1,1})_U}>_n \approx \frac{1.06}{\sqrt{n - 0.2}} \approx \frac{1.1}{\sqrt{n}} \tag{6.200}$$

The expected unbiased LS and WLS estimators $<(\log(a))_U>$ with their standard deviations have the same approximation in [27] (Note that there seems to be a typing error in Table 7 of [27] in the expression for the standard deviation of $\ln(a)$):

$$(\log a)_U \pm \sigma_{(\log a)_U} \approx \log a \pm \frac{1.1}{\sqrt{n}} \cdot \frac{1}{\beta} \tag{6.201}$$

The power function provides a convenient unbiasing method, estimation of the standard deviation and a way to quantify the efficiency of the estimators.

6.3.2.7 Comparison of Bias and Scatter in LS, WLS and ML

The ML, LS and WLS estimate the Weibull parameters. In Sections 6.2.3.4 and 6.3.2.6 quantitative representations for both estimator families were discussed. An overview of the results is shown in Table 6.5.

In all cases particularly the bias in the shape parameter is larger than that of the scale parameter and has the greatest impact on the reliability estimation at low probability of failure (see Section 6.2.3.5). The present section compares the bias and scatter results for the shape parameters b and $1/b$. The results are obtained for the standard Weibull-2 distribution (i.e. with $\alpha = \beta = 1$) that can be used for unbiasing through the pivotal functions (see Sections 6.2.3.3 and 6.3.2.5). The estimator of the standard Weibull-2 distribution is indicated as $b_{1,1}$ again.

The estimator of the Weibull shape parameter $b_{1,1}$ is compared in Figure 6.3 for the bias and in Figure 6.4 for the standard deviation of the unbiased estimator. The smallest bias occurs for WLS, but the LS bias is only slightly greater. The ML estimator is significantly larger for small sample size n.

After removing the systematic error by unbiasing, the random error remains (i.e. the standard deviation). As shown in Figure 6.4, the smallest standard deviation is achieved with WLS. However, the scatter for the ML estimator is about $0.1/2.9 \approx 3.4\%$ higher and it is not sure that this is significant except at very small n. For very small n (up to about $n = 4$) the LS scatter is smaller than that of the ML estimator, while at larger n the scatter

Table 6.5 Parameters of the power function D_n (Eq. (6.185)) for ML, LS and WLS estimators [27].

	$<b_{1,1}>_n - <b_{1,1}>_\infty$			$<\sigma_{(b_{1,1})_U}>_n$			$<1/b_{1,1}>_n - <1/b_{1,1}>_\infty$			$<\sigma_{(1/b_{1,1})_U}>_n$		
	ML	LS	WLS	ML	LS	WLS	ML	LS	WLS	ML	LS	WLS
P	-1	-0.9	-1	-0.5	-0.45	-0.5	-1	n.a.	n.a.	-0.5	-0.47	-0.5
Q	1.32	0.68	0.66	0.8	0.8	0.8	0.76	0	0	0.8	0.9	0.8
R	2	2	2	3	2.9	2.9	0.18	n.a.	n.a.	1.2	0.9	1.2

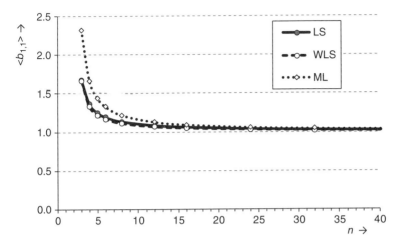

Figure 6.3 Expected estimator of the Weibull shape parameter b in case of $\alpha = \beta = 1$ based on the power functions in Eqs (6.52), (6.195) and (6.194).

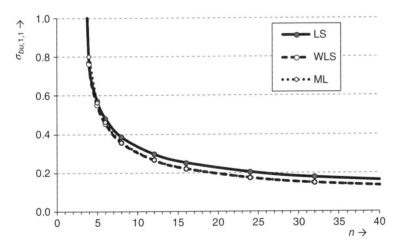

Figure 6.4 Standard deviation of the unbiased Weibull shape parameter in case of $\alpha = \beta = 1$ based on the power functions in Eqs (6.52), (6.195) and (6.194). The WLS and the ML curve largely overlap.

of the LS estimator decays slower due to its smaller negative power P of 0.45 instead of 0.5 for WLS and ML.

What if unbiasing is not carried out? In that case the standard deviation of the biased estimator is obtained by multiplying the standard deviation of the unbiased estimator by $<b_{1,1}>$. The random error of the biased WLS estimator remains the smallest for every n. For small n up to 8 the biased ML estimator has the largest standard deviation, while the scatter of the biased LS estimator takes over from $n = 9$ due to its smaller negative power P compared to WLS and ML.

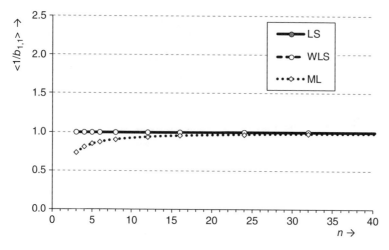

Figure 6.5 Expected estimator of the inverse Weibull shape parameter $1/b$ in case of $\alpha = \beta = 1$ based on the power functions in Eqs (6.56) and (6.186). The WLS and LS lines overlap.

One may wonder what is the virtue of correcting for a bias when it is equal to or smaller than the standard deviation. It is arguable, but particularly in case of a need for early decision-making based on small data sets, that there are reasons in favour of unbiasing. Firstly, accepting the reality of a significant random error is not an excuse for also accepting a systematic error that can be corrected for. Secondly, if such cases repeatedly occur, then the standard deviation of the average decreases with the root mean square of the number of cases, while the systematic error remains the same. At some point in time this standard deviation as a random error may no longer overshadow the systematic error. If best decision-making with small data set cases is a repeated part of the job, then unbiasing should pay off in the long run.

The estimator of the Weibull shape parameter $1/b_{1,1}$ is compared in Figure 6.5 for the bias and in Figure 6.6 for the standard deviation of the unbiased estimator. The WLS and LS are both unbiased and therefore have the smallest absolute bias (namely 0). The ML estimator has a significant negative bias for small sample sizes n to the downside (i.e. the ML estimator of $1/b$ is systematically too low).

After removing the systematic error by unbiasing (only necessary for the ML estimator), the standard deviation remains as the random error. As shown in Figure 6.6 and Table 6.5, the standard deviations of WLS and ML are the same. The standard deviation for the unbiased LS estimator decays slower with power $P = 0.47$ against 0.5 for WLS and ML.

Without unbiasing, the error in the biased estimator $1/b$ is obtained by multiplying the standard deviation of the unbiased estimator $1/b_U$ with $<1/b_{1,1}>$, which equals 1 for LS and WLS and < 1 for ML. Therefore, and for all n, the standard deviation of the biased ML estimator is smaller than for WLS by a factor that equals the ML bias in the $1/b_{1,1}$ estimator. The standard deviation of the biased LS estimator will be 65% higher for $n = 2$, which in the range from $n = 4$ to beyond 50 hovers between 34% and 27% (with a minimum at $n = 20$), as can be calculated based on the parameters in Table 6.5. It may seem an asset to have a lower standard deviation when working with biased ML

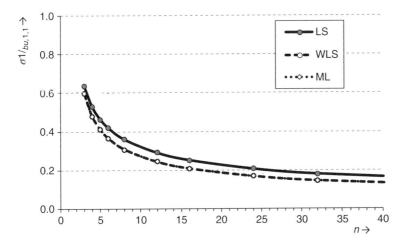

Figure 6.6 Standard deviation of the unbiased inverse Weibull shape parameter $1/b$ in case of $\alpha = \beta = 1$ based on the power functions in Eqs (6.52), (6.195) and (6.194). The WLS and ML curve practically overlap.

estimators, but it may also produce biased decisions, therefore the behaviour of unbiased estimators remains important.

The power function D_n proves to be very useful in such comparisons.

What is better: ML, LS or WLS? It depends. The LS method is the easiest to implement and is transparent with analytical solutions even for censored data. ML is often promoted for its merits for large sample sizes, but WLS performs equally well. With unbiased estimators WLS performs equally or even better than ML in terms of bias and scatter. One of the advantages of linear regression (LS and WLS) is that it is consistent with Weibull plots, provided that the expected plotting position is used (rather than the median plotting position). The best fit in the graph has exactly the same parameters as those determined by linear regression.

6.3.3 LR for the Exponential Distribution

The exponential distribution was introduced in Section 4.2.4. It describes the statistics of random failure (i.e. processes that have no memory of past operational life). The cumulative function shows as (Eq. (4.33)):

$$F_{\exp}(t; \theta) = 1 - e^{-\frac{t}{\theta}} \tag{6.202}$$

The parameter θ is called the characteristic lifetime and equals the mean time to failure. The hazard rate $h(t)$ of the exponential distribution is a constant (Eq. (4.34)):

$$h(t; \theta) = \frac{1}{\theta} \tag{6.203}$$

The exponential distribution can also be regarded as a Weibull distribution, with shape parameter fixed as $B = 1$ (see Section 4.2.3). In terms of a Weibull distribution, θ is the scale parameter.

Applying linear regression to the exponential distribution calls for a linear relationship, as in Eq. (6.75):

$$Y = A_0 + A_1 \cdot X \tag{6.204}$$

In Section 5.5 such a relationship was also needed to produce a linear exponential plot and formulated in Eq. (5.42) as a linear relationship between t and $\ln(1 - F)$:

$$Z(F_{\exp}) = -\ln(1 - F_{\exp}) = \frac{1}{\theta} \cdot t \tag{6.205}$$

This linear relationship between $\ln(1 - F_{\exp})$ and t can also be used for linear regression.

Alternatively, the exponential distribution can be regarded as a special case of the Weibull distribution with a fixed shape parameter $B = 1$, as mentioned above. The linear relationship for Weibull plots take the log of Eq. (6.205), which produces (cf. Eq. (5.43)):

$$\log(-\ln(1 - F_{\exp})) = \log t - \log \theta \tag{6.206}$$

Therefore, the exponentially distributed data can also be analysed as Weibull data and it can be evaluated whether the data are indeed exponentially distributed by checking whether the shape parameter estimator b equals 1. Linear regression analysis of Weibull data was discussed in Sections 6.2.3 through 6.3.2.7. In the following sections, linear regression of the exponentially distributed data based on the linear relationship of Eq. (6.205) will be discussed.

The expected plotting position $<Z_{i,n}>$ becomes (cf. Eq. ((5.44):

$$<Z_{i,n}> = <-\ln(1 - F_{\exp,i,n})> = -\int_0^1 \ln(1 - p) \cdot i \binom{n}{i} \cdot p^{i-1} \cdot (1 - p)^{n-i} dp \tag{6.207}$$

The solution is (see Eq. (5.45)):

$$<Z_{i,n}> = i \binom{n}{i} \sum_{j=0}^{i-1} \binom{i-1}{j} \frac{(-1)^j}{(n-i+j+1)^2} \tag{6.208}$$

The corresponding exact $F(<Z_{i,n}>)$ is:

$$F(<Z_{i,n}>) = 1 - \exp(-<Z_{i,n}>) \tag{6.209}$$

As with the Weibull distribution, this function appears almost linear with i. It can be approximated by (see Section 5.5.1 and Eq. (5.49)):

$$F(<Z_{i,n}>) \approx F(<Z_{1,n}>) + \frac{i-1}{n-1} \cdot (F(<Z_{n,n}>) - F(<Z_{1,n}>)) \tag{6.210}$$

Here:

$$F(<Z_{1,n}>) = 1 - \exp(-(1/n)) \tag{6.211}$$

$$F(<Z_{n,n}>) \approx 1 - \frac{0.562047488}{n + 0.518922085} \tag{6.212}$$

This approximation calculates $F(<Z_{i,n}>)$ within typically better than 0.7%

Finally, as t is the covariable, the linear relationship for linear regression reads:

$$t = \theta \cdot Z(F_{\text{exp}}) \tag{6.213}$$

Comparing Eqs (6.204) and (6.213) reveals that the location parameter a_0 equals 0 and the regression coefficient a_1 corresponds to the coefficient θ. Equation (6.213) is a single-coefficient relationship of the type that was discussed in Section 6.3.1.4 (cf. Eq. (6.120)):

$$Y = a_1 \cdot X \tag{6.214}$$

6.3.3.1 LR by Unweighted LS for the Exponential Distribution

Assume n observation pairs $(<Z_i>, t_i)$ are available. The averages are defined as (cf. Eq. (6.81)):

$$\overline{<Z>} = \frac{1}{n} \sum_{i=1}^{n} <Z_i> \tag{6.215}$$

$$\bar{t} = \frac{1}{n} \sum_{i=1}^{n} t_i \tag{6.216}$$

The unweighted LS estimator η_{LS} of the coefficient θ can be found by applying Eq. (6.123) in terms of sums or Eq. (6.124) in terms of averages:

$$\eta_{LS} = \frac{\sum_{i=1}^{n} t_i \cdot <Z_i>}{\sum_{i=1}^{n} <Z_i>^2} = \frac{\overline{<Z> \cdot t}}{\overline{<Z>^2}} \tag{6.217}$$

The internal error $s_{\text{int},\eta}^2$ follows from Eq. (6.125):

$$s_{\text{int},\eta}^2 = \frac{1}{\sum_{i=1}^{n} (<Z_i> - \overline{<Z>})^2} \cdot \frac{\sum_{i=1}^{n} (t_i - \tau_{LS} \cdot <Z_i>)^2}{n-1} \tag{6.218}$$

The correlation coefficient $r_{Z,t}$ can be analytically calculated as in Section 6.3.1.1 (cf. Eq. (6.94)):

$$r_{Z,t} = \frac{\sum_{i=1}^{n} ((<Z_i> - \overline{<Z>}) \cdot (t_i - \bar{t}))}{\sqrt{\sum_{i=1}^{n} (<Z_i> - \overline{<Z>})^2} \cdot \sqrt{\sum_{i=1}^{n} (t_i - \bar{t})^2}} = \frac{s_{Z,t}}{s_Z \cdot s_t} \tag{6.219}$$

The external error $s_{\text{ext},\eta}^2$ is found as in Eq. (6.98):

$$s_{\text{ext},\eta}^2 = \sum_{i=1}^{n} \left(\frac{(<Z_i> - \overline{<Z>})}{\sum_{i=1}^{n} (<Z_i> - \overline{<Z>})^2} \right)^2 \cdot (\Delta t_i)^2 \tag{6.220}$$

Again, these errors can be processed with Eqs (6.96) or (6.97) to find estimators of the corresponding errors in functions of η_{LS}.

The confidence intervals can be determined with the beta distribution (Sections 5.2.4, 5.2.5 and 5.5.4) if the data are considered to be randomly drawn from a large population (cf. Section 5.2.6). Alternatively, they can be based on the variance if the scatter can be considered normally distributed (cf. Sections 6.3.1.1 and 6.3.1.2).

6.3.3.2 LR by Weighted LS for the Exponential Distribution

Linear regression by WLS follows the approach of Section 6.3.1.2. The weights w_i are defined as (n is omitted as suffix below):

$$w_i = \frac{1}{<Z_i^2> - <Z_i>^2} \tag{6.221}$$

The expected value $<Z_i^2>$ is found as:

$$
\begin{aligned}
<Z_i^2> &= -\int_0^1 [\ln(1-p)]^2 \cdot i \binom{n}{i} \cdot p^{i-1} \cdot (1-p)^{n-i} dp \\
&= -\int_0^1 [\ln q]^2 \cdot i \binom{n}{i} \cdot (1-q)^{i-1} \cdot q^{n-i} dq \\
&= -i \binom{n}{i} \cdot \sum_k^{i-1} \binom{i-1}{k} (-1)^k \left[\int_0^1 [\ln q]^2 \cdot q^{n-i+k} dq \right]
\end{aligned}
\tag{6.222}
$$

which can be solved with integral 496 in [44] (realising that the limit $q \ln q = 0$ for q approaching 0):

$$
\begin{aligned}
\int_0^1 q^{n-i+k} \cdot (\ln q)^2 dq &= \frac{q^{n-i+k+1}(\ln q)^2}{n-i+k+1} \Big|_0^1 - \frac{2}{n-i+k+1} \int_0^1 q^{n-i+k} \cdot (\ln q) dq \\
&= \left(\frac{q^{n-i+k+1}(\ln q)^2}{n-i+k+1} - \frac{2 \cdot q^{n-i+k+1} \cdot \ln q}{(n-i+k+1)^2} + \frac{2 \cdot q^{n-i+k+1}}{(n-i+k+1)^3} \right) \Big|_0^1 \\
&= \frac{2}{(n-i+k+1)^3}
\end{aligned}
\tag{6.223}
$$

The expected plotting position value $<Z_i>$ was defined in Eq. (6.208). With these expressions, the weights w_i can be calculated.

Assume n observation pairs (Z_i, t_i) are available. The weighted averages are defined as (cf. Eq. (6.81)):

$$\overline{(<Z>)}_w = \frac{\sum_{i=1}^n (w_i \cdot <Z_i>)}{\sum_{i=1}^n w_i} \tag{6.224}$$

$$\bar{t}_w = \frac{\sum\limits_{i=1}^{n} w_i t_i}{\sum\limits_{i=1}^{n} w_i} \tag{6.225}$$

The weighted LS estimator η_{LS} of the coefficient θ can be found by applying Eq. (6.123) in terms of sums or Eq. (6.124) in terms of averages:

$$\eta_{LS} = \frac{\sum\limits_{i=1}^{n} w_i t_i <Z_i>}{\sum\limits_{i=1}^{n} w_i <Z_i>^2} = \frac{\overline{(<Z> \cdot t)_w}}{\overline{(<Z>^2)_w}} \tag{6.226}$$

The internal error $s_{\text{int},\eta}^2$ follows from Eq. (6.128):

$$s_{\text{int},\eta}^2 = \frac{1}{\sum\limits_{i=1}^{n} w_i (<Z_i> - \overline{<Z>})^2} \cdot \frac{\sum\limits_{i=1}^{n} w_i (t_i - \eta_{LS} \cdot <Z_i>)^2}{n-1} \tag{6.227}$$

The denominator contains a factor $n-1$ instead of $n-2$ due to the fact that with one parameter only one degree of freedom is lost. The correlation coefficient $r_{Z,t}$ can be calculated analytically as in Section 6.3.1.1 (cf. Eq. (6.94)):

$$r_{Z,t} = \frac{\sum\limits_{i=1}^{n} (w_i (<Z_i> - \overline{<Z>}_w) \cdot (t_i - \bar{t}_w))}{\sqrt{\sum\limits_{i=1}^{n} w_i (<Z_i> - \overline{<Z>}_w)^2} \cdot \sqrt{\sum\limits_{i=1}^{n} w_i (t_i - \bar{t}_w)^2}} = \frac{s_{Z,t}}{s_Z \cdot s_t} \tag{6.228}$$

and the external error $s_{\text{ext},\eta}^2$ as in Eq. (6.116):

$$s_{\text{ext},\eta}^2 = \sum\limits_{i=1}^{n} w_i \left(\frac{(<Z_i> - \overline{<Z>}_w)}{\sum\limits_{i=1}^{n} w_i (<Z_i> - \overline{<Z>}_w)^2} \right)^2 \cdot (\Delta t_i)^2 \tag{6.229}$$

Again, these errors can be processed with Eqs (6.96) or (6.97) to find estimators of the corresponding errors in functions of η_{LS}.

The confidence intervals can be determined with the beta distribution (Sections 5.2.4, 5.2.5 and 5.5.4) if the data are considered to be randomly drawn from a large population (cf. Section 5.2.6). Alternatively, they can be based on the variance if the scatter can be considered normally distributed (cf. Sections 6.3.1.1 and 6.3.1.2).

6.3.3.3 Processing Censored Data with the Adjusted Rank Method

An introduction to censored data and the application of linear regression was given in Section 6.3.1.3. It is assumed that the population under investigation has size n, of which r objects failed at known times t_i ($i = 1, \ldots, r$). This means that the failed fraction has sample size r and the surviving fraction has sample size $n - r$. It was explained that two methods are in use to find the ranking possibilities for the r observations: the adjusted rank method and the adjusted plotting position method. The approach to process censored data is the same as with the graphical analysis in Section 5.5.4.

The variables for the linear regression are the $<Z_{i,n}>$:

$$<Z_{i,n}> = -\ln(1 - F_{\exp,i,n}) = -\int_0^1 \ln(1 - p) \cdot i \binom{n}{i} \cdot p^{i-1} \cdot (1 - p)^{n-i} dp \qquad (6.230)$$

but in the above the index i is defined within the subgroup of failed objects, which has sample size r, and not the entire population. Within the total population with sample size n, the ranking index is yet to be determined, since the ranking is likely subject to change until all objects have ultimately failed. This is particularly true if some objects have a shorter operational life than the $r < n$ already observed times to failure at the evaluation moment, as explained in Section 6.3.1.3. As also applied to the Weibull distribution (Section 6.3.2.3), a way to solve this is to inventory all possible rankings and evaluate for each failed sample with index i what its average ranking index $I(i)$ or plotting position $Z_{I,n}$ is. The latter is discussed in the next section.

With the adjusted rank method first all uncensored data t_i ($i = 1, \ldots, r$) and censored data τ_j ($j = 1, \ldots, n - r$) are collected as in Table 6.2. The uncensored data are attributed to rank index i and subsequently an adjusted rank index $I(i)$ is calculated through the equation (see Eq. (5.3) in Section 5.2.2):

$$I(i) = I(i - 1) + \frac{n + 1 - I(i - 1)}{n + 2 - C_i} \qquad (6.231)$$

C_i is the sum of the number of censored and uncensored observations including the failure i. $I(0)$ is zero by definition. In case all censored observations are greater than the uncensored observations (e.g. when all objects were simultaneously put to test from the beginning), $C_i - 1$ equals $I(i - 1)$ and all $I(i)$ are simply equal to i ($i = 1, \ldots, r$).

Generally, the $I(i)$ are neither necessarily integers themselves nor are their differences necessarily integers. The summation in Eq. (6.230) is therefore hard to carry out. However, since $Z_{i,n}$ is a function of $F_{i,n}$ and the cumulative function $F(<Z_{i,n}>)$ for the exponential distribution can be approximated with Eqs (6.210)–(6.212), the cumulative function $F(<Z_{I,n}>)$ can conveniently be approximated as:

$$F(<Z_{I,n}>) \approx F(<Z_{1,n}>) + \frac{I - 1}{n - 1} \cdot (F(<Z_{n,n}>) - F(<Z_{1,n}>)) \qquad (6.232)$$

Here:

$$F(<Z_{1,n}>) = 1 - \exp(-(1/n)) \qquad (6.233)$$

$$F(<Z_{n,n}>) \approx 1 - \frac{.562047488}{n + .518922085} \qquad (6.234)$$

It is noteworthy that this expression does not require I to be an integer. From this the expected plotting position with the adjusted ranking I follows as:

$$<Z_{I,n}> \approx -\ln(1 - F(<Z_{I,n}>)) \qquad (6.235)$$

As for $F(<Z_{I,n}>)$, it is not required for the approximation of $<Z_{I,n}>$ that I is an integer.

The observed failure data t_i are now attributed to the adjusted rank, that is:

$$t_I = t_{I(i)} = t_i \tag{6.236}$$

With the calculated $<Z_{I(i),n}>$ and the observed failure data $t_{I(i)}$, unweighted linear regression is fairly straightforward for a set of r pairs of variables and covariables $(x_{I(i)}, y_{I(i)})$ with:

$$x_I = x_{I(i)} = <Z_{I(i),n}> \tag{6.237}$$

$$y_I = y_{I(i)} = t_{I(i)} \tag{6.238}$$

It should be noted that the analysis is based on r pairs and therefore n must be replaced by r when summations over $I(i)$ (basically over i) are carried out in the equations of Section 6.3.1.1. From that perspective, the averages are defined as (cf. Eq. (6.81)):

$$\overline{<Z>} = \frac{1}{r} \sum_{i=1}^{r} <Z_{I(i)}> \tag{6.239}$$

$$\overline{\log t} = \frac{1}{r} \sum_{i=1}^{r} t_{I(i)} \tag{6.240}$$

For censored data with r uncensored observations the unweighted LS estimators η_{LS} according to Section 6.3.1.4 are now found as (cf. Eq. (6.217)):

$$\eta_{LS} = \frac{\sum\limits_{i=1}^{r} t_{I(i)}<Z_{I(i)}>}{\sum\limits_{i=1}^{r} <Z_{I(i)}>^2} = \frac{\overline{(<Z> \cdot t)}}{\overline{(<Z>^2)}} \tag{6.241}$$

with the averages as in Eqs (6.239) and (6.240), $I(i)$ from Eq. (6.22), $<Z_{I(i),n}>$ from Eqs (6.232)–(6.234) and $t_{I(i)}$ from Eq. (6.236). The correlation coefficient and the errors are found in the same way through the equations in Section 6.3.3.1, by replacing n with r.

Unfortunately, a convenient approximation for $<Z_I^2>$ is not yet available, which makes this method not suitable for weighted linear regression as yet. But if a suitable expression is found, the estimator can be worked out with Eq. (6.226), where n is replaced by r, and i in the suffices is replaced by $I(i)$.

6.3.3.4 EXTRA: Processing Censored Data with the Adjusted Plotting Position Method

As mentioned previously, the censored data and the application of linear regression were introduced in Section 6.3.1.3. Again it is assumed that the population under investigation has size n, of which r objects failed at known times t_i ($i = 1, \ldots, r$). This means that the failed fraction has sample size r and the surviving fraction has sample size $n - r$. It was explained that two methods are in use to find the ranking possibilities for the r observations: the adjusted rank method and the adjusted plotting position method. The latter is briefly discussed here with reference to the Weibull case. The approach to process censored data is the same as with the graphical analysis in Section 5.4.4.

The method is elaborated for the Weibull distribution in Section 6.3.2.4 and can be applied to the exponential distribution in a similar fashion, albeit that the estimation of the coefficient follows the equations as in Section 6.3.3.3.

6.3.3.5 Expected LS and WLS Estimator for the Exponential Distribution

Again, first pivotal quantities and functions are discussed. Let u be distributed with the normalized exponential distribution (i.e. $\theta = 1$):

$$F(u; \theta = 1) = 1 - e - \left(\frac{u}{1}\right)^1 = 1 - e^{-u} \tag{6.242}$$

Therefore:

$$Z = -\ln(1 - F(u; \theta = 1)) = -\ln(1 - 1 + e^{-u}) = u \tag{6.243}$$

The LS estimators η_{LS} of the characteristic lifetime parameter $\theta = 1$ are indicated as η_1. Any exponential distribution can be produced by substituting:

$$u = \frac{t}{\theta} \tag{6.244}$$

The variable u is a pivotal quantity. The model does not change with the choice of the θ parameter. The LS estimator η_1 for the uncensored case is (cf. Eq. (6.217)):

$$\eta_1 = \frac{\sum\limits_{i=1}^n u_i \cdot <Z_i>}{\sum\limits_{i=1}^n <Z_i>^2} = \frac{\overline{(<Z> \cdot u)}}{\overline{(<Z>^2)}} \tag{6.245}$$

with:

$$\overline{(<Z> \cdot u)} = \frac{1}{n} \sum_{i=1}^n (u_i \cdot <Z_i>) \tag{6.246}$$

$$\overline{(<Z>^2)} = \frac{1}{n} \sum_{i=1}^n (<Z_i>^2) \tag{6.247}$$

The LS estimator of the general characteristic parameter θ is indicated as η_θ. Substitution of Eq. (6.244) in Eq. (6.245) with subsequent comparison with Eq. (6.217) shows that:

$$\eta_\theta = \theta \cdot \eta_1 \tag{6.248}$$

By substituting Eq. (6.243)), the expected LS estimator $<\eta_1>$ of $\theta = 1$ appears unbiased:

$$<\eta_1> = <\frac{\sum\limits_{i=1}^n u_i \cdot <Z_i>}{\sum\limits_{i=1}^n <Z_i>^2}> = \frac{\sum\limits_{i=1}^n <u_i> \cdot <Z_i>}{\sum\limits_{i=1}^n <Z_i>^2} = \frac{\overline{(<Z>^2)}}{\overline{(<Z>^2)}} = 1 \tag{6.249}$$

The expected LS estimator $<\eta_\theta>$ of θ is also unbiased:

$$<\eta_\theta> = \theta \cdot <\eta_1> = \theta \tag{6.250}$$

Also, for WLS the estimator $<\eta_1>$ of $\theta = 1$ is unbiased:

$$<\eta_1> = <\frac{\sum\limits_{i=1}^n w_i u_i \cdot <Z_i>}{\sum\limits_{i=1}^n w_i <Z_i>^2}> = \frac{\overline{(<Z>^2)}_w}{\overline{(<Z>^2)}_w} = 1 \tag{6.251}$$

The expected WLS estimator $<\eta_\theta>$ of θ is therefore also unbiased:

$$<\eta_\theta> = \theta \cdot <\eta_1> = \theta \tag{6.252}$$

With censoring and the adjusted ranking, the (W)LS estimator $<\eta_1>$ comes close to being unbiased, while for the adjusted plotting position the averaging over Z also seems to produce an unbiased estimator.

6.3.4 LR for the Normal Distribution

The normal distribution was introduced in Section 4.3.1. The distribution density function is defined in Eq. (4.50) as:

$$f(x) = \frac{1}{\sigma\sqrt{2\pi}} \cdot e^{-\frac{1}{2}\left(\frac{x-\mu}{\sigma}\right)^2} \tag{6.253}$$

The two parameters are the mean μ and the standard deviation σ. The cumulative distribution F cannot be derived analytically, but most mathematical software has a normal distribution module built in; there are numerical tables (cf. Appendix G) and approximations (Section 10.6) that yield values for F.

Applying linear regression to the normal distribution calls for a linear relationship as in Eq. (6.75):

$$T = A_0 + A_1 \cdot Z \tag{6.254}$$

The use of the characters 'Y' and 'X' in Eq. (6.75) is avoided here in order to prevent confusion with the use of 'x' and 'y' in previous sections with discussions on the normal distribution. However, Eq. (6.254) is the same kind of linear expression needed for LS and WLS parameter estimation as with the discussion on the Weibull and exponential distributions.

For producing normal plots a linear relationship was also required, as discussed in Section 5.6. This was achieved by plotting observed covariable values x_i ($i = 1, \ldots, n$) and expected values $<y_i>$ of the harmonized variable y against each other. The harmonized variable is defined as (see Eq. (5.57)):

$$y(x) = \frac{x - \mu}{\sigma} \tag{6.255}$$

The expected values $<y_{i,n}>$ are the variables $<Z_{i,n}>$ similar to the cases of the Weibull and the exponential distribution. However, due to the fact that there are no analytical expressions for the normal F and R, there is also no analytical expression for the expected values $<y_{i,n}>$ (see Section 5.6). However, the $<y_{i,n}>$ can be determined numerically or through approximations. The (W)LS will be elaborated in terms of $<Z_{i,n}>$, but it may be kept in mind that the plotting position $Z_{i,n}$ is just the same as $y_{i,n}$.

With the Weibull and the exponential distributions the expected plotting positions $<Z_{i,n}>$ could be calculated analytically, but it was more convenient to estimate them through approximations for $F(<Z_{i,n}>)$. These $F(<Z_{i,n}>)$ appeared practically linear with i (cf. Sections 5.4.1 and 5.5.1). For the normal distribution approximation also appears almost linear, as discussed in Section 5.6.1. Table 5.18 in that section gives an overview of two types of linear approximation for $F_{i,n}$ with accuracy in terms of a squared sum of residuals as in Eq. (5.70).

The most accurate approximation, particularly for small sample sizes n ($n < 10$), as discussed in Section 5.6.1, is (see Eq. (5.69) and Table 5.18):

$$F(<Z_{i,n}>) = F(<y_{i,n}>) \approx \frac{1.01}{n + 0.36} \cdot i - \frac{1}{200} \cdot \frac{n + 65}{n + 0.36} \tag{6.256}$$

A widely used approximation and second best for $n > 3$ is the formula of Blom [30] (see Eq. (5.61) and Table 5.18):

$$F(<Z_{i,n}>) = F(<y_{i,n}>) \approx \frac{i - 0.375}{n + 0.25} \tag{6.257}$$

The approximations of $<Z_{i,n}>$ are obtained by applying the inverse normal distribution:

$$<Z_{i,n}> \approx F^{-1}(F(<Z_{i,n}>)) \tag{6.258}$$

This can be achieved with most standard mathematical software and with the approximations in Section 10.6.

Finally, the variables in the linear relationship for linear regression are the $<y_{i,n}>$ and the covariables are the observed $x_{i,n}$. The linear relationship is therefore:

$$x = \mu + \sigma \cdot Z(F) \tag{6.259}$$

The coefficients a_0 and a_1 are therefore:

$$a_0 = \mu \tag{6.260}$$

$$a_1 = \sigma \tag{6.261}$$

6.3.4.1 LR by Unweighted LS for the Normal Distribution

Assume n observation pairs ($<Z_i>$, x_i) are available ($i = 1, \ldots, n$). The variables are the expected plotting positions $<Z_i>$, which are identical to the expected standard variable (i.e. $<Z_i> = <y_i>$). The covariables are the observed x_i ($i = 1, \ldots, n$). The averages are defined as (cf. Eq. (6.81)):

$$\overline{<Z>} = \frac{1}{n} \sum_{i=1}^{n} <Z_i> \tag{6.262}$$

$$\bar{x} = \frac{1}{n} \sum_{i=1}^{n} x_i \tag{6.263}$$

The unweighted LS estimators m_{LS} and s_{LS} according to Section 6.3.1.1 are now found as (note that $<Z_i>$ are the variables and x_i the covariables in the following equations):

$$m_{LS} = \bar{x} - s_{LS} \cdot \overline{<Z>} \tag{6.264}$$

$$s_{LS} = \frac{\overline{<Z> \cdot x} - \bar{x} \cdot \overline{<Z>}}{\overline{<Z>^2} - \overline{<Z>}^2} = \frac{\overline{(<Z> - \overline{<Z>}) \cdot x}}{\overline{(<Z> - \overline{<Z>})^2}}$$

$$= \frac{\overline{(<Z> - \overline{<Z>}) \cdot (x - \bar{x})}}{\overline{(<Z> - \overline{<Z>}) \cdot (<Z> - \overline{<Z>})}} \tag{6.265}$$

Alternatively, Eq. (6.85) can be expressed in terms of summations in various ways:

$$s_{LS} = \frac{\sum\limits_{i=1}^{n}(<Z_i> \cdot x_i) - \sum\limits_{i=1}^{n} x_i \cdot \frac{1}{n}\sum\limits_{i=1}^{n}<Z_i>}{\sum\limits_{i=1}^{n}(<Z_i>^2) - \frac{1}{n}\left(\sum\limits_{i=1}^{n}<Z_i>\right)^2} = \frac{\sum\limits_{i=1}^{n}(<Z_i> \cdot x_i) - n \cdot \bar{x} \cdot \overline{<Z>}}{\sum\limits_{i=1}^{n}(<Z_i>^2) - n \cdot \overline{<Z_i>}^2}$$

$$= \frac{\sum\limits_{i=1}^{n}((<Z_i> - \overline{<Z>}) \cdot (x_i - \bar{x}))}{\sum\limits_{i=1}^{n}((<Z_i> - \overline{<Z>}) \cdot (<Z_i> - \overline{<Z>}))} = \frac{\sum\limits_{i=1}^{n}((<Z_i> - \overline{<Z>}) \cdot x_i)}{\sum\limits_{i=1}^{n}((<Z_i> - \overline{<Z>})^2)} \tag{6.266}$$

The seven expressions in Eqs (6.265) and (6.266) for s_{LS} are all equivalent alternatives. Let $\sigma^2_{int,m_{LS}}$ and $\sigma^2_{int,s_{LS}}$ be the variances due to the internal error of m_{LS} and s_{LS}. As before, variances are functions of expected values, which are generally unknown. Approximations are achieved by summations and averages of the observations. The variance $\sigma^2_{int,s_{LS}}$ of the regression coefficient or slope s_{LS} is approximated by $s^2_{int,s_{LS}}$:

$$\sigma^2_{int,s_{LS}} \approx s^2_{int,s_{LS}} = \frac{1}{\sum\limits_{i=1}^{n}(<Z_i> - \overline{<Z>})^2} \cdot \frac{\sum\limits_{i=1}^{n}(x_i - m_{LS} - s_{SL} \cdot <Z_i>)^2}{n-2} \tag{6.267}$$

The variance $\sigma^2_{int,m_{LS}}$ of the location parameter m_{LS} is approximated by $s^2_{int,m_{LS}}$:

$$\sigma^2_{int,m_{LS}} \approx s^2_{int,m_{LS}} = \left(\frac{1}{n} + \frac{\overline{<Z>}^2}{\sum\limits_{i=1}^{n}(<Z_i> - \overline{<Z>})^2}\right) \cdot \frac{\sum\limits_{i=1}^{n}(x_i - m_{LS} - s_{SL} \cdot <Z_i>)^2}{n-2}$$

$$\tag{6.268}$$

The correlation coefficient $r_{Z,x}$ can be analytically calculated as in Section 6.3.1.1 (cf. Eq. (6.94)):

$$r_{Z,x} = \frac{\sum\limits_{i=1}^{n}((<Z_i> - \overline{<Z>}) \cdot (x_i - \bar{x}))}{\sqrt{\sum\limits_{i=1}^{n}(<Z_i> - \overline{<Z>})^2} \cdot \sqrt{\sum\limits_{i=1}^{n}(x_i - \bar{x})^2}} = \frac{s_{Z,x}}{s_Z \cdot s_x} \tag{6.269}$$

The external errors $s^2_{ext,s_{LS}}$ and $s^2_{ext,m_{LS}}$ are found as in Eqs (6.98) and (6.99):

$$s^2_{ext,s_{LS}} = \sum\limits_{i=1}^{n}\left(\frac{(<Z_i> - \overline{<Z>})}{\sum\limits_{i=1}^{n}(<Z_i> - \overline{<Z>})^2}\right)^2 \cdot (\Delta x_i)^2 \tag{6.270}$$

$$s^2_{ext,m_{LS}} = \left(\frac{1}{n}\right)^2 \sum\limits_{i=1}^{n}(x_i)^2 \tag{6.271}$$

These errors can be processed with Eqs (6.96) or (6.97) to find estimators of functions of the parameters s and m.

The confidence intervals can be determined with the beta distribution (Sections 5.2.4, 5.2.5 and 5.6.2) if the data are considered to be randomly drawn from a large population (cf. Section 5.2.6). Alternatively, they can be based on the variance if the scatter can be considered normally distributed (cf. Sections 6.3.1.1 and 6.3.1.2).

6.3.4.2 Processing Censored Data with the Adjusted Rank Method

An introduction to censored data and the application of linear regression was given in Section 6.3.1.3. It is assumed that the population under investigation has size n, of which r objects failed at known x_i $(i = 1, ..., r)$. This means that the failed fraction has sample size r and the surviving fraction has sample size $n - r$. It was explained that two methods are in use to find the ranking possibilities for the r observations: the adjusted rank method and the adjusted plotting position method. The approach to process censored data is the same as with the graphical analysis in Section 5.5.4.

The variables for the linear regression are the $<Z_{i,n}>$:

$$<Z_{i,n}> = <y_{i,n}> = \int_{-\infty}^{\infty} y \cdot f(y) \cdot i \binom{n}{i} \cdot [F(y)]^{i-1} \cdot [R(y)]^{n-i} dy \qquad (6.272)$$

This cannot be solved analytically because there are no analytical expressions for F and R. They can be approximated either with the expressions in Section 10.6, or more conveniently by using the method where $F(<Z_{i,n}>)$ are determined with Eqs (6.256) or (6.257), after which the inverse cumulative function F^{-1} is used to obtain the $<Z_{i,n}> = <y_{i,n}>$ (cf. Eq. (6.258)).

As in the case of censored Weibull and exponential data, one should be aware that with the censored data set the index i is defined within the subgroup of failed objects which has sample size r, and not the entire population. Within the total population with sample size n, the ranking index is yet to be determined, since the ranking is likely subject to change until all objects have ultimately failed. This is particularly true if some objects have a shorter operational life than the $r < n$ already observed times to failure at the evaluation moment, as explained in Section 6.3.1.3. As also applied to the Weibull distribution (Section 6.3.2.3), a way to solve this is to inventory all possible rankings and evaluate for each failed sample with index i what its average ranking index $I(i)$ or plotting position $Z_{I,n}$ is. The latter is discussed in the next section.

With the adjusted rank method, first all uncensored data x_i $(i = 1, ..., r)$ and censored data ξ_j $(j = 1, ..., n - r)$ are collected as in Table 6.2. The uncensored data are attributed to rank index i and subsequently an adjusted rank index $I(i)$ is calculated through the equation (see Eq. (5.3) in Section 5.2.2):

$$I(i) = I(i-1) + \frac{n + 1 - I(i-1)}{n + 2 - C_i} \qquad (6.273)$$

C_i is the sum of the number of censored and uncensored observations including the failure i. $I(0)$ is zero by definition. In case all censored observations are greater than the uncensored observations (which happens if all objects were simultaneously put to test from the beginning), $C_i - 1$ equals $I(i-1)$ and all $I(i)$ are simply equal to i $(i = 1, ..., r)$.

Generally, the $I(i)$ are neither necessarily integers themselves nor are their differences necessarily integers. The approximations in Eqs (6.256) or (6.257) allow us to work with non-integer $I(i)$. The inverse normal cumulative function F^{-1} also does not require integer values $I(i)$.

From this, the expected plotting position with the adjusted ranking I follows as:

$$<Z_{I,n}> \approx F^{-1}(F(<Z_{I,n}>)) \tag{6.274}$$

The observed failure data x_i $(i = 1, \ldots, r)$ are now attributed to the adjusted rank, that is:

$$x_I = x_{I(i)} = x_i \tag{6.275}$$

With the calculated $<Z_{I(i),n}>$ and the observed failure data $x_{I(i)}$, unweighted linear regression is fairly straightforward for a set of r pairs of variables and covariables $(<Z_{I(i),n}>, x_{I(i)})$.

It should be noted that the analysis is based on r pairs, and therefore n must be replaced by r when summations over $I(i)$ (basically over i) are carried out in the equations of Section 6.3.1.1. From that perspective, the averages are defined as (cf. Eq. (6.81)):

$$\overline{<Z>} = \frac{1}{r} \sum_{i=1}^{r} <Z_{I(i)}> \tag{6.276}$$

$$\overline{x} = \frac{1}{r} \sum_{i=1}^{r} x_{I(i)} \tag{6.277}$$

The unweighted LS estimators m_{LS} and s_{LS} according to Section 6.3.1.1 are now found as (note that $<Z_i>$ are the variables and x_i the covariables in the following equations):

$$m_{LS} = \overline{x} - s_{LS} \cdot \overline{<Z>} \tag{6.278}$$

$$s_{LS} = \frac{\overline{(<Z> - \overline{<Z>}) \cdot (x - \overline{x})}}{\overline{(<Z> - \overline{<Z>}) \cdot (<Z> - \overline{<Z>})}}$$

$$= \frac{\sum_{i=1}^{r}((<Z_i> - \overline{<Z>}) \cdot (x_i - \overline{x}))}{\sum_{i=1}^{r}((<Z_i> - \overline{<Z>}) \cdot (<Z_i> - \overline{<Z>}))} \tag{6.279}$$

Any of the seven expressions for s_{LS} in Eqs (6.265) and (6.266) can be adjusted by replacing n by r. Also, linear regression can be carried out with WLS for the uncensored and for the censored case. This is not elaborated here.

6.3.4.3 EXTRA: Processing Censored Data with the Adjusted Plotting Position Method

As mentioned in the previous subsection, the censored data and the application of linear regression were introduced in Section 6.3.1.3. Again it is assumed that the population under investigation has size n, of which r objects failed at known times t_i $(i = 1, \ldots, r)$. This means that the failed fraction has sample size r and the surviving fraction has sample size $n - r$. It was explained that two methods are in use to find the ranking possibilities for the r observations: the adjusted rank method and the adjusted plotting position method. The latter is briefly discussed in the present section with reference to the normal case. The approach to process censored data is the same as with the graphical analysis in Section 5.6.2.

The method is elaborated for the Weibull distribution in Section 6.3.2.4 and can be applied to the normal distribution in a similar fashion.

6.3.4.4 Expected LS Estimators for the Normal Distribution

Again, first pivotal quantities and functions are discussed. Let u be the normalized variable as defined in Eq. (6.255):

$$u = \frac{x - \mu}{\sigma} \tag{6.280}$$

The variable u is a pivotal quantity. It is noteworthy that u is the same as y and Z. The model does not change with the choice of μ and σ. The LS estimators for the standard distribution (i.e. with $\mu = 0$ and $\sigma = 1$) are $m_{0,1}$ and $s_{0,1}$. According to Eqs (6.264) and (6.265), these are:

$$m_{0,1} = \overline{u} - s_{0,1} \cdot \overline{<Z>} \tag{6.281}$$

$$s_{0,1} = \frac{\overline{(<Z> - \overline{<Z>}) \cdot (u - \overline{u})}}{\overline{(<Z> - \overline{<Z>}) \cdot (<Z> - \overline{<Z>})}} \tag{6.282}$$

with averages as defined in Eq. (6.81). Their expected values are:

$$\begin{aligned}
<s_{0,1}> &= <\frac{\overline{(<Z> - \overline{<Z>}) \cdot (u - \overline{u})}}{\overline{(<Z> - \overline{<Z>}) \cdot (<Z> - \overline{<Z>})}}> \\
&= \frac{\overline{(<y> - \overline{<y>}) \cdot <(u - \overline{u})>}}{\overline{(<y> - \overline{<y>}) \cdot (<y> - \overline{<y>})}} = \frac{\overline{(<y> - \overline{<y>})^2}}{\overline{(<y> - \overline{<y>})^2}} = 1
\end{aligned} \tag{6.283}$$

$$<m_{0,1}> = <\overline{u}> - s_{0,1} \cdot \overline{<Z>} = \overline{<y>} - 1 \cdot \overline{<y>} = 0 \tag{6.284}$$

Since the mean and standard deviation of the standard normal distribution are 0 (respectively 1), the expected LS estimators $<m_{0,1}>$ and $<s_{0,1}>$ are equal to the distribution parameters and therefore unbiased.

The LS estimators of the general parameters μ and σ are indicated as $m_{\mu,\sigma}$ and $s_{\mu,\sigma}$. Substitution of Eq. (6.280) in Eqs (6.281) and (6.282) with subsequent comparison with Eqs (5.61) and (5.62) shows that:

$$<s_{0,1}> = \frac{\overline{(<Z> - \overline{<Z>}) \cdot <\left(\frac{x - \mu}{\sigma} - \frac{\overline{x} - \mu}{\sigma}\right)>}}{\overline{(<Z> - \overline{<Z>}) \cdot (<Z> - \overline{<Z>})}} = \frac{1}{\sigma} \cdot <s_{\mu,\sigma}> \tag{6.285}$$

$$<m_{0,1}> = <\frac{x - \mu}{\sigma} - s_{0,1} \cdot \overline{<Z>}> = \frac{1}{\sigma}(<m_{\mu,\sigma}> - \mu) \tag{6.286}$$

Substitution of Eqs (6.283) and (6.284) in these equations yields:

$$<s_{\mu,\sigma}> = \sigma \tag{6.287}$$

$$<m_{\mu,\sigma}> = \mu \tag{6.288}$$

which means that the LS estimators m_{LS} and s_{LS} are unbiased.

6.3.5   LR Applied to Power Law Reliability Growth

The Duane and Crow AMSAA plots in Sections 5.7.1 and 5.7.2 are linear plots. The variables can be number with covariable time, or time with covariable MTBF (either cumulative or instantaneous). Since these are linear plots, LR with the unweighted LS method as described in Section 6.3.1.1 can be applied.

In the literature, confidence intervals are also provided and these can be applied to carry out LR with WLS as described in Section 6.3.1.2.

For further reading, see for example [45].

6.4   Summary

This chapter discussed distribution parameter estimation: general aspects of parameter estimation (Section 6.1); maximum likelihood estimators (Section 6.2); linear regression estimators (Section 6.3).

As for general aspects of parameter estimation (Section 6.1), the basic properties bias (Section 6.1.1), efficiency (Section 6.1.3) and consistency (Section 6.1.1.3) are discussed. Bias is the systematic error between the expected estimator and the true value. This difference is an expected value as well. Subtracting the bias from the estimator produces an unbiased estimator. Efficiency concerns the random error, which reduces with increasing sample size n. Consistency is required to obtain better estimation in terms of both bias and efficiency with increasing sample size n. Traditionally, consistency is associated with sample sizes approaching infinity. However, with asset management and incident management small sample sizes n often occur and special attention is paid to asymptotic behaviour due to small sample size n. A power function D_n is introduced that describes such asymptotic behaviour. It has the shape of a three-parameter power function: $D_n = Q \cdot (n - R)^P$ (Eq. (6.18)).

As for ML estimators (Section 6.2), these are applied to the Weibull-2 (Section 6.2.3), exponential (Section 6.2.4) and normal distributions (Section 6.2.5). For the Weibull distribution there is no analytical solution for the ML estimators. Generic expressions are possible due to pivotal functions (Section 6.2.3.3). Next, the power function D_n appears able to fit the bias and standard deviation, which leads to effective unbiasing formulas (Section 6.2.3.4). For censored data the shape parameter can be unbiased. The bias leads to overestimation of the quality with respect to early failures (Section 6.2.3.5). The ML estimator for the characteristic lifetime of an exponential distribution has an analytical expression, namely the average time, which is unbiased for uncensored data. The ML estimators for the normal distribution are the well-known mean and variance with analytical expressions and only unbiased if the data are uncensored. With censored data no analytical expression is available, because there is no analytical expression for the normal cumulative distribution and the reliability.

As for LR estimators (Section 6.3), after a general introduction on unweighted and weighted LS and censored data (Section 6.3.1), these are also applied to the Weibull-2 (Section 6.3.2), exponential (Section 6.3.3) and normal distributions (Section 6.3.4). LR works with ranked plotting positions quite similar to graphical analysis with parametric plots in Chapter 5. With censored data these ranked plotting positions must be manipulated. Two methods are discussed: the adjusted ranking method (e.g. Section 6.3.2.3) and the adjusted plotting position method (e.g. Section 6.3.2.4). Though the latter seems to be the most correct from the viewpoint of averaging possibilities, the differences seem quite small and the former is recommended in standards IEEE 930 and IEC 62539. For the uncensored case of Weibull, the power function D_n is used again to describe the bias, unbiasing and standard deviation. An advantage of the LR methods is that they can be fully consistent with graphical analysis (i.e. the best fit in graphs is also the best fit in parameter estimation, provided the same plotting position is used).

Questions

6.1 With reference to maximum likelihood and linear regression:
Consider the uncensored data in Table 6.6 (copied from Table 6.1).
a. Assuming these are exponential data, estimate the characteristic lifetime θ through LR, WLR and ML.
b. Assuming these are Weibull data, estimate the scale parameter α and the shape parameter β through LR, WLR and ML.
c. Unbias the estimated parameters (as far as possible).
d. Assuming these are lognormal data, convert these to normal data (as discussed in Section 5.6.3).
e. Estimate the normal and lognormal parameters.
f. Determine the correlation coefficient for the Weibull, exponential and normal/lognormal cases.
g. Determine the 90% confidence interval for Weibull, exponential and normal distributions with LR for the weakest object (i.e. $i = 1$).
h. Compare and discuss what distribution is most appropriate in your opinion.

6.2 With reference to maximum likelihood and linear regression:
In Table 6.7 the data of Table 6.1 are shown in a presumably earlier stage (i.e. part of the data is censored as yet). Perform the same analysis as above on these data:
a. Assuming these are exponential data, estimate the characteristic lifetime θ through LR, WLR and ML.
b. Assuming these are Weibull data, estimate the scale parameter α and the shape parameter β through LR, WLR and ML.

Table 6.6 Data for question 1.

Operational t (yr)	2.00	2.27	3.47	4.52	5.19	15.1	16.3	36.7	40.6	48.7	55.9
Failed?	Yes	Yes	Yes	Yes	Yes	Yes	Yes	Yes	Yes	Yes	Yes
Index i	1	2	3	4	5	6	7	8	9	10	11

Table 6.7 Data for question 2.

Operational t (yr)	2.00	2.27	3.47	4.52	5.19	15.1	16.3	16.3	16.3	16.3	16.3
Failed?	Yes	Yes	Yes	Yes	No	Yes	Yes	No	No	No	No
Index i	1	2	3	4	—	6	7	—	—	—	—

Table 6.8 Data for question 3.

Index	1	2	3	4	5	6	7	8	9	10	11	12
Observed t_i	0.22	0.38	0.47	0.52	0.61	0.71	>0.71	>0.71	>0.71	>0.71	>0.71	>0.71

    c. Unbias the estimated parameters (as far as possible).

    d. Assuming these are lognormal data, convert these to normal data (as discussed in Section 5.6.3).

    e. Estimate the normal and lognormal parameters.

    f. Determine the correlation coefficient for the Weibull, exponential and normal/lognormal cases.

    g. Determine the 90% confidence interval for Weibull, exponential and normal distributions with LR.

    h. Compare and discuss what distribution is most appropriate in your opinion.

6.3 With reference to parameter-free plotting, Kaplan–Meier, Weibull plot and Weibull LR with confidence intervals:

Consider the data in Table 6.8.

    a. What is this type of censoring called?

    b. Produce a parameter-free graph.

    c. Produce a Kaplan–Meier graph.

    d. Produce a Weibull plot and estimate the Weibull parameters by ML, LR and WLR.

    e. Unbias the parameters as far as possible.

    f. Estimate the 90% confidence interval in case of the parameter-free plot and the Weibull plot for the weakest object (i.e. $i = 1$).

7

System and Component Reliability

A utility grid is built up with many components and subgrids (parts of a grid). The reliability of a grid depends on the reliabilities of the underlying parts and the way these are interconnected. In more general terms, grids are systems built up from units and the system reliability depends on the reliability of those units plus the way the units are built together. As an example, the system reliability can be enhanced by increasing the unit reliabilities, but also by increasing redundancy in the system. On the other hand, installation errors, connections or lack of cooling, and so on can reduce reliability.

This chapter discusses system dependencies on system configurations and unit performances. Concepts and techniques are discussed to evaluate the reliabilities in systems like connections, circuits and substation bays.

After the basics in Section 7.1, block diagrams are introduced in Section 7.2 as a method to represent system components and/or failure mechanisms in a total configuration. Next, special basic system configurations are discussed, namely series (Section 7.3) and parallel (Section 7.4) systems. In some cases a system seems redundant, but in reality suffers from a common cause that cancels the redundancy, which is discussed in Section 7.5. Subsequently, systems are discussed that require a minimal redundancy, the so-called k-out-of-n systems (Section 7.6). Simple systems can be analysed in the elementary series and parallel configurations, but more complex systems require a different approach, of which three methods are discussed in Section 7.7.

7.1 The Basics of System Reliability

A cumulative distribution $F(t)$ characterizes a failure mechanism that occurs in components. Section 2.4 discussed the relation between $F(t)$ and the reliability $R(t)$, as well as other functions such as the hazard rate $h(t)$. Once the system reliability R_S is known, the other distribution representations follow as discussed in Section 2.4 as:

$$F_S(t) = 1 - R_S(t) \tag{7.1}$$

$$f_S(t) = -\frac{dR_S(t)}{dt} \tag{7.2}$$

$$h_S(t) = -\frac{1}{R_S(t)} \cdot \frac{dR_S(t)}{dt} \tag{7.3}$$

These functions may be substituted by the specific distributions discussed in Chapter 4, including the uniform, Weibull, exponential, normal and lognormal distributions. Measures of performance like expected lifetime were discussed in Chapter 3. Confidence intervals form an interval in which a time to fail can be expected with a specified probability, as discussed in Sections 5.2.4 and 5.2.5. An enhanced stress may accelerate ageing according to the power law model, which was discussed in Section 2.6.

Distributions are characterized by graphical representations and/or estimated parameters, as discussed in Chapters 5 and 6 respectively. When data are collected and analysed, the underlying distribution reveals itself. Forensic analysis of failed objects may prove whether this is all due to the same mechanism (or not), and whether a single distribution or a mixed distribution applies. In this chapter it is assumed that the failure distribution of single units is known by definition or by the analysis methods described above.

Functions like the hazard rate may be investigated further to identify which part is most critical to the system integrity. For instance, if there are n components with reliabilities R_i ($i = 1,\ldots, n$) in the system that all are critical to the system, then the system hazard rate can be described as:

$$h_S(t) = -\frac{1}{R_S(t)} \cdot \frac{dR_S(t)}{dt} = -\frac{1}{R_S(t)} \cdot \sum_{i=1}^{n} \left[\frac{\partial R_S}{\partial R_i} \cdot \frac{dR_i(t)}{dt} \right] = \sum_{i=1}^{n} h_{S,i} \tag{7.4}$$

In electrical networks multiple units are built into (sub)systems that together form grids. So, what about the performance of such systems as a whole? The performance of these systems can be derived from the performance of the individual components and how they are connected, as discussed in the following sections.

7.2 Block Diagrams

Block diagrams are a method to determine the reliabilities of systems in which the contributions of subsystem and component failure are reflected. The block diagram visualizes the assembly of reliabilities. The diagram consists of a starting point (usually not indicated explicitly) that is often positioned on the left of the diagram. From this point a possibly branched network of blocks and interconnecting lines or arrows develops.

Figure 7.1 shows an example of a simple block diagram that describes a hybrid vehicle where the reliabilities of the tyres, the fuel tank and the battery are considered. The car is able to drive if all the tyres are intact and at least either fuel is in the tank or the battery is charged. If an uninterrupted path of functioning blocks exists that connects the start and end point, then this system is still functioning. If not, the system is interrupted (i.e. is down).

The blocks are shown as boxes that represent a relevant unit and/or failure mechanism. These can be regarded as switches for functioning properly or not. Failure mechanisms may concern a component such as a tyre that may run flat, or something intangible like the interface of two insulating materials in which electrical discharging and treeing may lead to breakdown. The resulting short-circuit may interrupt the electric power supply. The interface is a physical reality, but it is not an object by itself (see also Section 8.3.1).

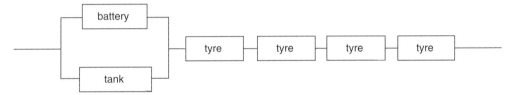

Figure 7.1 Simple block diagram for a hybrid vehicle where the reliability is represented by the integrity of the four tyres and the availability of the two energy storage facilities: the tank for fuel and the battery for electricity.

In the case of Figure 7.1, all four tyres must be intact, but of the energy sources one may be depleted without keeping the vehicle from functioning. The path runs along the energy source that does work. If both energy sources are available, this is even better. Two parallel paths provide redundancy. A flat tyre would break up all possible paths and end the functionality of the vehicle (assuming that a car with one or more flat tyres cannot be used).

In some cases it may be required for a system to have multiple functioning paths in order to meet the standards of reliability. An example is a regulator supervising the utilities, who may state that a grid must always be redundant except for a prescribed maximum time that the redundancy may be lifted. The loss of redundancy (during a period longer than the time constraint) may then be regarded as a failed system. This is an example of a *k*-out-of-*n* system where *k* units must work in order to keep the system of *n* units running. The next sections will elaborate the situations of series systems, parallel systems, combined series and parallel systems, common cause failures and various degrees of redundancy. Also, the analyses of complex systems with three methods are discussed.

7.3 Series Systems

A series system is a configuration in which each component is critical to the system integrity. If one component fails, the complete system fails. In other words, the system only functions if all components function.

Examples of a series system are: a chain that is as strong as its weakest link; a power cable that fails if breakdown occurs at any single insulation defect; a car that needs repair if one of the tyres runs flat. Systems with competing failure mechanisms are also examples of series systems. For instance, partial discharge and thermal degradation may simultaneously be active in one asset. The asset only survives if it survives both processes and fails if either of the processes leads to a breakdown of the asset.

Figure 7.2 shows the block diagram of a series system that consists of *n* components. The system reliability $R_S(t)$, failure distribution $F_S(t)$ and hazard rate $h_S(t)$ depend on

Figure 7.2 Block diagram of a series system.

the $R_i(t)$, $F_i(t)$ and $h_i(t)$ of the individual components with label i $(i = 1,...,n)$ as:

$$R_S(t) = R_1(t) \cdot R_2(t) \cdot \ldots \cdot R_n(t) = \prod_{i=1}^{n} R_i(t) \tag{7.5}$$

$$F_S(t) = 1 - R_S(t) = 1 - \prod_{i=1}^{n} R_i(t) = 1 - \prod_{i=1}^{n}(1 - F_i(t)) \tag{7.6}$$

$$h_S(t) = \frac{-1}{R_S(t)} \frac{dR_S(t)}{dt} = \frac{-1}{\prod_{j=1}^{n} R_j(t)} \sum_{i=1}^{n} \left(\frac{dR_i(t)}{dt} \frac{\prod_{j=1}^{n} R_j(t)}{R_i(t)} \right) = -\sum_{i=1}^{n} \frac{1}{R_i(t)} \frac{dR_S(t)}{dt} = \sum_{i=1}^{n} h_i(t) \tag{7.7}$$

If all components happen to have the same reliability $R(t)$, then Eqs (7.5)–(7.7) are simplified to:

$$R_S(t) = [R(t)]^n \tag{7.8}$$

$$F_S(t) = 1 - R_S(t) = 1 - [R(t)]^n = 1 - [1 - F(t)]^n \tag{7.9}$$

$$h_S(t) = n \cdot h(t) \tag{7.10}$$

If the context already makes clear that reliability R, failure distribution F and hazard rate h are functions of time t then this may no longer be expressed explicitly in the equations, which leads to a shorter notation:

$$R_S = R_1 \cdot R_2 \cdot \ldots \cdot R_n = \prod_{i=1}^{n} R_i \tag{7.11}$$

$$F_S = 1 - R_S = 1 - \prod_{i=1}^{n} R_i = 1 - \prod_{i=1}^{n}(1 - F_i) \tag{7.12}$$

$$h_S = \sum_{i=1}^{n} h_i \tag{7.13}$$

If all component reliabilities are the same (i.e. each $R_i = R$), then the equations become:

$$R_S = R^n \tag{7.14}$$

$$F_S = 1 - R_S = 1 - R^n = 1 - (1 - F)^n \tag{7.15}$$

$$h_S = n \cdot h \tag{7.16}$$

If component reliabilities are known either as a function or as a value, then these can be substituted in the equations above to find the system reliability, failure distribution and hazard rate.

For example, if it is known that each of the three single-phase cables that together form a 150-kV cable circuit have reliability of 92% after 36 years, then the reliability of the circuit is $R_S(36 \text{ years}) = [R(36 \text{ years})]^3 = 0.92^3 = 77.8\%$.

It should be noted that these three single-phase cables physically lie in parallel, but if any of the three single-phase cables breaks down, then the whole circuit fails. As the circuit integrity relies on the integrity of all three single-phase cables (despite the physical appearance), this is a series system in terms of reliability!

If the failure times of the n components are distributed according to the same exponential distribution with parameter θ, then the system also has an exponential reliability function R_S, but with a parameter $\theta_S = \theta/n$ as shown in Eqs (7.17)–(7.19). So, the characteristic lifetime θ_S of a series system is n times smaller than the characteristic lifetime θ of each of the n individual components:

$$R_S = R^n = [\exp(-t/\theta)]^n = \exp(-nt/\theta) = \exp\left(-\frac{t}{(\theta/n)}\right) = \exp\left(-\frac{t}{\theta_S}\right) \quad (7.17)$$

$$F_S = 1 - R_S = 1 - \exp\left(-\frac{t}{(\theta/n)}\right) = 1 - \exp\left(-\frac{t}{\theta_S}\right) \quad (7.18)$$

$$h_S = \frac{1}{\theta_S} = \frac{n}{\theta} = n \cdot h \Rightarrow \theta_S = \frac{\theta}{n} \quad (7.19)$$

The reliability, failure probability and hazard rate of a series system that consists of components that fail according to the same Weibull distribution with scale parameter α and shape parameter β are:

$$R_S = \left[\exp\left(-\left(\frac{t}{\alpha}\right)^\beta\right)\right]^n = \exp\left(-n\left(\frac{t}{\alpha}\right)^\beta\right) = \exp\left(-\left(\frac{t}{(\alpha/n^{1/\beta})}\right)^\beta\right)$$

$$= \exp\left(-\left(\frac{t}{\alpha_S}\right)^\beta\right) \quad (7.20)$$

$$F_S = 1 - R_S = 1 - \exp\left(-\left(\frac{t}{(\alpha/n^{1/\beta})}\right)^\beta\right) = 1 - \exp\left(-\left(\frac{t}{\alpha_S}\right)^\beta\right) \quad (7.21)$$

$$h_S = n \cdot \frac{\beta}{\alpha^\beta} t^{\beta-1} = \frac{\beta}{\alpha_S{}^\beta} t^{\beta-1} \quad (7.22)$$

The equations show that a series system consisting of components that fail according to a Weibull distribution will also feature a Weibull distribution, but with a smaller $n^{1/\beta}$ scale parameter. When Weibull explored the distribution that was named after him, he basically used this property as a precondition to deduce a suitable function [46].

As the exponential distribution can be regarded as a special case of the Weibull distribution (namely with $\beta = 1$ and $\alpha = \theta$) and the shape parameter β of the system reliability equals the shape parameter of the components according to Eq. (7.20), it becomes self-evident that the system reliability in that case also follows an exponential distribution.

Equations (7.20)–(7.22) describe the so-called 'volume effect'. If a chain consists of n links that each have the same Weibull distribution with parameters α and β, then the system will exhibit a Weibull distribution with parameters α_S and β_S that are related as:

$$\alpha_S = \frac{\alpha}{n^{1/\beta}} \quad (7.23)$$

$$\beta_S = \beta \quad (7.24)$$

This is an important finding that has implications for testing and defining reliability demands for practical applications. For instance, if the Weibull distribution of a certain cable type is investigated by testing single-phase cable lengths of, for example, 10 m, then the parameters α and β may be found. If this cable type is used for a 10-km circuit then actually 30 km is involved, namely 10 km for each of the three single-phase cables that form the circuit. The length of the cable involved is therefore a factor 30 km/10 m = 3000 times the length of the test cables. Equation (7.23) then predicts that the scale parameter of the circuit is a factor $3000^{1/\beta}$ smaller than that of the test cables.

For example, if $\beta = 4$, the expected lifetime of the circuit is a factor $3000^{1/4} = 7.4$ times smaller than the average lifetime of the test cables. Long cable circuit lengths therefore require a higher-quality cable than short circuit lengths because of this volume effect. The concept is that a longer piece of cable has a greater probability of containing a weaker spot to break down than a shorter piece.

7.4 Parallel Systems and Redundancy

A parallel system is a redundant system. The system fails if all components fail (for k-out-of-n systems that require k units to function, see Section 7.6). Examples of parallel systems are: a double circuit where each of the circuits is capable of carrying the full load; the earlier mentioned hybrid vehicle with a fuel tank and a battery that are both capable of powering motion; a team of colleagues that can take over specific tasks if necessary; and so on.

Figure 7.3 shows the block diagram of a parallel system that consists of n parallel components. The system reliability $R_S(t)$, failure distribution $F_S(t)$ and hazard rate $h_S(t)$ depend on the $R_i(t)$, $F_i(t)$ and $h_i(t)$ of the individual components with label i ($i = 1, \ldots, n$) as (using the short notation):

$$F_S = \prod_{i=1}^{n} F_i \tag{7.25}$$

$$R_S = 1 - F_S = 1 - \prod_{i=1}^{n} F_i = 1 - \prod_{i=1}^{n}(1 - R_i) \tag{7.26}$$

$$h_S = \frac{1}{R_S}\frac{\partial F_S}{\partial t} = \frac{1}{1 - \prod_{i=1}^{n} F_i} \cdot \frac{\partial\left(\prod_{i=1}^{n} F_i\right)}{\partial t} \tag{7.27}$$

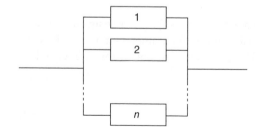

Figure 7.3 Block diagram of a parallel system.

These equations bear a close resemblance to those for the series system, but note that R and F are interchanged. Again, both functions of time as well as values can be substituted in the equations. If all component reliabilities and so on are the same, then the equations become:

$$F_S = F^n \tag{7.28}$$

$$R_S = 1 - F_S = 1 - F^n = 1 - (1-R)^n \tag{7.29}$$

$$h_S = \frac{1}{R_S}\frac{\partial F_S}{\partial t} = \frac{F^{n-1}}{1-F^n} \cdot \frac{\partial(F^n)}{\partial t} = \frac{n}{1-F^n} \cdot \frac{\partial F}{\partial t} = \frac{n \cdot F^{n-1} \cdot f}{1-F^n} \tag{7.30}$$

with f the distribution density.

If the components all exhibit the same exponential distribution with parameter θ, then the resulting system does not exhibit an exponential distribution:

$$F_S(t) = \prod_{i=1}^{n} F_i = [1 - \exp(-t/\theta)]^n \tag{7.31}$$

$$R_S = 1 - F_S = 1 - [1 - \exp(-t/\theta)]^n \tag{7.32}$$

$$h_S = \frac{1}{R_S}\frac{\partial F_S}{\partial t} = \frac{n \cdot [1 - \exp(-t/\theta)]^{n-1} \cdot \exp(-t/\theta) \cdot 1/\theta}{1 - [1 - \exp(-t/\theta)]^n} \tag{7.33}$$

It is noteworthy that components which form an electrical series network can still be a reliability parallel system. If, for instance, two switchgears are installed at both ends of a connection, then a short-circuit current can be interrupted by either one of the switchgears. As the function of interrupting a short-circuit current is facilitated as long as at least one of the switchgears is operational, this system design is a parallel system.

7.5 Combined Series and Parallel Systems, Common Cause

Series and parallel systems can be combined, of which Figure 7.1 was already an example. In that case a simple parallel subsystem (battery and tank) was put in series with a series subsystem of four tyres. This section will explore a few basic configurations and explain when redundancy becomes less effective due to a phenomenon called 'common cause' failure.

Figure 7.4 shows two series subsystems (AB and CD) that are combined in a parallel system. Figure 7.5 shows two parallel subsystems (AC and BD) that are combined in a series system. These two system combinations are evaluated and compared first.

The reliability of the system in Figure 7.4 is determined as:

$$R_{S,AB} = R_A \cdot R_B \Rightarrow F_{S,AB} = 1 - R_A \cdot R_B \tag{7.34}$$

Figure 7.4 Block diagram of a parallel system of series subsystems.

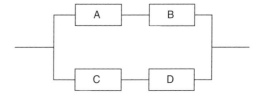

$$R_{S,CD} = R_C \cdot R_D \Rightarrow F_{S,CD} = 1 - R_C \cdot R_D \tag{7.35}$$

$$F_S = F_{S,AB} \cdot F_{S,CD} = (1 - R_A \cdot R_B) \cdot (1 - R_C \cdot R_D) \tag{7.36}$$

$$R_S = 1 - F_S = 1 - (1 - R_A \cdot R_B) \cdot (1 - R_C \cdot R_D) = R_A R_B + R_C R_D - R_A R_B R_C R_D \tag{7.37}$$

The reliability of the system in Figure 7.5 is determined as:

$$F_{S,AC} = F_A \cdot F_C \Rightarrow R_{S,AC} = 1 - F_A \cdot F_C = 1 - (1 - R_A) \cdot (1 - R_C) = R_A + R_C - R_A \cdot R_C \tag{7.38}$$

$$F_{S,BD} = F_B \cdot F_D \Rightarrow R_{S,BD} = 1 - F_B \cdot F_D = 1 - (1 - R_B) \cdot (1 - R_D) = R_B + R_D - R_B \cdot R_D \tag{7.39}$$

$$R_S = R_{S,AC} \cdot R_{S,BD} = (R_A + R_C - R_A R_C) \cdot (R_B + R_D - R_B R_D) = R_A R_B + R_A R_D \dots$$
$$\dots + R_C R_B + R_C R_D - R_A R_B R_D - R_C R_B R_D - R_A R_C R_B - R_A R_C R_D + R_A R_C R_B R_D \tag{7.40}$$

$$F_S = 1 - R_S = 1 - (R_A + R_C - R_A \cdot R_C) \cdot (R_B + R_D - R_B \cdot R_D) \tag{7.41}$$

The reliabilities of the two systems differ considerably, although they are constructed from the same components. The difference in configuration has a profound effect. If, for instance, components C and B fail, then the parallel system of series subsystems (Figure 7.4) fails, while the series system of parallel systems (Figure 7.5) remains operational. The degree of redundancy in the latter case is larger, because the paths AD and CB are added to the paths AB and CD. It can be shown that the R_S of Figure 7.5 is equal to or greater than the R_S of Figure 7.4.

When increasing redundancy by adding parallel systems, care must be taken that no series relationships remain that reduce the effectiveness of the redundancy. Such (sometimes hidden) series relationships are referred to as common cause problems. Below, the common cause phenomenon for a secure radio communication is elaborated in Example 7.1. Due to this effect, the radio system becomes less reliable than anticipated.

Example 7.1 *Redundancy and Common Cause* For repair missions a radio communication system is specified that must function at least 30 days with a reliability of $R(30\,\text{days}) = 96\%$. Experience shows that under the expected conditions during such

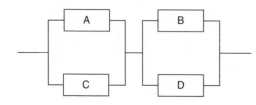

Figure 7.5 Block diagram of a series system of parallel subsystems.

Figure 7.6 The aimed for redundant system of two radios.

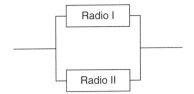

a mission the failure rate appears as $\lambda = 0.007/\text{day}$ and the failure distribution can be regarded as exponential. The reliability can then be calculated as:

$$R(30 \text{ days}) = e^{-0.007 \cdot 30} = 81\% \tag{7.42}$$

Because this reliability is too low, it is decided to increase the redundancy by using two radios (Figure 7.6). If one radio fails, the other can take over. The reliability can be calculated and indeed proves to fulfil the requirement of $R(30 \text{ days}) > 96\%$:

$$R(30 \text{ days}) = 1 - (1 - R)^2 = 1 - (1 - e^{-0.007 \cdot 30})^2 = 96.4\% > 96\% \tag{7.43}$$

However, in practice the failure probability appears as 5.6% (i.e. the reliability is 94.4%, which is less than the prescribed 96%). Apparently, the system of two radios is not as redundant as expected and the failure modes are investigated. Three failure modes labelled A, B and C are found, as shown in Table 7.1. The sum of the hazard rates appears to be 0.007/day, but the table also shows that 0.006/day is due to the radio itself, while $\lambda_V = 0.001/\text{day}$ is caused by the (external) power supply.

When the external power supply fails, this phenomenon C appears to interrupt both radios at the same time, and the presence of a second radio does not help in that case. The outage due to a failing power supply is a *common cause* problem. The redundancy does not work for that phenomenon. Apparently, the true system is not that of Figure 7.6 but of Figure 7.7, which is a parallel subsystem of radio reliabilities in series with a power-supply reliability.

Using the data of Table 7.1, Eqs (7.5) and (7.26), the reliability of this system can indeed be calculated as:

$$R_S(30 \text{ days}) = R_{power} \cdot (1 - (1 - R_{radio})^2) = 94.4\% < 96\% \tag{7.44}$$

Some solutions to the common cause problem are: providing a redundant power supply (e.g. an uninterrupted power supply (UPS)) or using a third radio. The latter does

Table 7.1 Example of failure modes with rate and cause.

Hazard rate		Phenomenon
h_A	0.002 failures/day	Defect in electronics of the radio
h_B	0.004 failures/day	Mechanical damage to the radio
h_C	0.001 failures/day	Failing external power supply
h_{tot}	0.007 failures/day	Total number of outages

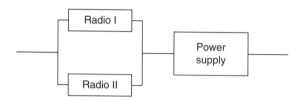

Figure 7.7 The true redundant system of the two radios.

not solve the power supply issue, but does increase the overall availability of the radio after, which the 96% may still be reached.

7.6 EXTRA: Reliability and Expected Life of *k*-out-of-*n* Systems

The previous sections described situations where at least one path through the diagram was necessary to keep the system functional. For a series system of n components, this means that n out of n components have to function in order to have a functional system. For a parallel system of n components, it means that at least 1 out of n components must remain functional in order to have a functional system.

There are systems where a specific number of k out of the total number n of components must function in order to keep the system functional. Examples are: a team of 10 persons of which at least three persons must be present in order to carry out certain tasks; a network of three redundant systems where it is obliged by regulations to have at least two systems intact; and so on. Such systems are referred to as k-out-of-n systems. These systems can also be represented in block diagrams with a notation that at least k out of n components must function.

Let us assume that at some moment j out of n components work, with $j \geq k$. The number of possible combinations of j out of n components being intact is equal to j permutations of n. The probability that j components work is R^j and the probability that $n - j$ failed is $F^{n-j} = (1 - R)^{n-j}$. The probability that j out of n components work and $n - j$ failed is therefore:

$$R_{j\text{-out-of-}n} = \binom{n}{j} \cdot R^j \cdot F^{n-j} = \binom{n}{j} \cdot R^j \cdot (1 - R)^{n-j} \tag{7.45}$$

The reliability $R_{S,n,k}$ of a k-out-of-n system, where at least k out of n components must function, is the sum of probabilities of working systems with $k \leq j \leq n$:

$$R_{S,n,k} = \sum_{j=k}^{n} \binom{n}{j} \cdot R^j \cdot F^{n-j} = \sum_{j=k}^{n} \binom{n}{j} \cdot R^j \cdot (1 - R)^{n-j} \tag{7.46}$$

In the following it will be useful to have a relationship between $R_{S,n,k}$ and $R_{S,n,k+1}$. A recurrent relation follows from Eq. (7.46):

$$R_{S,n,k} = \binom{n}{k} \cdot R^k \cdot (1 - R)^{n-k} + \sum_{j=k+1}^{n} \binom{n}{j} \cdot R^j \cdot (1 - R)^{n-j}$$

$$= \binom{n}{k} \cdot R^k \cdot (1 - R)^{n-k} + R_{S,n,k+1} \tag{7.47}$$

The expected lifetime $\theta_{S,n,k}$ can be found with Eq. (7.46) using Eq. (3.8) in Section 3.1:

$$\theta_{S,n,k} = \int_0^\infty R_{S,n,k}\, dt = \int_0^\infty \left(\sum_{j=k}^n \binom{n}{j} \cdot R^j \cdot (1-R)^{n-j} \right) dt \tag{7.48}$$

A recurrent relation follows for $\theta_{S,n,k}$ from Eqs (7.47) and (3.8):

$$\theta_{S,n,k} = \int_0^\infty \binom{n}{k} \cdot R^k \cdot (1-R)^{n-k}\, dt + \theta_{S,n,k+1} \tag{7.49}$$

The integral in Eq. (7.49) can be rewritten with Eq. (2.21) as:

$$\int_0^\infty \binom{n}{k} \cdot R^k \cdot (1-R)^{n-k}\, dt = \int_0^1 \binom{n}{k} \cdot R^k \cdot (1-R)^{n-k} \frac{-1}{h \cdot R}\, dR$$

$$= \int_0^1 \binom{n}{k} \cdot (1-F)^{k-1} \cdot F^{n-k} \frac{1}{h}\, dF \tag{7.50}$$

Substitution then yields an expression for the expected lifetime $\theta_{S,n,k}$.

In the special case of the exponential distribution, the hazard rate h is constant with $h = \lambda = 1/\theta$. In that case the last part of Eq. (7.50) using Eq. (4.52) of the incomplete beta function with $A\% = 100\%$ can be rewritten and we find:

$$\theta_{S,n,k} - \theta_{S,n,k+1} = \theta \cdot \int_0^1 \binom{n}{k} \cdot (1-F)^{k-1} \cdot F^{n-k}\, dF = \theta \cdot \frac{B(k, n+k-1)|_{A\%=100\%}}{k} = \frac{\theta}{k} \tag{7.51}$$

From this recurrent relation the expected lifetime of a k-out-of-n system of which the failure behaviour of all components is described by the same exponential distribution with parameter θ is:

$$\theta_{S,n,k} = \frac{\theta}{k} + \theta_{S,n,k+1} = \frac{\theta}{k} + \frac{\theta}{k+1} + \theta_{S,n,k+2} = etc. \Rightarrow \theta_{S,n,k} = \sum_{j=k}^n \frac{\theta}{j} \tag{7.52}$$

As a check, a series system only works if all n components work, which means that it is an n-out-of-n system by definition. From Eq. (7.52) the expected life of the series system $\theta_{S,n,n}$ corresponds to the result in Eq. (7.19):

$$\theta_{S,n,n} = \sum_{j=n}^n \frac{\theta}{j} = \frac{\theta}{n} \tag{7.53}$$

As another check, a parallel system is a 1-out-of-n system and Eq. (7.52) then yields, for the expected lifetime $\theta_{S,n,1}$:

$$\theta_{S,n,1} = \sum_{j=1}^n \frac{\theta}{j} = \theta \cdot \sum_{j=1}^n \frac{1}{j} = \theta \cdot \left(1 + \frac{1}{2} + \frac{1}{3} + .. + \frac{1}{n} \right) \tag{7.54}$$

7.7 Analysis of Complex Systems

Combinations of series and parallel systems were discussed in Section 7.5. However, many systems cannot be split into series and parallel subsystems, which makes it difficult to derive the system reliability R_S. The example shown in Figure 7.8 is called a

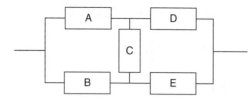

Figure 7.8 A bridge system looks simple but cannot be split into series and parallel subsystems.

bridge system. Although the configuration seems simple, it cannot be solved in terms of a combination of series and parallel subsystems.

This looks like a combination of two parallel systems in series (Figure 7.5), but the connections from A to E and from B to D run through a block C representing a component C that can fail. Because of C this system is called a bridge system. As mentioned above, this system cannot be analysed with the methods in the previous chapters. The three following sections present methods that are suitable to treat more complex systems: the conditional method; the up-table method; and the method of minimal paths and blockades.

These methods yield a solution for R_S. Once R_S is known, the other appearances of the distribution, namely F_S, f_S and h_S, can be found through the relations in Eqs (7.1)–(7.3).

7.7.1 Conditional Method

The conditional method divides the possible states of the system into a set of two (or more) complementary conditions. The last is not a prerequisite, but prevents corrections for possibilities that otherwise would be counted multiple times.

The concept is elaborated for the example of the bridge system of Figure 7.8. Component C is identified as a suitable component for defining complementary conditions. Let F_C be the failure probability and R_C the probability of survival of component C. The two alternative cases of C being failed and not-failed form a set of two complementary conditions. If C has failed the system C_0 is the result (see Figure 7.9); if C remains intact the system C_1 remains (see Figure 7.10).

The system C_0 corresponds to the parallel series system of Figure 7.4. Note that the components are labelled differently in Figure 7.9. The expression for the reliability under the condition that C is failed, $R_S|_{C0}$, becomes:

$$R_S|_{C0} = 1 - F_S|_{C0} = 1 - (1 - R_A \cdot R_D)(1 - R_B \cdot R_E) = R_A R_D + R_B R_E - R_A R_B R_D R_E$$

$$(7.55)$$

The probability that system C_0 applies is equal to the failure probability $F_C = 1 - R_C$. The Bayesian probability (Section 2.5.2.1) of a functioning system $R_{S,C0}$ under the

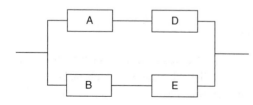

Figure 7.9 The C_0 system: the bridge system where C has failed, which breaks up the connection AE and BD.

Figure 7.10 System C_1: the bridge system where C is intact and the connection AE and BD is preserved.

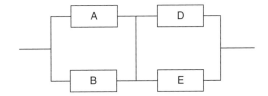

condition that C failed is therefore:

$$R_{S,C0} = F_C \cdot R_S|_{C0} = F_C \cdot (1 - (1 - R_A \cdot R_D)(1 - R_B \cdot R_E))$$
$$= (1 - R_C) \cdot (R_A R_D + R_B R_E - R_A R_B R_D R_E) \quad (7.56)$$

In case all components have the same reliability R, then Eqs (7.55) and (7.56) become:

$$R_S|_{C0} = 1 - (1 - R^2)^2 = 2R^2 - R^4 \quad (7.57)$$

$$R_{S,C0} = (1 - R) \cdot (2R^2 - R^4) = 2R^2 - 2R^3 - R^4 + R^5 \quad (7.58)$$

The condition of C being failed has now been analysed. The complementary condition that C is intact yields a different system, namely C_1, as shown in Figure 7.10. Now the paths AE and BD remain and the system corresponds to the series parallel system of Figure 7.5. The parallel subsystems are denoted 'subS,AB' and 'subS,DE', respectively. The system reliability under the condition that C functions, $R_S|_{C1}$, then becomes:

$$F_{subS,AB} = F_A \cdot F_B \Rightarrow R_{subS,AB} = 1 - (1 - R_A) \cdot (1 - R_B) = R_A + R_B - R_A \cdot R_B \quad (7.59)$$

$$F_{subS,DE} = F_D \cdot F_E \Rightarrow R_{subS,DE} = 1 - (1 - R_D) \cdot (1 - R_E) = R_D + R_E - R_D \cdot R_E \quad (7.60)$$

$$R_S|_{C1} = R_{subS,AB} \cdot R_{subS,DE} = [R_A + R_B - R_A R_B] \cdot [R_D + R_E - R_D R_E] \quad (7.61)$$

The probability that the system C_1 applies is equal to the probability that C functions (i.e. the reliability R_C). The Bayesian probability $R_{S,C1}$ that C_1 applies and that the system works becomes:

$$R_{S,C1} = R_C \cdot R_S|_{C1} = R_C \cdot [R_A + R_B - R_A R_B] \cdot [R_D + R_E - R_D R_E] \quad (7.62)$$

If all components have the same reliability R, then Eqs (7.61) and (7.62) become:

$$R_S|_{C1} = R_{subS}^2 = [2R - R^2]^2 = 4R^2 - 4R^3 + R^4 \quad (7.63)$$

$$R_{S,C1} = R \cdot R_S|_{C1} = 4R^3 - 4R^4 + R^5 \quad (7.64)$$

To obtain the complete system reliability R_S, the probabilities of the working system under the two complementary conditions (i.e. $R_{S,C0}$ and $R_{S,C1}$) are added:

$$R_S = R_{S,C0} + R_{S,C1} = F_C \cdot R_S|_{C0} + R_C \cdot R_S|_{C1}$$
$$= (1 - R_C) \cdot (R_A R_D + R_B R_E - R_A R_B R_D R_E)$$
$$+ R_C \cdot [R_A + R_B - R_A R_B] \cdot [R_D + R_E - R_D R_E]$$
$$= R_A R_D + R_B R_E + R_A R_C R_E + R_B R_C R_D - R_A R_B R_D R_E - R_A R_C R_D R_E - R_B R_C R_D R_E$$
$$- R_A R_B R_C R_D - R_A R_B R_C R_E + 2 \cdot R_A R_B R_C R_D R_E \quad (7.65)$$

In case the two conditions have not excluded each other, some possibilities will have been counted twice or more, which must be corrected for. Here, the conditions of C failed or intact are exclusive and no conditions are counted twice or more.

Again, if all components have the same reliability R, Eq. (7.65) yields:

$$R_S = F_C \cdot R_S|_{C0} + R_C \cdot R_S|_{C1} = 2R^2 + 2R^3 - 5R^4 + 2R^5 \tag{7.66}$$

The conditional method requires finding one or more components whose failure or survival simplifies the system in such a way that it can be represented as combinations of series and parallel systems (or alternatively, easily solved with the methods of the next two subsections).

7.7.2 Up-table Method

Another approach to derive the reliability R_S is the up-table method. This requires a systematic inventory of all combinations of failed and surviving components that keep the system operational or in its 'up-state'. If the system is not operational anymore due to a specific combination of failed components, then the system is in its 'down-state'. It is practical to work with logic values. Component survival and system up-state are indicated by 1 and component failure and system down-state are indicated by 0. In fact, the up-table is a so-called 'truth table' indicating the operability of components and system.

An up-table represents all operational combinations. Often the inventory is done by systematically grouping combinations by the number of surviving components.

As for the bridge system in Figure 7.8, the total number of components is 5. The number of operational components can be counted down from 5 to 0. In the latter situation every component has failed and so the system will have failed also.

The columns of Table 7.2 show all combinations for the bridge system in Figure 7.8. The seventh row shows the resulting system state and the eighth row shows the system reliability R_S for the case that all components have the same reliability R. Although Table 7.2 shows all combinations, often groups of combinations are left out if the system will obviously be down. For instance, if the number of surviving components is simply too small to make the system operational. For the bridge system it is obvious that with 0 or 1 surviving component the system can never be operational, and those combinations could therefore be skipped. The purpose of the up-table is to inventory all operational system states. The complete inventory of down-states is not necessary.

The system reliability in the eighth row is obtained by a logic AND operation on all component states in the same column. This can be written as the product of reliabilities for logic 1 and failure probability for logic 0. For instance, in the third column there are one failed and four surviving components (0,1,1,1,1), that is the reliability state for component A is FALSE (logic value 0) and the failure state is therefore TRUE while for all other components the reliability state is TRUE (logic value 1). The corresponding expression with failure of A and survival of B, C, D, E is:

$$R_S = F_A \cdot R_B \cdot R_C \cdot R_D \cdot R_E = (1 - R_A) \cdot R_B \cdot R_C \cdot R_D \cdot R_E \tag{7.67}$$

If all components have the same reliability R, then Eq. (7.67) with one failure and four survivors becomes:

$$R_S = F \cdot R^4 = (1 - R) \cdot R^4 \tag{7.68}$$

Table 7.2 All combinations of failed and surviving components for the bridge system in Figure 7.8.

# Okay	5	4					3										2										1					0
A	1	0	1	1	1	1	0	0	0	0	1	1	1	1	1	1	0	0	0	1	0	0	1	0	1	1	0	0	0	0	1	0
B	1	1	0	1	1	1	0	1	1	1	0	0	0	1	1	1	0	0	1	0	0	1	0	1	0	1	0	0	0	1	0	0
C	1	1	1	0	1	1	1	0	1	1	0	1	1	0	0	1	0	1	0	0	1	0	0	1	1	0	0	0	1	0	0	0
D	1	1	1	1	0	1	1	1	0	1	1	0	1	0	1	0	1	0	0	0	1	1	1	0	0	0	0	1	0	0	0	0
E	1	1	1	1	1	0	1	1	1	0	1	1	0	1	0	0	1	1	1	1	0	0	0	0	0	0	1	0	0	0	0	0
System	1	1	1	1	1	1	0	1	1	1	1	1	1	1	1	0	0	0	1	0	0	0	1	0	0	0	0	0	0	0	0	0
R_S	R^5	$5R^4(1-R)$					$8R^3(1-R)^2$										$2R^2(1-R)^3$										0					0

The first row indicates the number of surviving components. The second to sixth rows systematically inventory the component states. The seventh row shows the system state for each column. The eighth row shows the system reliability, given the same reliability R for all components.

If all states are inventoried, the system reliability is found by summing all probabilities of survival. Only the system up-states are identified with the system reliability, identified in the seventh row by 1. Table 7.2 subsequently shows the result for the situation that all components have the same reliability R in the eighth row. The sum of all system reliability probabilities is the system reliability R_S:

$$R_S = R^5 + 5 \cdot R^4 \cdot (1-R) + 8 \cdot R^3 \cdot (1-R)^2 + 2 \cdot R^2 \cdot (1-R)^3$$
$$= 2R^2 + 2R^3 - 5R^4 + 2R^5 \tag{7.69}$$

This result agrees with the result of Eq. (7.66) that was found with the conditional method. If the components do not have the same reliability, the expressions must be inventoried in terms of the individual reliabilities R_A, R_B, R_C, R_D and R_E as in Eq. (7.67). The resulting 16 expressions in the case of the bridge system can then be summed, which leads to the same result as in Eq. (7.65), derived with the conditional method.

It can be convenient to reduce the up-table. This is possible if the system reliability is independent of the state of one or more components. For instance, in case components A and D in Figure 7.8 are both surviving, then the states of B, C and E do not matter. All combinations where A and D both survive will yield the same result: the system is up.

Such combinations can be searched for in the truth table. Figure 7.11 shows only the combinations of Table 7.2 that yield a system up-state. Comparing the first two columns, the columns only appear to differ in the state of component A, but both yield the system up-state. Apparently, the system reliability does not depend on the state of component A, given the other four components are up. Both columns can then be replaced by a single column where the state of component A is indicated with 'X', meaning a 'do not care' value. This value for R_A will fall out of the equation when the reliabilities are summed:

$$R_{column\ 1\&2} = R_A \cdot R_B \cdot R_C \cdot R_D \cdot R_E + (1-R_A) \cdot R_B \cdot R_C \cdot R_D \cdot R_E$$
$$= R_B \cdot R_C \cdot R_D \cdot R_E \tag{7.70}$$

Because $R_A + (1-R_A) = 1$, the reliability R_A falls out of the equation indeed. The other columns can be swept in the same manner, as shown in Figure 7.11, which may reduce the truth table considerably and may make it easier to find an expression for R_S. In the

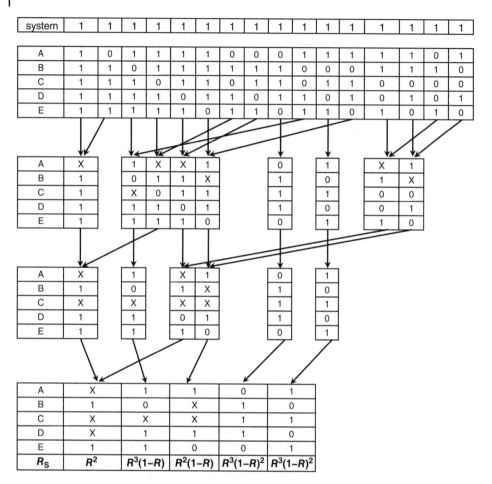

Figure 7.11 Example of reducing a truth table. Only the system up-states of Table 7.2 are shown. Do not care states are identified by comparing columns and marked by X (the do-not-care symbol). Such states fall out of the expressions for the system state reliability. For simplicity, the reliabilities R_A, R_B, R_C, R_D and R_E are taken equal, namely R. The full expression is in Eq. (7.71).

case of the truth table of the bridge system of Figure 7.8, the number of columns is reduced from 16 to 5 (columns and expressions).

The sequence of sweeping and grouping columns has an influence on the precise expressions on the bottom row, but after summation the result is the same.

Particularly if the components have distinct reliabilities, the sweeping method can be very useful in obtaining a reduced table and simpler expressions. The resulting expression for the system reliability R_S after the sweeping method with distinguishable component reliabilities becomes:

$$
\begin{aligned}
R_S &= R_B R_E + R_A(1 - R_B)R_D R_E + R_A R_D(1 - R_E) + (1 - R_A)R_B R_C R_D(1 - R_E) \\
&\quad + R_A(1 - R_B)R_C(1 - R_D)R_E \\
&= R_B R_E + R_A R_D + R_B R_C R_D + R_A R_C R_E - R_A R_B R_D R_E - R_A R_B R_C R_D \\
&\quad - R_B R_C R_D R_E - R_A R_B R_C R_E - R_A R_C R_D R_E + 2 \cdot R_A R_B R_C R_D R_E
\end{aligned}
\tag{7.71}
$$

If all components have the same reliability R, the system reliability R_S is derived from the reduced table as:

$$R_S = R^2 + R^3 \cdot (1-R) + R^2 \cdot (1-R) + R^3 \cdot (1-R)^2 + R^3 \cdot (1-R)^2$$
$$= 2R^2 + 2R^3 - 5R^4 + 2R^5 \tag{7.72}$$

This is the same result as was obtained with the conditional method (Eq. (7.66)) and with the up-table method without sweeping the table (Eq. (7.69)).

The method of the up-table requires fewer calculations than the conditional method, but the table quickly grows with the number of components involved. Sweeping by identifying 'do not care' situations and reducing the table can reduce the number of columns significantly, which then leads to simpler expressions. The conditional method and the up-table method are equivalent alternatives.

7.7.3   EXTRA: Minimal Paths and Minimum Blockades

The third method to deduce the system reliability R_S inventories the minimal paths with which the system remains operational and/or the minimum blockades at which the system fails. This method does not easily lead to the system reliability, but does provide insight into critical combinations. Care must be taken with the block diagrams used, which have a slightly different meaning from the block diagrams in previous sections.

The term 'minimal paths' refers to paths through the reliability diagram that ensure the functioning of the system. Other components may provide additional redundancy, but the minimal path is a combination from which all redundancies are skimmed.

The minimal path method inventories all minimal paths. For instance, the path AD in Figure 7.8 is such a minimal path. If all components but A and D are failed, the system is just functioning, but there is no redundancy left. If one of these remaining components fails, the system is down. Table 7.3 shows all four minimal paths of the bridge system in Figure 7.8. Also, the probability of functioning (i.e. the reliability) of those paths is given for the case of distinguished reliabilities as well as the case of components all having the same reliability R.

The minimal paths themselves may also be put into a block diagram to provide an overview of the possible combinations. This minimal path diagram takes the appearance of a parallel system, but should not be confused with a reliability diagram. One should be aware that the minimal path diagram usually has a high common cause content. Figure 7.12 shows such a minimal path diagram. If the individual component blocks are made visible, then the representation of Figure 7.8 follows, which brings us back to the starting point. Minimal path diagrams must be clearly indicated as such.

Table 7.3 Minimal paths for the bridge system in Figure 7.8.

Minimal path	Reliability of path	Reliability of path if all $R_i = R$
AD	$R_A \cdot R_D$	R^2
ACE	$R_A \cdot R_C \cdot R_E$	R^3
BCD	$R_B \cdot R_C \cdot R_D$	R^3
BE	$R_B \cdot R_E$	R^2

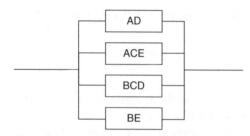

Figure 7.12 Minimal path diagram. *Note*: this has the appearance of a reliability parallel system but should not be confused with it. The seeming redundancy has a considerable common cause content. Eliminating the common cause leads back to Figure 7.8.

The common causes can be eliminated by elaborating the system reliabilities in terms of the component reliabilities and eliminating duplications. This can be done by introducing a state vector $\vec{X}$ that contains all state values x_i of the n components ($i = 1,\dots,n$) as logic values ($1 = $ up or working; $0 = $ down or failed):

$$\vec{X} = (x_1, x_2, \dots, x_m) \tag{7.73}$$

Subsequently, the function $\varphi(\vec{X})$ is defined that evaluates the system state based on the configuration and the state values of the components. This function is also called the structure function. If the system is up, then $\varphi(\vec{X}) = 1$ and if the system is down, $\varphi(\vec{X}) = 0$. For instance, if in the case of the bridge system $x_A = 1$ and $x_D = 1$, then $\varphi(\vec{X}) = 1$.

The system reliability is the expected value of $\varphi(\vec{X})$, which is found by averaging this function over all possible combinations:

$$R_S = \langle \phi(\vec{X}) \rangle \tag{7.74}$$

Using Eq. (7.26) for the reliability of a parallel system, the structure function of the bridge system can be derived from Figure 7.13 as:

$$\varphi(\vec{X}) = 1 - (1 - x_A x_B) \cdot (1 - x_A x_C x_E) \cdot (1 - x_B x_E) \cdot (1 - x_B x_C x_D) \tag{7.75}$$

The expected value is found by elaborating Eq. (7.75). Duplications are eliminated because of the properties of logic variables. The product of a logic variable with itself (and higher powers similarly) results in the logic variable:

$$x_i^2 = x_i \tag{7.76}$$

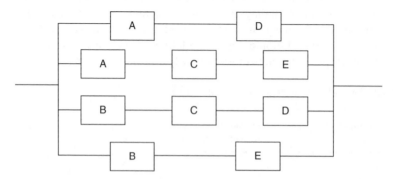

Figure 7.13 The minimal path diagram elaborated by expanding the paths as series systems in the parallel system. *Note*: the common cause content remains at this stage. For instance, a single failure of component B takes out two of the four minimal paths.

Table 7.4 Minimal blockades for the bridge system in Figure 7.8.

Minimal blockade	Probability of blockade	Probability of blockade if all $F_i = F$
AB	$F_A \cdot F_B$	F^2
DE	$F_D \cdot F_E$	F^2
ACE	$F_A \cdot F_C \cdot F_E$	F^3
BCD	$F_B \cdot F_C \cdot F_D$	F^3

The expected value of $\varphi(\vec{X})$ in case of the bridge system now becomes:

$$\langle \phi(\vec{X}) \rangle = \langle 1 - (1 - x_A x_B) \cdot (1 - x_A x_C x_E) \cdot (1 - x_B x_E) \cdot (1 - x_B x_C x_D) \rangle$$
$$= \langle x_A x_B \rangle + \langle x_B x_E \rangle + \langle x_A x_C x_E \rangle + \langle x_B x_C x_D \rangle - \langle x_A x_C x_D x_E \rangle - \langle x_A x_B x_C x_D \rangle$$
$$- \langle x_A x_B x_C x_E \rangle - \langle x_A x_B x_D x_E \rangle - \langle x_B x_C x_D x_E \rangle + 2 \cdot \langle x_A x_B x_C x_D x_E \rangle \quad (7.77)$$

Powers of x_i were eliminated with Eq. (7.76). The system reliability is equal to the expected structure function value, which in turn is the sum of expected values of sub-structure functions, which in turn are equal to subsystem reliabilities. The system reliability follows from Eq. (7.77) as:

$$R_S = R_A R_D + R_B R_E + R_A R_C R_E + R_B R_C R_D - R_A R_C R_D R_E - R_A R_B R_C R_D$$
$$- R_A R_B R_C R_E - R_A R_B R_D R_E - R_B R_C R_D R_E + 2 \cdot R_A R_B R_C R_D R_E \quad (7.78)$$

This result corresponds to the results with the conditional method (Eq. (7.65)) and the up-table method (Eq. (7.71)). In case all components have the same reliability R, and if there is no need to distinguish the reliabilities, then R_S can again be written as:

$$R_S = 2R^2 + 2R^3 - 5R^4 + 2R^5 \quad (7.79)$$

An alternative to the minimal path is to inventory the minimal blockades that could make the system fail. If one of those blocks occurs then the system fails and otherwise it remains operational. The minimal blockades are presented in Table 7.4.

Also, this table can be represented in a blockade diagram which takes the form of a series system. If any of these blockades occurs, the system will be down. Again, duplications exist, so care must be taken as with the minimal path diagram. This representation gives a quick overview of which combined failures will lead to system failure.

The elimination of duplications follows the same route as with the minimal path method, after which the system reliability follows as in Eq. (7.78).

7.8 Summary

This chapter discussed the reliability of systems based on units and their configuration: basics of system reliability (Section 7.1); block diagrams (Section 7.2); series systems (Section 7.3); parallel systems (Section 7.4); combined series and parallel systems including common cause (Section 7.5); reliability of k-out-of-n systems (Section 7.6); and analysis of complex systems (Section 7.7).

As for the basics (Section 7.1), the fundamental statistical functions for systems are defined in line with Section 2.4, where the representations of distributions were discussed. Block diagrams (Section 7.2) are a graphic representation of systems that give an overview of a system in terms of units, reliabilities and/or failure mechanisms. These diagrams are useful for composing reliability expressions.

As for fundamental system configurations, series (Section 7.3) and parallel systems (Section 7.4) are discussed. With series systems all units must work in order to have the system work and the system reliability R_S is therefore the product of the individual reliabilities R_i of the units: $R_S = \Pi R_i$. With parallel systems at least one unit must work to have the system work or alternatively: the system fails if all units have failed and the system failure distribution F_S is therefore the product of the individual failure distribution F_i of the units, $F_S = \Pi F_i$. Parallel systems provide redundancy in the system. Series and parallel system reliability configurations can differ from the geometrical appearance of units. For example, two switchgears in series provide redundancy in switching and form a parallel system in terms of reliability. Similarly, three parallel single-phase cables together form a three-phase AC circuit and because the circuit fails if any one single-phase cable fails, this is a series system in terms of reliability.

As for combined series and parallel systems including common cause (Section 7.5), a system often consists of combinations. In case of redundancy, it is possible that seemingly redundant systems still have a failure mode in common, which is called the common cause. Additionally, in some cases a minimum redundancy is required or a minimum of parallel units are required to have the system work. These are the so-called k-out-of-n systems (Section 7.6). More complex systems (Section 7.7) like bridge systems sometimes cannot be split into a combination of series and parallel systems. Three methods are discussed to analyse such systems: the conditional method using a Bayesian approach (Section 7.7.1); the up-table method or truth table method with possible reduction due to so-called 'do not care' states (Section 7.7.2); and the minimal paths method or minimal blockades method (Section 7.7.3).

Questions

7.1 With reference to block diagrams, series systems and parallel systems:
Consider the system of identical components in Figure 7.14. Let the reliabilities of each component equal R_1. Let the lifetime of each component be exponentially distributed with a characteristic lifetime $\theta = 500$ hours.
a. Show the relation between dR/dt and R for an exponential distribution of which the mean lifetime is θ_1.
b. Give an expression of the system reliability R_{S1} in terms of R_1.

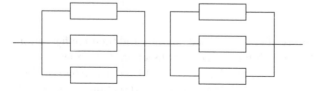

Figure 7.14 Block diagram for questions 1(a)–(d).

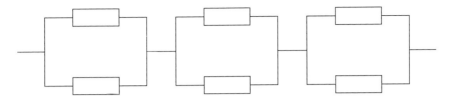

Figure 7.15 Block diagram for question 1(e).

   c. Give an expression for the system distribution density f_S in terms of R_1 and θ_1.

   d. Calculate for the system of Figure 7.14 at time $t = 100$ hours:

      i. The reliability of a component.

      ii. The reliability of the system.

      iii. The distribution density f_S of the system.

      iv. The hazard rate h_S of the system.

   e. Cheaper components become available with a higher characteristic lifetime, namely $\theta_2 = 800$ hours. A disadvantage is that these components require another configuration, namely as in Figure 7.15. The question is whether this is an improvement. Give an expression of this alternative system reliability R_{S2} in terms of R_2.

   f. Check the reliability change by calculating the ratio of R_{S2} and R_{S1} at times:

      i. $t = 100$ hours.

      ii. $t = 500$ hours.

      iii. $t = 800$ hours

7.2 With reference to block diagrams, series systems, parallel systems and bridge systems:

Consider the system of identical components in Figure 7.16. This is a parallel system of three series systems. Assume the component reliabilities are equal and indicated as R. The lifetime of each component is exponentially distributed and equal to θ.

   a. Express the system reliability R_S in terms of R.

   b. Determine for a single series system the hazard rate and the mean time to failure. Express these in terms of θ.

   c. Determine for the complete systems the hazard rate and the mean time to failure. Express these in terms of θ.

   d. An alternative configuration with the same units is shown in Figure 7.17. Determine whether the configuration of Figure 7.17 is an improvement in comparison

Figure 7.16 Block diagram for questions 2(a)–(c).

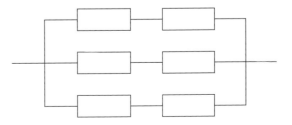

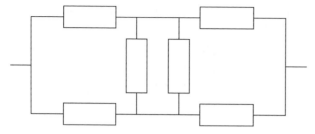

Figure 7.17 Block diagram for question 2(d).

to Figure 7.16 if $R = 0.9$. (*Hint*: a possible solution approach is to first determine the reliability R_{par} of the central blocks and next apply the conditional method.)

7.3 With reference to block diagrams, series systems, parallel systems and common cause:

Consider the configuration of parallel series systems in Figure 7.18. This system consists of two pieces of equipment with their own power supply. The system works as long as one of the apparatus works. The reliability of each apparatus R_A with their power supply R_P is assumed to follow an exponential distribution. The mean lifetime θ_A of an apparatus is 1000 days, while that of a power supply θ_P is 700 days.

a. Give an expression of the system reliability R_S in terms of the reliabilities of (single) apparatus R_A and R_P.

b. Calculate the system reliabilities at two moments, namely:
  i. $t_1 = 200$ days
  ii. $t_1 = 700$ days.

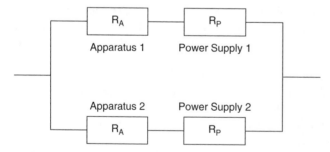

Figure 7.18 Block diagram for questions 3(a), (b).

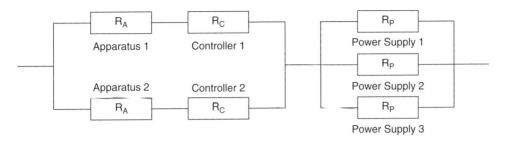

Figure 7.19 Block diagram for questions 3(c), (d).

c. A new configuration is designed in order to increase the reliability R_{new} to beyond 95% at t_1 and beyond 50% at t_2. The new configuration design is shown in Figure 7.19. This new system uses apparatus with a controller that can connect the apparatus to any of the working power supplies. The new system works as long as one apparatus works in combination with its dedicated controller and at least one power supply works. The reliabilities of the apparatus and power supplies follow an exponential distribution as described above, but the controllers age and their reliability R_C follows a Weibull distribution with parameters $\alpha = 4000$ days and $\beta = 2$. Give an expression for the new system reliability in terms of R_A, R_P and R_C.

d. Check whether the targets of $R_{new}(t_1) > 95\%$ and $R_{new}(t_2) > 50\%$ are achieved with the new design.

8

System States, Reliability and Availability

In previous chapters, components and systems have been analysed in terms of reliability as a function of time (or other performance variables). Each system had two possible states: working or failed. The working or operational state was the initial state and the failed state the permanent final state.

However, the grid and parts of the grid (like cable circuits) can often be repaired after failure. Individual components and parts may fail and be removed, but by corrective maintenance systems can come back into operation. As a last part of the theory, the discussion is taken beyond the reliability and lifetime of systems, namely to the availability as the equilibrium between failure and repair. This is the purpose of the present chapter.

There may be a balance between the working or up-state and the failed or down-state of a system. The availability A is defined as the ratio between the up-time and the total time (i.e. up-time plus down-time). If it is stressed that availability is a percentage up-time of the total time, then availability can also be indicated as $A\%$.

This chapter is about techniques to determine the availability of systems that can come back into service after at least one down-state. Also, the effect of spare parts is taken into account. If a system can nevertheless fall into a state of failure where repair is not possible, then the reliability of this (largely) repairable asset or system can be determined as well. Markov chains offer an elegant method to determine availability, reliability and average down-time.

As an introduction to Markov chains, first the most simple system consisting of one component is studied. Concepts such as states, state value, transition rates, availability and steady state will be defined and discussed. Subsequently, the general approach with Markov chains and Laplace transforms is introduced. The final subjects will be redundant systems and concepts like MTBF).

8.1   States of Components and Systems

Components can be either working or failed. This is indicated as the up-state (respectively down-state) of the component. Systems are composed of one or more components. In the following, the number of distinguishable components in the system is m. Due to their redundancy, systems can contain failed components, but may nevertheless still remain in an up-state. If no redundancy is left and another component fails, then the system will be in a down-state.

Reliability Analysis for Asset Management of Electric Power Grids, First Edition. Robert Ross.
© 2019 John Wiley & Sons Ltd. Published 2019 by John Wiley & Sons Ltd.
Companion website: www.wiley.com/go/ross/reliabilityanalysis

As in Section 7.7.3, the state variable x is introduced:

- $x = 1$ means the up-state (i.e. the component or system is working).
- $x = 0$ means the down-state (i.e. the component or system is failed).

Again, the state vector $\overrightarrow{X}$ is introduced which represents the state variables of all m individual components. The state vector is one-dimensional with variables that have binary values (i.e. 1 (working) or 0 (failed)):

$$\overrightarrow{X} = (x_1, x_2, \ldots, x_m) \tag{8.1}$$

A system can be in various states (e.g. with various degrees of redundancy). These states S_j are functions of the state vector:

$$S_j = S_j(\overrightarrow{X}) \tag{8.2}$$

For example, if a connection between two substations consists of two parallel circuits c1 and c2, then the state vector $\overrightarrow{X} = (x_{c1}, x_{c2})$ can describe four cases: (1,1), (1,0), (0,1), (0,0). In the first case both circuits are in an up-state, in the second and third cases one of two circuits is up and the other is down, while in the last case both circuits are down.

Whether a system state S_j is a working or failed state is expressed with the state function φ_j:

$$\phi_j = \phi(S_j) = \phi(S_j(\overrightarrow{X})) \tag{8.3}$$

The first three cases yield three distinct system states S_j ($j = 0,1,2$), all of which are system up-states (i.e. $\varphi(S_j) = 1$ for $j = 0,1,2$). The last case yields a system down-state S_3 (i.e. $\varphi(S_3) = 0$).

Some assumptions are used in the following:

1) The set of possible states S_j is the complete sample space of states (i.e. the system is always in one of those states).
2) The system can be in only one state at a time.
3) Only one transition can take place at a time.

The last assumption applied to the above-mentioned connection consisting of two parallel circuits means that the state vector cannot jump from (1,1) to (0,0) in one transition. The transition from (1,1) to (0,0) will always be through (1,0) or (0,1) and thus requires two subsequent transitions, however fast the second transition may follow the first one.

8.2 States and Transition Rates of One-Component Systems

The simplest case of states and transitions is the situation of a single component that can only fail without the possibility of repair. If such a component is maintained as described in Sections 1.3.2, 1.3.3 and 1.3.4, then the failure distribution is exponential (at least in approximation). In addition, if a bath tub model applies (see Section 2.4.5) and the study is limited to the period when a random failure process dominates, then again the exponential distribution applies (Section 4.2.4). For that reason we will focus on the exponential distribution. This is not a prerequisite for the Markov method, but the exponential distribution features a constant hazard rate which simplifies many calculations.

After describing the one-component system with failure and no repair, the situation is extended to the case where not only failure but also repair is possible. An equilibrium will then be established between the working and the failed (repairing) state.

8.2.1 One-Component System with Mere Failure Behaviour

First we will study the failure behaviour of a one-component system. At the start of operation the system is working (i.e. is in an up-state). According to a distribution $F(t)$ it will fail, after which it stays in a down-state.

Figure 8.1 shows two representations of the one-component system: a block diagram and a state diagram. The block diagram visualizes the coherence of reliabilities (see Section 7.2). The state diagram shows how the states S_j are related to each other. In the case of a one-component system without repair, there is only one relation: the system can transit from the initial state S_0 to the final state S_1. Note that the indices 0 and 1 do not refer to the down-state and the up-state, but are merely indices of states counting from zero.

In the special case of a one-component system, the state function φ happens to have the same value as the component state variable x:

$$\text{One-component system:} \quad \phi = \phi(S_j(x)) = x \tag{8.4}$$

That is, if the component is in its up-state, then the state function $\varphi = 1$ and if the component is failed, then the state function $\varphi = 0$. Table 8.1 shows the component and system states of the one-component system.

The probability that a system is in a state S_j at time t is denoted $P_j(t)$. The sum of all probabilities $P_j(t)$ is always equal to 1 (i.e. the system is always in one of the possible states that together form the complete sample space Ω (cf. Eq. (2.4)). In the present case of a one-component system the working state is S_0 and the failed state is S_1. The probability that the system is in the non-failed state S_0 is the reliability $R(t) = P_0(t)$ and the probability that the system is in the failed state S_1 is $F(t) = P_1(t)$. Indeed, the sum of $R(t)$ and $F(t)$ is always equal to 1 (Eq. (2.12)). Repeating Eq. (2.21), the hazard rate is defined as:

$$h(t) = \frac{f(t)}{R(t)} = \frac{1}{R(t)} \frac{dF(t)}{dt} = -\frac{1}{R(t)} \frac{dR(t)}{dt} \tag{8.5}$$

Figure 8.1 Two representations of a one-component system denoted by its hazard rate h_{01}. The suffix '01' indicates a transition from the initial state S_0 to the final state S_1. The left diagram is a block diagram (see Section 7.2) and the right diagram is a state diagram.

Table 8.1 Component and system states for a one-component system.

Component state x	Vector $\vec{X}$	System state	System state function ϕ
$x = 1$	$\vec{X} = (1)$	S_0	$\phi(S_0) = 1$
$x = 0$	$\vec{X} = (0)$	S_1	$\phi(S_1) = 0$

In terms of P_0 and P_1 in relation to the transition rate $h_{01}(t)$ from S_0 to S_1 we find:

$$h_{01}(t) = \frac{1}{P_0(t)} \frac{dP_1(t)}{dt} = -\frac{1}{P_0(t)} \frac{dP_0(t)}{dt} \tag{8.6}$$

This can be written in the form of two differential equations:

$$\frac{dP_0(t)}{dt} = \dot{P}_0(t) = -P_0(t) \cdot h_{01}(t) \tag{8.7}$$

$$\frac{dP_1(t)}{dt} = \dot{P}_1(t) = P_0(t) \cdot h_{01}(t) \tag{8.8}$$

These expressions are in line with Eq. (2.21) where the hazard rate was introduced.

As mentioned before, the sum of the probabilities is 1 and therefore constant. An increase or decrease in P_0 means a decrease (respectively increase) in P_1. The time derivatives of P_0 and P_1 have opposite signs, as shown in Eqs (8.7) and (8.8). Again, this was also stated for $R(t)$ and $F(t)$ when $f(t)$ was introduced in Eq. (2.16).

In case the failure distribution is exponential, the transition rate $h_{01}(t)$ is a constant λ_{01} (see Section 4.2.4). Equations (8.7) and (8.8) then become:

$$\dot{P}_0(t) = -P_0(t) \cdot \lambda_{01} \tag{8.9}$$

$$\dot{P}_1(t) = P_0(t) \cdot \lambda_{01} \tag{8.10}$$

The increase in P_1 and the decrease in P_0 are both proportional to the probability that the system resides in state S_0 and with λ_{01}. In this simple case it is apparent that the system will also fail according to an exponential distribution if the component does. However, in general, even with components all failing according to an exponential distribution, it will depend on the exact configuration whether the system follows an exponential distribution or not. In general, the system will probably have a more complicated distribution. The distributions that describe the system transitions from one state to another, as well as the failure behaviour, then follow from solving the differential equations. This will be shown in the next section.

With the boundary condition that the initial state is working ($P_0(0) = 1$ and $P_1(0) = 0$), the solution of Eq. (8.9) is the well-known expression for the exponential reliability:

$$P_0(t) = e^{-\lambda_{01} \cdot t} \tag{8.11}$$

The solution of Eq. (8.10) is the exponential failure distribution:

$$P_1(t) = 1 - e^{-\lambda_{01} \cdot t} \tag{8.12}$$

8.2.2 One-Component System with Failure and Repair Behaviour

How will the situation change if the component can be repaired and the system can therefore get back into operation? The block diagram and the state diagram for that case are shown in Figure 8.2. In this case it is assumed that transitions all follow an exponential distribution. As a consequence, the failure process is characterized by a constant failure rate λ_{01} (where the suffix 01 denotes transition from S_0 to S_1). The repair process is characterized by a repair rate of μ_{10} (where the suffix 10 indicates a transition from S_1

Figure 8.2 The block diagram (left) and state diagram (right) of a repairable one-component system with constant hazard (or failure) rate λ_{01} and repair rate μ_{10}.

to S_0). In the following, failure rates will be indicated by λ and repair rates by μ just for clarity. There is no theoretical reason for such a distinction; these are all just transition rates from one state to another.

Repair from state S_1 to S_0 might be regarded as 'failure' from the failed state to the working state. Based on symmetry and mutuality, the following equations can be deduced from Eqs (8.9) and (8.10):

$$\dot{P}_0(t) = -P_0(t) \cdot \lambda_{01} + P_1(t) \cdot \mu_{10} \tag{8.13}$$

$$\dot{P}_1(t) = P_0(t) \cdot \lambda_{01} - P_1(t) \cdot \mu_{10} \tag{8.14}$$

The derivative of P_0 is negatively proportional to both P_0 and λ_{01}. It is also positively proportional to both P_1 and μ_{10}. This can be interpreted as: the more assets are in state S_0 and the higher the failure rate λ_{01}, the more assets will fail. Likewise, the more assets are in state S_1 and the higher the repair rate μ_{10}, the more assets will be in the transition to get back into operation.

Equations (8.13) and (8.14) can be written in matrix notation:

$$\begin{pmatrix} \dot{P}_0(t) \\ \dot{P}_1(t) \end{pmatrix} = (P_0(t), P_1(t)) \cdot \begin{pmatrix} -\lambda_{01} & \lambda_{01} \\ \mu_{10} & -\mu_{10} \end{pmatrix} = (P_0(t), P_1(t)) \cdot \vec{M} \tag{8.15}$$

Here, $\vec{M}$ is the transition matrix that characterizes the complete system behaviour. This matrix follows directly from the state diagram. It may be noted that the sum of the row elements always equals 0, which means that the sum of the probabilities is a constant equal to 1. In practice, the differential equations can be deduced from this matrix (so the analysis procedure will then be reversed to the introduction above, where the matrix followed as an alternative notation).

Subsequently, the differential equations can be solved, from which probabilities $P_i(t)$ of residing in state S_i follow. This establishes how the system develops from its starting state to other states in time.

Linear differential equations can be solved with Laplace transforms. A set of such Laplace transforms is listed in Appendix B. The Laplace transform of a function $P(t)$ is defined as the following complex integral:

$$P(s) = L\{P(t)\} = \int_0^\infty e^{-st} \cdot P(t)dt \tag{8.16}$$

Here, $L\{\}$ is the notation for the Laplace transform and s is a complex number. The inverse Laplace transform $L^{-1}\{\}$ is again a complex integral:

$$P(t) = L^{-1}\{P(s)\} = \int_{\gamma-j\infty}^{\gamma+j\infty} e^{st} \cdot P(s)ds \tag{8.17}$$

Here, j denotes the imaginary unit: $j^2 = -1$.

The Laplace transform is used to find probabilities P_0 and P_1 (i.e. to solve Eqs (8.13) and (8.14)), which were taken together in Eq. (8.15).

The Laplace transform of a derivative $dP(t)/dt$ is:

$$L\left\{ \frac{dP(t)}{dt} \right\} = L\{\dot{P}(t)\} = s \cdot P(s) - P(t = 0) \tag{8.18}$$

Here, $P(t = 0)$ is the initial value of $P(t)$ at the start of operating the system. Therefore, the Laplace transform of a derivative requires knowledge of the initial value. In the present case $P_0(t = 0) = 1$ and therefore $P_1(t = 0) = 0$. The Laplace transform of Eqs (8.13) and (8.14) then yields:

$$sP_0(s) - P_0(t = 0) = sP_0(s) - 1 = -P_0(s) \cdot \lambda_{01} + P_1(s) \cdot \mu_{10} \tag{8.19}$$

$$sP_1(s) - P_1(t = 0) = sP_1(s) - 0 = P_0(s) \cdot \lambda_{01} - P_1(s) \cdot \mu_{10} \tag{8.20}$$

From Eq. (8.20) it follows that:

$$P_1(s) = \frac{P_0(s) \cdot \lambda_{01}}{s + \mu_{10}} \tag{8.21}$$

Combining Eqs (8.19) and (8.21) leads to an expression for $P_0(s)$:

$$P_0(s) = \frac{1}{s + \lambda_{01} - \left(\dfrac{\lambda_{01}}{s + \mu_{10}} \right) \cdot \mu_{10}} = \frac{s + \mu_{10}}{(s + \lambda_{01})(s + \mu_{10}) - \lambda_{01}\mu_{10}}$$

$$= \frac{\left(\dfrac{\mu_{10}}{\lambda_{01} + \mu_{10}} \right)}{s} + \frac{\left(\dfrac{\lambda_{01}}{\lambda_{01} + \mu_{10}} \right)}{s + \lambda_{01}} \tag{8.22}$$

Inverse Laplace transformation yields an expression for $P_0(t)$ (see Appendix B):

$$P_0(t) = \frac{\mu_{10}}{\lambda_{01} + \mu_{10}} + \frac{\lambda_{01}}{\lambda_{01} + \mu_{10}} \cdot \exp[-(\lambda_{01} + \mu_{10}) \cdot t] \tag{8.23}$$

This establishes how the reliability $R(t) = P_0(t)$ of the one-component system evolves in time. In the long run an equilibrium will be established between failure and repair. For large t the exponential term will approach 0 and then we find the equilibrium for P_0:

$$\lim_{t \to \infty} P_0(t) = P_{0,\infty} = \frac{\mu_{10}}{\lambda_{01} + \mu_{10}} \tag{8.24}$$

The faster the repair, the higher μ_{10} and the greater the probability that the system works. It is noteworthy that P_0 does not approach 0 for infinitely large t. There will always be a balance between failure and repair in the present case.

The ratio between the probability of a working system and the sum of probabilities (i.e. 1) is the availability $A(t)$. This is also the ratio between up-time and the total time.

In the case of a one-component system there is only one working state (namely S_0) and therefore $A(t)$ is equal to $P_0(t)$. In the general case there may be more working states and the availability $A(t)$ is then equal to the sum of all probabilities for a working system:

$$A(t) = \sum_{i,\text{working}} P_i(t) \tag{8.25}$$

If the interest is limited to determining the final availability $A_\infty = A(t \to \infty)$ rather than $A(t)$ for all t, then there is a more convenient solution without Laplace transforms. The crux is that in the long run all time derivatives of the probabilities $dP_i(t)/dt$ approach 0 (i.e. all probabilities $P_{i,\infty}$ become constant and are called 'steady states'). Therefore, for large t the derivatives in Eqs (8.13) and (8.14) approach 0:

$$\lim_{t\to\infty} \dot{P}_0(t) = \lim_{t\to\infty}(-P_0(t) \cdot \lambda_{01} + P_1(t) \cdot \mu_{10}) \Rightarrow 0 = -P_{0,\infty} \cdot \lambda_{01} + P_{1,\infty} \cdot \mu_{10} \quad (8.26)$$

$$\lim_{t\to\infty} \dot{P}_1(t) = \lim_{t\to\infty}(P_0(t) \cdot \lambda_{01} - P_1(t) \cdot \mu_{10}) \Rightarrow 0 = P_{0,\infty} \cdot \lambda_{01} - P_{1,\infty} \cdot \mu_{10} \quad (8.27)$$

This set of equations is dependent, but by adding the requirement that the sum of the probabilities equals 1, a set of independent equations occurs in which $P_{0,\infty}$ can be solved:

$$\left.\begin{array}{l} 0 = -P_{0,\infty} \cdot \lambda_{01} + P_{1,\infty} \cdot \mu_{10} \\ P_{0,\infty} + P_{1,\infty} = 1 \end{array}\right\} \Rightarrow \left.\begin{array}{l} P_{1,\infty} = P_{0,\infty} \cdot \lambda_{01}/\mu_{10} \\ P_{0,\infty} = 1 - P_{1,\infty} \end{array}\right\}$$

$$\Rightarrow P_{0,\infty} = \frac{\mu_{10}}{\lambda_{01} + \mu_{10}} = A_\infty \quad (8.28)$$

The steady-state approach yields the same solution as in Eq. (8.24) as it should, but it is usually easier because the Laplace transforms between s-domain and t-domain can be avoided.

By determining $P_0(t)$ and/or $P_{0,\infty}$ the behaviour of the system is established. With repair an equilibrium is reached as in Eqs (8.24) and (8.28). If no repair is carried out (or it is stated that $\mu_{10} = 0$) then the down-state S_1 is the permanent final state. The reliability of the system $R(t)$ equals $P_0(t)$ for the one-component system and an expected lifetime can be derived. Without repair the long-term availability A_∞ will be 0. The average availability will also be 0, because once failed the system will reside in its failed state for an infinitely long time.

Forecasts and requirements with respect to failure, expected lifetime and availability become possible once the probabilities of a working system are known. The Markov–Laplace method as well as the steady-state approach are powerful tools for determining these probabilities.

8.3 System State Probabilities via Markov Chains

Knowledge of the reliability $R(t)$ and availability $A(t)$ of components and systems supports balancing risks and decision-making about requirements, utilization, investments, and so on. The Markov–Laplace method enables the analysis of complex configurations with multiple failure processes, redundancy, spare parts and repairs. Section 8.2 discussed the example of a one-component system. As a start for the Markov–Laplace method, the present section will discuss Markov chains for the general case of multiple components and processes. Subsequently, the next section will apply the method to redundant configurations and spare parts.

The following steps are discussed in Sections 8.3.1 to 8.3.4:

1. Establishing the relevant system states (Section 8.3.1).
2. Drawing the state diagram (Section 8.3.2).
3. Deriving differential equations based on the state diagram (Section 8.3.3).
4. Deriving differential equations based on the transition matrix (Section 8.3.4).

The differential equations define the system behaviour. From these equations follow functions and quantities like the reliability $R(t)$, availability $A(t)$, hazard rate $h(t)$, expected lifetime θ and MTBF.

8.3.1 Component and System States

A system is built up from components. Knowledge of the configuration of the components is necessary to evaluate the system behaviour. This can be less trivial than it seems.

For instance, Figure 8.3 shows a failed rod made of glass fibre-reinforced epoxy resin featuring an electrical tree along its surface. In undamaged condition such components are typically used for mechanical support of conductors or transfer of mechanical force for switching. It is usually immersed in another insulation medium like oil, gas or vacuum. The insulating system is a combination of this rod and the insulating medium in which the rod is immersed. At first glance it might seem that the material with the lowest breakdown strength determines the reliability of the system. As both insulating materials must remain intact, this could be regarded as a series system of two components like in Section 7.3. In practice, however, the interface between the two insulating materials is often electrically weaker than each of the insulating materials. The failure behaviour of this interface is influenced by both materials. Knowledge of the bulk behaviour is not sufficient. The interface appears to be a very crucial third component. The system may therefore be described as a series system of three components: epoxy rod, surrounding insulating medium and interface. One must be aware that such effects may be hidden, but nevertheless can be most crucial to the system failure behaviour.

A block diagram can be drawn once the configuration is known sufficiently, including the behaviour of intangible parts like surfaces and interfaces. This defines the structure of the system. If there are various failure mechanisms then various block diagrams may be necessary as the combined action of components may be different.

State variables x_i can be assigned to the components. These state variables can be combined to the state vector $\overrightarrow{X}$ (repeating Eq. (8.1) for the case of m components):

$$\overrightarrow{X} = (x_1, x_2, \ldots, x_m) \tag{8.29}$$

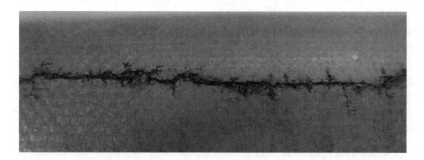

Figure 8.3 Electric failure along the surface of a solid insulator at the interface with air. The interface between air and solid insulator features its own failure mechanism and can be regarded as an additional component to insulator and air.

The system states S_j and the state function φ_j can then be established:

$$\phi_j = \phi(S_j) = \phi(S_j(\overrightarrow{X})) = \phi(S_j(x_1, x_2, \ldots, x_m)) \qquad (8.30)$$

In the example of a one-component system only two states existed: on the one hand the working or up-state and on the other hand the failed or down-state. With larger systems, more states exist. In order to reduce the complexity of the evaluation, it can be practical to try to combine states if they lead to the same value for the state function ϕ_j and only differ in details that are not relevant to the failure mechanism. This will be illustrated first.

Let us look again at the case of the two parallel circuits that form a redundant connection. Assume the circuits are coded A and B (Figure 8.4). The first step is to determine the states and state variables, as shown in Table 8.2.

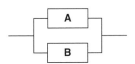

Figure 8.4 Block diagram of a redundant connection of two parallel circuits A and B.

The states and state variables are all listed in Table 8.2. If a state is assigned to each distinct vector $\overrightarrow{X}$, then four states $\widetilde{S}_j$ can be distinguished. In the last column, the relevant characteristics are summarized and it is noteworthy that states $\widetilde{S}_1$ and $\widetilde{S}_2$ have the same characteristics: 1 working and 0 redundant component. If it is irrelevant for the analysis which of the two components has failed, then states $\widetilde{S}_1$ and $\widetilde{S}_2$ can be combined into one state, leading to the simplified representation in Table 8.3.

It is recommended to reduce the number of distinguishable states S_j in order to simplify further analysis. The reduction of the number of states from four to three reduces the transition matrix from a 4×4 to a 3×3 matrix and reduces the equations to be solved from four to three.

Table 8.2 System states and state variables for the system in Figure 8.4.

Component state variables		Vector $\overrightarrow{X}$	System states $\widetilde{S}_j$	System state function $\widetilde{\phi}_j$	System characteristics
x_A	x_B				
1	1	(1,1)	$\widetilde{S}_0$	1	2 components OK, 1 redundant
1	0	(1,0)	$\widetilde{S}_1$	1	1 component OK, 0 redundant
0	1	(0,1)	$\widetilde{S}_2$	1	1 component OK, 0 redundant
0	0	(0,0)	$\widetilde{S}_3$	0	0 component OK, 0 redundant

Table 8.3 Revised system states S_j for the system in Figure 8.4.

Vector $\overrightarrow{X} = (x_A, x_B)$	System states S_j	State function ϕ_j	System characteristics
(1,1)	S_0	1	2 components OK, 1 redundant
(1,0) and (0,1)	S_1	1	1 component OK, 0 redundant
(0,0)	S_2	0	0 component OK, 0 redundant

8.3.2 System States and Transition Rates for Failure and Repair

After the system states have been established, the state diagram can be drawn with the transition rates. This requires an inventory of possible transitions and their rates. In general, such rates can be functions of time like the hazard rate $h(t)$, but often it is assumed that exponential distributions rule the transitions, in which case all transition rates are constant. Where possible and relevant, the failure rates will be indicated with a λ and the repair rates with a μ.

There are two main reasons why exponential distributions are applicable. Firstly, the analyses concern maintained (sub)systems where the hazard rate $h(t)$ can often be approximated with a constant average hazard rate h_{ave} (see Section 4.2.4). Secondly, with reference to the bath tub model (see Section 2.4.5), most components are operated for the largest part of their service life during the period of random failure, which follows an exponential distribution (Section 4.2.4).

Often there are also arguments why this is not applicable for all transitions. For instance, if the components start to enter the stage where wear-out dominates (the last part in the bath tub model, see Section 2.4.5) then the hazard rate is increasing. Often this increase is relatively low compared to the scale of planning and the hazard rate may then be regarded as semi-constant (i.e. the approximation of a constant rate is reasonable, but at each date the exact hazard rate has to be determined).

Another example is the situation where the type of repair is known to take a fixed amount of time with a narrow bandwidth. This is not an exponentially distributed transition then either. Nevertheless, the choice is often to use the Markov–Laplace method with convenient constant rates as a reasonable approximation or to solve the differential equations, acknowledging the time aspect in the transition rate. In the remaining subsections, the elaborations also employ the Markov–Laplace method with constant transition rates.

A state diagram shows the possible states, possible transitions and transition rates. Figure 8.2 showed the diagram of a repairable one-component system. Figure 8.5 shows the state diagram of the redundant system in Figure 8.4, for which the states are detailed in Table 8.3.

The diagram in Figure 8.5 shows that a component can fail whereby the system transits from state S_0 to S_1 or from state S_1 to S_2. The diagram also shows no possibility of two components failing at exactly the same time, with the system transiting from S_0 to S_2 directly. This is not a fundamental condition, because it would be possible to draw and calculate with this transition, but usually this option is left out of the scope as long as the failures are independent events. The argument is that at least an infinitesimally small time will elapse between the two subsequent events. The same applies to the repair process.

The transition rates in the diagram are indicated without value as yet, but that information could be included. The arrows already make clear in which direction the transition takes place. The constant rate can also be expressed in terms of characteristic lifetime θ, because $\theta = 1/\lambda$ (see Section 4.2.4).

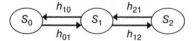

Figure 8.5 State diagram for a repairable system with three states like a redundant cable connection as detailed in Table 8.3.

Table 8.4 Transition rates for the case of Figure 8.5.

Transition rate	Value	Explanation
h_{01}	2λ	Two components are in operation; each can fail, whereby the system transits from S_0 to S_1, so the transition rate is twice the failure rate of each
h_{12}	λ	One component is in operation, so once the failure rate
h_{10}	μ	One component is failed and is getting repaired
h_{21}	μ	Two components are failed, but only one is repaired at the time

Figure 8.6 State diagram of Figure 8.5 with the transition rates specified.

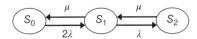

A typical description that could accompany the diagram is: the components have a failure rate of λ and there is one repairer who can repair one component at a time with a repair rate of μ. There could be other descriptions, like the possibility of being able to repair two failed components simultaneously (which doubles the repair rate from state S_2 to S_1, but is not the same as a direct transition from S_2 to S_0). However, here the choice is made to have only one repairer at a time.

The transition rates for the case of Figure 8.5 are specified in Table 8.4 and used in Figure 8.6.

8.3.3 Differential Equations Based on the State Diagram

The state diagram in Figure 8.6 shows states S_j with transitions to and from other states. The probability that the system resides in state S_j:

- Decreases with transitions from S_j to other states S_k ($k \neq j$). These transitions are recognizable by outbound arrows at state S_j.
- Increases with transitions to S_j from other states S_k ($k \neq j$). These transitions are recognizable by inbound arrows at state S_j.

For $R(t)$ and $h(t)$ the following holds (cf. Eq. (2.21)) in case of an exponential distribution:

$$\dot{R}(t) = -R(t) \cdot h(t) = -R(t) \cdot \lambda \tag{8.31}$$

Likewise, each transition from S_j to S_k has a negative contribution looking from S_j:

$$\text{For } k \neq j : \dot{P}_j(t) = -P_j(t) \cdot h_{jk} \tag{8.32}$$

For $F(t)$ and $h(t)$ (cf. Eq. (2.21)) it then holds that, for exponential distributions:

$$\dot{F}(t) = R(t) \cdot h(t) = R(t) \cdot \lambda \tag{8.33}$$

Likewise, each transition from S_j to S_k has a positive contribution looking from S_k:

$$\text{For } k \neq j : \dot{P}_k(t) = P_j(t) \cdot h_{jk} \tag{8.34}$$

The total derivative of the probability that the system will reside in state S_j is the sum of all transition contributions:

$$\dot{P}_j(t) = -P_j(t) \cdot \sum_{k \neq j} h_{jk} + \sum_{k \neq j} P_k(t) \cdot h_{kj} \tag{8.35}$$

The complete set of differential equations for the system arises by applying Eq. (8.35) to all states S_j. Whether contributions are positive or negative follows directly from the state diagram by evaluating the direction of the arrows.

For the example of the redundant cable connection (see Figure 8.6), the following set of differential equations is found for $j = 0,1,2$:

$$\dot{P}_0(t) = -P_0(t) \cdot 2\lambda + P_1(t) \cdot \mu$$
$$\dot{P}_1(t) = +P_0(t) \cdot 2\lambda - P_1(t) \cdot (\lambda + \mu) + P_2(t) \cdot \mu$$
$$\dot{P}_2(t) = +P_1(t) \cdot \lambda - P_2(t) \cdot \mu \tag{8.36}$$

8.3.4 Differential Equations Based on the Transition Matrix

An alternative method for determining the differential equations employs the transition matrix $\overrightarrow{M}$ that consists of elements h_{jk}. The set of differential equations can be written in matrix notation:

$$(\dot{P}_0(t), \dot{P}_1(t), \ldots, \dot{P}_m(t)) = (P_0(t), P_1(t), \ldots, P_m(t)) \begin{pmatrix} h_{00} & h_{01} & \cdots & h_{0m} \\ h_{10} & h_{11} & \cdots & h_{1m} \\ \cdots & \cdots & \cdots & \cdots \\ h_{m0} & h_{m1} & \cdots & h_{mm} \end{pmatrix} \tag{8.37}$$

The non-diagonal elements h_{jk} represent a transition from state S_j to another state S_k. In a state diagram like Figure 8.5 this concerns all transitions with outbound arrows from state S_j. The sum of all outbound transition rates is the total rate of transitions from state S_j to other states S_k. In other words, the sum of all transition rates h_{jk} is the total decrease per unit time that leaves S_j and does not come back to S_j. Therefore. it is also the negative change of the transition from S_j to itself. As a consequence. each h_{jj} is equal to the negative sum of all h_{jk} on the same row r_j for all $k \neq j$:

$$h_{jj} = -\sum_{\substack{k=0, \\ k \neq j}}^{\substack{k \neq j \\ m}} h_{jk} \tag{8.38}$$

This is also the way the matrix can be filled out:

1. Determine, per row r_j, first all h_{jk} with $k \neq j$.
2. Take h_{jj} equal to the negative sum of the h_{jk} with $k \neq j$ (as in Eq. (8.38)).

In the case of the redundant cable connection (see Figure 8.6), the matrix $\overrightarrow{M}$ is found as:

$$\overrightarrow{M} = \begin{pmatrix} -(2\lambda + 0) & 2\lambda & 0 \\ \mu & -(\mu + \lambda) & \lambda \\ 0 & \mu & -(0 + \mu) \end{pmatrix} = \begin{pmatrix} -2\lambda & 2\lambda & 0 \\ \mu & -(\mu + \lambda) & \lambda \\ 0 & \mu & -\mu \end{pmatrix} \tag{8.39}$$

The differential equations are found by the product:

$$(\dot{P}_0(t), \dot{P}_1(t), \dot{P}_2(t)) = (P_0(t), P_1(t), P_2(t)) \cdot \begin{pmatrix} -2\lambda & 2\lambda & 0 \\ \mu & -(\mu + \lambda) & \lambda \\ 0 & \mu & -\mu \end{pmatrix} \tag{8.40}$$

It is noteworthy that the matrix is filled out row by row, but used column by column in the product. Elaborating Eq. (8.40) yields the set of differential equations that defines the system behaviour:

$$\begin{aligned}
\dot{P}_0(t) &= -P_0(t) \cdot 2\lambda + P_1(t) \cdot \mu \\
\dot{P}_1(t) &= +P_0(t) \cdot 2\lambda - P_1(t) \cdot (\lambda + \mu) + P_2(t) \cdot \mu \\
\dot{P}_2(t) &= \qquad\qquad + P_1(t) \cdot \lambda \qquad - P_2(t) \cdot \mu
\end{aligned} \tag{8.41}$$

This is the same result as in Eq. (8.36), which was achieved directly from the state diagram. The methods are equivalent alternatives. The matrix seems more laborious, because first the matrix is filled out and then the differential equations are derived. On the other hand, the matrix provides a means to keep an overview that may help to detect possible mistakes.

The differential equations also offer a simple, quick check for errors. If the probabilities P_k on the right-hand side of the equations are written under each other in an orderly manner as in Eq. (8.41), then the sum of the rates per P_k is equal to 0. For instance, in Eq. (8.41) the coefficient for $P_0(t)$ appears as -2λ in the first equation and 2λ in the second equation, and so the sum of the coefficients is $-2\lambda + 2\lambda = 0$. Similarly, the sum of the coefficients for P_1 appears as $\mu - (\lambda + \mu) + \lambda = 0$ and the sum of the coefficients for P_2 appears as $\mu - \mu = 0$. This check may also help to quickly detect errors. The background of this check is a consequence of the fact that the sum of all probabilities constantly remains 1 and therefore the sum of the derivatives remains 0 for all $P_k(t)$.

8.4 Markov–Laplace Method for Reliability and Availability

In the previous section differential equations were derived that define the behaviour of a system of components. The present section shows how the equations are solved to find the reliability and availability of such systems. The complete method uses both Markov chains of states and Laplace transforms from the t-domain to the s-domain and vice versa.

For special cases shortcuts are possible to determine expected lifetimes, such as the MTBF. These methods will be dealt with in Section 8.5. As mentioned before, in all cases it is assumed that the exponential distribution applies to all transitions and therefore all transition rates h_{jk} are constant.

The set of differential equations is written in matrix notation in Eq. (8.37), from which each differential equation follows as in Eq. (8.35):

$$\dot{P}_j(t) = -P_j(t) \cdot \sum_{k \neq j} h_{jk} + \sum_{k \neq j} P_k(t) \cdot h_{kj} \tag{8.42}$$

Laplace transforms provide a convenient way to solve these linear differential equations. The Laplace transform $L\{P(t)\}$ of a time function $P(t)$ yields $P(s)$:

$$P(s) = L\{P(t)\} = \int_0^\infty e^{-st} \cdot P(t)dt \tag{8.43}$$

In the literature, Laplace-transformed functions are also written with lowercase characters, but this notation might lead to confusion as in statistics lowercase $f(t)$ is used for the derivative of $F(t)$. In other notations asterisks are used (e.g. P for the time function and P^* for the Laplace-transformed function). This symbol may, however, also be in use for conjugated complex numbers. $LP(s)$ and $L(P(t))(s)$ are also used. Because various mathematical disciplines merge, notation conflicts may arise. In the present context the distinction is made by explicitly indicating the functions by their variable: functions of t as $P(t)$ and functions of s as $P(s)$. If no confusion is possible, time functions $P(t)$ may be indicated just by P.

The transform of Eq. (8.42) yields:

$$sP_j(s) - P_j(t = 0) = -P_j(s) \cdot \sum_{k \neq j} h_{jk} + \sum_{k \neq j} P_k(s) \cdot h_{kj} \tag{8.44}$$

For the complete set:

$$(sP_0(s) - P_0(t = 0), \ldots, sP_m(s) - P_m(t = 0)) = (P_0(s), \ldots, P_m(t)) \cdot \overrightarrow{M} \tag{8.45}$$

For this set, expressions can be found for all $P_j(s)$. The time functions $P_j(t)$ are found by an inverse Laplace transform:

$$P(t) = L^{-1}\{P(s)\} = \int_{\gamma - j\infty}^{\gamma + j\infty} e^{st} \cdot P(s)ds \tag{8.46}$$

The availability is equal to the sum of all probabilities of working states:

$$A(t) = \sum_{j, working} P_j(t) \tag{8.47}$$

In the case of the redundant cable connection of Figure 8.6 with the differential equations of Eq. (8.41), while assuming that the system was fully functioning initially (meaning $P_0(t = 0) = 1$ and $P_1(t = 0) = P_2(t = 0) = 0$), the transformed set is:

$$sP_0(s) - P_0(t = 0) = sP_0(s) - 1 = -2\lambda \cdot P_0(s) \qquad + \mu \cdot P_1(s)$$
$$sP_1(s) - P_1(t = 0) = sP_1(s) - 0 = \quad 2\lambda \cdot P_0(s) - (\lambda + \mu) \cdot P_1(s) + \mu \cdot P_2(s)$$
$$sP_2(s) - P_2(t = 0) = sP_2(s) - 0 = \qquad\qquad \lambda \cdot P_1(s) - \mu \cdot P_2(s) \tag{8.48}$$

The solution of this set in terms of $P_j(s)$ and subsequent inverse Laplace transform yields the probabilities $P_j(t)$ that define the system behaviour. If no permanent final state exists (because presumably repair is always possible), then an equilibrium will be reached between working and failed states. With known $P_j(t)$, the system availability can be determined as the sum of the probabilities of working states $S_{j,working}$:

$$A(t) = \sum_{j,working} P_j(t) = 1 - \sum_{j,failed} P_j(t) \tag{8.49}$$

The system of the redundant cable connection does not have a permanent final state, because even from the down-state S_2 repair is still possible. In that case the availability $A(t)$ is used to describe the system behaviour. Then, the system has two up-states (S_0 and S_1) and the availability thus becomes:

$$A(t) = P_0(t) + P_1(t) = 1 - P_2(t) \tag{8.50}$$

If one or more permanent final states, also referred to as 'absorbing states', exist then the system ends in a down-state and the probability of system failure $F(t)$ approaches 100% with increasing time. With known $P_j(t)$ the system reliability $R(t)$ can be determined as the sum of probabilities of working states $S_{j,working}$:

$$R(t) = \sum_{j,working} P_j(t) = 1 - \sum_{j,failed} P_j(t) \tag{8.51}$$

So, if repair from state S_2 would not be possible in the case of the redundant cable connection, then the system would be permanently failed once state S_2 is reached. All matrix elements in the bottom row of the matrix in Eq. (8.39) would then be equal to 0. In that case the system would be described by the system reliability $R(t)$:

$$R(t) = P_0(t) + P_1(t) = 1 - P_2(t) \tag{8.52}$$

The complete Markov–Laplace method to find the probabilities can be summarized with the steps in Table 8.5. Whichever of the two is applicable, Eq. (8.49) or (8.51), establishes the system behaviour in time.

Table 8.5 Summary of the Markov–Laplace method.

Step	Action	Sections
1	Determine the relevant components in a block diagram	8.1
2	Determine the (minimum) set of states S_j	8.3.1
3	Draw a state diagram with states S_j, transition rates h_{jk} and the direction of the transitions	8.3.2
4	Option 4a: derive the differential equations for the probabilities P_j based on the state diagram	8.3.3
	Option 4b: derive the differential equations for the probabilities P_j based on the transition matrix	8.3.4
5	Achieve the Laplace transforms of the differential equations	8.4
6	Solve the set of Laplace-transformed differential equations which yield the Laplace-transformed probabilities $P_j(s)$	8.4
7	Achieve the inverse Laplace transform which produces the $P_j(t)$ that defines the system behaviour	8.4

8.5 Lifetime with Absorbing States and Spare Parts

As mentioned in the previous section, states that do not allow repair cause permanent system failure. Such states are called 'absorbing states'. The present section will show how the expected lifetime of a system with one or more absorbing states can be calculated.

The expected lifetime θ was introduced in Section 3.1 and Eq. (3.8) as:

$$\theta = \langle t \rangle = \int_0^\infty t \cdot f(t)dt = \int_0^\infty R(t)dt \tag{8.53}$$

An expected lifetime can be determined not only for a system with a permanent final state, but also for a repairable system with a down-state. In that case the expected lifetime can be the MTTFF (i.e. the mean time to first failure) and the MTBF. These two expected times will be addressed in Section 8.6. The discussion largely follows the procedure for the expected lifetime of the present section.

Expected lifetimes can be calculated with Eq. (8.53), but the Markov–Laplace method offers a simpler and quicker alternative. At first, we take a closer look at the Laplace transform $R(s)$ of the reliability $R(t)$:

$$L\{R(t)\} = \int_0^\infty e^{-st}R(t)dt = R(s) \tag{8.54}$$

The variable s can be any complex number. If s approaches 0, then $R(s)$ approaches θ:

$$\lim_{s \downarrow 0} R(s) = \lim_{s \downarrow 0} \int_0^\infty e^{-st}R(t)dt = \int_0^\infty 1.R(t)dt = \int_0^\infty R(t)dt = \theta \tag{8.55}$$

According to Eq. (8.51), $R(t)$ is equal to the sum of the probabilities $P_j(t)$ of working states (and also 1 minus the sum of the probabilities of the failed states). Likewise, $R(s)$ is also the sum of probabilities $P_j(s)$ of working states and 1 minus the probabilities of failed states:

$$R(s) = \int_0^\infty e^{-st}R(t)dt = \int_0^\infty e^{-st} \sum_{j,working} P_j(t)dt = \sum_{j,working} \int_0^\infty e^{-st}P_j(t)dt = \sum_{j,working} P_j(s) \tag{8.56}$$

From Eqs (8.55) and (8.56) it follows that if the $P_j(s)$ of all the working (or all the failed) states are known, then $R(s)$ is produced as the sum of the probabilities of the working states. If we let s approach 0, then $R(s \rightarrow 0)$ is equal to the expected lifetime θ.

There is no need for an inverse Laplace transform, but it is sufficient to take steps 1 to 6 in the Markov–Laplace method (Table 8.5) and then take the limit $s \rightarrow 0$ as in Eq. (8.55).

For the time domain it holds that the sum of all probabilities $P_j(t)$ is equal to 1. For that reason, $R(t)$ can also be determined as $1 - F(t)$ (cf. the right-hand side of Eq. (8.51)). This does not hold for the s-domain. The Laplace transform of the sum of the probabilities $P_j(t)$ is the Laplace transform of 1:

$$L\left(\sum_j P_j(t)\right) = L(1) = \int_0^\infty e^{-st} \cdot 1 \, dt = \frac{-e^{-st}}{s}\bigg|_0^\infty = \frac{1}{s} \tag{8.57}$$

Taking the limit $s \rightarrow 0$ causes the Laplace transform to approach infinity. The expected lifetimes of the working states are finite, which is equivalent to the sum of the transformed probabilities of the working states being finite if $s \rightarrow 0$. But if the system transits to an absorbing state, it resides permanently in this failed state. The Laplace transform of the sum of probabilities of failed states (particularly absorbing states) is therefore infinite:

$$\lim_{s\downarrow0} \sum_{j,working} P_j(s) = \lim_{s\downarrow0} R(s) = 0 \tag{8.58}$$

$$\lim_{s\downarrow0} \sum_{j,failed} P_j(s) \rightarrow \infty \tag{8.59}$$

For that reason, the transformed system reliability $R(s)$ cannot be obtained from (the transformed) 1 minus the transformed failure probabilities. It is necessary to calculate the probabilities of the working states.

As an example, consider the situation of three components that each permanently fail according to an exponential distribution with failure rate λ. The questions are whether it makes a difference employing the three components in parallel as a double-redundant system or keeping one or two components as a spare unit to replace a failing component immediately.

Let it be justified to assume that the time to replace is neglected. And also, if a component is kept as a spare unit it is not in service and cannot fail during that time either in this model.

Table 8.6 shows the analysis steps as listed in Table 8.5, except that step 7 is replaced by taking the limit $s \rightarrow 0$ rather than using the inverse Laplace transform. The three cases are:

A. All three components are simultaneously employed from the beginning and provide the maximum redundancy initially.
B. Just two components are simultaneously employed from the beginning and one is kept as a spare to be employed if either of the other two fails.
C. Only one component is employed from the beginning. There is no redundancy, but there is a maximum stock of spares.

No repair is possible. The question is which of the three systems will provide the longest expected system lifetime. What is the best strategy?

After step 7 in Table 8.6 it becomes apparent that keeping stock is more effective. The third configuration provides a 63.6% longer expected system lifetime than operating all components in parallel. However, the redundancy is completely lifted in the third option and applying corrective maintenance (Section 1.3.1) must be acceptable indeed.

Keeping the spare parts out of operation does have a significant contribution to the expected lifetime of the system, but it must be kept in mind that in the absence of redundancy, the system is down after each failure. In practice, the question will be whether the consequences of the interruption in system operation and the required time for exchanging a component are negligible indeed and do not impose an unacceptable risk that is presently left out of the equation.

The result of one component in operation and corrective maintenance after failure bears a resemblance to non-rechargeable batteries that are replaced after exhaustion. Using three batteries subsequently then also leads to a total expected lifetime of three times the lifetime of a single battery. This battery example does not fully apply, because the exhaustion of batteries will not follow an exponential distribution usually. Still, keeping spares in combination with fast repair can be a good strategy in maintenance as an alternative to redundancy.

Table 8.6 Comparison of three-component systems with redundant and/or spare components. At step 6, $P_3(s)$ is not elaborated as it approaches infinity. In all cases, system reliability $R_S(s)$ is the sum of $P_0(s)$, $P_1(s)$ and $P_2(s)$, from which the expected system lifetime θ_S is derived and expressed in terms of the expected component lifetime $\theta = 1/\lambda$. Initially, the system works fully (i.e. $P_0(t=0) = 1$).

Step	3 parallel	2 parallel, 1 spare	1 employed, 2 spares
1			
2	S_0: 3 OK, system OK S_1: 2 OK, system OK S_2: 1 OK, system OK S_3: 0 OK, system fails	S_0: 2 OK, 1 spare, system OK S_1: 1 OK, 1 spare, system OK S_2: 1 OK, 0 spare, system OK S_3: 0 OK, 0 spare, system fails	S_0: 1 OK, 2 spare, system OK S_1: 1 OK, 1 spare, system OK S_2: 1 OK, 0 spare, system OK S_3: 0 OK, 0 spare, system fails
3			
4	$\dot{P}_0(t) = -3\lambda P_0(t)$ $\dot{P}_1(t) = 3\lambda P_0(t) - 2\lambda P_1(t)$ $\dot{P}_2(t) = 2\lambda P_1(t) - \lambda P_2(t)$ $\dot{P}_3(t) = \lambda P_2(t)$	$\dot{P}_0(t) = -2\lambda P_0(t)$ $\dot{P}_1(t) = 2\lambda P_0(t) - 2\lambda P_1(t)$ $\dot{P}_2(t) = 2\lambda P_1(t) - \lambda P_2(t)$ $\dot{P}_3(t) = \lambda P_2(t)$	$\dot{P}_0(t) = -\lambda P_0(t)$ $\dot{P}_1(t) = \lambda P_0(t) - \lambda P_1(t)$ $\dot{P}_2(t) = \lambda P_1(t) - \lambda P_2(t)$ $\dot{P}_3(t) = \lambda P_2(t)$

5

$sP_0(s) - 1 = -3\lambda P_0(s)$
$sP_1(s) = 3\lambda P_0(s) - 2\lambda P_1(s)$
$sP_2(s) = 2\lambda P_1(s) - \lambda P_2(s)$
$sP_3(s) = \lambda P_2(s)$

$sP_0(s) - 1 = -2\lambda P_0(s)$
$sP_1(s) = 2\lambda P_0(s) - 2\lambda P_1(s)$
$sP_2(s) = 2\lambda P_1(s) - \lambda P_2(s)$
$sP_3(s) = \lambda P_2(s)$

$sP_0(s) - 1 = -\lambda P_0(s)$
$sP_1(s) = \lambda P_0(s) - \lambda P_1(s)$
$sP_2(s) = \lambda P_1(s) - \lambda P_2(s)$
$sP_3(s) = \lambda P_2(s)$

6

$P_0(s) = \dfrac{1}{s+3\lambda}$

$P_1(s) = \dfrac{3\lambda}{s+2\lambda}\,\dfrac{1}{s+3\lambda}$

$P_2(s) = \dfrac{2\lambda}{s+\lambda}\,\dfrac{3\lambda}{s+2\lambda}\,\dfrac{1}{s+3\lambda}$

$P_0(s) = \dfrac{1}{s+2\lambda}$

$P_1(s) = \dfrac{2\lambda}{s+2\lambda}\,\dfrac{1}{s+2\lambda}$

$P_2(s) = \dfrac{2\lambda}{s+\lambda}\,\dfrac{2\lambda}{s+2\lambda}\,\dfrac{1}{s+2\lambda}$

$P_0(s) = \dfrac{1}{s+\lambda}$

$P_1(s) = \dfrac{\lambda}{s+\lambda}\,\dfrac{1}{s+\lambda}$

$P_2(s) = \dfrac{\lambda}{s+\lambda}\,\dfrac{\lambda}{s+\lambda}\,\dfrac{1}{s+\lambda}$

7

$\lim_{s\downarrow 0} P_0(s) + P_1(s) + P_2(s) = \dfrac{11}{6\lambda}$
$\Rightarrow \theta_5 = \dfrac{11}{6}\,\theta$

$\lim_{s\downarrow 0} P_0(s) + P_1(s) + P_2(s) = \dfrac{2}{\lambda}$
$\Rightarrow \theta_5 = 2\cdot\theta$

$\lim_{s\downarrow 0} P_0(s) + P_1(s) + P_2(s) = \dfrac{3}{\lambda}$
$\Rightarrow \theta_5 = 3\cdot\theta$

8.6 Mean Lifetimes MTTFF and MTBF

If a system has no absorbing state (i.e. no permanent final state) then it will come back into operation due to repair after failure. Section 8.5 dealt with such systems and it was found that the sum of the Laplace-transformed probabilities of working states was finite (see Eq. (8.58)) and equal to the expected lifetime θ. With a repairable system, the sums of the Laplace-transformed probabilities of both the working and the failed states will be infinite. For infinite time the system will balance between the working and failed states and the times to reside in either type of state will be infinite.

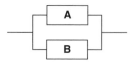

Figure 8.7 Block diagram of a redundant connection of two parallel circuits A and B (repeated).

Still two types of expected lifetimes exist that are finite and can be determined with an adapted Markov–Laplace method: the MTTFF and the MTBF. They are found in the same way as expected lifetimes in the case of absorbing states after adapting the state diagram of the repairable system.

This is illustrated by studying the redundant cable connection again (Figure 8.7). That system consists of two parallel cable circuits A and B, which can be repaired after failure. Both the MTTFF and the MTBF are illustrated in Figure 8.8.

The MTTFF is the mean time that passes between commissioning and the first failure. The initial state is S_0 and therefore $P_0(t = 0) = 1$ and all other probabilities $P_j(t = 0) = 0$. The simple adaptation to the model is to make state S_2 absorbing, as shown on the left in Figure 8.9. The corresponding state diagram is shown in Figure 8.10. The expected or mean time of that system produces the MTTFF.

Although the differential equations can be derived from Figure 8.10 using Section 8.3.3, it is illustrative to regard the transition matrix of Eq. (8.39). Absorbing states S_j have transition rates $h_{jk} = 0$ from S_j to other states S_k (with $k \neq j$). The rates h_{jk} are found on row r_j of the transition matrix. As the rate h_{jj} is the negative of all the other rates on that row and all $h_{jk} = 0$, also $h_{jj} = 0$. The adaptation of Figure 8.10 leads to a simple replacement in the transition matrix by setting all h_{2k} to 0 in the third row:

$$\vec{M}_{original} = \begin{pmatrix} -2\lambda & 2\lambda & 0 \\ \mu & -(\lambda + \mu) & \lambda \\ 0 & \mu & -\mu \end{pmatrix} \rightarrow \vec{M}_{S_2\ absorbing} = \begin{pmatrix} -2\lambda & 2\lambda & 0 \\ \mu & -(\lambda + \mu) & \lambda \\ 0 & 0 & 0 \end{pmatrix} \quad (8.60)$$

It is very easy to effectuate absorbing states in the transition matrix and they are conveniently recognizable by the rows filled with zeros. From the new matrix the differential equations can be derived following Eq. (8.37). However, the original set of differential

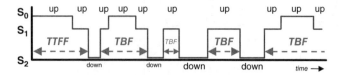

Figure 8.8 Example of up- and down-times of the system in Figure 8.7. S_0 and S_1 are up-states and S_2 is the down-state. TTFF is time to first failure and TBF is time between failures. The means MTTFF and MTBF can be found with the Markov–Laplace method.

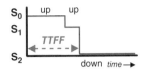 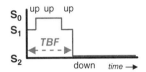

Figure 8.9 The MTTFF and the MTBF can be determined by turning state S_2 into an absorbing state and taking S_0 (respectively S_1) as the initial state. On the left is the setup to determine MTTFF and on the right is the setup to determine MTBF.

Figure 8.10 Adaptation of the original repairable system turns state S_2 into an absorbing state, after which the Markov–Laplace method can produce the MTTFF and the MTBF with the appropriate initial state.

equations can also be conveniently adapted by eliminating all terms with $P_j(t)$ if state S_j is taken as absorbing. The new set of differential equations to solve the MTTFF and the MTBF can be achieved both from the new matrix as well as from eliminating all terms with $P_2(t)$ from the set of equations in Eq. (8.41). The result is:

$$\dot{P}_0(t) = -2\lambda \cdot P_0(t) + \mu \cdot P_1(t)$$
$$\dot{P}_1(t) = 2\lambda \cdot P_0(t) - (\lambda + \mu) \cdot P_1(t)$$
$$\dot{P}_2(t) = \lambda \cdot P_1(t) \tag{8.61}$$

The Laplace transform produces, in case of the MTTFF (i.e. $P_0(t = 0) = 1$):

$$sP_0(s) - P_0(t = 0) = sP_0(s) - 1 = -2\lambda \cdot P_0(s) + \mu \cdot P_1(s)$$
$$sP_1(s) - P_1(t = 0) = sP_1(s) - 0 = 2\lambda \cdot P_0(s) - (\lambda + \mu) \cdot P_1(s)$$
$$sP_2(s) - P_2(t = 0) = sP_2(s) - 0 = \lambda \cdot P_1(s) \tag{8.62}$$

The transformed probabilities of the two working states are S_0 and S_1:

$$P_0(s) = \cfrac{1}{s + 2\lambda - \cfrac{2\lambda\mu}{s + \lambda + \mu}} \tag{8.63}$$

$$P_1(s) = \frac{2\lambda}{s + \lambda + \mu} P_0(s) \tag{8.64}$$

Expected lifetimes follow from s approaching 0 (Section 8.5). The MTTFF is achieved as the sum of the two expected times:

$$\text{MTTFF} = \lim_{s\downarrow0}(P_0(s) + P_1(s)) = \frac{\lambda + \mu}{2\lambda^2} + \frac{2\lambda}{\lambda + \mu}\frac{\lambda + \mu}{2\lambda^2} = \frac{3\lambda + \mu}{2\lambda^2} \tag{8.65}$$

The MTBF is achieved in a similar way. The system, state diagram and transition matrix are the same as with the procedure to determine the MTTFF. The difference between the MTBF and the MTTFF is the initial state. At $t = 0$ the system of Figure 8.7 starts at state S_0 and it takes at least two steps to reach the down-state S_2. When determining the MTTFF the probability $P_0(t = 0) = 1$.

For the MTBF the starting point is different. If the system is repaired after a failure, then the first up-state is only one step from falling back into the down-state (so S_1 in the present example). If further repair can also be carried out before the next failure occurs, then the system can get back into a new state. In the case of the system of Figure 8.7, the initial state after repair from the down-state is then S_1 instead of S_0. Therefore, with determining the MTBF, the probability $P_1(t = 0) = 1$. The Laplace transform yields a different set of equations:

$$sP_0(s) - P_0(t = 0) = sP_0(s) - 0 = -2\lambda \cdot P_0(s) + \mu \cdot P_1(s)$$
$$sP_1(s) - P_1(t = 0) = sP_1(s) - 1 = 2\lambda \cdot P_0(s) - (\lambda + \mu) \cdot P_1(s)$$
$$sP_2(s) - P_2(t = 0) = sP_2(s) - 0 = \lambda \cdot P_1(s) \tag{8.66}$$

The solutions of the transformed probabilities of the working states become:

$$P_0(s) = \frac{\mu}{s + 2\lambda} P_1(s) \tag{8.67}$$

$$P_1(s) = \frac{1}{s + \lambda + \mu - \frac{2\lambda\mu}{s+2\lambda}} \tag{8.68}$$

Similarly, the MTBF follows as the sum of the expected time from s approaching 0:

$$\text{MTBF} = \lim_{s\downarrow 0}(P_0(s) + P_1(s)) = \frac{\mu}{2\lambda^2} + \frac{1}{\lambda} = \frac{2\lambda + \mu}{2\lambda^2} \tag{8.69}$$

The expected lifetimes MTTFF and MTBF can be found without the inverse Laplace transform. The MTTFF will be larger than or equal to the MTBF, because the time to first failure generally starts from a state that is more than one step away from the down-state, whereas the time between failures starts from a state that is only one step away from the down-state. Also, in this case the numerator of the MTTFF in Eq. (8.65) appears to be λ larger than the numerator of the MTBF in Eq. (8.69).

8.7 Availability and Steady-State Situations

With repairable systems an equilibrium is established between failure and repair in the long run. In this section a convenient method is discussed to determine the availability in the long run.

The average time of functioning and of failure will both become infinite, so determining the expected lifetimes as in Sections 8.5 and 8.6 is not an option. However, it is possible to determine the ratio of up-time and total time (sum of up-time and down-time), which is the availability $A(t)$. This ratio is equal to the sum of probabilities of up-states. In the long run the average $A(t)$ will approach a constant value A_∞.

The availability $A(t)$ at $t = 0$ is equal to the probability of the initial state. Usually this means $P_0(t = 0) = 1$. In practice, commissioning tests may be carried out and operation starts if such tests were successful. In practice, there may be concerns about the full compliance of a new, even warranted product. In such a case, $P_0(t = 0) < 1$ and therefore the sum of probabilities is > 0. This situation can be dealt with through the Markov–Laplace method (Section 8.4), given the $P_j(t = 0)$ are known for all j.

Again, the case of the redundant cable circuit consisting of two parallel repairable cable circuits is used as an example to illustrate the method. As a start, the initial state is taken as S_0 and so $A(t = 0) = P_0(t = 0) = 1$. Next we study how A_∞ can be determined.

In the long run the system will develop from its initial state S_0 into the situation where the system will find an equilibrium between the probabilities $P_j(t)$. These probabilities $P_j(t)$ approach constants π_j and their derivatives $\dot{P}_j(t)$ become 0. This is the so-called 'steady state' situation in which the availability $A(t)$ also approaches the constant value A_∞.

The calculation of A_∞ follows the Markov–Laplace method up to step 4 in Table 8.5 (i.e. deriving the differential equations; cf. Eq. (8.35)). However, now the derivatives are set to 0 and the probabilities $P_j(t)$ are replaced by their asymptotic average π_j. For each j:

$$0 = -\pi_j(t) \cdot \sum_{k \neq j} h_{jk} + \sum_{k \neq j} \pi_k(t) \cdot h_{kj} \tag{8.70}$$

This set of equations is not independent and the solution requires one more condition. This condition is the fact that the sum of all probabilities must be equal to 1 at all times, which also applies to the π_j:

$$\sum_j \pi_j = 1 \tag{8.71}$$

The steady-state probabilities π_j can be solved from Eqs (8.70) and (8.71), after which follows the sum of all probabilities of working states (cf. Eq. (8.49)):

$$A_\infty = \sum_{j,working} \pi_j = 1 - \sum_{j,failed} \pi_j \tag{8.72}$$

In the case of the redundant cable circuit consisting of two parallel repairable cable circuits (see Figure 8.7), the following differential equations apply:

$$\dot{P}_0(t) = -P_0(t) \cdot 2\lambda + P_1(t) \cdot \mu$$
$$\dot{P}_1(t) = +P_0(t) \cdot 2\lambda - P_1(t) \cdot (\lambda + \mu) + P_2(t) \cdot \mu$$
$$\dot{P}_2(t) = +P_1(t) \cdot \lambda - P_2(t) \cdot \mu \tag{8.73}$$

In the steady-state situation with inclusion of Eq. (8.71), this becomes:

$$0 = -2\lambda \cdot \pi_0 + \mu \cdot \pi_1$$
$$0 = 2\lambda \cdot \pi_0 - (\lambda + \mu) \cdot \pi_1 + \mu \cdot \pi_2$$
$$0 = \lambda \cdot \pi_1 - \mu \cdot \pi_2$$
$$1 = \pi_0 + \pi_1 + \pi_2 \tag{8.74}$$

One of the first three equations could be left out, because the set is not independent. The fourth equation is the third independent equation that is required in order to solve this set. Elaboration yields:

$$\left. \begin{array}{l} \pi_0 = \dfrac{\mu}{2\lambda} \cdot \pi_1 \\[2mm] \pi_2 = \dfrac{\lambda}{\mu} \pi_1 \\[2mm] 1 = \left(\dfrac{\mu}{2\lambda} + 1 + \dfrac{\lambda}{\mu} \right) \cdot \pi_1 \end{array} \right\} \Rightarrow \begin{array}{l} \pi_1 = \dfrac{2\lambda\mu}{\mu^2 + 2\lambda\mu + 2\lambda^2} \\[2mm] \pi_0 = \dfrac{\mu^2}{\mu^2 + 2\lambda\mu + 2\lambda^2} \\[2mm] \pi_2 = \dfrac{2\lambda^2}{\mu^2 + 2\lambda\mu + 2\lambda^2} \end{array} \tag{8.75}$$

The asymptotic availability A_∞ now follows from Eq. (8.72) with S_0 and S_1 as working states:

$$A_\infty = \pi_0 + \pi_1 = 1 - \pi_2$$

$$\Leftrightarrow A_\infty = \frac{\mu^2 + 2\lambda\mu}{\mu^2 + 2\lambda\mu + 2\lambda^2} \tag{8.76}$$

It is noteworthy that determination of A_∞ does not require Laplace transforms at all.

If the repair rate is much larger than the failure rate (i.e. $\mu \gg \lambda$) then A_∞ approaches 1 (Eq. (8.76)). This is one of the strategies to increase availability: fast repair (i.e. short repair times) compared to the MTTFF and the MTBF significantly improves the availability. Where redundancy is limited (possibly because of costs and/or mass or volume of critical components), the strategy of a stock of spare parts combined with fast repair can provide relieve.

8.8 Summary

This chapter discussed the reliability and availability of systems, particularly taking repair and spare parts into account, addressing: states of components and systems (Section 8.1); states and transition rates for a one-component system (Section 8.2); state probabilities via Markov chains (Section 8.3); Markov–Laplace method for reliability and availability (Section 8.4); lifetime with absorbing states and spare parts (Section 8.5); mean lifetimes MTTFF and MTBF (Section 8.6); availability and steady-state situations (Section 8.7).

As for component and system states (Section 8.1), rather than characterizing a system by reliabilities as in Chapter 7, systems are analysed by their states (i.e. being failed or non-failed). In Section 8.2 this is elaborated firstly for a single component as a non-repairable (Section 8.2.1) and next as a repairable (Section 8.2.2) system. Section 8.3 discusses a system composed of two components where various transitions become possible. Grouping of system states is introduced to reduce calculations. The transition matrix is explained as an alternative to working with separate equations.

The Markov–Laplace method (Section 8.4) combines Markov chains and Laplace transforms and an overview of the successive steps is listed. Absorbing states (Section 8.5) are states from which repair is not possible. A method to determine the reliability and expected lifetime of a system with both repairable and absorbing states is presented. Also, the comparison of spare parts versus redundancy is discussed, which leads to the conclusion that spare parts with fast repair can be a good alternative to redundancy, provided that uncertainties associated with unforeseen failure and sudden need for repair are acceptable.

As for mean times to failures (Section 8.6), the MTTFF and the MTBF are distinguished. The MTTFF runs from the initial state while the MTBF starts from the first state that is reached by repair after failure. When failure requires at least two steps from the initial state, the MTTFF is generally longer than the MTBF. Finally, Section 8.7 discusses behaviour in the long run where an equilibrium is reached between working and failed states. The probabilities of the states become constant and the states are called steady states. The analysis can be simplified in that case.

Questions

Note: A table of Laplace transforms can be found in Appendix B.

8.1 With reference to component and system states, transition matrix, differential equations and availability:

A connection consists of two DC circuits. The power supply is sufficient as long as at least one circuit is operational. In case of failure(s) one circuit can be repaired at a time. The mean repair time is one circuit per week (assume an exponential distribution). As for reliability and availability, only the circuits are considered in this exercise.

   a. Describe the states, the transition matrix and the differential equations for the probabilities of operation of the system.

   b. What is the long-term availability of the connection?

8.2 With reference to parallel systems, availability A_∞, MTTFF and MTBF:

Consider a parallel system of two components with each an exponentially distributed lifetime with failure rate λ. When a component fails, the surviving component endures a heavier stress. The failure rate λ_2 during this enhanced stress is $M \cdot \lambda$ with $M > 1$. The system remains operational though. If the second component fails before the other one is repaired, the system goes down but remains repairable. There is only one repair crew working with a repair rate μ and therefore no more than one repair is carried out at a time. Let $\lambda = 8 \times 10^{-3}$ failures/hour and $\mu = 5 \times 10^{-2}$ repairs/hour.

   a. Give an inventory of the system states and produce the transition matrix.

   b. Calculate the mean lifetime until first failure (MTTFF) if $M = 1.7$.

   c. Calculate the mean lifetime between failures (MTBF) if $M = 1.7$.

   d. Determine the availability A_∞ of this system.

   e. There is an option to install cheaper components. An advantage is that the repair rate then doubles (i.e. to 10×10^{-2} repairs/hour). A disadvantage is that if a single component carries the full load then the failure rate increases more, namely with $M = 3$. Show whether the implementation of the cheaper components is an improvement with respect to: availability A_∞, MTTFF and MTBF.

8.3 With reference to availability, repair, non-repair, spare parts, mean time to failure and reliability:

Consider a system consisting of two components that are parallel in operation. The system works as long as at least one component works. Additionally, one spare part is in stock. The components feature a failure rate λ if they are operational. A repair crew is available that can repair with a repair rate μ. The moment a component fails, it is immediately replaced by the spare component (so the time for replacement can be ignored) and the repair crew commences the work on the failed component.

   a. Make an inventory of the system states and compose the transition matrix.

   b. Determine the availability A_∞ of this system.

   c. Consider now the impact of the repair crew itself being unable to work from the start. As a consequence, no repairs are possible but replacement by a spare is possible. The question is how the reliability of the system decays as a function of time. The initial state is again a parallel system of two working components plus one spare component. Compose the transition matrix for this new situation.

d. Determine the mean time to failure for the system if no repair is possible.

e. Determine the reliability $R(t)$ as a function of time.

8.4 With reference to system states, repair and MTTFF:

Consider a system of three identical components. The configuration consists of one unit in series with the other two units in parallel, as shown in Figure 8.11. All components have a lifetime that is distributed according to the same exponential distribution with failure rate λ. If the system is down, none of the surviving components can fail. There is one repair crew with a repair rate μ. If component 1 fails while one of the other components has already failed, then the component that is repaired first will take the position of component 1 (and the second repaired component takes another position in order to complete the system).

a. Give the system states and compose a state diagram (*Tip*: reduce the number of states).

b. Identify the system state(s) where the system is down.

c. Compose the transition matrix and give the corresponding differential equations.

d. Determine the availability of the system.

e. Determine the mean time to first failure (MTTFF).

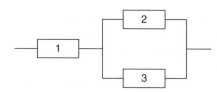

Figure 8.11 Block diagram for question 4.

9

Application to Asset and Incident Management

In this chapter, various cases are presented that illustrate the application of the previous chapters. Most of these examples are inspired by actual cases that occurred in practice. The applications are grouped in four categories:

1. Maintenance styles and grid design in Section 9.1.
2. Health index in Section 9.2.
3. Testing and quality assurance in Section 9.3.
4. Incident management dealing with typical fault situations in Section 9.4.

9.1 Maintenance Styles

In the following subsections, various maintenance cases are discussed with reference to the theory presented in Chapters 1–8.

9.1.1 Period-Based Maintenance Optimization for Lowest Costs

PBM (see Section 1.3.2) typically follows directives on maintenance actions irrespective of the condition of the assets involved.

The upfront investment costs are followed by periodic costs during operational life. With PBM, such costs are made periodically. The idea is that such costs lower the hazard rate and therefore prolong asset life, which reduces asset depreciation. Maintenance costs are weighed against replacement costs. The case of Section 9.1.1.1 describes how costs can be minimized by optimizing the maintenance frequency. The one and only criterion is cost in this approach. Usually, other considerations are also involved in planning and decision-making, as will be discussed in later cases, but here a method for cost minimizing is presented as a start. After the case description, an overview of references to book sections with introductory material and theoretical background is provided. Next, the case is elaborated and finally some remarks on the case are discussed.

9.1.1.1 Case Description

PBM is a maintenance style where, at fixed periods, the considered asset is brought back into a new state. This means that the hazard rate is reset to 0 (see Figure 9.1). PBM is only useful for wear-out processes (cf. Examples 4.1–4.3 in Section 4.2.4.1.).

Reliability Analysis for Asset Management of Electric Power Grids, First Edition. Robert Ross.
© 2019 John Wiley & Sons Ltd. Published 2019 by John Wiley & Sons Ltd.
Companion website: www.wiley.com/go/ross/reliabilityanalysis

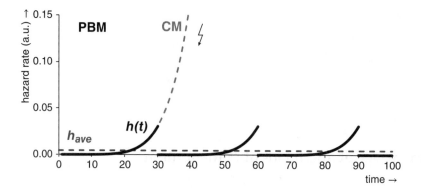

Figure 9.1 If the hazard rate $h(t)$ is repeatedly reset due to maintenance (here PBM), an average hazard rate h_{ave} can be determined that approximates the original hazard rate $h(t)$. This constant h_{ave} defines an exponential distribution that approximates the original distribution. If the hazard rate is not reset, the object runs to failure as in CM.

The period can be in terms of calendar time, but also in terms of operational hours, switching operations, number of lightning strikes, and so on. In the present example the ageing and maintenance period will be in terms of time T.

Replacement comes with investments (capital expenses, CAPEX) and servicing comes with periodic operational costs (operational expenses, OPEX). Over the total lifetime, the costs are called the total cost of ownership (total expenses, TOTEX) or lifecycle cost (LCC). The question is how to optimize the servicing period T for lowest total cost.

9.1.1.2 References to Introductory Material
The following sections provide introductory material:

- Section 1.3.2: on period-based maintenance.
- Section 2.4.5: on the hazard rate and the three types of failure behaviour: child mortality, random failure and wear-out failure.
- Section 2.5.3: on the alternative meaning of child mortality with early failures.
- Section 4.2.4.1: on the exponential distribution and average hazard rate.
- Section 4.2.1.1: on the Weibull-2 distribution and failure behaviour.
- Section 4.2.1.2: on the Weibull-2 mean.

9.1.1.3 PBM Cost Optimization Analysis
The approach for finding the period T with the lowest total cost is:

1. Find an average hazard rate as a function of the period T irrespective of the distribution.
2. Define the relevant cost rates and relate these to period T.
3. Minimize the total cost rate in general by optimization of T.
4. Illustrate the procedure for the exponential and Weibull-2 distributions.

Assume an asset has a hazard rate $h(t)$ as a function of time t. After a period $t = T$, the asset is serviced and it becomes as good as new. The servicing is repeated after every period T and every time the hazard rate h is reset to its initial value $h(0)$, as shown in Figure 9.1 (copied from Figure 4.10 in Section 4.2.4.1). As a consequence, the hazard rate cycles about an average hazard rate h_{ave}. The exact nature of $h(t)$ itself does not matter, and therefore this holds for any distribution within reasonable boundaries (it would not work if, for example, the cycling period was too long to prevent run to failure). The resulting constant average h_{ave} is, however, characteristic for an exponential distribution, even if the underlying distribution is lognormal, Weibull, and so on.

Section 4.2.4.1 discusses how the average h_{ave} is obtained. As a result, the average hazard rate h_{ave} can be calculated over the first period T as (see Eq. (4.38)):

$$h_{ave}(T) = -\frac{-1}{T} \cdot \ln(R(t))|_0^T = \frac{-\ln(R(T))}{T} \tag{9.1}$$

As mentioned before, it only makes sense to reset the hazard rate in case of wear-out failure mechanisms, because these have an increasing hazard rate and therefore a reset is an improvement. With random failure and child mortality processes, servicing efforts do not improve and may even deteriorate the situation (except when child mortality is in fact a fast wear-out process).

A refinement can be made to include the effect of the operational time $t - kT$ at the end by adjusting Eq. (4.38), but whether or not this is desirable is left to the reader for individual cases. Further elaboration will be based on the average hazard rate h_{ave} as a function of T, as in Eq. (9.1) above.

The hazard rate h of an exponential distribution is the inverse of the mean lifetime θ. Therefore:

$$h_{ave}(T) = \frac{1}{\theta(T)} \tag{9.2}$$

The next step is to define the cost rates. There are two types of costs, namely investment costs or CAPEX K_c on the one hand and maintenance costs or OPEX K_m on the other hand. The related cost rates concern: the capital cost rate C_c, which is taken as the cost K_c per lifetime (so $h_{ave} \cdot K_c$) and the maintenance cost rate C_m, which is taken as the cost K_m per period T. Assuming K_m constant, the total cost rate C_t is the sum of C_m and C_c:

$$C_t = C_m + C_c = \frac{K_m}{T} - K_c \cdot h_{ave}(T) = \frac{1}{T} \cdot (K_m - K_c \cdot \ln R(T)) \tag{9.3}$$

The total cost can be optimized for T to find the PBM at the lowest cost rate:

$$0 = \frac{\partial C_t}{\partial T} = \frac{-1}{T^2} \cdot (K_m - K_c \cdot \ln R(T)) + \frac{-1}{T} \cdot K_c \cdot \frac{d \ln R(T)}{dT} \tag{9.4}$$

The underlying distribution has to be substituted in Eq. (9.4), after which the solution gives the optimum T with lowest total cost rate.

This is illustrated for the exponential and the Weibull distributions. As for the exponential distribution with $R(T) = \exp(-T/\theta)$, Eq. (9.4) becomes:

$$0 = \frac{\partial C_t}{\partial T} = -\frac{K_m}{T^2} \Rightarrow |T| \to \infty \tag{9.5}$$

For the exponential distribution, the optimum cycle period T is infinite. This makes sense; the exponential distribution has a constant hazard rate h. Resetting the hazard rate to $h(0)$ is not an improvement. The average hazard rate $h_{ave} = h$. This means that the efforts and maintenance expenses do not improve the asset condition, while maintenance costs depend on the period T between servicing actions. Naturally, optimization leads to infinitely long periods T between servicing to minimize the (useless) costs.

In case of a Weibull-2 distribution (cf. Section 4.2.1.1), the optimization of T through Eq. (9.4) leads to:

$$T_{opt,Weibull} = \alpha \cdot \left(\frac{K_m}{(\beta - 1) \cdot K_c} \right)^{1/\beta} = \alpha \cdot \left(\frac{1}{(\beta - 1)} \right)^{1/\beta} \cdot \left(\frac{K_m}{K_c} \right)^{1/\beta} \tag{9.6}$$

In case β approaches 1, the Weibull distribution corresponds to the exponential distribution and again T becomes infinite. In case of $\beta < 1$, there are solutions for Eq. (9.4) such as certain values (like the inverses of even numbers). However, these are not local minima, so the total cost rate has not been minimized. (Note that it is good practice to not only determine the points where the derivative becomes 0, but also to check whether those points are indeed minima.) For child diseases where the hazard rate decays, maintenance adversely restores the hazard rate at its highest value, namely $h(0)$. For $\beta > 1$, the Weibull-2 distribution concerns a wear-out process and the solution in Eq. (9.4) is the true cost rate minimum.

This procedure can be carried out once a model for the hazard rate is available (i.e. $h(t)$ is known).

Example 9.1 *Ageing, Maintenance and Optimized Cycle Period* **T** Assume certain equipment contains a component that is dominating the failure behaviour of the equipment. The failure distribution is a Weibull-2 distribution with parameters $\alpha = 230$ days and $\beta = 2.5$. As $\beta > 1$, this is a wear-out process. The replacement cost is €5000. If the component is serviced, the cost is €331.46. With each servicing there are maintenance costs, but the failure rate decreases. As a result, the depreciation is spread over a longer period.

The optimum servicing cycle T_{opt} can be determined with Eq. (9.6) based on the costs and the Weibull parameters. The average hazard rate h_{ave} follows from Eq. (9.1). The maximum hazard rate can be found as $h_{max} = h(T)$. The characteristic lifetime θ can be estimated as $1/h_{ave}$. The cost rates are found as in Eq. (9.3). The cost rates can also be determined by a chosen, not optimized cycle period T. For example, $T_{set} = 200$ days. The result is shown in Table 9.1 and Figure 9.2.

9.1.1.4 Remarks
It may be noted that the cost rate optimization method does not involve any risk appreciation. In a more holistic approach, this would be an important driver in decision-making.

Table 9.1 Comparison of CM, PBM with optimized T and PBM with chosen example of $T = 200$ days.

Servicing	T[days]	h_{ave} [days^{-1}]	h_{max}[days^{-1}]	θ[days]	C_c[€/day]	C_m[€/day]	C_t[€/day]
None (i.e. CM)	∞	∞	∞	204	24.50	0.00	24.50
Optimum Eq. (9.6)	102.5	0.001 3	0.003 2	773	6.47	9.70	16.17
Chosen example	200.0	0.003 2	0.008 8	284	17.63	4.97	22.60

Note: θ is determined for CM from the Weibull distribution with the gamma function (see Section 4.2.1.2, Eq. (4.28)). With servicing, θ is determined from the resulting exponential distribution with h_{ave}.

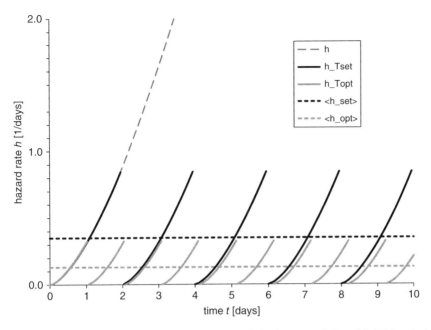

Figure 9.2 Comparison of hazard rate [12] in case of: CM (i.e. run to failure, 'h'); PBM optimized T for lowest cost rate ('h_Topt'); chosen example of $T = 200$ days ('h_Tset'). Also, the average hazard rates are drawn in case of: PBM optimized T for lowest cost rate ('<h_opt>'); chosen example of $T = 200$ days ('<h_set>'). For PBM with optimized T, not only the average hazard rate is lower than that of the chosen period $T = 200$ days, but also the cost rate is lower (see Table 9.1).

However, for comparison and optimization of maintenance, the optimization of T may nevertheless be a convenient method.

If desired, the case can be extended to other lifetime costs or period-based costs like interest and inflation, but the present case aims at illustrating the methodology of optimizing PBM. The optimization equations may become more complicated with other factors involved, but the principle remains the same: find an average hazard rate and use it to find expressions for balancing the cost rates.

In practice, PBM may not be able to fully reset the hazard rate. There may be another irreversible ageing process active or perfect servicing may just not be possible. As shown in Figure 9.3, despite the servicing the asset may age and the average hazard rate is no longer a constant, but is pushed up by the hazard rate of a process that apparently cannot be solved by servicing [12].

9.1.2 Corrective versus Period-Based Replacement and Redundancy

Corrective maintenance (CM, see Section 1.3.1) means that no action is taken unless a failure or an unacceptable shortcoming occurs. With this style, assets run to fail, after which the grid is corrected by repair or replacement of the assets. An advantage is that the full lifetime has been used; disadvantages are that assets may fail without prior notice, collateral damage as well as safety issues may occur in the vicinity of the assets and unplanned mitigation may cause emergency situations. The damage can be physical

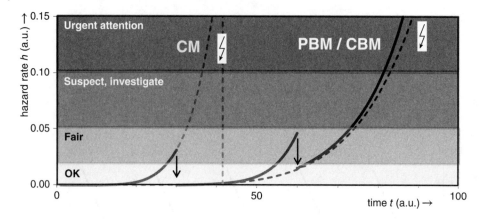

Figure 9.3 The effect of an extra ageing process that comes on top of a maintainable issue (cf. [12]). The grey tones reflect various HI indications (see Section 1.3.3.4). It is an example of servicing that is partly successful to reset the hazard rate down to the extra hazard rate, which may be the ultimate reason for failure of the asset. In practice, the hazard rate scale may be rather logarithmic instead of the used arbitrary units (a.u.).

damage to other assets, but can also concern intangible matters such as reputation, non-compliance with regulations, and so on.

Apart from the risks on site, the integrity of the grid performance is also jeopardized by failures. A strategy can be to apply redundancy by parallel systems, as discussed in Section 7.4 and Chapter 8.

As an alternative, PBM (see Section 1.3.2) may be employed to prevent failures by timely replacement. Advantages of successful PBM are that the number of surprise failures can be drastically reduced and that replacement actions can be planned well in advance; a disadvantage is that service life is sacrificed and therefore depreciation of assets is higher than with CM.

9.1.2.1 Case Description

CM and PBM have their characteristic advantages and disadvantages. This means that under certain conditions CM is favoured and under other conditions PBM.

CM is taken as the reference in this case. The two strategies of PBM and redundancy come with a price, but are able to overcome some disadvantages of CM.

The question is: how does CM compare to PBM and redundancy? Furthermore, how can we illustrate when one would be favoured over the others?

9.1.2.2 References to Introductory Material

The following sections provide introductory material:

- Section 1.3.1: on corrective maintenance.
- Section 1.3.2: on period-based maintenance.
- Section 2.6: on ageing dose, power law and accelerated and multi-stress ageing.
- Chapter 5: on plotting.
- Chapter 6: on parameter estimation.
- Section 7.4: on parallel systems and redundancy.
- Sections 8.3–8.7: on various examples for parallel and repairable systems.

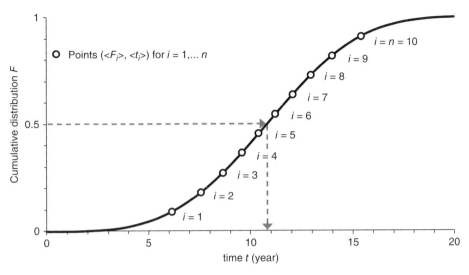

Figure 9.4 Example of a batch of $n = 10$ assets which are ranked by their sequence of time to failure. The distribution in this example is Weibull-2 with parameters $\alpha = 12$ and $\beta = 3.5$. The mean and the median (indicated in the figure) are each about 10.8 years.

9.1.2.3   Analysis of Corrective versus Period-Based Replacement and Redundancy

The approach for comparing CM, redundancy and PBM is:

1. Define a CM reference for the discussion.
2. Discuss the background and effect of redundancy.
3. Discuss the PBM approach and compare the CM and PBM in terms of referenced service life.

First, an example of the case is defined. Figure 9.4 shows the distribution of a batch of assets. It is assumed that the distribution is known from earlier experience and in this case it is a Weibull distribution with scale parameter $\alpha = 12$ and shape parameter $\beta = 3.5$. As often in CM there is knowledge neither about the individual conditions nor about the times to breakdown nor about the individual rankings. So, each of the 10 assets is a black box in that sense, but the group is a sample of a presumably known distribution. What is known is that the *expected* times are as plotted in Figure 9.4.

If these assets are operated with CM, then the mean as well as the median lifetime θ will be about 10.8 years (that these happen to be almost the same is an irrelevant coincidence). All will fail in service. There will be 10 unplanned outages, possibly with collateral damage, and safety issues plus other consequences.

As a next step, the grid performance may be protected by redundancy. Then, one of the assets may fail, but the grid as a system keeps functioning (Section 7.4). For example, instead of a single circuit, a double circuit can be installed, and so on. In the following it is assumed that redundancy is carried out by installing double assets that can each take over the full load if the other asset fails. A second failure will down the system though, and the probability of a black-out following the first failure depends on the required time for repair (cf. Section 8.7). A block diagram of a redundant system was discussed in Section 8.3.1 and is shown in Figure 9.5.

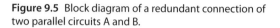

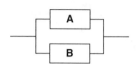

The reliability and lifetime of an asset can depend on the ageing conditions, as discussed in Section 2.6. Experience may be built up with how the distribution in Figure 9.4 is influenced. For now, it is assumed that the ageing conditions do not influence the distribution. This means that the hazard rate of an asset is influenced neither by faults in a neighbouring asset, nor by increasing the full load on one asset after another asset fails, nor by running at double load, nor by reducing the load after the failed asset is repaired or replaced. These may prove to be different in practice, and the present case may be adjusted if required in the underlying Markov models by failure and repair rates (cf. Section 8.3.2). For now, the focus is just on the effect of redundancy itself.

The mean lifetime per asset remains the same, which means that the depreciation of such a grid doubles, since two assets are simultaneously ageing instead of one. However, the system MTTFF and MTBF will increase, as described in Section 8.6. The long-term availability A_∞ is described in Section 8.7. With given asset failure rate λ and asset repair rate μ, the performance of a parallel system can be given in terms of the MTTFF as in Eq. (9.7), the MTBF as in Eq. (9.8) and the long-term availability A_∞ as in Eq. (9.9). The underlying assumption is that exponential distributions (at least in sufficient approximation) apply:

$$\text{MTTFF} = \frac{3\lambda + \mu}{2\lambda^2} \tag{9.7}$$

$$\text{MTBF} = \frac{2\lambda + \mu}{2\lambda^2} \tag{9.8}$$

$$A_\infty = \frac{\mu^2 + 2\lambda\mu}{\mu^2 + 2\lambda\mu + 2\lambda^2} \tag{9.9}$$

In case an asset is used under the regime of CM without redundancy, the lifetime of assets is $<t> = <1/\lambda>$, which becomes $\theta = 1/\lambda$ in case of an exponential distribution, where θ and λ are constants. The availability of this single asset becomes $A_{1,\infty} = \mu/(\lambda + \mu)$. The lifetime without interruption of a system compared to a non-redundant asset, while assuming a much larger repair rate than the failure rate (i.e. $\mu \gg \lambda$), is:

$$\frac{\text{MTTFF}}{\theta} \approx \frac{\mu}{2\lambda} \approx \frac{\text{MTBF}}{\theta} \tag{9.10}$$

The system availability grows:

$$A_\infty - A_{1,\infty} \approx \frac{\lambda}{\mu}\left(1 - \frac{2\lambda}{\mu}\right) \tag{9.11}$$

It may also be informative to compare the respective unavailabilities $U = 1 - A$. This yields the same ratio as with the times to failure in Eq. (9.10):

$$\frac{1 - A_\infty}{1 - A_{1,\infty}} = \frac{U_\infty}{U_{1,\infty}} \approx \frac{2\lambda}{\mu} \tag{9.12}$$

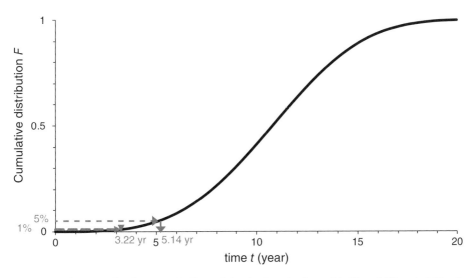

Figure 9.6 The same distribution as in Figure 9.4, selecting the time with 1% and 5% probability of breakdown.

With the redundancy strategy the investment costs K_c for the 10 assets will double to $2K_c$ (ignoring possible discount benefits if installing double). At about twice the cost, the mean periods without failure increase considerably. As a quantitative example, if repair took about two weeks and the mean time to asset failure was 10.8 years as mentioned above, then the redundancy would increase the MTTFF or MTBF (i.e. the period without system failure) from 10.8 years to 140.4×10.8 years ≈ 1516 years.

Another strategy to prevent system failure is to periodically replace the assets. Again, it is assumed that CM is the maintenance style, so the assets are indistinguishable and no information about the hazard rate of the individual assets is known. If only knowledge about the distribution in general is available, when should the assets be replaced? This depends on the hazard appetite (or risk appetite) of the utility. In the present example there are 10 assets. The question might be asked: with what probability is it acceptable for an asset to fail and the system to go down? From the sample size $n = 10$, it follows that with replacement at a failure probability of 10% it is almost certain that a failure will be observed. So, a lower probability must be selected. Figure 9.6 shows an example where 5% or 1% probability is selected with the time at which these probabilities are reached. In practice, lower probabilities might be required to maintain an acceptable service level, since the number of assets may be much higher and still no incidents could be afforded.

Another approach would be to specify the level of the hazard rate h. A 0.01 (respectively 0.05) hazard rate is reached at $t(h = 0.01) = 3.11$ years (respectively $t(h = 0.05) = 5.93$ years) with parameters $\alpha = 12$ and $\beta = 3.5$ using the reverse expression for the Weibull-2 hazard rate in Eq. (4.22):

$$t = \left(\frac{h \cdot \alpha^{\beta}}{\beta} \right)^{\frac{1}{\beta - 1}} \tag{9.13}$$

The approach of replacing at low probabilities or hazard rates of 1% and 5% yields replacement intervals of about 3 and 5 years. While redundancy requires more simultaneous investments, preventive periodic replacement requires more sequential investments. Since the mean lifetime is 10.8 years, the cost rate increases in first approximation with a factor of about 10.8/3 and 10.8/5. Actually, the interest is in the mean cost rate $C_c = K_c/\theta$ and therefore not the mean time to failure but its inverse should be averaged (cf. Eq. (9.3)). As an estimate, the cost rate will increase by a factor of about 2 to 3 in order to have a 5% (respectively 1%) probability that an asset fails. It also means that with these strategies, in about 5 years/5% = 100 years (respectively 3 years/1% = 300 years) such an event is very likely to occur.

9.1.2.4 Remarks

With the redundancy approach at lower cost rate a higher system reliability and availability is achieved. The difference though is that with CM and redundancy the assets will start to fail after typically 6 years with a frequency of more than once a year (see Figure 9.4). The impact on the system may be negligible (as assumed here, but may not be true in practice), however, there may be consequences of these failures that should be weighed as well. For instance, if the assets can explode with failure and jeopardize safety or cause much higher repair costs than replacement alone, then the price and risk may become unacceptable. Allowing assets to fail will probably also require a more robust design and installation to keep collateral damage acceptable.

In [12] it was stated that CM is appropriate when preventive efforts and costs do not outperform (even unexpected) failure with its consequences and repair. There may just be no realistic alternative to CM (e.g. if inspections and servicing are impossible with respect to certain types of ageing and failure modes). Practice shows that a considerable part of the grid connections follow CM until there is a reason for closer attention. For example, many polymer cable joints are operated largely with CM.

The most applied strategy to overcome unexpected failure is redundancy. It leads to installing double circuits or even higher multiples. Care must be taken that connections are not prone to common-cause faults (i.e. the system design must be such that a fault in one circuit does not lead to losing the parallel circuits too) [12]. Redundancy saves grid integrity, but it does not protect against collateral damage and safety threats. Timely replacement and robust installation methods can.

As a final remark, the present case assumed the failure distribution to be known. This may not be so, particularly with transmission grids where the assets are more capital-intensive and failures may be scarce due to the more robust design. Periodic replacement is hard to plan without knowledge of the distribution. As long as no failures occur, one does not know how much remaining life is sacrificed by preventive replacement unless failures do occur. For that reason, redundancy is generally chosen to build a robust grid where the impact is high. The lower the impact, the more one might rely on fast repair.

9.1.3 Condition-Based Maintenance

CBM (see Section 1.3.3) aims to keep the asset hazard rate below a target limit. With preventive periodic replacement (see Section 9.1.2), the weak and strong assets are treated the same. The individual assets get a tailor-made approach with CBM. Each asset is inspected for its condition, to receive dedicated maintenance. In principle, CBM can be more cost-efficient than PBM, but is usually also more technologically advanced, because it requires adequate diagnostics and proper expert rules to interpret the diagnostics. Sometimes one can make clever use of already available data.

9.1.3.1 References to Introductory Material

The following sections provide introductory material:

- Section 1.3.2: on period-based maintenance.
- Section 1.3.3: on condition-based maintenance.
- Section 1.3.3.4: on health index.
- Section 2.4.5: on hazard rate.
- Chapter 5: on plotting.
- Sections 5.7.1 and 5.7.2: on Crow/AMSAA plots.
- Chapter 6: on parameter estimation.

9.1.3.2 Analysis of Condition versus Period-Based Replacement

The approach for comparing PBM and CBM:

1. Define an illustrative example for CBM.
2. Illustrate in a graph how the individual warning signals work.
3. Distinguish the population failure distribution and the individual asset failure probability.
4. Estimate and compare the lifetimes with PBM and CBM.

A shortcoming of PBM is treating all assets alike. Preventive replacement prevents even the weakest asset from failing. CBM is more efficient in that respect. Assume that the same cumulative distribution F as in the previous sections (Weibull-2 with $\alpha = 12$ and $\beta = 3.5$) characterizes the failures of a population of $n = 10$ assets.

Assume that all assets can be inspected and some time before an asset fails it always gives a signal that a failure is imminent. This warning signal can consist of detectable levels of corrosion, leaking oil, high oil humidity level, presence of contamination on an insulator surface, partial discharge level, and so on. At some point in time this signal appears and marks the start of a growing hazard rate for that particular asset. Ideally, a quantitative relation between the condition indicator and the hazard rate is known (e.g. through testing or experience). Such a detectable signal would help to plan timely remedial action like servicing or replacement.

Figure 9.7 shows the cumulative distribution F as a solid line. Small black diamonds mark the moment for each asset that a warning signal of imminent failure emerges. Open diamonds mark a sequence of confidence levels that failure may occur with probabilities 1%, 5% and 10%, ..., 70%. In fact, now two failure distributions show: the overall cumulative distribution F and the individual asset failure distribution characterized by the sequence of diamonds, which is just another representation of a distribution.

The first distribution (i.e. F) is hardly informative. It shows what percentage of the population fails, but does not have information about the individual assets. It may be based on earlier experience and is useful for PBM, as discussed in Sections 9.1.1 and 9.1.2 where the weak and strong assets are treated alike. It does not help to plan tailor-made actions for individual assets.

However, the second distribution represented by the diamonds indicates the confidence levels (i.e. probabilities) for each individual asset based on observed condition

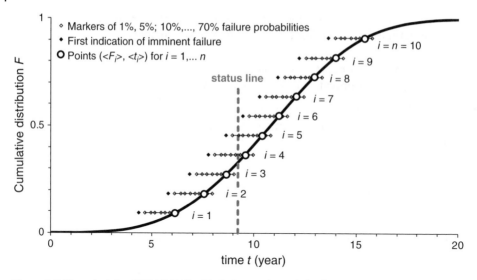

Figure 9.7 The principle of CBM [12]. The black diamonds mark the first signals of imminent failure of the asset, while the open diamonds show the failure probability deciles for that asset. The moving status line at $t = 9.2$ years shows that five assets ($i = 1,...,5$) have signalled imminent failure, whereas the other five have not. Depending on the adopted confidence level, remedial action should be taken for the first five (if not yet failed or treated).

(e.g. oil quality in a transformer, etc.). That second distribution is the one that is used for CBM. The likelihood per asset of imminent failure is part of the expert rules mentioned in Section 1.3.3. For example, having assessed a certain oil quality, an expert rule should quantify the probability of failure in the near future. Developing such rules based on experience is a great challenge, but once successfully developed, they can open the door to considerable savings on asset depreciation.

As an illustration, Figure 9.7 shows a status line that marks an arbitrarily chosen evaluation moment. It can be seen that three assets (labelled $i = 1,2,3$) most likely have failed, two assets signalled imminent failure ($i = 4,5$) and the remaining five assets ($i = 6, ..., 10$) have given no signal for imminent failure and are not suspect as yet. If confidence levels can be assigned to inspection results, then guidelines can be developed for timely replacement of assets. If, for instance, the confidence level for condition-based replacement is set at 1% (see Figure 9.8), then the average lifetime is $t = 9.33$ years. This means that each asset is replaced with 99% probability of not failing until replacement. Similarly, for condition-based replacement with 5% failure probability (the second open diamond for each asset) the average lifetime to replacement is $t = 9.62$ years. Considering the status line at the moment $t = 9.2$ years again, regardless of whether 1% or 5% has been used, the assets $i = 1, ..., 5$ would all have been replaced.

With PBM the confidence levels of 1% and 5% led to replacement after 3.22 and 5.14 years. This means that CBM employs the operational lifetime of the assets better with a factor $9.33/3.22 = 2.9$ (respectively $9.62/5.14 = 1.9$). It is a considerable saving on replacement costs, to be earned with adequate diagnostics and expert rules. These do not come for free usually, which requires weighing the costs against the savings and other aspects like operational impact.

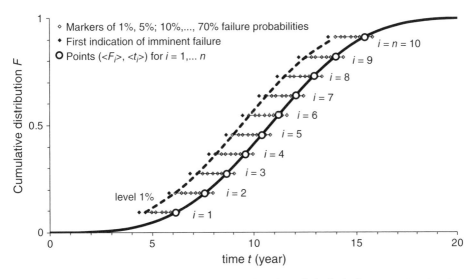

Figure 9.8 The principle of CBM where assets are replaced or refurbished when a warning signal predicts an imminent failure. The black dotted line connects the 1% failure probabilities of the individual assets and indicates when an asset would be replaced with 99% probability of not failing.

9.1.3.3 Remarks

A hazard rate plot gives another perspective (see Figure 9.9). As each asset i has its individual failure probability, it also has an individual hazard rate h_i following from that individual distribution. These individual hazard rates are based on a supposedly known relationship between inspection results and failure probability (see Figure 9.9). As mentioned, these hazard rates are just another representation of the failure probability. The dotted line representing the population hazard rate h_F indicates the overall hazard rate (the solid line in Figures 9.7 and 9.8). The hazard rate h_F is useful for PBM to trigger periodic replacement, whereas the individual hazard rates h_i are indicative of the condition-based urgency to replace the individual assets. This provides a link to the concept of the HI (see Sections 1.3.3.4 and 9.1.1.4), which quantifies a rating of individual asset health (i.e. condition). One way to rate the condition is by the hazard rate (although there are other definitions in use as well).

CBM aims at maintaining a hazard rate level below a critical hazard rate level h_c. This may apply to single assets as in Figure 9.9, but generally applies to systems such as circuits and connections. The number of failures is proportional to the average hazard rate and the number of assets in use. It is not uncommon to aim for average asset hazard rates in the order of 10^{-3}/year, 10^{-4}/year or better. With thousands of assets in a grid and an aim to keep the annual number of failures limited to a minimum, such average hazard rates are needed. Small subgroups of assets may occasionally have higher hazard rates if this does not jeopardize the overall performance of the grid.

The annual failure rate within asset groups is often monitored continuously. A steady or declining hazard rate for assets is an indicator of successful CBM. In a balanced situation, the hazard rate is constant (i.e. the population follows an exponential distribution). The hazard rate is generally a function of time, but steady for an exponential distribution. The overall hazard rate can thus easily be calculated by evaluating the total number

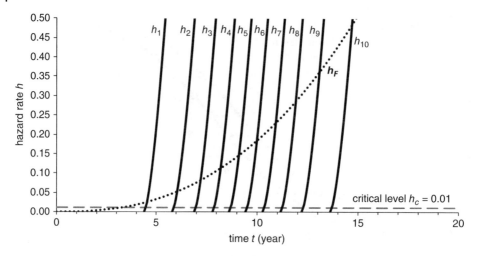

Figure 9.9 Hazard rate plot of the same case as in Figures 9.7 and 9.8 [12]. The hazard rates h_i based on the failure probability functions (indicated by diamonds) represent the hazard rates of the individual assets i. The idea is that monitor signals reflect the asset conditions and that signals can be interpreted in terms of failure probability (i.e. hazard rate). The population hazard rate h_F is plotted as a dotted curve (as if CM was applied). The individual hazard rates might be linked to HI as in Figure 9.3. Also indicated is a critical hazard level $h_c = 0.01$ that might be taken as a trigger to interfere. This level corresponds to about 1% failure probability.

of failures per unit time (disregarding the individual asset ages). Adequate CBM (and/or PBM) detects assets that do wear out by, for example, an increase in the hazard rate, subsequently reducing the hazard rate of such assets selectively by servicing, refurbishment or replacement.

Deviation from a flat hazard rate upwards by, for example, new ageing mechanisms or downwards by successful improvement programmes can be identified with Duane or Crow plots (see Sections 5.7.1, 5.7.2 and [47]).

9.1.4 Risk-Based Maintenance

RBM (see Section 1.3.4) aims at controlling the risk in terms of corporate business values (CBVs) below a target limit. Examples of CBVs are performance, safety, finances, and so on (see Section 1.3.4.1). With CBM the failure rate of assets was the ground for remedial action, but with RBM it is the damage that might be caused by an event like failure. RBM bears a close resemblance to CBM, because it requires estimating the probability of imminent asset failure but gives different weights to the importance of failures based on to the extent to which CBVs are violated. RBM therefore requires not only the availability of diagnostics and expert rules, but also the ability to forecast the impact of events.

9.1.4.1 References to Introductory Material
The following sections provide introductory material:

- Section 1.3.3: on condition-based maintenance.
- Section 1.3.4: on risk-based maintenance.
- Section 1.3.3.4: on health index.

- Section 1.3.4.1: on risk index.
- Section 2.4.5: on hazard rate.
- Chapter 5: on plotting.
- Chapter 6: on parameter estimation.

9.1.4.2   Analysis of Risk versus Condition-Based Maintenance

The approach for comparing RBM and CBM:

1. Define an illustrative example for RBM.
2. Discuss the risk plane.
3. Discuss the two axes of the risk plane: rate and impact.
4. Discuss two schools for the impact: monetizing versus scaling gravity.
5. Discuss the risk score for each CBV and for both single and multiple impacts.
6. Discuss how to compose the total risk index from the risk scores.

The risk concept with CBM is narrowed down to the likelihood of failure, but with RBM the risk concept is related to a series of business values like performance, safety and finances to name a few from the list in Section 1.3.4.1. Risk is often defined as the product of probability and impact or the product of occurrence rate and consequence (cf. Section 1.3.4):

$$risk = probability \times impact \qquad (9.14)$$

Risk can be plotted in a risk plane (see Figure 1.6) or, if split up into discrete categories, as a matrix (see Figure 1.7). The top right area of the plane is the sector of frequently occurring high-impact events which impose an unacceptable risk. The bottom left area is the sector of rarely occurring low-impact events which impose a negligible risk. The policy might be adopted to mitigate all medium and higher risks, while leaving the lower risks for later consideration.

The vertical axis is used for the likelihood that an event occurs. This can be probability, estimated lifetime or frequency, and so on. Figure 9.10 uses the hazard rate, also called the failure rate. Often the scale is taken as logarithmic, because the hazard rate can vary over multiple orders of magnitude. As the failure (un)likelihood is linked to the asset

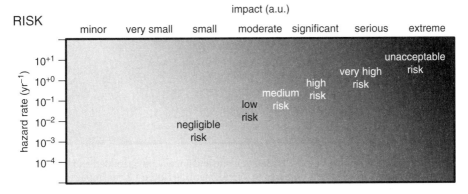

Figure 9.10 Example of risk plane. The frequency is expressed in terms of the hazard rate. The unit of the impact with respect to the various business values is yet to be determined.

health, the vertical axis may be directly associated with the HI (cf. Section 1.3.3.4), which is discussed in more depth in Section 9.2.

The horizontal axis is used for the impact that a failure has with respect to the various business values. As mentioned in Section 1.3.4.1, there are normally multiple business values involved in the evaluation. Using a single axis for the impact urges us to use a single impact unit that can be applied to the different business values such as performance, safety and finances. There are two main schools in quantifying the impact: monetizing and scaling gravity (in the sense of seriousness).

One way to quantify the impact is monetizing (i.e. all business value violations are translated into financial damage). The simplest way is to estimate the costs associated with the violations. For instance: if a high-pressure oil-filled cable leaks oil, the costs may concern repair and the cleaning of the soil; if a transformer burns down, the costs concern extinguishing the fire, repairing the damage to the substation and the loss of income due to the interrupted power transmission, and so on. Sometimes the financial damage may not sufficiently reflect the gravity of the incident, and a penalty may have to be added to the costs. The monetizing method supports decision-making conveniently if the concern is material damage. It becomes more tedious if appraisal of the impact is difficult or unethical, for instance if the above-mentioned high-pressure oil-filled cable leaks oil at a place where cleaning is not effective, like leaking oil into a river that washes the oil into the sea. If it cannot be cleaned, no costs are involved but is the impact then really nil? Or should a real or virtual penalty compensate this? An example that concerns ethics is loss of human life versus serious injury, where the costs may lead to very undesired outcomes. Again, the company may translate such damage into (virtual) penalties in order to take it into account with decision-making on timely investments.

Another method is scaling gravity. For each degree of damage per business value, it is evaluated how impacts scale from minor to extreme irrespective of costs as in Figure 9.11. Loss of human life is an extreme impact and will rank more gravely than serious injury, no matter the costs. Leaking oil into an environmentally important area can rank highly even if no cleaning expenses are made and whether penalties exist or not. The

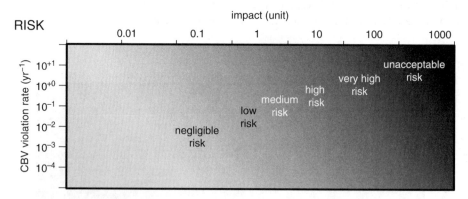

Figure 9.11 Example of risk plane. Both CBV violation rate and impact are quantified and Eq. (9.14) can be used to calculate a risk score. Compared to the risk matrix in Figure 1.7, the impact classes rank as: minor ≤ 0.01; very small 0.01–0.1; small 0.1–1; moderate 1–10; significant 10–100; serious 100–1000; extreme > 1000. Here, the risks rank as: negligible ≤ 0.01; low 0.1; medium 1; high 10; very high 100; extreme ≥ 1000.

advantage is that there is no need to keep track of market prices and inflation. However, assigning gravity to a range of impacts may be challenging. Though it is often argued that one way or another scaling gravity comes down to monetizing, there is a difference. The costs for mitigating a problem are principally considered after gravity considerations trigger investments. During the making of the investment plans costs may nevertheless be considered when choosing the most cost-efficient solution, but money plays no role in the fundamental decisions to start up the mitigation process.

Having an impact scale either by monetizing or by gravity scaling, the next step is to determine the risk score per CBV_i. The hazard rate is the rate at which failures occur. This is, however, not the same as the rate at which a CBV_i is violated. For instance, consider a termination that is estimated to fail with hazard rate h and then cause a black-out with a significant impact I_{CBVi}. If the redundancy is such that in less than $P_{CBVi} = 1\%$ of cases the failure will actually lead to a black-out, then the rate h_{CBVi} of violating the CBV_i 'quality of supply' is not h, but $h_{CBVi} = P_{CBVi} \cdot h = 1\% \cdot h$. So, there is a difference between the hazard rate and the CBV violation rate if not every failure event causes a CBV violation.

After having determined the CBV violation rate h_{CBVi} and with impact I_{CBVi}, the next step is to determine the risk score per CBV_i (i.e. RS_{CBVi}). The idea is to use Eq. (9.14) in an adapted form. The risk score for a CBV_i is the product of the rate h_{CBVi} and the impact I_{CBVi} of the CBV_i violation:

$$RS_{CBV_i} = h_{CBV_i} \cdot I_{CBV_i} \tag{9.15}$$

As a complication, it is possible that in a percentage P_{CBVi1} of cases a CBV_i is violated with impact I_{CBVi1}, while in another percentage P_{CBVi2} with impact I_{CBVi2}. These percentages P_{CBVij} can be used as weights for assessing a weighted risk score RS_{CBVi} for a CBV_i:

$$RS_{CBV_i} = \sum_j (h_{CBV_{ij}} \cdot I_{CBV_{ij}}) = h \cdot \sum_j (P_{CBV_{ij}} \cdot I_{CBV_{ij}}) \tag{9.16}$$

If multiple impacts are possible, one might take only the most severe impact for calculations or allow a higher h_{CBVi} than h.

At this stage there are risk scores for all CBVs. Having determined these risk scores RS_{CBVi} for each CBV_i, the final step is to determine the risk index RI that represents the total risk of the asset with respect to the investigated event (usually a failure). The RI is generally taken as the sum of the risk scores. If every CBV_i is considered equally important, then the sum will be unweighted (i.e. the terms have equal weights). If the company policy assigns different weights w_i to certain CBVs, then the collective RI for the asset risks can be taken as the weighted sum of the risk scores:

$$RI = \sum_i (w_i \cdot RS_{CBV_i}) \tag{9.17}$$

This RI can be used to prioritize the mitigation of the asset risk. If desired, not only the RI at present but also an RI in 3 or 10 years may be estimated to have an impression of the urgency of mitigating the situation.

9.1.4.3 Remarks

In practice, not every asset may be analysed for its risk index. It depends on the ease with which the risk scores can be determined. For instance, CBV 'safety' can be influenced by

RISK

CBV violation rate (yr⁻¹)	impact (unit)						
	0.01	0.1	1	10	100	1000	
10^{+1}	low	medium	high	very high	unacceptable	unacceptable	unacceptable
10^{+0}	negligible	low	medium	high	very high	unacceptable	unacceptable
10^{-1}	negligible	negligible	low	medium	high	very high	unacceptable
10^{-2}	negligible	negligible	negligible	low	medium	high	very high
10^{-3}	negligible	negligible	negligible	negligible	low	medium	high
10^{-4}	negligible	negligible	negligible	negligible	negligible	low	medium
	negligible	negligible	negligible	negligible	negligible	negligible	low

Figure 9.12 The risk plane of Figure 9.11 turned into discrete categories. The risks rank as: negligible ≤ 0.01; low 0.1; medium 1; high 10; very high 100; extreme ≥ 1000.

external decisions. If a bicycle path is diverted, which changes the distance between a substation and passers-by, it changes the safety risk as well (for better or worse). Keeping track of the *RI* continuously would imply that for all assets at the station, a re-evaluation might have to take place periodically. This may not be practical.

A different approach is to only consider asset issues that have been added to the portfolio of projects, because at least one risk score is equal to or exceeds 'medium risk' (i.e. $RS \geq 1$). This portfolio of issues can be evaluated by the sum of its associated *RI*. A steady sum of *RI* indicates a steady risk position. An increasing sum of *RI* indicates a deteriorating grid that is in need of significant remedial action (e.g. increase in budget or more efficient methods).

Often the risk plane is used in classes or categories, which turns it into a matrix as shown in Figure 9.12. The numerical qualification in this case is the same as in Figure 9.20.

9.2 Health Index

The following discussions concern a closer look at the use of the HI. The HI (see Section 1.3.3.4) is an indicator that is used by many utilities to represent the condition of assets [12, 48, 49]. The indicator and its use are not yet harmonized, and many views and methods are in use. The representation of the asset condition or health is what all methods have in common. However, whether this health is expressed in expected remaining lifetime, hazard rate, remaining quality categorization, measure for required mitigation or another measure is not determined.

9.2.1 General Considerations of Health Index

The HI as presented in Section 1.3.3.4 is a decision tree: based on a range of checks, health is expressed in terms of condition ('OK', 'fair', 'suspect') and recommended follow-up ('investigate', 'urgent attention'). In contrast, Figure 9.3 links the four classes to a hazard rate which is a statistical function as described in Section 2.4.5. The figure provides the concept of a translation, but the question is whether this is sufficient and workable. This section studies how the HI is developed.

9.2.1.1 References to Introductory Material on Health Index Concept Considerations

The following sections provide introductory material:

- Section 1.3.3: on condition-based maintenance.
- Section 1.3.3.4: on health index.
- Section 1.3.4.1: on risk index.
- Section 2.4.5: on hazard rate.

9.2.1.2 Analysis of the Health Index Concept

The approach for analysing the HI concept is:

1. Discuss HI in terms of hazard rate.
2. Indicate how HI (and RI) are defined in practice with techniques like FME(C)A.
3. Discuss crucial role of expert engineers and data.

The HI as an indicator of asset condition is usually discrete. This assists decision-making by assigning follow-up to each category (replace, investigate deeper, etc.). The outcomes can be based on a logic tree or discretization of a calculated result like the hazard rate. The prime question is: how can quantities like asset remaining life or hazard rate be assessed and categorized?

As mentioned in Section 1.3.3, CBM requires adequate diagnostics to assess the asset condition and expert rules to interpret the findings and take proper decisions with respect to maintenance. The expert rules are precisely the core of the HI.

There are various techniques for the set of expert rules, which comprise [50]: failure root cause analysis (FRCA); failure modes and effects analysis (FMEA); failure modes, effects and criticality analysis (FMECA); and so on. Though these techniques are usually applied in the design phase of a product, end-users like utilities may perform these also to define their maintenance strategies. The description of such techniques is beyond the scope of this book, but the concept is discussed here. Generally, the road to the expert rules in the HI involves:

1. *Define the functionalities of the asset.* For example, a line insulator must electrically separate the high-voltage line and the earthed tower, but also mechanically carry the line. There may be other functionalities thought of.
2. *Define the failure criteria of the asset.* Usually this is a total or partial stop of one of its functionalities. For example, a short circuit through or over the surface (flash-over) is a failure with respect to electrical separation; a physical fracture ceases the functionality of carrying the line. A flash-over due to a falling tree branch which can be re-energized immediately is not, because the asset still functions. If the falling tree branch were to unacceptably damage the insulator, then it is a failure.
3. *Analyse the asset subsystems or parts with their function(s).* For example, a polymer line insulator consists of a glass fibre-reinforced rod to provide electrical separation and mechanical support, a rubber housing and sheds around the rod to provide weather protection and prevent arcing (which in turn prevents radio disturbance), clamps on both sides of the rod to allow mounting to the tower and the line (while also clamping the rod), corona rings at both ends to control the electric stress, and so on.
4. *Describe the failure scenarios systematically.* For example, a line could drop for various reasons, like a clamp sliding off the rod or a fracture of the rod. The fracture can

be due to electrochemical degradation if the silicon rubber does not cover the full rod, for example. These inventories of scenarios can produce quite extensive fault trees. Also estimate the likelihood of occurrence.

5. *Evaluate the suitable diagnostics, inspections or remedies.* For example, tracking or damage to an insulator can be observed by camera; discharging may even be audible; and so on.

6. *Translate the diagnostic and inspection results into probabilities/hazard rates.* For example, given the observed deterioration, what would be the typical remaining life or what percentage of assets with these diagnostic findings are found to fail within, say, 1 month or 1 year? Is CBM possible and cost-efficient or should another maintenance style be chosen (cf. Section 1.3)?

7. *Evaluate the impact of a failure in terms of business values.* For example, minor to extreme (cf. Figure 9.10). The impact is also called criticality and this last step distinguishes FMECA from FMEA. The criticality also involves the impact and is related to the RI.

The outcomes of step 6 are a range of hazard rates or average remaining lifetimes. If diagnostics and inspection indicate that an asset is found to score on the early phases of a certain failure mode, then from step 6 the asset HI classification follows.

The process of inventorying failure modes can be extensive and requires the involvement of expert engineers. Considering the often thousands of installed assets, this can pay off the effort quite well. Furthermore, historic data about the deployed asset population, failures, numbers of removed assets for reasons other than failure, and so on are important. Particularly with long operational lives (on the order of decades), data over such periods are also required. Though these data may be hard to retrieve, it is recommended to set up and restore historic archives to objectively assess hazard rates.

9.2.1.3 Remarks

The HI as discussed is closely related to the hazard rate as a measure of how well the assets can perform here. In addition to imminent failure concerns, there are usually more considerations of importance to decision-making, such as:

- *Repair rate.* A surge arrestor can be replaced in weeks, but a 380-kV transformer takes years to be designed, built and installed. If the hazard rate reduces to the range [0.01/year,1/year], the stakes with regard to performance would become very high for transformers unless sufficient redundancy and/or spare units are provided.

- *Accessibility.* Not related to the repair rate, but rather due to planning. For example, preventive measures probably require the asset to be de-energized. Whether a planned outage is possible depends on the grid configuration and other (e.g. reconstruction) activities in the grid. Similarly, personnel must be available.

- *Obsoleteness.* If spare parts become obsolete, tailor-made and transition solutions are necessary. For example, the production of gas-pressurized cable has almost entirely ceased. Repair of failures in such cable systems requires taking out some cable length around the breakdown spot and inserting two cable joints with a piece of spare cable. If this cable is no longer produced and the stock of spare cable is depleted, other cable types must be inserted. The transition joints are large, costly and interrupt the system (e.g. oil pressure systems).

- *Operational stress.* Asset ageing can be accelerated or decelerated by its stress level or profile.

The HI is not yet standardized, and whether these aspects should be involved in the HI itself is a matter of debate. If the HI stays close to the concept of hazard rate or remaining life, then other factors may need to be considered separately, like reparability.

Another aspect that may require attention is confidence in the HI rating. On the one hand this may be related to the accuracy of the diagnostics, which will not be perfect. On the other hand the historic experience with a given failure mode for a given asset type may be scarce, which forms a weak basis for the expert rules. An approach can be to also build the expert rules on comparable but not identical assets. The resulting expert rules may therefore be less firm to apply, but may nevertheless provide a useful guideline even with possibly moderate confidence. The HI may be accompanied with a confidence interval or with a confidence score that indicates the firmness of the classification.

9.2.2 Combined Health Index

The HI has been used so far as an asset condition indicator. Circuits in the grid are built from groups of components that may each be characterized by their own asset HI. Similar to the condition assessment and HI assignment of individual assets, the condition and HI of groups of assets can be determined. Circuits or even higher integration levels in the grid, like connections (built from multiple circuits), substations, regional grids, and so on, may then be characterized by a combined (or composite) HI. This might be done with a so-called combined health index (CHI). Below, a model is discussed to define the CHI by hazard rate calculation rules for systems.

9.2.2.1 References to Introductory Material
The following sections provide introductory material:

- Section 1.3.3.4: on health index.
- Section 1.3.4.1: on risk index.
- Section 2.4.5: on hazard rate.
- Chapter 7: on system and component reliability.
- Chapter 8: on system states.

9.2.2.2 Analysis of the Combined Health Index Concept
The asset HI is expressed in terms of follow-up, estimated remaining life, estimated hazard rate, and so on. Most methods have in common that the HI categories reflect some kind of failure probability in a certain time. Two functions are adequate for that: the failure distribution density f (cf. Section 2.4.3) and the hazard rate h (cf. Section 2.4.5). As the HI is assigned to functioning components that might fail, the hazard rate concept is the most useful.

Circuits and connections are groups of aggregated components. These can be defined as systems, which were the subject of Chapter 7. The calculation rules for system hazard rates can be applied to find a CHI. The scales of HI classes can be numbered consecutively (e.g. class 1–4: OK, fair, suspect, urgent attention), which is generally not a good scale for the hazard rate itself. Therefore, the HI needs to be translated into hazard rates or hazard rate bandwidths as in Figure 9.3.

Once that is done, it becomes possible to calculate the CHI through the Markov–Laplace methods discussed in Chapter 8 and particularly Section 8.6, where the MTTFF and/or the MTBF are discussed. A common assumption with the use of the Markov–Laplace method is that transitions are exponentially distributed with constant hazard rates (Section 8.4). Particularly if wear-out occurs this is not true, because in that case the hazard rate is increasing. However, partly for convenience and partly because this increase is relatively low compared to the scale of planning, generally the hazard rate is still regarded as semi-constant as a reasonable approximation.

The inverses of these times until failure MTTFF and MTBF are hazard rates to first failure h_{TFF} (respectively between failures h_{BF}), which can be converted back to HI:

$$h_{TFF} \approx \frac{1}{\text{MTTFF}} \tag{9.18}$$

$$h_{BF} \approx \frac{1}{\text{MTBF}} \tag{9.19}$$

These are the hazard rates associated with the first (respectively the sequential) failures of a system that consists of multiple assets. Similar to the classification in Figure 9.3, classes of hazard rates can be associated with CHI scores. The method described in this section uses the Markov–Laplace method to produce a CHI for asset groups that is consistent with the HI methodology for single assets.

9.3 Testing and Quality Assurance

In the following sections various testing aspects are recapitulated.

9.3.1 Accelerated Ageing to Reduce Child Mortality

Two types of child mortality cases were discussed: as a competing simultaneous process in all objects (Section 2.4.5) and as a deviant subpopulation mixed through the remaining population (Section 2.5.3). Both types result in early failures.

A way to reduce the contribution of these two failure types is to burn them out of the population by accelerated ageing, as discussed in Example 2.10. It is a common part of factory quality control to apply an elevated stress level to let assets with a defect fail before delivery. Likewise, a test after installation in the field employs an elevated stress level like twice the operational voltage (e.g. the $2U_0$ test) to burn out installation errors.

The benefit of this type of testing is primarily preventing early failures during operation. A disadvantage is that the service life of healthy assets is also reduced by accelerated ageing. For that reason, older cable systems which have been repaired may be tested less severely.

The power law is used extensively as a model to describe the effect of accelerated ageing (cf. Section 2.6.2, Eq. (2.66)) and to convert the lifetimes τ_i from one stress level B_i to another:

$$\tau_1 = \tau_2 \cdot \left(\frac{B_2}{B_1} \right)^p \tag{9.20}$$

The power p in the power law is crucial and can be determined by performing a series of tests with a range of ramped voltages (see Example 2.11).

For cable and cable materials, various powers p for voltage-driven processes have been reported (cf. [51, 52]):

- water treeing: $2 \leq p \leq 4$;
- partial discharge-induced degradation: $p \sim 4$;
- contaminant effects in polymer insulation in real cable under dry conditions: $8 \leq p \leq 10$ (often $p \sim 9$ is taken as average);
- deterioration mechanisms in high-purity polymer cables with smooth extruded semi-conducting shields: $p \geq 15$.

If more than one mechanism occurs in the stressed assets, then the mechanism with the highest power p is accelerated the most. It should be evaluated whether the applied accelerated ageing burns out the right mechanism(s). Another point of consideration is that processes may deviate from what is happening during real operation. For instance, the phenomenon of water treeing can be accelerated by various parameters like frequency, salt content, contamination, and so on. The extent of water tree growth can be measured after staining with methylene blue [53, 54], but the type of acceleration clearly leaves its mark on the chemical composition and physical structure of the resulting water trees [55]. So, the question is justified: to what extent is the accelerated process identical with that encountered under normal operation? In research and product development, the statistics should also involve some fingerprinting (like material characterization) to check whether the right mechanism is accelerated in order to avoid fruitless improvement programmes.

Not all ageing processes follow this power law acceleration model. Processes that are temperature-driven (e.g. chemical reactions) are often accelerated by thermal excitation, which follows the Arrhenius law (Section 2.6.3, Eq. (2.74)) [14]:

$$t_{remaining} = A \cdot \exp(\Delta H / kT) \tag{9.21}$$

Here, A is a (time) constant, ΔH an activation energy, T the absolute temperature in Kelvin and k the Boltzmann constant.

9.3.2 Tests with Limited Test Object Numbers and Sizes

In quality control, destructive testing is performed to assess the limits of the asset strength. Additional lengths or pieces are produced to be sacrificed in testing. In this section it is assumed that the test data are Weibull-distributed, because each test object will fail at its weakest spot (cf. Section 4.2.1). Analysis of the data comprises estimation of the Weibull parameters. As a standard for asset quality, it may be required that (e.g. after a given number of years) the reliability is 99%. The test must check whether the asset qualifies itself.

Particularly when assets are capital-intensive, like large generators, there is a sensible drive to reduce costs by limiting both the number and size of the test objects. To what extent can this influence the test evaluation?

The following illustrates the possible effect of limited testing in the case of a large generator. The example focuses on the insulation reliability. Assume the generator contains 96 insulated copper bars, which are costly and labour-intensive to produce. For testing purposes one or two additional bars are produced. One bar is cut into four pieces for electric breakdown testing.

The manufacturer has built up a growing database of test results which lead to a reference Weibull distribution. All new test results are compared with this reference.

A client orders a generator with the agreement that it meets a particular test. After re-evaluation the manufacturer estimates the probability of passing this test to be only 50% and negotiations are started to find alternative tests and procedures to gain confidence that the generator is fit for purpose.

From the manufacturer's archives the claim is made that the average bar survives more than 400 years, which should be good enough for a generator that will be in operation for maybe 40 years. Is it?

The following steps were undertaken:

1. Evaluate the manufacturer data.
2. Evaluate the statistical analysis.
3. Evaluate the testing method to reach these data.
4. Evaluate the expected reliability of the generator.

As for point 1, the evaluation of the manufacturer data showed that the parameters of reference for the Weibull distribution are: $\alpha = 450$ and $\beta = 3.3$. The mean lifetime θ follows from Section 4.2.1.2, Eq. (4.28) and the claim seems confirmed:

$$q = \alpha \cdot \Gamma\left(1 + \frac{1}{\beta}\right) = 450 \cdot \Gamma\left(1 + \frac{1}{3.3}\right) \text{ yr} \approx 404 \text{ yr} \tag{9.22}$$

As for point 2, the statistical analysis procedure appeared to analyse the data by the maximum likelihood method (Section 6.2.3.1). This is where the first concerns arise (see Figure 9.13). Since only $n = 4$ test samples are put to the test, the estimated shape parameter has a significant bias. The manufacturer software did not compensate for this bias (which is standard). The unbiased shape parameter estimator b_U is discussed in Section 6.2.3.4 and shown in Eq. (6.53) as:

$$b_U = b_{\alpha,\beta} \cdot \frac{1}{\langle b_{1,1} \rangle} \approx b_{\alpha,\beta} \cdot \frac{n - 2}{n - 0.68} \tag{9.23}$$

With sample size $n = 4$ and $b_{\alpha,\beta} = 3.3$, the unbiased estimator yields $b_U = 1.99$. As mentioned before, the bias in the scale parameter is very small and can be ignored.

As for point 3, it was realized in evaluating the testing method that the four test objects were not full-size bars but only quarter sections (leaving the connection length out of the scope). The probability of finding a significant defect in a quarter section is smaller than that in a full-size bar. This is called the volume or length effect and was discussed in Section 7.3. The scale parameter is particularly sensitive to this effect. If a system with m times the length of the test object has Weibull parameter α_S and β_S versus the test objects that are Weibull-distributed with parameter α and β then their relationship is as shown in Eqs (7.23) and (7.24):

$$\alpha_S = \frac{\alpha}{m^{1/\beta}} \tag{9.24}$$

$$\beta_S = \beta \tag{9.25}$$

With $b_U = 1.99$ and $\alpha = 450$ years, the α_S for full-sized bars becomes $\alpha_S = 224$ years (see Figure 9.13). The mean lifetime θ of the full-sized generator bars therefore drops to $\theta = 199$ years.

As for part 4, in evaluating the expected reliability it is important to realize that the first bar to fail is the one that lets the generator fail. The mean is not of interest, but the

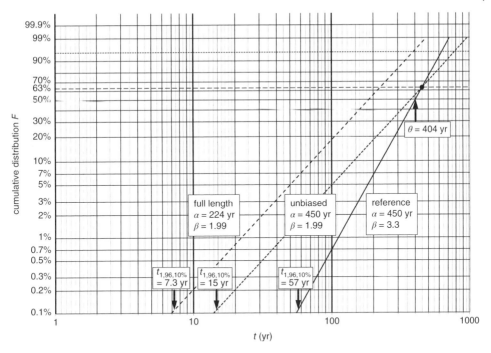

Figure 9.13 The course of the analysis. Shown are the reference graph (manufacturer test results), the unbiased graph (correction for small test set of $n = 4$) and the full-length graph (corrected for downscaled test objects). The initial expectation of $\theta = 404$ years obtained from the reference graph reduces to an expectation of 10% probability that a 96 bar generator fails within $t_{1,96,10\%} = 7.3$ years.

lifetime $\tau_{1,96,A\%}$ of the first failing bar out of 96 at probability $A\%$. This can be determined with the inverse beta distribution as the confidence limit $F_{1,n,A\%}$ (see Section 4.1.2.1 and Eq. (4.19)):

$$F_{i,n,A\%} = B^{inv}(A\%; i, n + 1 - i) \qquad (9.26)$$

The time at which this probability is reached with the corrected Weibull parameters is:

$$\tau_{i,n,A\%} = \alpha_S \cdot (-\ln(1 - F_{i,n,A\%}))^{\frac{1}{\beta_S}} \qquad (9.27)$$

For $A\% = 1\%$, 5% (respectively 10%) the time $\tau_{1,96,A\%}$ is 2.2 years, 5.1 years (respectively 7.3 years). So, after about 7.3 years there is already a 10% probability that this generator will fail (see Figure 9.13).

The client noticed the expected lifetime of 400 years for a bar (section) dropping to about 7 years with 10% failure probability for the generator. The seemingly too strict test appeared not so strict after all. The matter was solved by a higher-grade insulation and the test was passed accordingly.

It is interesting to consider what would have happened without the analysis. There probably would have been an unexpected failure. While the expectations for the lifetime were in hundreds of years, the median $\tau_{1,96,50\%}$ appears to be 18.8 years. Without the analysis, the client and the manufacturer would have believed that something odd happened and forensic studies might or might not have found a cause like transport to the site, lightning that might have damaged the insulation, sudden change in load, and so on.

Possibly nobody would have expected this generator to have failed intrinsically because it simply was not strong enough and had a 50% probability of failing before 18.8 years.

9.4 Incident Management (Determining End of Trouble)

Despite maintenance and testing, surprising failures may occur. Redundancy may have prevented a black-out, but if an issue is possible by any chance, then it may happen someday. If a failure occurs, generally a forensic investigation is conducted to reveal the cause. This investigation will probably document the status after the failure, consist of visual inspections, a systematic scan of possible scenarios, a check on production and installation records, a material investigation and a statistical study. The latter is the subject of the following subsections.

A number of questions will have to be answered:

- Is the observed failure behaviour normal?
- When might the next failure occur?
- How many next failures may be expected?

From a maintenance point of view, questions concern whether imminent failures can be detected in a timely manner and prevented by replacement or servicing. If so, a diagnostic with expert rules might be employed to carry out CBM (cf. Section 9.1.3).

Most studied cases below concern cable systems. The interesting part is that cable systems contain a countable number of accessories (cable terminations and cable joints) with lengths of cable in between that can house non-countable fault sites (anywhere in cable sections). The statistics of the accessories concern distributions with well-known sample sizes. What may not be known is whether or not all accessories suffer from the same degradation mechanism. The statistics of the cable are often less clear-cut than with countable accessories, because it is a continuum of possible breakdown spots. After a failure, some cable length around the fault is removed and replaced by a piece of repair cable plus two cable joints on both ends to connect with the remaining cable.

9.4.1 Component Failure Data and Confidence Intervals

The present case is discussed to compare confidence limits.

A series of components is supposed to stay in operation for about 30 years, but 11 are found to fail much earlier (at the times listed in Table 9.2).

As there are many more of these components in operation, the question is raised whether more failures can be expected. There are two types of confidence limit discussed: beta distribution-based for cases of random sampling from a large population; regression error-based for models based on samples. The beta distribution approach usually gives (much) wider confidence intervals. How does that influence decision-making?

Table 9.2 Failure data of components.

i	1	2	3	4	5	6	7	8	9	10	11
t_i (yr)	0.89	1.31	1.55	1.64	2.08	2.22	2.46	2.90	2.95	3.26	4.49

9.4.1.1 References to Introductory Material

The following sections provide introductory material:

- Sections 4.2.1, 4.2.1.1 and 4.2.2: on the Weibull distribution.
- Section 5.4.1: on Weibull plots.
- Section 5.4.5: on confidence intervals.
- Sections 6.3.1 and 6.3.2: on linear regression and Weibull parameter estimation.
- Section 10.3: on regression analysis and error estimation.

9.4.1.2 Analysis of the Case

The following steps were undertaken:

1. Plot the data in a Weibull plot.
2. Draw both the beta distribution-based and regression-based confidence limits.
3. Estimate the times before which the next failure might occur with 99% confidence.
4. Evaluate the meaning of both types of confidence limit.

Firstly, the available data are used to produce the Weibull plot shown in Figure 9.14. The expected plotting positions $<Z_{i,n}>$ are used for the Weibull-distributed data (cf. Eq. (5.24)):

$$\langle Z_{i,n}\rangle = \left[-\gamma + i\binom{n}{i}\sum_{j=0}^{i-1}\binom{i-1}{j}\frac{(-1)^{i-j}\cdot\ln(n-j)}{n-j}\right]\cdot\log e \tag{9.28}$$

which can be conveniently approximated with (cf. Eq. (5.28)):

$$\langle Z_{i,n}\rangle \approx \log\left(-\ln\left(1-\frac{i-0.44}{n+0.25}\right)\right) = \log\left(-\ln\left(\frac{n+0.69-i}{n+0.25}\right)\right) \tag{9.29}$$

The beta-based confidence intervals are produced as in Section 5.2.4, Eq. (5.14):

$$F_{i,n,A\%} = B^{inv}(A\%; i, n+1-i) \tag{9.30}$$

from which an $A\%$ confidence limit $t_{i,n,A\%}$ follows from Eq. (5.15) as:

$$t_{i,n,A\%} = F^{inv}(F_{i,n,A\%}(t)) \tag{9.31}$$

With a best fit with Weibull parameters α and β this becomes:

$$t_{i,n,A\%} = \alpha \cdot [-\ln(1 - F_{i,n,A\%}(t))]^{\frac{1}{\beta}} \tag{9.32}$$

The times $t_{i,n,A\%}$ are plotted at the same plotting positions $<Z_{i,n}>$ and indicated in Figure 9.14 with '+' marks for the confidence limits. The interpolation between the confidence limit $t_{i,n,1\%}$ marks and the confidence limit $t_{i,n,99\%}$ marks is produced as follows. The plotting position is determined using the approximation for $<Z_{i,n}>$ of Eq. (9.29) while using non-integer i values. The lower and upper confidence limits around these $<Z_{i,n}>$ are calculated as the confidence limits with Eq. (9.32) for the same non-integer i values.

The regression-based confidence limits characterize the LR fit rather than the possible scatter of the population (this is elaborated further in Sections 10.4.3 and 10.4.3.2). Such regression-based confidence limits are produced according to the regression analysis described in Section 10.3. The Weibull plot is regarded as a plot of variable $x_i = <Z_{i,n}>$ and covariable $y_i = \log(t_i)$, which are the observed data listed in Table 9.2. The variables $<Z_{i,n}>$ are assumed without error and the covariables $\log(t_i)$ scatter about

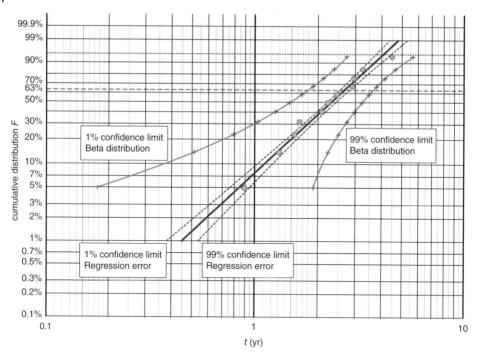

Figure 9.14 Weibull plot of the data in Table 9.2. The plot shows the data, the best fit surrounded by confidence limits based on regression error estimation and confidence limits based on the beta distribution (i.e. random sampling from a large population). The estimated parameters are $a = 2.7$ years and $b = 2.6$.

their theoretical values on the true line and about their projected values on the best fit. Note that in probability plots the variables are usually plotted along the vertical axis and the covariables along the horizontal axis.

The variance $(\Delta y)^2 = (\Delta \log(t))^2$ at an arbitrary Z value is derived in Section 10.3, Eq. (10.96) as:

$$(\Delta \log t)^2 \approx \frac{1}{n-2}\left(1 + \frac{n \cdot (Z - \overline{\langle Z_{i,n}\rangle})^2}{S_{ZZ}}\right) \cdot s^2 \tag{9.33}$$

With n the sample size, S_{ZZ} the squared sum of $\langle Z_{i,n}\rangle$ deviations from their mean (cf. Eq. (10.64)):

$$S_{ZZ} = \sum_{i=1}^{n}(\langle Z_{i,n}\rangle - \overline{\langle Z_{i,n}\rangle})^2 \tag{9.34}$$

and s^2 the quadratic sum of the residues (cf. Eq. (10.92)) with a_0 and a_1 the estimated parameters by regression analysis:

$$s^2 = \frac{1}{n}\sum_{i=1}^{n}\varepsilon_i^2 = \frac{1}{n}\sum_{i=1}^{n}(\log t_i - a_1 \cdot \langle Z_{i,n}\rangle - a_0)^2$$

$$= \frac{1}{n}\sum_{i=1}^{n}\left(\log t_i - \frac{1}{\beta} \cdot \langle Z_{i,n}\rangle - \log \alpha\right)^2 \tag{9.35}$$

The scatter of log(t) at arbitrary t is assumed to be normally distributed with mean <log(t)> and standard deviation Δlog(t). Section 4.3.1, Table 4.15 shows that the interval [$\mu - \sigma, \mu + \sigma$] corresponds to the ~68% confidence interval. Note that with given Z, the mean <log(t)> is defined by the linear relationship as in Eq. (5.19). In this case:

$$\mu = \langle \log t \rangle = \log \alpha + \frac{1}{\beta} \cdot Z \tag{9.36}$$

and:

$$\sigma = \Delta \log t \tag{9.37}$$

As the normal distribution is a symmetrical distribution, its mean is the median (i.e. the 50% limit) and half of the 68% interval is below (respectively above) the mean. Therefore, $\mu - \sigma = $ <log(t)>$ - $ Δlog(t) is the 16% confidence interval and $\mu + \sigma = $ <log(t)>$ + $ Δlog(t) the 84% confidence interval.

Using the standardized normal distribution other confidence limits can be determined by multiplying the standard deviation by a factor. Appendix G shows a table of cumulative standardized normal distribution values F as a function of multiples of the standard deviation. For instance, standardized variable $y = 1$ corresponds to $F = 0.8413447$, which is the above mentioned 84%. The $F = 99\%$ value corresponds to $y = F^{-1}(0.99) \approx 2.330$. The 99% confidence limit therefore corresponds to:

$$F^{-1}(99\%) = \frac{\log t_{99\%} - \langle \log t \rangle}{\Delta \log t} = \frac{\log t_{99\%} - \log \alpha - \frac{1}{\beta} \cdot Z}{\Delta \log t} \approx 2.33 \tag{9.38}$$

Therefore:

$$\log t_{99\%} = \left[\log \alpha - \frac{1}{\beta} \cdot Z \right] + F^{-1}(99\%) \cdot \Delta \log t$$

$$\approx \left[\log \alpha - \frac{1}{\beta} \cdot Z \right] + 2.33 \cdot \Delta \log t \tag{9.39}$$

Similarly, as the distribution is (assumed) symmetrical, subtracting 2.33·Δlog(t) gives the 100% to 99% = 1% confidence limit:

$$\log t_{1\%} \approx \left[\log \alpha - \frac{1}{\beta} \cdot Z \right] - 2.33 \cdot \Delta \log t \tag{9.40}$$

In this way, for arbitrary Z the regression-based $A\%$ confidence limits log($t_{A\%}$) follow as shown in Figure 9.14.

The two types of confidence interval are very different in width. Section 10.4.3 shows that beta-based and regression error-based confidence limits tend to differ in general. The beta-based confidence limits are informative about the scatter in the population, whereas the regression error-based confidence limits are a measure of the accuracy of the best fit of the data.

9.4.1.3 Remarks

It is very tempting to use the narrow regression confidence interval for deciding the reliability of the group. A method to evaluate whether more failures are likely is to hypothetically state that one more failure is to come and then check the upper 99% confidence limit. The evaluation moment is shortly after the last observed failure, say 4.5 years.

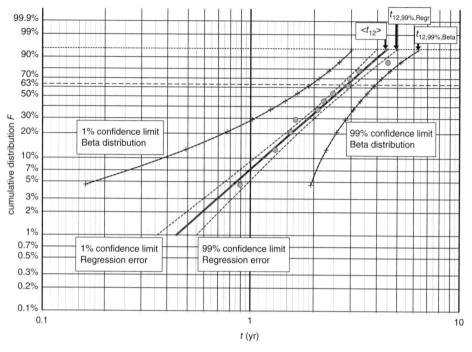

Figure 9.15 Weibull plot of the data from Table 9.2 with the assumption that a 12th failure is to be expected. The 99% confidence limit indicates before what time this failure would be expected to happen with 99% confidence.

Figure 9.15 shows again the Weibull plot of the data from Table 9.2, but now assumes a 12th failure occurs, which changes the plotting positions $<Z_{i,n}>$, because $n = 12$ instead of 11. The Weibull parameter estimators assuming $n = 12$ are $a = 2.9$ years and $b = 2.5$. The 12th observation is censored as yet (or may not happen at all). The 99% confidence limit marks a time before which this 12th failure would occur. The hypothesis that a 12th failure will follow may be rejected if the time passes the 99% confidence limit. The expected time of failure appears to be 4.38 years (which is before the 11th failure). Taking the regression-based 99% confidence limit, the time at which the hypothesis is rejected if no failure occurs before that is $t_{12,99\%,\text{regr}} = 4.91$ years, which is within about half a year. However, if the beta-based 99% confidence limit is applied, then $t_{12,99\%,\text{beta}} = 6.3$ years, which means almost two more years of uncertainty.

In a Bayesian approach the knowledge can be used that $t_{12} > 4.49$ years. The confidence limit corresponding to $t_{12,A\%} = 4.49$ years equals $A\% = 56\%$. The probability that the 12th failing joint survives $t_{A\%}$ is $1 - A\% = 44\%$. It follows for the 12th joint (see Section 2.5.2.1 and Eq. (2.42)):

$$P(survive\ t_{A\%}\ |\ survive\ 4.49\ y)$$

$$= \frac{P(survive\ 4.49\ y\ |\ survive\ t_{A\%}) \cdot P(survive\ t_{A\%})}{P(survive\ 107\ d)}$$

$$= \frac{1 \cdot (1 - A\%)}{R(4.49y)} = \frac{1 - A\%}{0.44} = 1 - \widetilde{A}\% \tag{9.41}$$

with $\tilde{A}\%$ the probability of the 12th joint failing before time $t_{A\%}$, which corresponds to the beta-based confidence limit $A\%$. Probability $\tilde{A}\%$ is related to $A\%$ as:

$$\tilde{A}\% = 1 - \frac{1 - A\%}{0.44} = \frac{A\% - 0.56}{0.44} \Leftrightarrow A\% = 0.44 \cdot \tilde{A}\% + 0.56 \tag{9.42}$$

The beta-based $A\% = 99\%$ confidence limit then corresponds to $\tilde{A}\% = 98\%$. Vice versa, a corrected confidence limit $\tilde{A}\% = 99\%$ corresponds to the beta-based confidence limit $A\% = 99.6\%$ and $t_{12,99.6\%} = 6.63$ years. Using the Bayesian approach and taking as the criterion $\tilde{A}\% = 99\%$ implies an additional 3.3 years to wait before the hypothesis of more failures to follow can be rejected.

The regression-based confidence limits provide a means to evaluate the scatter about the best fit of the observed data. The beta distribution describes the scatter of observations from a population with the estimated parameters. This means that it is recommended to take the larger beta-based confidence limit into consideration.

The fact that even the 11th failure is beyond the 99% regression-based confidence limit is a warning signal as such. As mentioned before, Section 10.4.3 discusses the comparison between beta and regression-based confidence limits in more detail.

9.4.2 Failures in a Cable with Multiple Problems and Stress Levels

Two circuits of three-core XLPE cable are installed: circuit A and circuit B (see Figure 9.16). Within 1 hour after energizing a fault occurs in circuit A and after repair a second fault occurs at 17 hours. The first was clearly a case of vandalism and the second was initially assumed to have the same cause, which however never became clear.

Within 4 days two failures occur in circuit B, which was initially thought to be vandalism again, but no external marks were found. A material investigation revealed a considerable extrusion problem (i.e. these were both due to a production error).

It was decided to perform a $2U_0$ test to detect and burn out possible defects. The result was seven more faults distributed over both circuits after which the test paused. After every fault, that cable was repaired (with unsuspected repair cable), after which the $2U_0$ test was resumed. Leaving A and B at U_0 stress after 7.48 hr@$2U_0$, one more fault occurred within a day. It was decided to resume the $2U_0$ test for 8.52 hours on both circuits. Now circuit A remained without fault, but circuit B produced another three faults. Subsequently the circuits were kept at U_0.

An overview of the energizing times at U_0 and the test at $2U_0$, together with an overview of applied stresses and faults, is shown in Table 9.3.

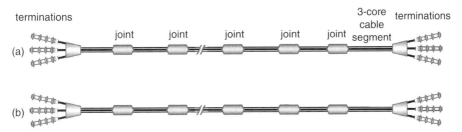

Figure 9.16 Circuits A and B each consist of a three-core XLPE-insulated cable (three phases within one cable). The terminations are generally mounted vertically, but here are drawn horizontally for schematic purposes. In the present case only the cable sections have failures, not the joints and terminations.

Table 9.3 Overview of applied stresses and periods to cables A and B. Also indicated is an overview of faults occurring in both cables with combined times to failure.

	Stress periods (hr)				Total stress periods (hr)	
	T_1 at U_0	T_2 at $2U_0$	T_3 at U_0	T_4 at $2U_0$	$T_{U0} = T_1 + T_3 + T_5$	$T_{2U0} = T_2 + T_4$
Cable A					2478	16
Fault no.						
A1	1				1	
A2	17				17	
A3	1093	2.21			1093	2.21
A4	1093	4.07			1093	4.07
A5	1093	5.12			1093	5.12
A6	1093	5.50			1093	5.50
A7	1093	7.48			1093	7.48
Cable B					3600	16
Fault no.						
B1	56				56	
B2	87				87	
B3	3364	4.50			3364	4.50
B4	3364	5.53			3364	5.53
B5	3364	7.48	18		3382	7.48
B6	3364	7.48	18	0.20	3382	7.68
B7	3364	7.48	18	1.56	3382	9.04
B8	3364	7.48	18	7.01	3382	14.49

This case consists of various types of fault and stressing the cables with two different voltages (U_0 and $2U_0$). What can be learnt from this case? What conclusions can be drawn?

9.4.2.1 References to Introductory Material
The following sections provide introductory material:

- Sections 2.6.1 and 2.6.2: on accelerated ageing and the power law.
- Sections 4.2.1, 4.2.1.1 and 4.2.2: on the Weibull distribution.
- Section 5.4.1: on Weibull plots.
- Section 5.4.5: on confidence intervals.
- Section 6.3.2: on Weibull parameter estimation by linear regression.

9.4.2.2 Analysis of the Case
The following steps were undertaken:

1. Investigate which subpopulations exist by applying the power law.
2. Find an appropriate power for the power law.
3. Evaluate the probability of future faults in these circuits.

Firstly, the existence of subpopulations is explored. From the onset it is clear that there are different failure causes that are most probably unrelated. The first fault was due to vandalism by driving a nail into the cable, the cause of the second fault was unclear, the third and fourth faults were due to an extrusion problem. The question is whether this shows in a plot of the data.

Interestingly, the times to failure are in many cases a blend of ageing at different voltage levels: U_0 and $2U_0$ (see Table 9.3). Section 2.6.1 explains the concept of the power law and Section 2.6.2 explains how ageing dose can be used to translate the ageing time at one stress level to an equivalent ageing time at another stress level. In particular, Eq. (2.68) can be used to translate the failure time of each fault to the equivalent ageing time θ_0 at U_0:

$$\theta_0 = \frac{T_{U_0} \cdot (U_0)^p + T_{2U_0} \cdot (2U_0)^p}{(U_0)^p} = T_{U_0} + 2^p \cdot T_{2U_0} \tag{9.43}$$

Here, t_{U0} and t_{2U0} are the times at the respective stress levels U_0 and $2U_0$; p is the power of the power law. Unfortunately, this power is unknown and performing the research as described in Section 2.6.1 by a series of ramped voltage tests is not very practical.

However, both the subpopulations and the power can be found with the following exercise. As a first step the failure data in Table 9.3 are used to calculate cumulative times to failure θ_0 with Eq. (9.43). After ranking the data for each power (see Table 9.4), the results are as plotted in Figure 9.17. A first observation from Table 9.4 and Figure 9.17 is that the faults A1, A2, B1 and B2 do not connect to the remaining faults and the choice is made to ignore these four faults for further analysis. For A1 the cause (vandalism) was different, but A2 never became clear. B1 and B2 had a problem related to the other faults, but appeared much bigger.

Another question is whether the two sets of faults in cables A and B should be regarded as two different subpopulations. In favour of separating the faults in A from those in B is the fact that the faults in cable A ceased to occur before half of the total stress period on that cable. However, on the other hand, the total stress time in cables A and B does not differ that much and in cable B faults are still occurring. As cables A and B have been produced in the same production batch, it is still possible that sections in A contain

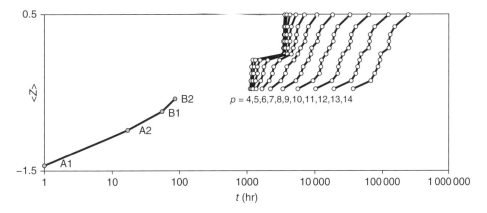

Figure 9.17 Weibull plot of the ranked data from Table 9.4.

Table 9.4 Overview of cumulative times to failure θ_0 (hr) for a range of assumed powers p in Eq. (9.43) applied to the data of Table 9.3.

	$p = 4$	$p = 5$	$p = 6$	$p = 7$	$p = 8$	$p = 9$	$p = 10$	$p = 11$	$p = 12$	$p = 13$	$p = 14$
Cable – operational time (hr)											
A	2 734	2 990	3 502	4 526	6 574	10 670	18 862	35 246	68 014	133 550	264 622
B	3 856	4 112	4 624	5 648	7 696	11 792	19 984	36 368	69 136	134 672	265 744
Fault – time to failure (hr)											
A1	1	1	1	1	1	1	1	1	1	1	1
A2	17	17	17	17	17	17	17	17	17	17	17
A3	1 128	1 164	1 234	1 376	1 659	2 225	3 356	5 619	10 145	19 197	37 302
A4	1 158	1 223	1 353	1 614	2 135	3 177	5 261	9 428	17 764	34 434	67 776
A5	1 175	1 257	1 421	1 748	2 404	3 714	6 336	11 579	22 065	43 036	84 979
A6	1 181	1 269	1 445	1 797	2 501	3 909	6 725	12 357	23 621	46 149	91 205
A7	1 213	1 332	1 572	2 050	3 008	4 923	8 753	16 412	31 731	62 369	123 645
B1	56	56	56	56	56	56	56	56	56	56	56
B2	87	87	87	87	87	87	87	87	87	87	87
B3	3 436	3 508	3 652	3 940	4 516	5 668	7 972	12 580	21 796	40 228	77 092
B4	3 452	3 541	3 718	4 072	4 780	6 195	9 027	14 689	26 015	48 666	93 968
B5	3 502	3 621	3 861	4 339	5 297	7 212	11 042	18 701	34 020	64 658	125 934
B6	3 505	3 628	3 874	4 365	5 348	7 314	11 246	19 111	34 839	66 297	129 211
B7	3 527	3 671	3 961	4 539	5 696	8 010	12 639	21 896	40 410	77 438	151 493
B8	3 614	3 846	4 309	5 237	7 091	10 801	18 220	33 058	62 733	122 084	240 786

errors that are similar to the sections in B that keep failing. It is possible that cable A will again start to fail. It was therefore decided to regard the faults (after exclusion of the first four) as one group for further analysis.

The next step is to find an appropriate power for further analysis. As a criterion, the correlation coefficient of the Weibull plot is optimized with power p. A complication is that it is unknown how many more production errors cables A and B contain. That is, the data are censored. With various assumptions of numbers r of right-censored data (cf. Section 5.2.2), the correlation coefficient $\rho<_Z>_{,\log t}$ was estimated (cf. Section 3.5.2). With assumptions of $r = 0,1,5,11,19$ in all cases the largest correlation is found for $p = 9.2$, which is used for further analysis. The resulting breakdown data are shown in Table 9.5. These are used for further analysis. Figure 9.18 shows the Weibull plot of the data in Table 9.5 with the hypothesis that (at least) one more failure is to follow.

In order to estimate the probability of next failures, it is assumed as a working hypothesis that one more failure will follow. With this hypothesis the Weibull parameters can be estimated. Knowing the failure distribution enables us to draw confidence limits using the beta distribution. The latter distribution can estimate probabilities and corresponding times of the 12th fault out of 12 faults. If this hypothetical last fault did not occur within that confidence interval, then the probability grows that it will not occur at all (i.e. that the problems would be over).

Table 9.5 The breakdown data in cables A and B calculated from Table 9.3 using Eq. (9.43) with $p = 9.2$. The first four faults are excluded as not belonging to this subpopulation.

	t_1	t_2	t_3	t_4	t_5	t_6	t_7	t_8	t_9	t_{10}	t_{11}
hr	2 486	3 659	4 320	4 560	5 808	6 201	6 850	8 097	8 223	9 080	12 516
day	104	152	180	190	242	258	285	337	343	378	521
yr	0.28	0.42	0.49	0.52	0.66	0.71	0.78	0.92	0.94	1.04	1.43

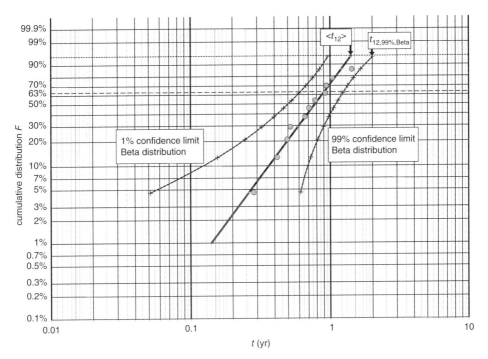

Figure 9.18 Weibull plot with the 1% and 99% confidence limits of the failure data of Table 9.5. It is assumed that one more failure is to come. The plot shows that this 12th failure should occur with 99% confidence before $t = 2$ years (with correction for $n = 12$ as a small data set, then $t = 2.15$ years).

The hypothesis is that more failures are to follow. The test criterion is that the next failure fails with 99% confidence before or at $t_{12,99\%}$. With $p = 9.2$ the total test time on cable A was 1.43 years and on cable B 1.56 years. For the linear regression analysis the Weibull parameters reach the result: $\alpha = 0.91$ years and $\beta = 2.47$. A correction might be applied for $n = 12$ with Eq. (6.198), with leads to $\beta = 2.27$. The estimated times $t_{12,99\%}$ for the estimated Weibull distribution and the corrected distribution are 2 years (respectively 2.15 years).

Again, a Bayesian approach could be used as discussed in Section 9.4.1.3. Taking $\tilde{A}\% = 99\%$ as criterion instead of $A\% = 99\%$ increases the time to reject the hypothesis of more failures: 2.1 and 2.26 years.

An additional $2U_0$ test could be applied to evaluate whether both cables remain failure-free up to the 2 years equivalent. However, if another failure occurs, the conclusion might be that enough is enough and the cables must be replaced without further testing.

9.4.2.3 Remarks

In practice, likely the conclusion would already be drawn that no further testing was worthwhile and total replacement was demanded. More than 11 failures related to production may be deemed enough evidence.

A refinement of the analysis can involve forensic investigation into the production for improved quality control. The aim is to evaluate whether possible correlations exist between the occurrence of errors leading to faults and distinguishable sub-batches of production. Can it be traced back in the production to which sub-batch faults B1 and B2 occurred in? Were there any special circumstances at the time of their production (humidity, extruder settings, temperature, etc.)? Can the other individual faults be traced back to a moment in production and what production parameters applied at those moments? Could there be contaminants in the materials, flaws in the thermal setting of the extruder, did air dryers and filters work well, were there possible influences from the weather (humidity, pressure, etc.), and so on? Such detailed information might also add to the understanding and probability estimation of future faults in each of these cables. Such lessons are also valuable to prevent future production problems.

9.4.3 Case of Cable Joints with Five Early Failures

A DC connection consisting of two parallel circuits each with 100 joints is taken into service (cf. Figure 9.19). The system is planned to operate for > 40 years, but five joints failed much earlier. The rounded-off failure times observed since commissioning were: 58, 78, 90, 100 and 107 days (i.e. times $t_i = 0.159, 0.208, 0.247, 0.274$ and 0.293 years; $i = 1, ..., 5$).

It was considered that these might be due to a child mortality issue, but fears grew that this could also be a systematic problem in the joints. If there were only one or two additional failures, the situation would be unwanted but under control. However, if many more failures followed, the system would require a major overhaul as soon as possible. The stakes are high and the company wants to know the urgency of taking countermeasures. The question is whether and how it can be decided that the trouble is over. What relation exists between time and certainty?

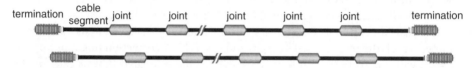

Figure 9.19 A scheme of the redundant connection consisting of two parallel circuits. For ease of schematics the terminations are drawn horizontally, but in real life these are normally mounted vertically.

9.4.3.1   References to Introductory Material

The following sections provide introductory material:

- Sections 2.4.3 and 2.4.4: on distribution density and mass function.
- Sections 2.4.5 and 2.5.3: on the bath tub model and child mortality.
- Section 3.6: on similarity of distributions.
- Sections 4.2.1, 4.2.1.1 and 4.2.2: on the Weibull distribution.
- Section 5.4.1: on Weibull plots.
- Section 5.4.5: on confidence intervals.
- Section 6.3.2: on Weibull parameter estimation by linear regression.
- Sections 7.4, 8.3, 8.6 and 8.7: on redundancy, reliability and availability.

9.4.3.2   Analysis of the Case

The following steps were undertaken:

1. An initial review and interpretation of the data.
2. Prognosis using a Weibull plot and confidence intervals (Section 9.4.3.3).
3. Estimation of sample size using the similarity index (Section 9.4.3.4).
4. Evaluating the urgency based on reliability and availability of the system (Section 9.4.3.5).

The data are shown in Table 9.6 with the intervals between consecutive failures. As a first impression, the time interval between the consecutive failure times is shortening, which suggests that the problem is accelerating as yet.

It is unknown how many joints feature the problem that leads to these early failures. Nevertheless, some initial conclusions may be drawn about the failure distribution at this stage. As each failure means that a fixed percentage ΔF of the population fails and the time between failures Δt is shortening according to Table 9.6, the distribution density $f(t) = dF/dt$ is apparently increasing. Likewise, the hazard rate $h(t)$ is increasing. Therefore, the present data do not support the idea of a child mortality process as in Section 2.4.5.

It is possible that only a limited subpopulation suffers from this phenomenon, and then a child mortality process in the sense of subpopulations that die out may apply, as described in Section 2.5.3. In a worst-case scenario the subpopulation may be a considerable fraction of the total population, if not the complete population. The main question is how many more failures can be expected and when.

Table 9.6 Data for the case of cable joints with early failures. The time between consecutive failures is shortening, indicating that the probability density has not reached its maximum as yet.

i	1	2	3	4	5
t_i (yr)	0.159	0.208	0.247	0.274	0.293
$t_i - t_{i-1}$ (yr)	0.159	0.049	0.038	0.027	0.019
$1/(t_i - t_{i-1})$ (yr^{-1})	6.3	20	26	37	52

Two approaches are applied here: the first is to evaluate the data with a Weibull plot and the likelihood that the problems are running to an end; the second is to use the similarity index to estimate the total number of failures.

9.4.3.3 Prognosis Using a Weibull Plot and Confidence Intervals

As for the first approach, a Weibull plot (cf. Section 5.4.1) is produced with various assumptions of failures to come. This means that there is an assumption about the sample size n of total affected joints, of which the first $r = 5$ observations are uncensored and the remaining $k = n - r$ observations are censored (i.e. unknown as yet). The results are shown in Figure 9.20 for $n = 5, 6, 15, 105$ and 255 (i.e. with $k = 0, 1, 10, 100$ and 250 censored observations). For each distribution the Weibull parameters are estimated.

The plotting positions $<Z_i>$ are given in Section 5.4.1, Eq. (5.24) and may alternatively be approximated as in Eqs (5.27) and (5.28). As the censored data are larger than the largest observed failure time (right-censored), there is no difference between adjusted ranking and adjusted plotting position (cf. Section 5.2.2).

For every n the data show very limited scatter. The regression based confidence intervals of the sample are certainly smaller than the confidence limits as provided by the beta distribution. As discussed before, this happens quite often. This population may just be regarded as a random sample from a larger population, but also as a complete population as such. As a criterion one might use the correlation coefficient as a measure for which n fits the data best. According to Figure 9.20, the data scatter less around the

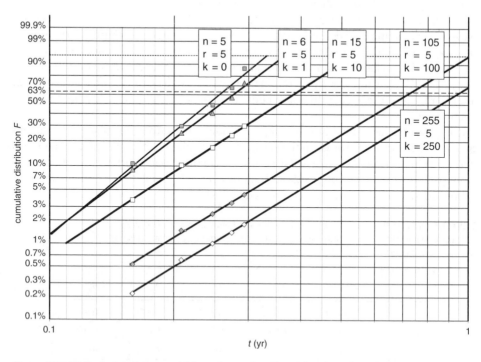

Figure 9.20 Weibull plot of observed failure times t_i ($r = 5$) and their best fits with various assumed number of expected failures n, of which r failed and $k = n - r$ are supposedly yet to come. The fits are obtained through unweighted linear regression.

best linear regression fit with higher assumed n. It was found that the correlation coefficient grew from 0.9932 for $n = 5$ to 0.999157 for $n = 11$ and from then on dropped to 0.9986 at $n = 305$, which may be taken as an indication that more failures can be expected since the correlation is better for $n > 5$. However, in all cases the fit is already quite good.

As a next step it may be evaluated with what probability the next failure might occur. Since the present number of failures $r = 5$, the next failure would be number $i = 6$. Based on the beta distribution, the confidence limits in Table 9.7 are calculated based on the Weibull parameters shown in Table 9.8 (as in Section 9.4.1.2 and Eq. (9.32)):

$$t_{i,n,A\%} = \alpha \cdot [-\ln(1 - F_{i,n,A\%}(t))]^{\frac{1}{\beta}} \tag{9.44}$$

The moment of evaluation was shortly after the last failure at $t = 0.293$ years (i.e. after $t = 107$ days). Table 9.7 shows for $n = 6$ and Weibull parameters $\alpha = 0.28$ and $\beta = 4.2$ that the next failure would occur with probability $A\%$ at or before $t_{6,6,A\%}$. Since it is already known that the sixth failure will not occur before day 107 while $A\%(107\ \text{days})$ is about 11%, the survival probability is 89%. The $A\%$ confidence limits can be corrected with

Table 9.7 Beta distribution-based probabilities and corresponding critical times $t_{i,n,A\%}$ that the sixth failure out of six occurred at or before.

A%	50%	66.7%	75%	80%	90%	95%	97.5%	99%	99.5%	99.9%
$t_{6,6,A\%}$ (yr)	0.34	0.36	0.37	0.37	0.39	0.41	0.42	0.44	0.45	0.47
$t_{6,6,A\%}$ (day)	124	131	134	137	143	149	154	160	164	172

Table 9.8 Estimated Weibull parameters for various sample sizes n. With these parameters the first five times to failure are estimated as expected times $<\tau_{i,n}>$ and median times $\tau_{M,i,n}$. The last two rows are the similarity indices according to Eqs (9.53) and (9.54), respectively.

n	5	6	7	8	9	10	20	30	50
a_n	0.262	0.282	0.300	0.316	0.330	0.342	0.435	0.497	0.579
b_n	4.6	4.2	4.0	3.9	3.8	3.7	3.5	3.5	3.4
$<\tau_{1,n}>$	0.168	0.167	0.167	0.167	0.167	0.167	0.168	0.168	0.168
$<\tau_{2,n}>$	0.210	0.212	0.213	0.214	0.215	0.215	0.217	0.217	0.217
$<\tau_{3,n}>$	0.241	0.244	0.246	0.246	0.247	0.247	0.249	0.250	0.249
$<\tau_{4,n}>$	0.270	0.273	0.273	0.274	0.274	0.274	0.275	0.275	0.274
$<\tau_{5,n}>$	0.307	0.303	0.301	0.300	0.299	0.298	0.297	0.296	0.294
$\tau_{M,1,n}$	0.170	0.169	0.168	0.168	0.168	0.168	0.168	0.168	0.167
$\tau_{M,2,n}$	0.211	0.213	0.214	0.215	0.215	0.216	0.217	0.217	0.217
$\tau_{M,3,n}$	0.242	0.245	0.246	0.247	0.248	0.248	0.250	0.250	0.249
$\tau_{M,4,n}$	0.271	0.273	0.274	0.274	0.275	0.275	0.275	0.275	0.274
$\tau_{M,5,n}$	0.306	0.303	0.301	0.300	0.299	0.299	0.297	0.297	0.295
$S_{<\tau n>}$	0.856	0.917	0.944	0.958	0.966	0.972	0.986	0.989	0.990
$S_{\tau M,n}$	0.862	0.920	0.946	0.959	0.967	0.972	0.986	0.989	0.990

the theorem of Bayes: the condition is that only joints are considered that survive day 107 (i.e. $R(107) = 100\%$ for those joints). It is assumed that only $A\%$ are considered that correspond to a time $t_{A\%} \geq 107$ days. This implies $R(107 \text{ days}|t_{A\%}) = 1$. The probability that the sixth joint survives, $t_{A\%}$, is $1 - A\%$. It follows for the sixth joint (see Section 2.5.2.1 and Eq. (2.42)) that:

$$
\begin{aligned}
&P(\text{survive } t_{A\%} \mid \text{survive } 107 \text{ days}) \\
&= \frac{P(\text{survive } 107 \text{ days} \mid \text{survive } t_{A\%}) \cdot P(\text{survive } t_{A\%})}{P(\text{survive } 107 \text{ days})} \\
&= \frac{1 \cdot (1 - A\%)}{R(107 \text{ days})} = \frac{1 - A\%}{0.89} = 1 - \tilde{A}\%
\end{aligned}
\tag{9.45}
$$

Here $\tilde{A}\%$ is the probability of the sixth joint failing before time $t_{A\%}$, which corresponds to the beta-based confidence limit $A\%$. Probability $\tilde{A}\%$ relates to $A\%$ as:

$$
\tilde{A}\% = 1 - \frac{1 - A\%}{0.89} = \frac{A\% - 0.11}{0.89}
\tag{9.46}
$$

If confidence limits are based on the scatter in the sample (cf. Section 10.3), the confidence limits will be much narrower. It depends on the interpretation of the sample and the population, which would be used as a criterion to evaluate whether the subpopulation with early failures is extinct or not. Using the beta distribution as in Table 9.7, the conclusion is that it would take weeks without failure before confidence grows sufficiently to justify the conclusion that the problems are over.

9.4.3.4 Estimation of Sample Size Using the Similarity Index

The number of affected joints n is unknown. It is known though that with each next failure a population fraction of $\Delta F = 1/n$ fails. The observed failure times t_i are given in Table 9.6. The probability density $f_{i,n}$ may be estimated as:

$$
f_{i,n}(t_i) = \frac{\Delta F_{i,n}}{\Delta t} = \frac{1}{n} \cdot \frac{1}{t_i - t_{i-1}}
\tag{9.47}
$$

The last row of Table 9.6 shows the values of $n \cdot f_{i,n}$.

On the other hand, by assuming a sample size n, the Weibull parameters a_n and b_n can be estimated. From a given sample size n and estimated Weibull parameters a_n and b_n the first five times to failure can be estimated. Table 9.8 shows the Weibull parameters estimated by unweighted linear regression (Section 6.3.2.1) for various n.

Subsequently, times to failure are estimated as the expected failures time $\langle \tau_{i,n} \rangle$ and median failure times $\tau_{M,i,n}$. Normally, only one of these methods is used, but here the two are determined in parallel for comparison. The expected times are obtained as in Section 4.1.2.1, Eq. (4.14):

$$
\begin{aligned}
\langle \tau_{i,n} \rangle &= \frac{\Gamma(n+1)}{\Gamma(i) \cdot \Gamma(n+1-i)} \int_0^1 \tau(p) \cdot p^{i-1} \cdot (1-p)^{n-i} dp \\
&= \frac{\Gamma(n+1)}{\Gamma(i) \cdot \Gamma(n+1-i)} \int_0^1 a_n \cdot [-\ln(1-p)]^{\frac{1}{b_n}} \cdot p^{i-1} \cdot (1-p)^{n-i} dp
\end{aligned}
\tag{9.48}
$$

The median times to failure $\tau_{M,i,n}$ are found by finding the median ranked probability $F_{M,i,n}$ as in Section 4.1.2.1, Eq. (4.19):

$$
F_{M,i,n} = F_{i,n,50\%} = B^{inv}(50\%; i, n+1-i)
\tag{9.49}
$$

Subsequently, the median times are found as:

$$\tau_{M,i,n} = F^{inv}(F_{M,i,n}) = a_n \cdot [-\ln(1 - F_{M,i,n})]^{\frac{1}{b_n}} \tag{9.50}$$

Similarly to approximating the probability density $f_{i,n}$ from the observed times t_i with Eq. (9.47), the estimated failure times $\langle \tau_{i,n} \rangle$ and the median failure times $\tau_{M,i,n}$ can be used to find probability densities $g_{i,n}$ as:

$$g_{i,n}(\langle \tau_{i,n} \rangle) = \frac{1}{n} \cdot \frac{1}{\langle \tau_{i,n} \rangle - \langle \tau_{i-1,n} \rangle} \tag{9.51}$$

and

$$g_{i,n}(\tau_{M,i,n}) = \frac{1}{n} \cdot \frac{1}{\tau_{M,i,n} - \tau_{M,i-1,n}} \tag{9.52}$$

The similarity index S_{fg} for discrete outcome sets (see Section 3.6.2) is a measure to evaluate how similar two sets of outcomes f_i and g_i are. In the present case it can be used to determine for which sample size n the highest similarity of $g_{i,n}$ with $f_{i,n}$ is found. The corresponding sample size n apparently describes the observations best and may be used as a measure of the number of affected joints. The similarity index S_{fg} follows from Eq. (3.105). With the expected times it is indicated as $S_{\langle \tau_{i,n} \rangle}$ and becomes:

$$
S_{\langle \tau_n \rangle} = \frac{\sum\limits_i f_{i,n} \cdot g_{i,n}}{\sum\limits_i f_{i,n} \cdot f_{i,n} + \sum\limits_i g_{i,n} \cdot g_{i,n} - \sum\limits_i f_{i,n} \cdot g_{i,n}}
$$
$$
= \frac{\sum\limits_i \frac{1}{t_i - t_{i-1}} \cdot \frac{1}{\langle \tau_{i,n} \rangle - \langle \tau_{i-1,n} \rangle}}{\sum\limits_i \left(\frac{1}{t_i - t_{i-1}} \right)^2 + \sum\limits_i \left(\frac{1}{\langle \tau_{i,n} \rangle - \langle \tau_{i-1,n} \rangle} \right)^2 - \sum\limits_i \frac{1}{t_i - t_{i-1}} \cdot \frac{1}{\langle \tau_{i,n} \rangle - \langle \tau_{i-1,n} \rangle}} \tag{9.53}
$$

With the median times it is indicated as $S_{\tau M,i,n}$ and becomes:

$$
S_{\tau_{M,n}} = \frac{\sum\limits_i f_{i,n} \cdot g_{i,n}}{\sum\limits_i f_{i,n} \cdot f_{i,n} + \sum\limits_i g_{i,n} \cdot g_{i,n} - \sum\limits_i f_{i,n} \cdot g_{i,n}}
$$
$$
= \frac{\sum\limits_i \frac{1}{t_i - t_{i-1}} \cdot \frac{1}{\tau_{M,i,n} - \tau_{M,i-1,n}}}{\sum\limits_i \left(\frac{1}{t_i - t_{i-1}} \right)^2 + \sum\limits_i \left(\frac{1}{\tau_{M,i,n} - \tau_{M,i-1,n}} \right)^2 - \sum\limits_i \frac{1}{t_i - t_{i-1}} \cdot \frac{1}{\tau_{M,i,n} - \tau_{M,i-1,n}}} \tag{9.54}
$$

The sample size n drops out of the equations. The similarity indices can be determined with just the observed failure times and the estimated failure times.

The last two rows of Table 9.8 show the similarity indices for a range of sample sizes n. Figure 9.21 shows the results of a wider range of sample sizes n. The results for the expected times and the median times resemble each other very well. Some details do differ though. For the expected times a maximum similarity index is found at $n = 90$, while for the median times the similarity index keeps increasing slightly but steadily with increasing n, up to at least 500. The question is how significant the increase is and how exact the maximum of the similarity index is. The relatively fast growth of the similarity index with n up to 20–30 is an indication that a significant number of failures can be expected.

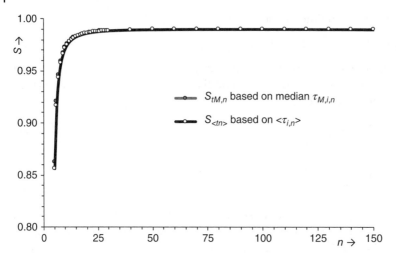

Figure 9.21 Similarity indices of approximated distribution densities $f_{i,n}$ and $g_{i,n}$.

The principle of the method with approximated distribution densities and the similarity index is illustrated in Figure 9.22. The observed distribution density $n \cdot f_{i,n}$ is still increasing. The estimated density $n \cdot g_{i,n}$ at $t \approx 0.3$ years is already decreasing for $n = 5$ and 6, which would mean that the distribution density is in its tails and the worst problems would be over. For $n = 10$ the distribution density tends to flatten, which would mean it is approaching its mode and the maximum of the problems. For $n = 20$ and 30, the estimated $n \cdot g_{i,n}$ is still rising and the shape starts to resemble that of $n \cdot f_{i,n}$. However, the increase is not yet as steep as found for the observed $n \cdot f_{i,n}$. This indicates that the distribution density is still at its onset and the worst is yet to come. It is concluded that

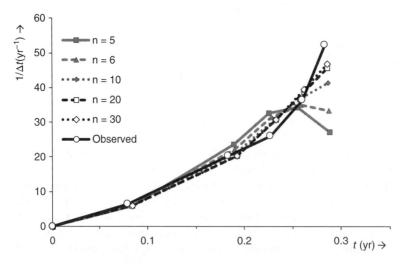

Figure 9.22 Approximations of the distribution densities $n \cdot f_{i,n}$ and $n \cdot g_{i,n}$ plotted against time, which is taken as half the interval (i.e. $1/(<t_i> - <t_{i-1}>)$ is plotted against $(<t_i> + <t_{i-1}>)/2$, etc.). Time t_i with $i \leq 0$ is taken as 0 and both $n \cdot f_{0,n}$ and $n \cdot g_{0,n}$ are also set to 0 for wear-out processes. With sample sizes $n > 30$ the similarity is still improving (see text).

the best similarity is found for high sample sizes and therefore many more failures are likely to occur.

9.4.3.5 Redundancy and Urgency

As discussed in Sections 7.4 and 8.3, installing parallel circuits provides a redundant system with a higher reliability and availability than a single circuit would provide. Whether acceptable levels of reliability and availability are achieved depends on the balance between failure rate and repair rate.

The time between failures MTBF dropped to a worrying 7 days (0.019 years) and the failure rate λ is therefore 0.14/day or 52.6/year. This is extremely high for grid components. Repair of a high-voltage cable typically takes 1–2 weeks depending on the applied cable technology. In special circumstances the repair time may be even (much) longer. Normally, for a given cable technology the repair time is fairly predictable. With the Markov–Laplace method a repair rate μ is used for ease of calculation. For the sake of simplicity, 1 week is assumed (i.e. $\mu = 0.14$/day), which equals λ in this case.

The MTBF is the time between failures of the system (i.e. the time between black-outs – total interruption of power supply). According to Section 8.6, Eq. (8.69)):

$$\text{MTBF} = \frac{2\lambda + \mu}{2\lambda^2} \tag{9.55}$$

Equation (9.55) yields MTBF = 10.5 days (i.e. every 1.5 weeks this system is down for about 1 week) and that is probably totally unacceptable. Likewise, the availability in the long run was given in Section 8.7, Eq. (8.76) as:

$$A_\infty = \frac{\mu^2 + 2\lambda\mu}{\mu^2 + 2\lambda\mu + 2\lambda^2} \tag{9.56}$$

Equation (9.56) yields an estimated availability of about 0.6, which means that the power supply is disturbed 40% of the time. The supply is severely hindered by the present problems.

9.4.3.6 Remarks

The present case initially discussed the likelihood of the next failure (Section 9.4.3.3). The hypothesis is that more failures will follow and by firstly assuming one more failure, it can be decided when that failure might occur and before which time the failure would occur with a certain confidence $A\%$. If no failure happens until that time, the hypothesis may be rejected and the conclusion is that no more failures of that kind come at all. On the other hand, if a failure does occur before that time, then the exercise is repeated. With ongoing failures the method confirms the hypothesis, but after each new failure the hypothesis is questioned again. The hypothesis method does not give an impression of how many failures can be expected to follow.

The method with approximations for the distribution density and use of the similarity index (Section 9.4.3.4) estimates the total sample size n. It is therefore a method to forecast how many more failures can be expected. It is recommendable to carry out simulations as discussed in Section 3.6.4 (see also Section 10.4) to evaluate the significance of the similarity index. Alternatively, the errors in the Weibull parameters may be estimated and from those the time errors and the similarity index errors may be determined to get a sense of accuracy.

Some refinements are possible in the similarity index method. The distribution densities might be estimated as continuous distributions, after which Section 3.6.3 may be applied.

The present case is characterized by relatively little scatter of the data. It is quite common that a limited data set features smaller confidence intervals based on the regressions than the confidence intervals that the beta distribution predicts. Section 10.4.3 shows how data sets from the n-dimensional distribution space produce less scatter. If the data are not independent and not randomly drawn, then there may be a reason why the sample is drawn from a subspace (cf. Section 5.2.6) and then the regression-based confidence limits apply following Section 10.3. However, if there appears no reason why the sample might be from a particular subspace, then the beta distribution-based confidence limits must be reckoned with.

In a similar situation in practice, a forensic investigation revealed a systematic error in the joints in that case. The similarity index method estimates a high number of failures yet to come, which was supported by the forensic evidence. However, Section 9.4.4 describes a case with much more scatter, which naturally affects the firmness of inferences.

9.4.4 Joint Failure Data with Five Early Failures and Large Scatter

Not every case is characterized with such a smooth data set as in Section 9.4.3. The case presented here concerns 30 joints installed in a cable bed of six single-phase cables that feed a production plant as in Figure 9.23.

The cables were designed to stay in service for at least 30 years (about 11 000 days), but joints failed after 1250, 1400, 1700, 1750 and 2150 days (i.e. 3.42, 3.84, 4.66, 4.79 and 5.89 years). The forensic studies remained inconclusive. Since the last failure at day 2150, a period of 550 days (1.51 years) has passed by without new failures. The date of evaluation is therefore day 2700. Hopes grew that the problems were over. However, outages were costly and a statistical analysis is required to assess the probabilities of more failures.

9.4.4.1 References to Introductory Material
The following sections provide introductory material:

- Sections 2.4.3 and 2.4.4: on distribution density and mass function.
- Sections 2.4.5 and 2.5.3: on the bath tub model and child mortality.
- Section 3.6: on similarity of distributions.

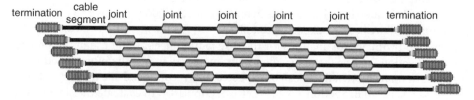

Figure 9.23 A cable bed consisting of six single-phase cables with five joints each. This connection can be split into two circuits of three single-phase cables each. For ease of schematics the terminations are drawn horizontally, but in real life these are normally mounted vertically.

Table 9.9 Data for the case of cable joints with early failures. The time between consecutive failures is varying due to the scatter of the times and there is no single maximum.

i	1	2	3	4	5
t_i (yr)	3.42	3.84	4.66	4.79	5.89
$t_i - t_{i-1}$ (yr)	3.42	0.42	0.82	0.13	1.10
$1/(t_i - t_{i-1})$ (yr^{-1})	0.29	2.38	1.22	7.69	0.91

- Sections 4.2.1, 4.2.1.1 and 4.2.2: on the Weibull distribution.
- Section 5.4.1: on Weibull plots.
- Section 5.4.5: on confidence intervals.
- Section 6.3.2: on Weibull parameter estimation by linear regression.

9.4.4.2 Analysis of the Case
The following steps were undertaken:

1. An initial review and interpretation of the data.
2. Prognosis using a Weibull plot and confidence intervals (Section 9.4.3.3).
3. Estimation of sample size using the similarity index (Section 9.4.3.4).

The data are shown in Table 9.9 with the intervals between consecutive failures. As a first impression, the time interval between the consecutive failure times varies considerably and it is uncertain whether the process is accelerating or not.

As in Section 9.4.3 it is unknown how many joints feature the problem that leads to these early failures. As mentioned, the intervals between the consecutive failures vary considerably. Using the approach again of each failure meaning that a fixed percentage ΔF of the population fails and realizing that with the time between failures Δt is varying, the distribution density $f(t) = dF/dt$ is apparently varying (i.e. featuring considerable scatter). The data may therefore be less conclusive than in the previous case.

The main question remains: how many more failures can be expected and when? Again, the two approaches of Section 9.4.3 are applied here: the first is to evaluate the data with a Weibull plot and the likelihood that the problems are running to an end; the second is to use the similarity index to estimate the total number of failures.

9.4.4.3 Prognosis Using a Weibull Plot and Confidence Intervals
As for the first approach, a Weibull plot (cf. Section 5.4.1) is produced with various assumptions of failures to come. This means that there is an assumption about the sample size n of total affected joints: the first $r = 5$ observations are uncensored and the remaining $k = n - r$ observations are censored (i.e. unknown as yet). The results are shown in Figure 9.24 for $n = 5$, 6, 14 and 30 (i.e. with $k = 0$, 1, 9 and 25 censored observations). For each distribution, the Weibull parameters are estimated.

Again, the plotting positions $<Z_i>$ are given in Section 5.4.1, Eq. (5.24) and may alternatively be approximated as in Eqs (5.27) and (5.28). As the censored data are larger than the largest observed failure time (right-censored), there is no difference between adjusted ranking and adjusted plotting position (cf. Section 5.2.2).

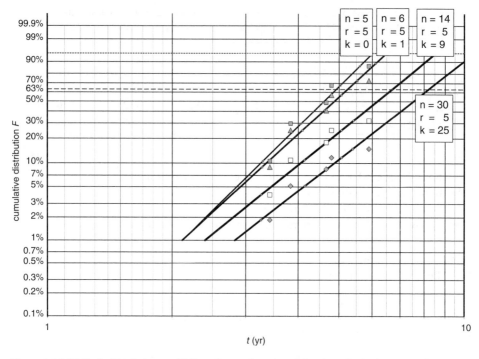

Figure 9.24 Weibull plot of observed failure times t_i ($r = 5$) and their best fits with various assumed number of expected failures n, of which r failed and $k = n - r$ are supposedly yet to come. The fits are obtained through unweighted linear regression.

Figure 9.24 shows quite some scatter for every n. The least scatter might be an indication for the best linear regression fit and consequently the best estimation of n. As a criterion, the correlation coefficient was first used as a measure for which n fits the data best. It was found that the correlation coefficient started as 0.973 for $n = 5$ and dropped off from there to 0.964 for $n = 6$ and further to 0.950 for $n = 14$ and 0.944 for $n = 30$, which may be taken as an indication that very few more failures can be expected since the correlation is best for $n = 5$. Although this conclusion cannot be firm due to the considerable scatter, even with lacking data it is still the best objective guess.

As a next step it can be evaluated with what probability the next failure might occur, similar to the approach in Sections 9.4.1.3 and 9.4.3.3. Having $r = 5$ observed times, the next failure would be $i = 6$. Based on the beta distribution the confidence limits in Table 9.10 are calculated. The Weibull plot is shown in Figure 9.25.

Table 9.10 Beta distribution-based confidence limits and corresponding critical times $t_{i,n,A\%}$ that the sixth failure out of six occurred at or before.

A%	50%	66.7%	75%	80%	90%	95%	97.5%	99%	99.5%	99.9%
$t_{6,6,A\%}$ (yr)	6.15	6.41	6.56	6.66	6.94	7.16	7.36	7.59	7.75	8.07
$t_{6,6,A\%}$ (day)	2246	2341	2395	2432	2531	2615	2687	2771	2829	2946

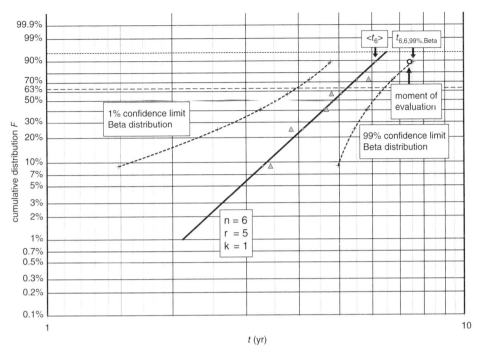

Figure 9.25 Weibull plot assuming a sample size of $n = 6$ with 1% and 99% confidence limits and indications of $<t_6>$ and $t_{6,6,99\%}$. The estimated Weibull parameters are: $a = 5.25$ years and $b = 5.04$.

The remark in Section 9.4.4 that a period of 550 days (1.51 years) passed by without new failures since the last failure at day 2150 is relevant. The evaluation date is day 2700 (i.e. year 7.40). This is very close to the time $t_{99\%}$ at which the hypothesis might be rejected. The expected time of a hypothetical sixth failure is 6.11 years.

There are two approaches that urge us to monitor the situation for a longer period. The first approach acknowledges that the sample size is small. Unbiasing for the parameter followed by estimating the time $t_{99\%}$ is not yet sorted out, but could be investigated with, for example, Monte Carlo simulation (see Section 10.4.4). As a guess, the bias of ML and LS is compared in Section 6.3.2.7, and only for the case of ML is a bias-enhancing factor for censoring known, namely $\sqrt{(n/r)}$ as discussed in Section 6.2.3.4, Eq. (6.62). If this factor also applies to the LS shape parameter, then the unbiased $b_{U,n=6}$ would be found by multiplying $b_{n=6}$ with a factor 0.843. The assumed unbiased shape parameter then becomes 4.2 instead of 5.04. As a consequence, $t_{6,6,99\%}$ would shift from 7.59 to 8.13 years (i.e. an additional 0.54 years). The expected time $<t_{6,6}>$ would rise from 6.11 to 6.29 years. This date is already passed at the moment of evaluation (7.4 years), which would now correspond to $A\% = 92\%$.

The second reason to evaluate for a longer period is based on the Bayesian approach when $\bar{A}\%$ confidence limits are used based on the observation that $t_6 > 5.89$ years (see Sections 9.4.1.2 and 9.4.3.3). This would increase $t_{6,6,A\%-99\%} = 7.59$ years to $t_{6,6,\bar{A}\%-99\%} = 7.68$ years. Involving the unbiasing effect would first decrease the shape parameter to $b_6 = 4.2$ years and subsequent calculations would yield $t_{6,6,\bar{A}\%-99\%} = 8.22$ years. This assumed correction is based on the assumption that censoring with LS also increases the bias with a factor $\sqrt{(n/r)}$.

9.4.4.4 Estimation of Sample Size Using the Similarity Index

Next, the similarity index exercise of Section 9.4.3.4 is also applied to this case. In that section the mathematics is explained. The data of the present case are shown in Table 9.9, of which the last row shows $1/(t_i - t_{i-1})$, the values that are also the values of $n \cdot f_{i,n}$ (cf. Eq. (9.29)):

$$\frac{1}{t_i - t_{i-1}} = n \cdot \frac{\Delta F_{i,n}}{\Delta t} = n \cdot f_{i,n}(t_i) \tag{9.57}$$

For various n the (as yet biased) Weibull parameters a_n and b_n are determined and listed in Table 9.11.

From Table 9.11 it can be seen that $S_{<\tau n>}$ and $S_{\tau M,n}$ agree quite well within typically 1%, but the most important conclusion is that the similarity with the reference remains very low at values even below 0.5 over a wide range (see Figure 9.26), which is due to the double peak in the frequency plot (see Figure 9.27), which in turn is the result of t_4 following t_3 relatively fast.

The similarity index method is not very conclusive. It correctly yields values <0.5 warning about the uncertainty in the data.

9.4.4.5 Remarks

As discussed in Section 6.1.2, small data sets are regularly encountered in asset management. It can be due to the need to limit samples for expensive tests or due to the occurrence of a crisis caused by a surprising number of failures in a short time. In order to get the situation under control after failures, a fast state update is required, possible scenarios have to be evaluated and early decisions have to be made based on the relatively scarce available data. Almost by definition, early or timely decision-making has

Table 9.11 Estimated Weibull parameters for various sample sizes n. With these parameters the first five times to failure are estimated as expected times $<\tau_{i,n}>$ and median times $\tau_{M,i,n}$. The calculation of the times $<\tau_{i,n}>$ and $\tau_{M,i,n}$ and the similarity indices $S_{<\tau n>}$ and $S_{\tau M,n}$ is discussed in Section 9.4.3.4.

n	5	6	7	8	10	12	14	20	30
a_n	4.94	5.25	5.52	5.74	6.13	6.46	6.74	7.43	8.27
b_n	5.4	5.0	4.8	4.7	4.6	4.5	4.4	4.3	4.3
$<\tau_{1,n}>$	3.387	3.383	3.384	3.386	3.389	3.392	3.394	3.398	3.403
$<\tau_{2,n}>$	4.098	4.130	4.149	4.161	4.176	4.185	4.191	4.201	4.209
$<\tau_{3,n}>$	4.597	4.643	4.664	4.677	4.692	4.700	4.705	4.714	4.720
$<\tau_{4,n}>$	5.064	5.091	5.099	5.103	5.106	5.107	5.108	5.109	5.110
$<\tau_{5,n}>$	5.636	5.553	5.517	5.496	5.473	5.461	5.453	5.441	5.432
$\tau_{M,1,n}$	3.431	3.424	3.423	3.422	3.423	3.424	3.424	3.426	3.427
$\tau_{M,2,n}$	4.125	4.157	4.175	4.186	4.201	4.209	4.215	4.224	4.231
$\tau_{M,3,n}$	4.616	4.662	4.683	4.696	4.710	4.719	4.724	4.732	4.738
$\tau_{M,4,n}$	5.075	5.104	5.113	5.117	5.121	5.122	5.123	5.124	5.124
$\tau_{M,5,n}$	5.635	5.560	5.527	5.507	5.485	5.473	5.465	5.453	5.445
$S_{<\tau n>}$	0.420	0.429	0.435	0.440	0.446	0.450	0.452	0.456	0.459
$S_{\tau M,n}$	0.427	0.434	0.440	0.444	0.450	0.453	0.456	0.459	0.462

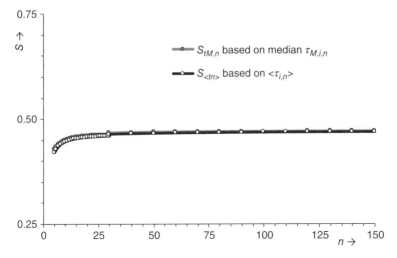

Figure 9.26 Similarity indices of approximated distribution densities $f_{i,n}$ and $g_{i,n}$. The values remain very low, which indicates a poor similarity.

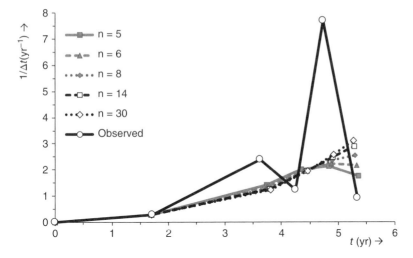

Figure 9.27 Approximations of the distribution densities $n \cdot f_{i,n}$ and $n \cdot g_{i,n}$ plotted against time, which is taken as half the interval (i.e. $1/(<t_i> - <t_{i-1}>)$ is plotted against $(<t_i> + <t_{i-1}>)/2$, etc.). Time t_i with $i \leq 0$ is taken as 0 and both $n \cdot f_{0,n}$ and $n \cdot g_{0,n}$ are also set to 0 for wear-out processes.

to reckon with censored data. If there were no worries about a significant number of imminent failures, then the decision-making would be unnecessary.

In Section 10.4.3 the effect of sampling small sets on the scatter and confidence limits is discussed. It is shown that many samples have a much smaller scatter than might be expected from the beta-based confidence limits at first sight, but the regression-based confidence and the beta-based confidence have different meanings. Some sampled sets do have large scatter though, as can be seen in Table 10.7 and Figure 10.21 in Section 10.4.3.2. The presently discussed case may just be one of those.

The method with the similarity indices (Section 9.4.4.4) is more sensitive to variations in the time between failures $(t_i - t_{i-1})$ than the confidence limit method (Section 9.4.4.3). The first aims at estimating the expected sample size n, but remained inconclusive, while the second aims at the probability of the next failure and its expected failure time.

To enlarge the data sets and increase the accuracy of the statistical analysis, it is advisable to evaluate whether data from comparable cases can be grouped. For instance, in Section 9.4.2.2 it was decided to regard the failure data of cables A and B as one group.

Finally, this case was also discussed in [70] using the ML method with unbiasing for estimating the Weibull parameters. The results deviate somewhat but are quite comparable to the findings in Table 9.10.

10

Miscellaneous Subjects

This chapter contains a variety of subjects in support of the matter discussed in previous chapters. The subjects discussed in this c1hapter are:

- Combinatorics (Section 10.1 ff.): permutations, combinations and the gamma function. These subjects enable counting possibilities in statistics.
- Power functions and asymptotic behaviour (Section 10.2 ff.): Taylor and Maclaurin series, polynomial fitting and power function fitting. The latter method is used extensively for bias and variance description.
- Regression analysis (Section 10.3): the unweighted and the weighted least-squares techniques are elaborated together with error analysis and regression-based confidence limits.
- Sampling from a population and simulations (Section 10.4 ff.): systematic sampling, numerical integration and expected values, ranked samples with size n and confidence limits, Monte Carlo experiments and random number generators. These subjects give background information for statistical experiments and illustrate the meaning of beta distribution-based and regression-based confidence limits.
- Hypothesis testing (Section 10.5): null hypothesis H_0 and alternative hypothesis H_1 together with Type I and Type II errors.
- Approximations for the normal distribution (Section 10.6) using Taylor series and the power function fitting.

10.1 Basics of Combinatorics

Combinatorics is a branch of mathematics that studies the counting of possible ways for combining and ranking elements from a group. In the next two subsections the concepts of permutations and combinations are discussed, followed by a description of the gamma function that can be used for both integer and non-integer numbers.

10.1.1 Permutations and Combinations

If a set consists of n distinct elements with n a natural number ($n = 0,1,2,\ldots$), then the factorial function $n!$ describes the number of ways to arrange the elements in sequences. These arrangements are also called the permutations of the set of elements. If the set is not empty, then $n > 0$. In that case, there are n elements to choose from for the first position. If there is only $n = 1$ element there is just one possible permutation. Provided

Reliability Analysis for Asset Management of Electric Power Grids, First Edition. Robert Ross.
© 2019 John Wiley & Sons Ltd. Published 2019 by John Wiley & Sons Ltd.
Companion website: www.wiley.com/go/ross/reliabilityanalysis

$n > 1$, there are $n - 1$ elements to choose from for the second position and so on until the nth element is also selected. Therefore, the number of permutations is:

$$n! = n \cdot (n-1) \ldots 2 \cdot 1 = \prod_{i=1}^{n} i \tag{10.1}$$

This can also be expressed as a recurrent relationship:

$$n! = n \cdot (n-1)! \tag{10.2}$$

If the set is empty, then $n = 0$. The number of permutations of this empty space is the factorial of 0. By definition, there is only one way to arrange an empty space, that is:

$$0! = 1 \tag{10.3}$$

If neither the full arrangements nor the sequence are of interest, but merely a smaller selection of k elements out of a set of n elements (so with $k < n$), these selections are called combinations. The number of combinations is given by the binomial coefficient, which is derived from the factorial function:

$$\binom{n}{k} = \frac{n!}{k! \cdot (n-k)!} = \frac{n \cdot (n-1) \cdot \ldots \cdot (n-k+1)}{k \cdot (k-1) \cdot \ldots \cdot 1} \tag{10.4}$$

This can be regarded as the number of ways that the total set can be divided into two subsets, namely the subset of selected elements and the subset of not selected elements. A set may also be divided into m subsets ($m < n$) with each subset i having x_i elements, which implies:

$$\sum_{i=1}^{m} x_i = n \tag{10.5}$$

The number of combinations is given by the multinomial coefficient, which is:

$$\frac{n!}{x_m! \cdot x_{m-1}! \cdot \ldots \cdot x_1!} = \frac{n!}{\prod_{i=1}^{m} x_i!} \tag{10.6}$$

10.1.2 The Gamma Function

The binomial coefficient and the factorial function can also be expressed in terms of the so-called gamma function.

The gamma function is an extension of the factorial function to real and complex numbers z. The gamma function is defined as an extension of the factorial function as in Eq. 10.1 for a complex number z with positive real part (i.e. $\text{Re}(z) > 0$):

$$\Gamma(z) = \int_{0}^{\infty} t^{z-1} e^{-t} dt \tag{10.7}$$

Partial integration of Eq. (10.7) yields a recurrent relationship for complex z with $\text{Re}(z) > 0$:

$$\Gamma(z+1) = z \cdot \Gamma(z) \tag{10.8}$$

For integers n this recurrent relationship of the gamma function is:

$$\Gamma(n+1) = n \cdot \Gamma(n) \tag{10.9}$$

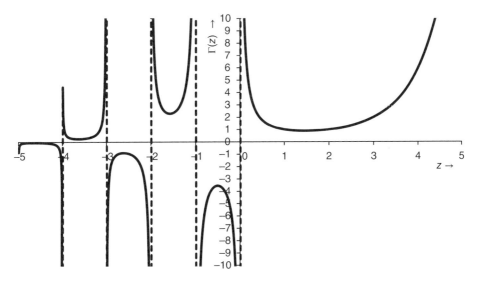

Figure 10.1 The gamma function for real numbers z. For natural numbers z > 0 it corresponds to the factorial function z! (Eq. 10.10).

Table 10.1 Gamma function values for some special cases.

x	0.5	1	1.5	2
$\Gamma(x)$	$\sqrt{\pi}$	1	$\frac{1}{2}\sqrt{\pi}$	1

Expansion of Eq. (10.9) gives an expression for the factorial $n!$:

$$\Gamma(n+1) = n! \tag{10.10}$$

The binomial coefficient in Eq. (10.4) can also be expressed in terms of the gamma function:

$$\binom{n}{k} = \frac{n!}{k! \cdot (n-k)!} = \frac{\Gamma(n+1)}{\Gamma(k+1) \cdot \Gamma(n-k+1)} \tag{10.11}$$

Since the gamma function can be applied not only to integers n, but also to non-integers z, Eq. (10.11) provides a means to generalize the binomial coefficient. With this extension the gamma function can be defined for all complex z with the exception of 0 and negative integers, where it has simple poles. The gamma function for real numbers is shown in Figure 10.1.

Appendix I contains a table for $\Gamma(x)$ with $x = 1.000,...,2.000$. Starting from those values, the recurrent relationship of Eq. (10.8) can be used to calculate $\Gamma(x)$ for x outside that range, except for 0 and negative integers. Some special values are given in Table 10.1.

10.2 Power Functions and Asymptotic Behaviour

In various situations it is very useful to approximate functions or number sequences with power functions. Examples are the use of the moment generating function

(Section 3.1.4), approximations of the normal distribution (Section 10.6) and approximations of the asymptotic behaviour of estimator bias and variance (Section 6.1.3). This section discusses the Taylor and Maclaurin series (Section 10.2.1), polynomial fitting (Section 10.2.2) and power function fitting (Section 10.2.3). The last method was applied in Chapter 6 to describe the bias and variance of parameter estimators.

10.2.1 Taylor and Maclaurin Series

Taylor series are used to represent a function $g(x)$ near a point $x = a$ by a series of power functions [56]. Generally, the function $g(x)$ is known and is required to be infinitely differentiable in an interval near $x = a$. The asymptote for $g(x)$ is $g(a)$ and the Taylor series describes how $g(x)$ approaches $g(a)$ when $x \to a$. How well $g(x)$ is represented for x remote from a depends on the specific case, but the Taylor series model aims specifically at a situation where $x \to a$. Appendix C provides a table of Taylor expansions.

The expression for Taylor series is:

$$g(x) = \sum_{m=0}^{\infty} \frac{g^{(m)}(a)}{m!} \cdot (x - a)^m \tag{10.12}$$

In case $a = 0$, these series are also called Maclaurin series.

In case the function is approximated with a finite instead of an infinite series, the expression becomes:

$$g(x) \approx \sum_{n=0}^{k} \frac{g^{(n)}(a)}{n!} \cdot (x - a)^n + o((x - a)^k) \tag{10.13}$$

Here again, $o((x - a)^k)$ denotes a function of $x - a$ that approaches zero faster than $(x - a)^k$ (cf. Eq. (4.83)).

Example 10.1 *Taylor Series of the Normal Distribution for $y \to 0$* The cumulative function of the normal distribution cannot be analytically determined, but some values as well as its first derivative are known. In Section 10.6 various approximations are discussed. The Taylor series can be used to generate one of those.

The Taylor series expression for the standard normal distribution $F(y)$ for $y \to 0$ (and $y \geq 0$) is:

$$F(y) = \sum_{m=0}^{\infty} \frac{F^{(m)}(0)}{m!} \cdot (y)^m \tag{10.14}$$

Although $F(y)$ cannot be expressed analytically, the value at $y = 0$ is known due to the symmetry of the normal distribution: $F(0) = 1/2$, which is associated with the $m = 0$ term.

The $m = 1$ term is the first derivative, which is the well-known normal density function $f(y)$ (see Table 4.17) with $f(0) = 1/\sqrt{(2\pi)}$:

$$F^{(1)}(y) = f(y) = \frac{1}{\sqrt{2\pi}} \cdot e^{-\frac{1}{2}y^2} \tag{10.15}$$

The higher derivatives of $F(y)$ can be obtained by differentiating $f(y)$. For example, for $m = 2$ the derivative $F^{(2)}(y) = f^{(1)}(y)$ is therefore:

$$F^{(2)}(y) = f^{(1)}(y) = \frac{-1}{\sqrt{2\pi}} \cdot y \cdot e^{-\frac{1}{2}y^2} \tag{10.16}$$

It follows that $F^{(2)}(0) = f^{(1)}(0) = 0$, which means that the $m = 2$ term becomes 0. For $m = 3$ the derivative $F^{(3)}(y)$ is:

$$F^{(3)}(y) = f^{(2)}(y) = \frac{-1}{\sqrt{2\pi}} \cdot \left(e^{-\frac{1}{2}y^2} - y^2 \cdot e^{-\frac{1}{2}y^2} \right) \tag{10.17}$$

It follows that $F^{(3)}(0) = f^{(2)}(0) = -1/\sqrt{(2\pi)}$. And so on:

$$\begin{aligned}
F(y) &= F(0) + \frac{f(0)}{1!}y + \frac{f^{(1)}(0)}{2!}y^2 + \frac{f^{(2)}(0)}{3!}y^3 + \dots \\
&= \frac{1}{2} + \frac{1}{\sqrt{2\pi}} \cdot \left(\frac{1}{1!}y + \frac{0}{2!}y^2 + \frac{-1}{3!}y^3 + \dots \right)
\end{aligned} \tag{10.18}$$

For all even m the derivative $F^{(m)}(0) = 0$ and for odd m the terms are $1/\sqrt{(2\pi)}$ where the sign flips with every next odd m. This can be rewritten as:

$$y \geq 0: \ F(y) = \frac{1}{2} + \frac{1}{\sqrt{2\pi}} \sum_{n=0}^{\infty} \frac{(-1)^n y^{2n+1}}{n! 2^n (2n+1)} \tag{10.19}$$

From the symmetry of the standard normal distribution it follows that $F(-y) = 1 - F(y)$ and from that the Taylor expansion for $y \leq 0$ follows. This result is presented in Section 10.6, in Eqs (10.123) and (10.124). However, the Taylor series do not need to be developed with an infinite number of terms. For large n the terms approach zero. The normal distribution can be calculated with any desired precision by tuning the maximum n.

10.2.2 Polynomial Fitting

Assume a set of $n+1$ observations (x_i, g_i) is known, where x_i is an independent variable, g_i the dependent variable and $i = 0, \dots, n$. A polynomial fit $p(x)$ can be used to describe $g(x)$ over the interval of known x_i (i.e. $[x_{i,min}, x_{i,max}]$).

Two approaches are discussed here: polynomial interpolation and polynomial regression. With polynomial interpolation the $n+1$ data points are described with an n-grade polynomial (powers 1 through n) plus one constant (power 0) (i.e. in total the coefficients of $n+1$ powers of x). The polynomial interpolation fit is exact at the data points (x_i, g_i).

With polynomial regression a lower-grade polynomial is used than the number of data points n and the procedure also differs from the interpolation method. The regression methods provide a best fit on average. The regression fit is not necessarily exact at the data points (x_i, g_i). The two methods are discussed in the next two sections.

10.2.2.1 Polynomial Interpolation
The n-grade polynomial to fit the $n+1$ data points (x_i, g_i) can be written as:

$$p(x) = \sum_{k=0}^{n} a_k x^k \tag{10.20}$$

The a_k are called the coefficients of the polynomial. For each data point (x_i, g_i):

$$p(x_i) = \sum_{k=0}^{n} a_k x_i^k = g_i \tag{10.21}$$

This yields a set of $n + 1$ equations with $n + 1$ coefficients a_k ($k = 0,\ldots,n$) to be solved. The equations written in matrix form are:

$$\begin{pmatrix} p(x_0) \\ p(x_1) \\ \vdots \\ p(x_n) \end{pmatrix} = \begin{pmatrix} x_0^0 & x_0^1 & \cdots & x_0^n \\ x_1^0 & x_1^1 & \cdots & x_1^n \\ \vdots & \vdots & \ddots & \vdots \\ x_n^0 & x_n^1 & \cdots & x_n^n \end{pmatrix} \cdot \begin{pmatrix} a_0 \\ a_1 \\ \vdots \\ a_n \end{pmatrix} = \begin{pmatrix} g_0 \\ g_1 \\ \vdots \\ g_n \end{pmatrix} \tag{10.22}$$

The matrix is a so-called Vandermonde matrix. It can be shown that the determinant of this matrix is non-zero, that the matrix is non-singular and that just one solution of the equations exists.

This set can be solved in various ways. The solution through the so-called Lagrange polynomials can readily be used and must produce the same result as other methods, since the solution is unique according to the Vandermonde matrix concept. Lagrange n-grade polynomials $L(x)$ are defined for each of the $n + 1$ variables x_m ($0 \leq m \leq n$) as:

$$L_m^n(x) = \prod_{\substack{j=0 \\ j \neq m}}^{n} \frac{(x - x_j)}{(x_m - x_j)} \tag{10.23}$$

For values $x = x_i$ the Lagrange polynomials equal the so-called Kronecker delta δ_{im}, which equals 1 if $i = m$ and 0 if $i \neq m$:

$$L_m^n(x_i) = \prod_{\substack{j=0 \\ j \neq m}}^{n} \frac{(x_i - x_j)}{(x_m - x_j)} = \delta_{im} \tag{10.24}$$

The polynomial must be such that $p(x_i) = g_i$ for each $i = 0,\ldots,n$. In terms of the Lagrange polynomials, the solution is:

$$p(x) = \sum_{i=0}^{n} g_i \cdot L_i^n(x) \tag{10.25}$$

The method becomes more accurate the more data points are used, but also more laborious.

An example of the method is provided in Example 10.2, where it is used to approximate the standard normal distribution.

As for the accuracy of the method, if $g(x)$ is known, at least $n + 1$ times differentiable and interpolated with an n-grade polynomial $p(x)$ over an interval $[x_0,\ldots,x_n]$, then the error $g(x) - p(x)$ is found to be:

$$|R_n(x)| = |g(x) - p(x)| \leq \frac{\left| \prod_{i=0}^{n}(x - x_i) \right|}{(n + 1)!} \cdot \max|f^{(n+1)}(\xi)| \quad \xi \in [x_0, x_n] \tag{10.26}$$

Table 10.2 Data of standard normal distribution used for polynomial interpolation.

i	0	1	2	3	4	5
y_i	0	0.5	1	1.5	2	2.5
$R(y_i)/f(y_i)$	$\sqrt{(\pi/2)}$	0.876364	0.655680	0.515816	0.421369	0.354265

Note: The interval for interpolation is $y \in [y_{min}, y_{max}] = [0,2.5]$.

Example 10.2 *Polynomial Interpolation of the Normal Distribution* Applying polynomial interpolation to the normal distribution requires a set of data and the interpolation can be carried out between the minimum and the maximum of the interval. The ratio $R(y)/f(y)$ is used for various approximations of the normal distribution. As an example, it will be carried out here with the polynomial interpolation method.

Assume the data in Table 10.2, where $g_i = R_i/f_i$ are obtained for the standard normal distribution.

The interpolation polynomial can now be found for [0,2.5], as can be seen from the y values in Table 10.2. The polynomial may be used outside this interval as well, but may (and will) prove to be considerably inaccurate.

The Lagrange polynomials are elaborated with Eq. (10.23):

$$L_0^5(y) = \frac{(y - 0.5) \cdot (y - 1) \cdot (y - 1.5) \cdot (y - 2) \cdot (y - 2.5)}{(-0.5) \cdot (-1) \cdot (-1.5) \cdot (-2) \cdot (-2.5)}$$

$$L_1^5(y) = \frac{y \cdot (y - 1) \cdot (y - 1.5) \cdot (y - 2) \cdot (y - 2.5)}{(0.5) \cdot (-0.5) \cdot (-1) \cdot (-1.5) \cdot (-2)}$$

$$L_2^5(y) = \frac{y \cdot (y - 0.5) \cdot (y - 1.5) \cdot (y - 2) \cdot (y - 2.5)}{(1) \cdot (0.5) \cdot (-0.5) \cdot (-1) \cdot (-1.5)}$$

$$L_3^5(y) = \frac{y \cdot (y - 0.5) \cdot (y - 1) \cdot (y - 2) \cdot (y - 2.5)}{(1.5) \cdot (1) \cdot (0.5) \cdot (-0.5) \cdot (-1)}$$

$$L_4^5(y) = \frac{y \cdot (y - 0.5) \cdot (y - 1) \cdot (y - 1.5) \cdot (y - 2.5)}{(2) \cdot (1.5) \cdot (1) \cdot (0.5) \cdot (-0.5)}$$

$$L_5^5(y) = \frac{y \cdot (y - 0.5) \cdot (y - 1) \cdot (y - 1.5) \cdot (y - 2)}{(2.5) \cdot (2) \cdot (1.5) \cdot (1) \cdot (0.5)} \tag{10.27}$$

The interpolation polynomial is found with Eq. (10.25):

$$\frac{R(y)}{f(y)} \approx p(y) = \sum_{i=0}^{5} \frac{R(y_i)}{f(y_i)} \cdot L_i^5(y) \tag{10.28}$$

The polynomial based on the data points of Table 10.2 becomes:

$$p(y) = \sqrt{\frac{\pi}{2}} - 0.98956 \cdot y + 0.574677 \cdot y^2 - 0.23367 \cdot y^3$$
$$+ 0.056976 \cdot y^4 - 0.00606 \cdot y^5 \tag{10.29}$$

This polynomial fits exactly at the data points of Table 10.2. The absolute and relative accuracy for R is shown in Table 10.3.

The solution is not as accurate as the six-parameter approximation of Zelen and Severo ([68], p. 932; see Eq. (10.131)), but considering the mere six data points is

Table 10.3 Accuracy for R (cf. Section 10.6 for other approximations and Table 10.12 for accuracies with two-parameter approximations).

Expression for $y \geq 0$:

$$R(y) \approx f(y) \cdot \left[\sqrt{\tfrac{\pi}{2}} - 0.98956 \cdot y + 0.574677 \cdot y^2 - 0.23367 \cdot y^3 + 0.056976 \cdot y^4 - 0.00606 \cdot y^5 \right]$$

Deviations over interval	[0,2.5]	[0,3]	[0,5]
Maximum positive ΔR	0.00027	0.00027	0.00027
Maximum negative ΔR	−0.00005	−0.00006	−0.00008
Maximum positive $\Delta R/R$	0.001	0.001	0.001
Maximum negative $\Delta R/R$	−0.00027	−0.04	−10.6

Note: The interpolation range is [0,2.5].

still quite effective within the interval [0,2.5]. Outside the interpolation interval the polynomial deviates significantly with respect to the relative error $\Delta R/R$ in the interval [0,5]. A relative error of 1 is first reached for $y > 3.8$, which lies roughly 50% outside the interval of the data points. The large error with the extrapolation illustrates the character of this method; it is indeed meant for interpolation.

10.2.2.2 Polynomial and Linear Regression

Polynomial regression is a method to estimate the coefficients of an n-grade polynomial from at least m observed data points (x_i, g_i) with $n < m$. Again, the independent variables are x_i and the dependent variables (i.e. the observations) are g_i. With polynomial interpolation the number of data points and the polynomial grade plus 1 (for the constant, i.e. the coefficient of x^0) are the same, but with polynomial regression generally there are more data points than the polynomial grade plus 1. As mentioned above, the polynomial does not necessarily fit the data points exactly.

The regression model yields an n-grade polynomial with a minimum error ε:

$$p(x) = a_0 + a_1 x + a_2 x^2 + \ldots + a_n x^n + \varepsilon = \sum_{j=0}^{n} a_j \cdot x^j + \varepsilon \tag{10.30}$$

Having m data points (x_i, g_i) with $(i = 1, \ldots, m)$, there are m equations that can be used to estimate the best regression coefficients a_j such that ε is minimal. In matrix form:

$$\begin{pmatrix} x_0^0 & x_0^1 & \cdots & x_0^n \\ x_1^0 & x_1^1 & \cdots & x_1^n \\ \vdots & \vdots & \ddots & \vdots \\ x_m^0 & x_m^1 & \cdots & x_m^n \end{pmatrix} \cdot \begin{pmatrix} a_0 \\ a_1 \\ \vdots \\ a_n \end{pmatrix} + \begin{pmatrix} \varepsilon_0 \\ \varepsilon_1 \\ \vdots \\ \varepsilon_m \end{pmatrix} = \begin{pmatrix} g_0 \\ g_1 \\ \vdots \\ g_m \end{pmatrix} \tag{10.31}$$

Note that n and m are different: a unique solution of the a_j $(j = 0, \ldots, n)$ requires $m > n$ (i.e. $m \geq n + 1$). Polynomial regression, as confusing as it may sound, is an example of linear regression, which is used to achieve the set of a_j that gives the best fit. Also note that the equations are not linear with x, but linear with each of the regression coefficients a_j. So, from the perspective of x it is a polynomial, but from the perspective of the regression coefficients a_j it is a linear relation, for which reason this is called linear regression.

In matrix notation:

$$\vec{\vec{X}} \cdot \vec{a} + \vec{\varepsilon} = \vec{g} \tag{10.32}$$

In this notation $\vec{\vec{X}}$ is the $(m+1)$-by-$(n+1)$ design matrix with elements $x_{ij} = (x_i)^j$ containing observed values of the variable x_i, $\vec{a}$ the parameter vector consisting of the regression coefficients to be estimated, $\vec{\varepsilon}$ the error vector to be minimized and $\vec{g}$ the dependent variable or response vector containing the observations.

The estimation of the $n+1$ regression coefficients a_i $(i = 0,\ldots,n)$ can be carried out by least-squares estimation, minimizing the error vector. In matrix notation the solution is:

$$\vec{a} = (\vec{\vec{X}}^T \vec{\vec{X}})^{-1} \vec{\vec{X}}^T \vec{g} \tag{10.33}$$

As mentioned above, this solution is unique if $m > n$. The matrix operations involved are transposition, multiplication and inversion (involving taking the adjugate), as explained below, after which the solution of Eq. (10.33) is worked out.

Transposition of an $(m+1)$-by-$(n+1)$ matrix $\vec{\vec{V}}$ with elements v_{ij} yields the $(n+1)$-by-$(m+1)$ transposed matrix $\vec{\vec{V}}^T$ with elements v_{ji} (i.e. the rows and columns are mirrored on the diagonal). In case of the transpose of the matrix in Eq. (10.31), we have the $(n+1)$-by-$(m+1)$ matrix:

$$\vec{\vec{X}}^T =
\begin{pmatrix}
x_0^0 & x_0^1 & \cdots & x_0^n \\
x_1^0 & x_1^1 & \cdots & x_1^n \\
\vdots & \vdots & \ddots & \vdots \\
x_m^0 & x_m^1 & \cdots & x_m^n
\end{pmatrix}^T
=
\begin{pmatrix}
x_0^0 & x_1^0 & \cdots & x_m^0 \\
x_0^1 & x_1^1 & \cdots & x_m^1 \\
\vdots & \vdots & \ddots & \vdots \\
x_0^n & x_1^n & \cdots & x_m^n
\end{pmatrix} \tag{10.34}$$

Multiplication of a p-by-q matrix $\vec{\vec{V}}$ with elements v_{ij} and a q-by-r matrix $\vec{\vec{W}}$ with elements w_{ij} yields a p-by-r matrix $\vec{\vec{V}}\vec{\vec{W}}$ with elements $(vw)_{i,j} = (v_{i,1}w_{1,j} + v_{i,2}w_{2,j} + \cdots + v_{i,n}w_{n,j})$. In case of multiplication between the brackets in Eq. (10.33), the result is an $(n+1)$-by-$(n+1)$ matrix:

$$\vec{\vec{X}}^T\vec{\vec{X}} =
\begin{pmatrix}
x_0^0 & x_1^0 & \cdots & x_m^0 \\
x_0^1 & x_1^1 & \cdots & x_m^1 \\
\vdots & \vdots & \ddots & \vdots \\
x_0^n & x_1^n & \cdots & x_m^n
\end{pmatrix}
\begin{pmatrix}
x_0^0 & x_0^1 & \cdots & x_0^n \\
x_1^0 & x_1^1 & \cdots & x_1^n \\
\vdots & \vdots & \ddots & \vdots \\
x_m^0 & x_m^1 & \cdots & x_m^n
\end{pmatrix}$$

$$=
\begin{pmatrix}
\sum\limits_{r=0}^{m} x_r^0 & \sum\limits_{r=0}^{m} x_r^1 & \cdots & \sum\limits_{r=0}^{m} x_r^n \\
\sum\limits_{r=0}^{m} x_r^1 & \sum\limits_{r=0}^{m} x_r^2 & \cdots & \sum\limits_{r=0}^{m} x_r^{n+1} \\
\vdots & \vdots & \ddots & \vdots \\
\sum\limits_{r=0}^{m} x_r^n & \sum\limits_{r=0}^{m} x_r^{n+1} & \cdots & \sum\limits_{r=0}^{m} x_r^{2n}
\end{pmatrix} \tag{10.35}$$

Finding the inverse of a matrix can be a tedious task. For relatively small matrices $\vec{V}$ the inverse $\vec{V}^{-1}$ can be constructed with the determinant $\det(\vec{V})$ and adjugate matrix $\mathrm{adj}(\vec{V})$:

$$\vec{V}^{-1} = \frac{\mathrm{adj}(\vec{V})}{\det(\vec{V})} \tag{10.36}$$

The determinant of a q-by-q matrix $\det(\vec{V})$ is found by a sequence of steps for larger dimensions. For $q = 2$ the determinant is found as:

$$\det(\vec{V}) = \det \begin{pmatrix} v_{11} & v_{12} \\ v_{21} & v_{22} \end{pmatrix} = v_{11}v_{22} - v_{12}v_{21} \tag{10.37}$$

For $q = 3$ the determinant is found as:

$$\det(\vec{V}) = \det \begin{pmatrix} v_{11} & v_{12} & v_{13} \\ v_{21} & v_{22} & v_{23} \\ v_{31} & v_{32} & v_{33} \end{pmatrix}$$

$$= v_{11} \cdot \det \begin{pmatrix} v_{22} & v_{23} \\ v_{32} & v_{33} \end{pmatrix} - v_{12} \cdot \det \begin{pmatrix} v_{21} & v_{23} \\ v_{31} & v_{33} \end{pmatrix} + v_{13} \cdot \det \begin{pmatrix} v_{21} & v_{22} \\ v_{31} & v_{32} \end{pmatrix}$$

$$= v_{11}v_{22}v_{33} - v_{11}v_{23}v_{32} - v_{12}v_{21}v_{33} + v_{12}v_{23}v_{31} + v_{13}v_{21}v_{32} - v_{13}v_{22}v_{31} \tag{10.38}$$

For q in general the Leibnitz formula is used, which is basically an extension of the procedure above.

As a next step to determine the inverse matrix, the adjugate matrix $\mathrm{adj}(\vec{V})$ is constructed. The adjugate of a q-by-q matrix is found by replacing each element v_{ij} by its cofactor c_{ij}, forming matrix $\vec{C}$, and next constructing the transpose matrix $\vec{C}^T$. The cofactor c_{ij} is found by removing row i and column j in $\vec{V}$ and calculating the determinant of the remaining matrix. Determinants where the sum $i + j$ is odd take a minus sign.

As an example, the adjugate of matrix $\vec{V}$ with $q = 2$ is:

$$\mathrm{adj}(\vec{V}) = \mathrm{adj} \begin{pmatrix} v_{11} & v_{12} \\ v_{21} & v_{22} \end{pmatrix} = \begin{pmatrix} \det(v_{22}) & -\det(v_{21}) \\ -\det(v_{12}) & \det(v_{11}) \end{pmatrix}^T = \begin{pmatrix} v_{22} & -v_{12} \\ -v_{21} & v_{11} \end{pmatrix} \tag{10.39}$$

The inverse of $\vec{V}$ is subsequently found by substitution of Eqs (10.37) and (10.39) in Eq. (10.36) for $q = 2$:

$$\vec{V}^{-1} = \frac{\mathrm{adj}(\vec{V})}{\det(\vec{V})} = \frac{1}{v_{11}v_{22} - v_{12}v_{21}} \cdot \begin{pmatrix} v_{22} & -v_{12} \\ -v_{21} & v_{11} \end{pmatrix} \tag{10.40}$$

As a second example, for $q = 3$ the adjugate of matrix $\vec{V}$ is:

$$\text{adj}(\vec{V}) = \vec{C}^T$$

$$= \begin{pmatrix} v_{22}v_{33} - v_{23}v_{32} & -v_{21}v_{33} + v_{23}v_{31} & v_{21}v_{32} - v_{22}v_{31} \\ -v_{12}v_{33} + v_{13}v_{32} & v_{11}v_{33} - v_{13}v_{31} & -v_{11}v_{32} + v_{12}v_{31} \\ v_{12}v_{23} - v_{13}v_{22} & -v_{11}v_{23} + v_{13}v_{21} & v_{11}v_{22} - v_{12}v_{21} \end{pmatrix}^T$$

$$= \begin{pmatrix} v_{22}v_{33} - v_{23}v_{32} & -v_{12}v_{33} + v_{13}v_{32} & v_{12}v_{23} - v_{13}v_{22} \\ -v_{21}v_{33} + v_{23}v_{31} & v_{11}v_{33} - v_{13}v_{31} & -v_{11}v_{23} + v_{13}v_{21} \\ v_{21}v_{32} - v_{22}v_{31} & -v_{11}v_{32} + v_{12}v_{31} & v_{11}v_{22} - v_{12}v_{21} \end{pmatrix} \qquad (10.41)$$

The inverse for $q = 3$ is found by substitution of Eqs (10.38) and (10.41) in Eq. (10.36), similarly to the case of $q = 2$.

Being able to carry out the operations of transposition, multiplication and inversion, at last the estimators for the regression coefficients follow from Eq. (10.33).

Finally, the procedure is applied to the case of $n = 1$, which assumes the $n + 1 = 2$ parameter model:

$$g(x) = a_0 + a_1 \cdot x \qquad (10.42)$$

This is an important case, as it is the foundation of linear regression parameter estimation (see Section 6.3). Having m data points (x_i, g_i) with $(i = 1, \ldots, m)$, the matrix equation becomes:

$$\begin{pmatrix} x_0^0 & x_0^1 \\ x_1^0 & x_1^1 \\ \vdots & \vdots \\ x_m^0 & x_m^0 \end{pmatrix} \cdot \begin{pmatrix} a_0 \\ a_1 \end{pmatrix} + \begin{pmatrix} \varepsilon_0 \\ \varepsilon_1 \\ \vdots \\ \varepsilon_m \end{pmatrix} = \begin{pmatrix} g_0 \\ g_1 \\ \vdots \\ g_m \end{pmatrix} \qquad (10.43)$$

The two regression coefficients follow from Eq. (10.33) and the procedures described above. The results are:

$$a_1 = \frac{m \cdot \sum\limits_{i=0}^{m} g_i x_i - \sum\limits_{i=0}^{m} g_i \cdot \sum\limits_{i=0}^{m} x_i}{m \cdot \sum\limits_{i=0}^{m} x_i^2 - \left(\sum\limits_{i=0}^{m} x_i \right)^2} \qquad (10.44)$$

and

$$a_0 = \frac{\sum\limits_{i=0}^{m} g_i \cdot \sum\limits_{i=0}^{m} x_i^2 - \sum\limits_{i=0}^{m} g_i x_i \cdot \sum\limits_{i=0}^{m} x_i}{m \cdot \sum\limits_{i=0}^{m} x_i^2 - \left(\sum\limits_{i=0}^{m} x_i \right)^2} \qquad (10.45)$$

The expression for a_0 can be rewritten by expressing $\sum g_i x_i$ by means of Eq. (10.44) in terms of a_1 and substituting in (10.45). This yields:

$$a_0 = \frac{1}{m} \sum_{i=0}^{m} g_i - a_1 \cdot \frac{1}{m} \sum_{i=0}^{m} x_i = \langle g \rangle - a_1 \cdot \langle x \rangle \tag{10.46}$$

These are the well-known equations for determining the slope a_1 and g-intercept a_0 (i.e. g at $x = 0$) for the assumed model of (10.42). The coefficients a_0 and a_1 are used in Section 6.3 to estimate distribution parameters. Linear regression analysis is discussed further in Section 10.3.

10.2.3 Power Function Fitting

A third type of fitting abandons the idea of integer powers in approximations. A three-parameter power function [57] was introduced that had proven useful for describing asymptotic behaviour of the bias of the maximum likelihood (ML) shape parameter estimator and later for other biases and standard deviations as well [42] (see Sections 6.1.3 and 6.2.3.4).

The method aims at situations where a function or sequence of values E_n monotonically approaches an asymptote E_∞ with increasing variable n. In contrast to the Fisher consistency defined for $n \to \infty$ (cf. Section 6.1.1.3), the method focuses on the behaviour at the other extreme, namely the behaviour of E_n at small n. The character 'E' is chosen as the values are normally the expected values of the function at given variable value n. Though initially developed in the context of a discrete variable, the method also applies to continuous variable x and function $g(x)$ with asymptote $g_\infty = g(x \to \infty)$. Through data manipulation the method can be applied to other limits as well. Here the description focuses on functions that approach an asymptote, with variable n approaching infinity.

As an assumption, the method uses the power function $D_n = D(n;P,Q,R)$ to fit the difference between E_n and E_∞:

$$D_n = E_n - E_\infty = Q \cdot (n - R)^P \tag{10.47}$$

where n is the variable and P, Q and R are parameters. If the difference D_n is zero, then $Q = 0$ and the parameters P and R may have any arbitrary finite value. With $Q \neq 0$ and n approaching infinity, the difference D_n is required to approach 0, which implies that $P < 0$. As a consequence, there is a singularity at $n = R$. Therefore, the power function is generally used for $n > R$. Another consequence of the power function model is that higher derivatives alternate in sign if $P < 0$.

The parameters can be found by, for example, regression analysis. A convenient way is to evaluate the logarithmic function:

$$\log D_n = \log(E_n - E_\infty) = \log Q + P \cdot \log(n - R) \tag{10.48}$$

That is, $\log(D_n)$ has a linear regression relationship with $\log(n - R)$.

As an example, Table 10.4 shows data of the Weibull parameter ML estimator $b_{1,1}$ as a function of sample size n, as published in [27]. The notation $b_{1,1}$ means that random samples with sample size n are drawn from the Weibull distribution with parameters $\alpha = \beta = 1$. The observed data $(n, \langle b_{1,1}(n) \rangle)$ are used to assess the asymptotic behavior of the expected ML shape parameter estimator (cf. Section 6.2.3.3). The theoretical value and asymptote for infinitely large sample sizes is $E_\infty = 1$, since $\beta = 1$.

Table 10.4 The bias in the ML estimator of the Weibull shape parameter $b_{1,1}$ (i.e. original distribution has parameters $\alpha = 1$ and $\beta = 1$) [27].

n	3	4	5	6	8	12	16	24	32	64	100	128
$<b_{1,1}>$	2.264	1.661	1.442	1.332	1.227	1.1368	1.0967	1.0622	1.0454	1.0222	1.0134	1.0112

Note: The asymptote $E_\infty(b_{1,1})$ is known to be $\beta = 1$.

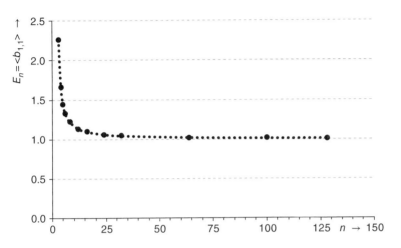

Figure 10.2 Data of Table 10.4 showing asymptotic behaviour.

Figure 10.2 shows the data of Table 10.4 and the asymptote for $n \to \infty$ is clearly 1. Figure 10.3 shows the same data in a log–log plot after the asymptote E_∞ is subtracted. Varying R has a large effect, particularly on the lower $\log(n - R)$ values. At $R = 2$ the initially convex graph becomes straight, while at $R > 2$ the graph becomes concave. With a straight line the correlation coefficient $\rho_{\log(n-R),\log(Dn)}$ is highest, which can be used to optimize R. Least-squares estimation of the graph subsequently yields P as the slope and Q as the intercept at $\log(n - R) = 0$ (i.e. $n - R = 1$). Substitution of P, Q and R in Eq. (10.47) defines E_n as:

$$E_n = E_\infty + D_n = E_\infty + Q \cdot (n - R)^P \tag{10.49}$$

Alternatively, P might be believed to have a certain value (e.g. -1 instead of a value like -0.9875; for theoretical reasons or convenience), after which Q and R can be found by regression analysis (see Section 10.3). For that purpose, Eq. (10.47) can be rewritten as:

$$(D_n)^{1/P} = (E_n - E_\infty)^{1/P} = -R \cdot Q^{1/P} + Q^{1/P} \cdot n \tag{10.50}$$

with given P and n being the variable, $(-R \cdot Q^{1/P})$ is the intercept and $Q^{1/P}$ the regression coefficient. From these follow R, Q and with given P follows ultimately Eq. (10.49).

The power function model was also applied in Section 10.6 to approximate the reliability function $R(y)$ (not to be confused with the R-parameter in the power function) of the normal distribution. Close examination shows that this is a good approximation in the case of the normal distribution, but not an exact expression.

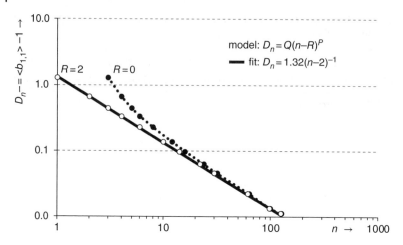

Figure 10.3 Data of Table 10.4 on a log–log scale. If R can be adjusted such that the line becomes straight, then P and Q follow from the slope (respectively the intercept) at $\log(n - R) = 0$ (i.e. $n - R = 1$).

Nevertheless, in other cases the power function appears to be just the exact solution in parameter estimation. As a first (and trivial) example, an unbiased parameter estimator means $E_n = E_\infty$ and then $D_n = 0$, which can be regarded as following the power function with $Q = 0$, as mentioned above. It is trivial, but an exact solution.

A second (non-trivial) example is the standard deviation of the mean. The expression for the standard deviation $\sigma_{MY,n}$ of the mean M_Y is (Section 4.3.1.3, Eq. (4.74)):

$$\sigma_{M_{Y,n}} = \frac{1}{\sqrt{n}} = n^{-\frac{1}{2}} \tag{10.51}$$

The asymptote for $E_\infty(\sigma_{MY,n}) = 0$. Therefore, D_n is equal to $E_n = \sigma_{MY,n}$ and the power function parameters are $P = -1/2$, $Q = 1$ and $R = 0$. The power function is an exact representation in that case.

10.3 Regression Analysis

The parameters that characterize a distribution or other function can often be found (directly or indirectly) by assuming a linear relationship between a variable X and a covariable Y (both of which have yet to be defined):

$$Y = A_0 + A_1 \cdot X \tag{10.52}$$

The parameter A_0 is the location parameter or Y-intercept (i.e. $Y(X = 0)$) and A_1 is the regression coefficient or slope (i.e. derivative dY/dX). This section will discuss the method of linear regression, including the variances of the estimated parameters and the variance of an arbitrary $Y(X)$. The analysis is discussed for unweighted as well as for weighted data.

Section 6.3.1.1 discussed how to obtain the best regression parameter estimators a_0 and a_1 through the least-squares method from a set of variable and covariable pairs

(x_i, y_i) with $i = 1, \ldots, n$ and the two boundary conditions of Eqs (6.79) and (6.80):

$$\frac{\partial S}{\partial a_0} = -2 \cdot \sum_{i=1}^{n} (y_i - a_0 - a_1 \cdot x_i) = -2 \cdot \left(\sum_{i=1}^{n} y_i - n \cdot a_0 - a_1 \cdot \sum_{i=1}^{n} x_i \right) = 0 \quad (10.53)$$

$$\frac{\partial S}{\partial a_1} = -2 \cdot \sum_{i=1}^{n} x_i \cdot (y_i - a_0 - a_1 \cdot x_i) = -2 \cdot \left(\sum_{i=1}^{n} x_i y_i - a_0 \cdot \sum_{i=1}^{n} x_i - a_1 \cdot \sum_{i=1}^{n} x_i^2 \right) = 0$$
$$(10.54)$$

Recalling that a mean $\bar{u}$ is defined as:

$$\bar{u} = \frac{1}{n} \sum_{i=1}^{n} u_i \quad (10.55)$$

the solutions are found in Eq. (6.84) for a_0 and for a_1 in Eq. (6.85) in terms of means and Eq. (6.86) in terms of sums:

$$a_0 = \bar{y} - a_1 \cdot \bar{x} \quad (10.56)$$

$$a_1 = \frac{\overline{x \cdot y} - \bar{y} \cdot \bar{x}}{\overline{x^2} - \bar{x}^2} = \frac{\overline{(x - \bar{x}) \cdot y}}{\overline{(x - \bar{x})^2}} = \frac{\overline{(x - \bar{x}) \cdot (y - \bar{y})}}{\overline{(x - \bar{x}) \cdot (x - \bar{x})}} \quad (10.57)$$

Alternatively, Eq. (6.85) can be expressed in terms of summations in various ways:

$$a_1 = \frac{\sum_{i=1}^{n} (x_i \cdot y_i) - \sum_{i=1}^{n} y_i \cdot \frac{1}{n} \sum_{i=1}^{n} x_i}{\sum_{i=1}^{n} (x_i^2) - \frac{1}{n} \left(\sum_{i=1}^{n} x_i \right)^2} = \frac{\sum_{i=1}^{n} (x_i \cdot y_i) - n \cdot \bar{y} \cdot \bar{x}}{\sum_{i=1}^{n} (x_i^2) - n \cdot \bar{x}^2}$$

$$= \frac{\sum_{i=1}^{n} ((x_i - \bar{x}) \cdot (y_i - \bar{y}))}{\sum_{i=1}^{n} ((x_i - \bar{x}) \cdot (x_i - \bar{x}))} = \frac{\sum_{i=1}^{n} ((x_i - \bar{x}) \cdot y_i)}{\sum_{i=1}^{n} ((x_i - \bar{x})^2)} \quad (10.58)$$

As an overview, there are (cf. Figure 10.4):

- Observed pairs (x_i, y_i) with $i = 1, \ldots, n$.
- A true linear relationship $\tilde{y}(x)$ with exact, but unknown, parameters A_0 and A_1:

$$\tilde{y} = A_0 + A_1 \cdot x \quad (10.59)$$

- A best-fit linear relationship $\hat{y}(x)$ based on the observed pairs with best estimated parameters a_0 and a_1:

$$\hat{y} = a_0 + a_1 \cdot x \quad (10.60)$$

The question is: how can the accuracies (i.e. the variance of a_0, a_1 and y) be determined from the observed pairs? The equations to be solved are:

$$(\Delta a_0)^2 = \langle (a_0 - A_0)^2 \rangle \quad (10.61)$$

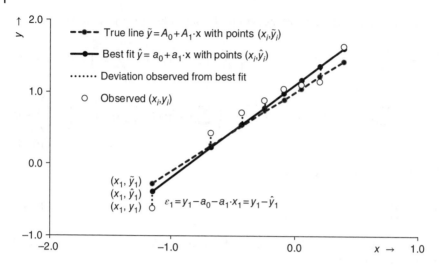

Figure 10.4 Illustration of the observed pairs (x_i, y_i), the true line $\tilde{y}(x)$, the best fit $\hat{y}(x)$ and the residual $\varepsilon_i = (y_i - \hat{y}_i)$.

$$(\Delta a_1)^2 = \langle (a_1 - A_1)^2 \rangle \tag{10.62}$$

$$(\Delta y)^2 = \langle (y - \tilde{y})^2 \rangle \tag{10.63}$$

First, a set of three sums S_{XX}, S_{YY} and S_{XY} is introduced for ease of notation. The squared sum S_{XX} of X is defined with various equivalent expressions as:

$$S_{XX} = \sum_{i=1}^{n} (x_i - \bar{x})^2$$

$$= \sum_{i=1}^{n} (x_i - \bar{x}) \cdot x_i = \sum_{i=1}^{n} x_i^2 - \frac{1}{n} \left(\sum_{i=1}^{n} x_i \right)^2 = \sum_{i=1}^{n} x_i^2 - n \cdot \bar{x}^2 \tag{10.64}$$

Likewise, the squared sum S_{YY} of Y is defined with various equivalent expressions as:

$$S_{YY} = \sum_{i=1}^{n} (y_i - \bar{y})^2$$

$$= \sum_{i=1}^{n} (y_i - \bar{y}) \cdot y_i = \sum_{i=1}^{n} y_i^2 - \frac{1}{n} \left(\sum_{i=1}^{n} y_i \right)^2 = \sum_{i=1}^{n} y_i^2 - n \cdot \bar{y}^2 \tag{10.65}$$

Likewise, the cross-product sum S_{XY} is defined with various equivalent expressions as:

$$S_{XY} = \sum_{i=1}^{n} (x_i - \bar{x})(y_i - \bar{y})$$

$$= \sum_{i=1}^{n} (x_i - \bar{x}) \cdot y_i = \sum_{i=1}^{n} x_i \cdot (y_i - \bar{y})_i$$

$$= \sum_{i=1}^{n} x_i y_i - \frac{1}{n} \left(\sum_{i=1}^{n} x_i \right) \cdot \left(\sum_{i=1}^{n} y_i \right) = \sum_{i=1}^{n} x_i y_i - n \cdot \bar{x} \cdot \bar{y} \tag{10.66}$$

As a consequence, a_1 can also be written as:

$$a_1 = \frac{S_{XY}}{S_{XX}} \tag{10.67}$$

The idea is that y_i is distributed about the true $\tilde{y}_i$. It is assumed that the residual d_i about the true value has mean 0 and variance σ^2:

$$\langle d_i \rangle = \langle y_i - \tilde{y}_i \rangle = 0 \tag{10.68}$$

$$\langle d_i^2 \rangle = \langle (y_i - \tilde{y}_i)^2 \rangle = \sigma^2 \tag{10.69}$$

Furthermore, it is assumed that the residuals d_i and d_j are uncorrelated if $i \neq j$, that is:

$$\forall i \neq j : \langle d_i \cdot d_j \rangle = 0 \tag{10.70}$$

The variance of estimators a_0 and a_1 is given by:

$$(\Delta a_0)^2 = \langle (a_0 - A_0)^2 \rangle \tag{10.71}$$

$$(\Delta a_1)^2 = \langle (a_1 - A_1)^2 \rangle \tag{10.72}$$

For the variance of y it is helpful to take one other step first, namely defining a parameter b. As mentioned in Section 6.3.1.1, a_0 and a_1 appear not independent:

$$\langle (a_0 - A_0) \cdot (a_1 - A_1) \rangle = -\bar{x} \cdot (\Delta a_1)^2 \tag{10.73}$$

A set of independent parameters a_1 and b is obtained if the best fit is written as:

$$\hat{y} = a_0 + a_1 \cdot x = b + a_1 \cdot (x - \bar{x}) \tag{10.74}$$

with b (noting that the average of both y and $\hat{y}$ is equal to $\bar{y}$):

$$b = a_0 + a_1 \cdot \bar{x} = \bar{y} = \frac{1}{n} \sum_{i=1}^{n} y_i \tag{10.75}$$

The best fit can then be written with independent errors Δa_1 and Δb:

$$\hat{y} = (b \pm \Delta b) + (a_1 \pm \Delta a_1) \cdot (x - \bar{x}) \tag{10.76}$$

This can be conceived as the best fit that varies about the point $(\bar{x}, \bar{y})$ with independent variations of a rotation and a shift along the y-axis. As a consequence, y will vary around the true $\tilde{y}$ with variance $(\Delta y)^2$:

$$(\Delta y)^2 = (\Delta b)^2 + (x - \bar{x})^2 \cdot (\Delta a_1)^2 \tag{10.77}$$

Having basic expressions for Δa_0, Δa_1, Δb and Δy, the question is how to estimate these from the observed pairs (x_i, y_i) with $i = 1, \ldots, n$. From Eq. (10.58) it follows that:

$$a_1 = \sum_{i=1}^{n} \left(\frac{(x_i - \bar{x})}{S_{XX}} \right) \cdot y_i \tag{10.78}$$

This is a linear function of the y_i with given coefficients that are constant for given i. The variance of a_1 relates to the variances of the y_i as:

$$(\Delta a_1)^2 = \sum_{i=1}^{n} \left(\frac{(x_i - \bar{x})}{S_{XX}} \right)^2 \cdot (\Delta y_i)^2 \tag{10.79}$$

If it is assumed that all variances $(\Delta y_i)^2$ are equal to σ^2, then the variance $(\Delta a_1)^2$ can be expressed as:

$$(\Delta a_1)^2 = \sum_{i=1}^{n} \left(\frac{(x_i - \bar{x})}{S_{XX}} \right)^2 \cdot \sigma^2 = \frac{1}{S_{XX}} \cdot \sigma^2 \tag{10.80}$$

Similarly, $(\Delta b)^2$ can be achieved as:

$$(\Delta b)^2 = \left(\frac{1}{n} \right)^2 \sum_{i=1}^{n} (\Delta y_i)^2 = \frac{1}{n} \cdot \sigma^2 \tag{10.81}$$

Using Eq. (10.75) the variance $(\Delta a_0)^2$ can be expressed as:

$$(\Delta a_0)^2 = (\Delta b)^2 + \bar{x}^2 \cdot (\Delta a_1)^2$$
$$= \left(\frac{1}{n} + \frac{\bar{x}^2}{S_{XX}} \right) \cdot \sigma^2 \tag{10.82}$$

The variances $(\Delta a_1)^2$, $(\Delta b)^2$ and $(\Delta a_0)^2$ are all expressed in terms of σ^2, which parameter is generally unknown, but can be estimated from the observed data as follows. To solve σ^2 the true line must be considered with the model in Eq. (10.59). For the true parameters A_1 and B, the same expressions hold as for the best estimated parameters a_1 and b in Eq. (10.78) (respectively Eq. (10.75)):

$$A_1 = \sum_{i=1}^{n} \left(\frac{(x_i - \bar{x})}{S_{XX}} \right) \cdot \tilde{y}_i \tag{10.83}$$

$$B = A_0 + A_1 \cdot \bar{x} = \frac{1}{n} \sum_{i=1}^{n} \tilde{y}_i \tag{10.84}$$

The deviation of the best estimators and the true parameters is:

$$a_1 - A_1 = \sum_{i=1}^{n} \left(\frac{(x_i - \bar{x})}{S_{XX}} \right) \cdot (y_i - \tilde{y}_i) = \sum_{i=1}^{n} \left(\frac{(x_i - \bar{x})}{S_{XX}} \right) \cdot d_i \tag{10.85}$$

$$b - B = \frac{1}{n} \sum_{i=1}^{n} (y_i - \tilde{y}_i) = \frac{1}{n} \sum_{i=1}^{n} d_i \tag{10.86}$$

The error from the true line is d_i:

$$d_i = y_i - \tilde{y}_i = y_i - A_1 \cdot (x_i - \bar{x}) - B \tag{10.87}$$

The error from the best fit is $\varepsilon_i = y_i - \hat{y}_i$. (see Eq. (6.77)). This can be written as:

$$\varepsilon_i = y_i - \hat{y}_i = y_i - a_1 \cdot (x_i - \bar{x}) - b$$
$$= d_i - [(a_1 - A_1) \cdot (x_i - \bar{x}) + (b - B)] \tag{10.88}$$

The squared sum of ε_i can be elaborated as:

$$\sum_{i=1}^{n} \varepsilon_i^{\,2} = \sum_{i=1}^{n} d_i^{\,2} - \frac{1}{S_{XX}} \left(\sum_{i=1}^{n} (x_i - \bar{x}) \cdot d_i \right)^2 - \frac{1}{n} \sum_{i=1}^{n} d_i^{\,2} \qquad (10.89)$$

The expected value of this sum is the sum of estimated variances s^2. Recalling that d_i and d_j are uncorrelated if $i \neq j$ (Eq. (10.70)) and all x terms remain constant:

$$\left\langle \sum_{i=1}^{n} \varepsilon_i^{\,2} \right\rangle = n \cdot \langle s^2 \rangle$$

$$= \left\langle \sum_{i=1}^{n} d_i^{\,2} \right\rangle - \left\langle \frac{1}{S_{XX}} \left(\sum_{i=1}^{n} (x_i - \bar{x}) \cdot d_i \right)^2 \right\rangle - \left\langle \frac{1}{n} \sum_{i=1}^{n} d_i^{\,2} \right\rangle$$

$$= n \cdot \sigma^2 - \sigma^2 - \sigma^2 \qquad (10.90)$$

This yields a relation between the observable variance s^2 and the unknown variance σ^2:

$$\sigma^2 = \frac{n}{n-2} \cdot \langle s^2 \rangle \qquad (10.91)$$

The estimate for $<s^2>$ is s^2:

$$s^2 = \frac{1}{n} \sum_{i=1}^{n} \varepsilon_i^{\,2} = \frac{1}{n} \sum_{i=1}^{n} (y_i - a_1 \cdot x_i - a_0)^2 \qquad (10.92)$$

Now, it is possible to give expressions for the variances based on the observed pairs (x_i, y_i):

$$(\Delta a_1)^2 \approx \frac{n}{n-2} \cdot \frac{s^2}{S_{XX}} = \frac{1}{n-2} \cdot \frac{\displaystyle\sum_{i=1}^{n} (y_i - a_1 \cdot x_i - a_0)^2}{\displaystyle\sum_{i=1}^{n} (x_i - \bar{x})^2} \qquad (10.93)$$

Similarly, $(\Delta b)^2$ can be achieved:

$$(\Delta b)^2 \approx \frac{1}{n-2} \cdot s^2 = \frac{1}{n-2} \cdot \frac{1}{n} \sum_{i=1}^{n} (y_i - a_1 \cdot x_i - a_0)^2 \qquad (10.94)$$

Using Eq. (10.75), the variance $(\Delta a_0)^2$ can be obtained as:

$$(\Delta a_0)^2 \approx \left(\frac{1}{n} + \frac{\bar{x}^2}{S_{XX}} \right) \cdot \frac{n}{n-2} \cdot s^2$$

$$= \left(\frac{1}{n} + \frac{\bar{x}^2}{\displaystyle\sum_{i=1}^{n} (x_i - \bar{x})^2} \right) \cdot \frac{1}{n-2} \cdot \sum_{i=1}^{n} (y_i - a_1 \cdot x_i - a_0)^2 \qquad (10.95)$$

And finally:

$$(\Delta y)^2 = (\Delta b)^2 + (x - \bar{x})^2 \cdot (\Delta a_1)^2$$

$$\approx \frac{1}{n-2} \left(1 + \frac{n \cdot (x - \bar{x})^2}{S_{XX}} \right) \cdot s^2 \qquad (10.96)$$

The smallest error in y can thus be expected at $x = \bar{x}$.

In case of weighted linear regression, the error estimate for a_1 becomes:

$$(\Delta a_1)^2 \approx \frac{1}{n-2} \cdot \frac{\sum\limits_{i=1}^{n} w_i(y_i - a_1 \cdot x_i - a_0)^2}{\sum\limits_{i=1}^{n} w_i(x_i - \bar{x})^2} \tag{10.97}$$

Similarly, the variance of b is determined as:

$$(\Delta b)^2 \approx \frac{1}{n-2} \cdot \frac{\sum\limits_{i=1}^{n} w_i(y_i - a_1 \cdot x_i - a_0)^2}{\sum\limits_{i=1}^{n} w_i} \tag{10.98}$$

Using Eq. (10.75) the variance of a_0 is obtained as:

$$(\Delta a_0)^2 \approx \left(\frac{1}{\sum\limits_{i=1}^{n} w_i} + \frac{(\bar{x}_w)^2}{\sum\limits_{i=1}^{n} w_i(x_i - \bar{x}_w)^2} \right) \cdot \frac{1}{n-2} \cdot \sum\limits_{i=1}^{n} w_i(y_i - a_1 \cdot x_i - a_0)^2 \tag{10.99}$$

And finally:

$$(\Delta y)^2 = (\Delta b)^2 + (x - \bar{x}_w)^2 \cdot (\Delta a_1)^2$$

$$\approx \left(\frac{1}{\sum\limits_{i=1}^{n} w_i} + \frac{(x - \bar{x}_w)^2}{\sum\limits_{i=1}^{n} w_i(x_i - \bar{x}_w)^2} \right) \cdot \frac{\sum\limits_{i=1}^{n} w_i(y_i - a_1 \cdot x_i - a_0)^2}{n-2} \tag{10.100}$$

The smallest error in y can thus be expected at $x = \bar{x}_w$.

More information about regression analysis can be found, for example, in [58].

10.4 Sampling from a Population and Simulations

If an expected property like the lifetime of an asset population must be determined, then this property must be assessed systematically for all assets and the mean must be taken of this set. If the expected value is not physically observed but calculated, then the systematic evaluation for all assets generally is done with the use of the cumulative distribution.

As mentioned in Section 2.4.1, the cumulative distribution F describes the sequence of object events with respect to a given measure. Each asset has a certain ranking in terms of F. Running from $F = 0\%$ to 100% systematically addresses each asset. If the property is represented by a variable X that is a function of F, then the expected value $<X>$ is found with discrete F through Eq. (3.2):

$$E(X) = \langle X \rangle = \frac{\sum\limits_{j=1}^{n} f(x_j)x_j}{\sum\limits_{j=1}^{n} f(x_j)} \tag{10.101}$$

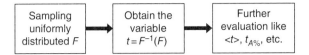

Figure 10.5 Process of sampling from a population, obtaining the distributed variable and carrying out further analysis.

and for continuous F through Eq. (3.3):

$$E(X) = \langle X \rangle = \frac{\int_0^1 x(F)dF}{\int_0^1 dF} = \int_0^1 x(F)dF \qquad (10.102)$$

In this approach the cumulative distribution value is in fact treated as a variable F that is uniformly distributed (see Section 4.1.1 and following).

Simulations like scenario tests, but also calculations like evaluating the expected lifetime or a confidence limit like $t_{1\%}$ or $t_{50\%}$, follow the procedure shown in Figure 10.5. First, samples from the uniformly distributed F population are drawn. Subsequently, these F values are converted into the variable of interest (i.e. $X(F)$), which can be time to failure t, $\log(t)$, reciprocal $1/t$, and so on. Finally, the analysis of the generated data is carried out, which yields output like the expected value of lifetime t, $A\%$ ranking time $t_{A\%}$, and so on.

The following subsections describe the sampling of cumulative distribution values F with various sample sizes n, how expected values can be determined and the relationship between confidence limits. Section 10.4.1 discusses systematic sampling from an n-dimensional F-space. Numerical integration of F and $X(F)$ is the subject of Section 10.4.2. The relation between the sample space and confidence intervals is explored in Section 10.4.3. Section 10.4.4 discusses Monte Carlo simulations with random number generators that form an efficient alternative to numerical integration. Finally, Section 10.4.5 introduces alternative sampling like geometrical and fractal patterns, aimed at increased efficiency.

10.4.1 Systematic Sampling

This section shows how sampling in the space of cumulative distribution values can be done. An n-sized sample drawn from a distribution consists of a series of n values. If the sample size n is larger than 1, then the distribution values F_i are usually ranked in order of magnitude. Figure 10.5 may be adjusted as in Figure 10.6.

As a start, it is assumed that $n = 1$. Samples are drawn from a one-dimensional space of cumulative distribution values F. If only one sample is selected to represent the distribution (i.e. not randomly selected), the mean might be used. This is indicated in Figure 10.7 by the dot. The $<F>$ is at the mean 'ranked' position $<F_{i=1,n=1}> = 1/(1+1) = 0.5$ (cf.

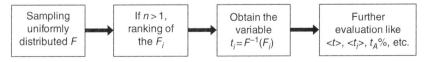

Figure 10.6 Process of sampling from a population, ranking the samples in case of $n > 1$, obtaining the distributed variable(s) and carrying out further analysis.

Figure 10.7 The dot indicates the selection of the mean <F> that represents the distribution.

Figure 10.8 The dots represent the selection of the means of the quartiles that represent the distribution in each quartile.

Section 4.1.2.1, Eq. (4.13)). This can be used to calculate the time to failure $t(<F>)$ as a first-order approximation of $\langle t \rangle$.

As a next step in sampling, a representative F may be selected for subintervals of the range. For instance, the interval $[0,1]$ may be split into m equal subintervals. In case $m = 4$ the subintervals correspond to the quartiles (see Section 3.2). The subintervals are projections of the interval $[0,1]$ and are indexed by j ($j = 1,\ldots,m$). Each subinterval can be represented by the mean $<F_{i,n,j,m}>$ ($i = 1,2,3,4$), as indicated in Figure 10.8. These means are also the medians or means of each subinterval and have values:

$$\langle F_{i,n,j,m} \rangle = \frac{i}{n+1} \cdot \frac{1}{m} + \frac{j-1}{m} \tag{10.103}$$

Note that the subinterval means $<F_{i,n,j,m}>$ should not be confused with the mean ith ranked cumulative distribution $<F_{i,n}>$ of the full interval $[0,1]$, having values $i/(n+1)$. These subinterval means are not regarded as an m-sized sample, but as m 1-sized samples. The sampling is in a one-dimensional F-space.

Dividing the interval into m subintervals has an application in numerical integration to determine expected values like $<t>$, as will be discussed in Section 10.4.2.

If the sample size $n > 1$, more than one cumulative probability value is sampled. These values are ranked and indexed F_i such that $F_i < F_k$ if $i < k$. For the case of $n = 2$, Figure 10.9 shows the mean $(<F_1>,<F_2>) = (1/3,2/3)$ that represents the cumulative distribution pairs.

Again, subintervals may be defined with m the divisional factor. Figure 10.10 shows how the mean $(<F_1>,<F_2>)$ can be selected in subintervals with $m = 2$ and $m = 3$, respectively. This figure shows that the subspaces consist of triangles with the same orientation as that in Figure 10.9 (diagonal faces to the lower right) or with a mirrored orientation (diagonal faces to the upper left).

The construction of the means can be carried out by calculating all possible $<F_{i,n,j,m}>$ for $i = 1,\ldots,n$ and $j = 1,\ldots,m$. All permutations of $i = 1,\ldots,n$ must be considered and, if necessary, re-ranked. To clarify this, first the mean of the lower subspace is determined, similar to (10.103):

$$\langle F_{i,n,=2j=1,m} \rangle = \frac{i}{2+1} \cdot \frac{1}{m} + \frac{1-1}{m} = \frac{i}{3 \cdot m} \tag{10.104}$$

So, $m = 2$ yields $<F_{1,2,1,2}> = 1/6$ and $<F_{2,2,1,2}> = 2/6$, which gives point $(F_1,F_2) = (1/6,2/6)$ and similarly $m = 3$ yields $<F_{1,2,1,3}> = 1/9$ and $<F_{2,2,1,3}> = 2/9$ giving point $(F_1,F_2) = (1/9,2/9)$. Next, all these points are elevated by all possible combinations of j/m. In case of $m = 2$, four points $(1/6,2/6)$, $(1/6,2/6 + 1/2)$, $(1/6 + 1/2,2/6)$ and $(1/6 + 1/2,2/6 + 1/2)$ are obtained, that is $(1,2)/6$, $(1,5)/6$, $(4,2)/6$ and $(4,5)/6$. The third point must be re-ranked to become $(2,4)/6$, which is the centre of mass of the triangle facing up-left.

Similarly, the points for the case $m = 3$ can be found. Also, their re-ranking is necessary in the triangles facing up-left. Apparently, two families of subspaces are produced, subdividing the $n = 2$ dimensional space. These families have their origin in

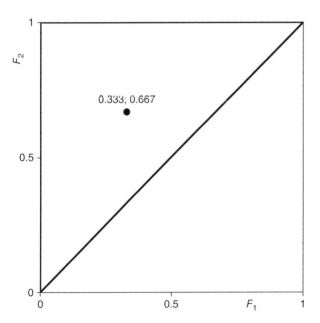

Figure 10.9 The dot represents the selection of the mean ($<F_1>,<F_2>$) = (1/3,2/3) that represents the distribution. The ranked two-dimensional F-space comprises the triangle above the diagonal (i.e. the part where $F_2 > F_1$). The mean ($<F_1>,<F_2>$) is the centre of mass of that triangle.

the permutations of $i = 1,\ldots,n$. Therefore, $n!$ families of subspace orientations can be expected in case of an arbitrary n. The number of subspaces is m^n.

Table 10.5 shows overviews of cumulative distribution values for various sample sizes n and divisional factors m.

In the following sections the means of the subintervals/subspaces will be used for numerical integration and to compare confidence limits based on the beta distribution and regression analysis.

10.4.2 Numerical Integration and Expected Values

When an integral of a function $g(x)$ over an interval $[a,b]$ is not known as an elementary function, it may still be possible to determine the integral through numerical integration or historically 'quadrature'. This historic term refers to calculating the surface under a plotted function.

There are various ways to perform numerical integration. As an example, let $g(x)$ be defined as:

$$g(x) = 430 - 25 \cdot x - 5 \cdot (x - 2)^2 + 3 \cdot (x - 6)^3 \tag{10.105}$$

The integration interval in this example is taken as $[0,10]$. Typically, numerical integration would be used when there is no analytical solution for $G(x) = \int g(x)dx$, while in this case the solution can be calculated as:

$$G(x)|_0^{10} = \int_0^{10} g(x)dx = \left[-238 \cdot x + \frac{319}{2} \cdot x^2 - \frac{59}{3} \cdot x^3 + \frac{3}{4} \cdot x^4 \right]\Big|_0^{10}$$

$$= \frac{4210}{3} \tag{10.106}$$

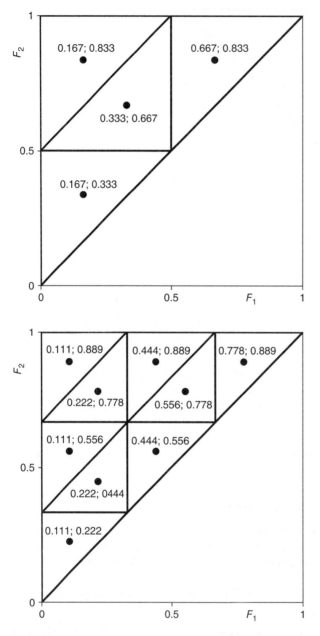

Figure 10.10 Sampling from the cumulative distribution space with sample size $n = 2$. Subspaces are shown with factor $m = 2$ (top) and $m = 3$ (bottom). The dots are the means that represent the distribution in each subspace.

Table 10.5 Mean cumulative distribution values for various sample sizes n and divisional factors m.

$n = 3, m = 3$ (×1/12)			$n = 3, m = 2$ (×1/8)			$n = 4, m = 2$ (×1/10)			
F_1	F_2	F_3	F_1	F_2	F_3	F_1	F_2	F_3	F_4
1	2	3	1	2	3	1	2	3	4
1	2	7	1	2	7	1	2	3	9
1	2	11	1	3	6	1	2	4	9
1	3	6	1	6	7	1	2	8	9
1	3	10	2	3	5	1	3	4	7
1	6	7	2	5	7	1	3	7	9
1	6	11	3	5	6	1	4	7	8
1	7	10	5	6	7	1	7	8	9
1	10	11				2	3	4	6
2	3	5				2	3	6	9
2	3	9				2	4	6	8
2	5	7				2	6	8	9
2	5	11				3	4	6	7
2	7	9				3	6	7	9
2	9	11				4	6	7	8
3	5	6				6	7	8	9
3	5	10							
3	6	9							
3	9	10							
5	6	7							
5	6	11							
5	7	10							
5	10	11							
6	7	9							
6	9	11							
7	9	10							
9	10	11							

In this case that knowledge can be used to compare the numerical integration result with the true value. The function is shown in Figure 10.11.

A very straightforward method is to split the integration interval $[a,b]$ into a series of m subintervals, just as was done in the previous section with one-sized sampling. In the example of $g(x)$ the interval boundaries are $a = 0$ and $b = 10$. The means x_j of the jth subintervals are:

$$x_j = \frac{j - .5}{m} \cdot (b - a) = \frac{j - .5}{m} \cdot 10 \tag{10.107}$$

The width Δx or w_j of each subinterval equals $(1/m) \cdot 10$. The function $g(x)$ is sampled at the mean or median of each subinterval:

$$g_j = g(x_j) = g\left(\frac{j - .5}{m} \cdot (b - a)\right) = g\left(\frac{j - .5}{m} \cdot 10\right) \tag{10.108}$$

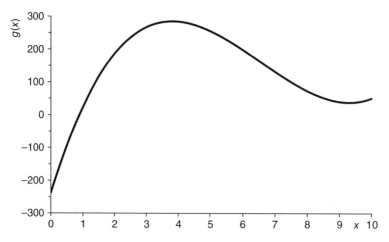

Figure 10.11 The example function g(x).

The numerical integral then becomes:

$$\int_0^{10} g(x)dx \approx \sum_{j=1}^{m} \frac{g_j \cdot 10}{m} \tag{10.109}$$

The result of Eq. (10.109) is 1403.45 for $m = 100$ and 1403.3345 for $m = 1000$, while the true value is 1403.3333. This method is also called the midpoint rule or rectangle rule.

The interval width $1/m$ is the same for all intervals, but there may be reasons to vary the width. For instance, where $g(x)$ varies greatly the interval width w_j may therefore be chosen smaller, while the function may remain more constant in other parts and using larger w_j there saves calculations. Equation (10.109) then becomes:

$$\int_0^{10} g(x)dx \approx \sum_{j=1}^{m} w_j g_j \tag{10.110}$$

where w_j is the width of the interval and g_j is sampled at the mean of the jth interval. This is illustrated in Figure 10.12.

An important application is determining the expected value of a population of assets. The cumulative distribution F covers the full population. The expected value of a function g is assessed by calculating the mean value $<g_F>$ as the integral:

$$\langle g_F \rangle = \int_0^1 g(F)dF \tag{10.111}$$

Example 10.3 *Determining Mean Lifetime and Screening of Products* As an example of using a numerical integration, the mean time to breakdown $<t>$ of a product population with a mixed distribution of a child mortality process, a random failure mechanism and a wear-out mechanism is discussed. In this example the child mortality is considerable and may be tackled by screening the products through an endurance test to burn out the inferior products and improve the quality of the delivered products. Numerical integration is used to quantify the effect.

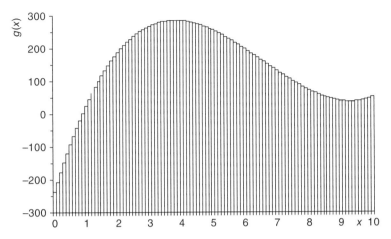

Figure 10.12 Splitting up the integration interval into n subintervals with width $10/m$. The function $g(x)$ is sampled at the median or mean of each subinterval. In this case, for all intervals, $w_j = 1/m$.

Table 10.6 Three simultaneous processes are active in each asset.

Process	Distribution F	Scale parameter α (yr)	Shape parameter β
Child mortality failure	F_{CMF}	0.03	0.001
Random failure	F_{RF}	1500	1
Wear-out failure	F_{WF}	140	4.5

Assume these three processes have Weibull parameters as specified in Table 10.6. The hazard rate of the processes is shown in Figure 10.13. The distribution of lifetimes is shown in Figure 10.14.

The mean lifetime can be calculated with the integral:

$$\langle t_F \rangle = \int_0^1 t(F)dF \tag{10.112}$$

Distribution F is built up with three processes. If an asset fails due to either one of the processes, the asset fails. In terms of systems, this is a series system that survives if it survives all processes (see Section 7.3). The total reliability $R = 1 - F$ is equal to the product of the individual process reliabilities $R_k = 1 - F_k$ (k = CMF, RF or WF) (cf. Eq. (7.11)):

$$R(t) = R_{CMF}(t) \cdot R_{RF}(t) \cdot R_{WF}(t) \tag{10.113}$$

This yields:

$$F(t) = 1 - R_{CMF}(t) \cdot R_{RF}(t) \cdot R_{WF}(t)$$
$$= 1 - \exp\left[-\left(\frac{t}{\alpha_{CMF}}\right)^{\beta_{CMF}} - \left(\frac{t}{\alpha_{RF}}\right)^{\beta_{RF}} - \left(\frac{t}{\alpha_{WF}}\right)^{\beta_{WF}} \right] \tag{10.114}$$

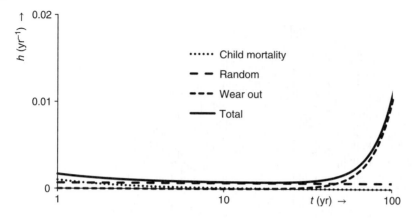

Figure 10.13 Hazard rate of the processes in Table 10.6.

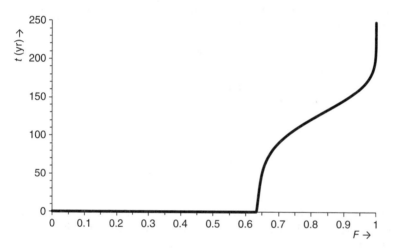

Figure 10.14 Lifetime distribution of the processes in Table 10.6. This figure was actually produced by calculating F(t), but putting F along the horizontal axis and t along the vertical axis.

This function is inconvenient to invert to an analytical expression $t(F)$. Rather than using Eq. (10.112), a more convenient expression for the mean lifetime was presented in Section 3.1, Eq. (3.8):

$$\langle t_F \rangle = \int_0^\infty t \cdot f(t)dt = \int_0^\infty R(t)dt \qquad (10.115)$$

The reliability follows from Eqs (10.113) and (10.114) and is shown in Figure 10.15. The expected lifetime <t> can be found by numerical integration of $R(t)$. As the child mortality process particularly gives large fall-out at small t, the subinterval width w_j may be chosen smaller at low t_j in order to increase accuracy. The subintervals should neither overlap nor have gaps in between. A numerical integration over [0,400] yields <t> = 44.6 years. Details of the applied numerical integration in that case are: $w_j = 0.001$ for $t_1 = 0.0005$, ..., $t_{20} = 0.0195$; $w_j = 0.01$ for $t_{21} = 0.025$, ..., $t_{38} = 0.195$; $w_j = 0.1$ for $t_{39} = 0.25$, ..., $t_{56} = 1.95$; $w_j = 1$ for $t_{57} = 2.5$, ..., $t_{454} = 399.5$.

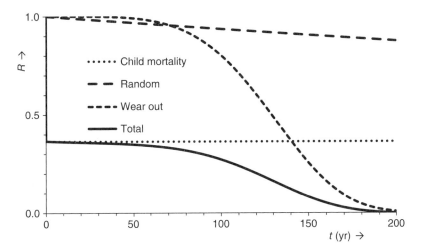

Figure 10.15 Reliability of the processes in Table 10.6.

The accuracy of the result may be tested in order to evaluate whether firstly the interval is sufficient large and secondly the subinterval is sufficiently small. In the above case the numerical integration did not increase after 300 years with an accuracy of 0.00001 (checked up to 600 years). As discussed below this example, it is recommended to avoid integrals over an infinite interval by translating into an integral over a finite interval if possible.

The reliability R drops to about 0.37 almost instantaneously after $t = 0$ (see Figure 10.15), which means that about 63% of the products fail immediately after applying stress.

Instead of integration in the t domain, a finite integral interval $[0,1]$ can be used if F is taken as the variable. That requires Eq. (10.115) to be rewritten as Eq. (10.112).

An alternative to improving production quality can be to apply an endurance test to the products for about 9 hours (or if a power for the power law is known, to carry out an accelerated ageing test as discussed in Section 2.6.2, Example 2.10). In this case 63% of the product batch will fail, after which 37% remains. That surviving part of the population will feature a mean lifetime $<t>$ of almost 121 years, as can be calculated by setting the starting reliability to 1 from $t = 0.001$ years (=8.8 hours). The test wastes 63% of the production but enhances the quality of the delivered products significantly and may save considerably on after-sales servicing.

The numerical integration shown in Example 10.3 can be applied after sampling the cumulative distribution as described in Section 10.4.1, not only for $n = 1$ but also for $n > 1$. Where a higher precision is desired, the subdivision into subspaces may be chosen higher. Of course, the weight of subspaces in the integral must remain proportional to their volume.

As a final remark, Example 10.3 shows both a finite integration interval for F over $[0,1]$ in Eq. (10.112) and an infinite integration interval for t over $[0,\infty)$ in Eq. (10.115). In case of a finite interval and integration of a function without singularities, a finite outcome is expected. Singularities may or may not lead to finite integration outcomes, and this has to be evaluated case by case. In case of infinite intervals, the function that is integrated should reduce sufficiently fast to zero for the variable (say t) approaching ∞ in order to yield a finite integration outcome. With statistics, often an expected value for the population must be determined (e.g. expected lifetime). It may have advantages to translate an integral such that it integrates over cumulative probabilities F because of its finite interval. Section 3.6.3 discussed expected values $<f_g>$, $<f_f>$, and so on in terms of both integrals over t (see Eqs (3.114)–(3.116)) and F (see Eqs (3.117)–(3.120)).

10.4.3 Ranked Samples with Size n and Confidence Limits

Section 5.2.6 discussed the common finding that confidence intervals based on the beta distribution generally tend to be much wider than those based on regression analysis. In the following it is shown, for a homogeneous population, how these are related and why regression analysis seems to produce smaller errors.

In the following, examples of the cumulative distribution spaces in Section 10.4.1 are used to generate samples with sample size n: $F_1, \ldots, F_n$. It is assumed that a true distribution with its parameters is known that defines $F(t)$ and $t(F)$. The following steps are undertaken:

- Since the distribution with its parameters is known, each sample $F_1, \ldots, F_n$ corresponds to a sample $t_1, \ldots, t_n$.
- In Section 10.4.1 the cumulative distribution space was subdivided, which leads to a range of samples $F_1, \ldots, F_n$ and therefore also to a range of samples $t_1, \ldots, t_n$. These various samples are all equally probable to occur.
- The samples $t_1, \ldots, t_n$ are plotted in a graph together with the confidence intervals around the true distribution. This compares the scatter of lines with the beta distribution-based confidence limits.
- For each sample the confidence limits based on the regression analysis can also be calculated and compared with the beta distribution-based confidence limits.

To illustrate the argument a Weibull distribution with parameters $\alpha = 1$ a.u. and $\beta = 2$ is used in all cases. The Weibull distribution is therefore:

$$F(t) = 1 - \exp\left[-\left(\frac{t}{\alpha}\right)^\beta\right] = 1 - \exp[-t^2] \tag{10.116}$$

The times to breakdown follow from F as the inverse Weibull distribution:

$$t(F) = F^{-1}(F) = \alpha \cdot [-\ln(1-F)]^{\frac{1}{\beta}} = [-\ln(1-F)]^{\frac{1}{2}} \tag{10.117}$$

The Weibull distribution $F(t)$ and the inverse Weibull distribution $t(F)$ are shown in Figures 10.16 and 10.17, respectively.

10.4.3.1 Behaviour of Population Fractions

Figure 10.10 showed two-dimensional spaces of ranked cumulative distribution values (F_1, F_2) in which sample pairs are marked with a dot. Each subspace contains sets (F_1, F_2) and with random selection each subspace has the same probability of being drawn, given the areas for $n = 2$ or volumes for $n \geq 3$ are equal. If they are not equal, the probabilities differ and are proportional to the surface or volume of the subspace.

The pairs (F_1, F_2) can now be converted into pairs of failure times (t_1, t_2) using Eq. (10.117) and the given parameters $\alpha = 1$ a.u. and $\beta = 2$. A set of two observations is just

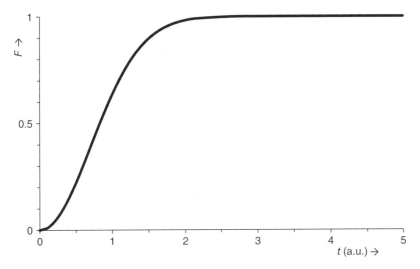

Figure 10.16 The Weibull distribution of Eq. (10.116) with $\alpha = 1$ a.u. and $\beta = 2$.

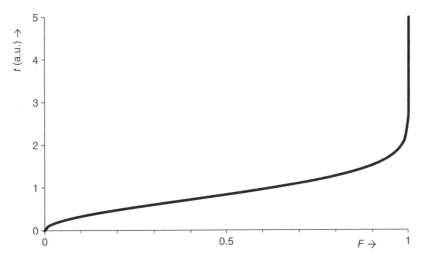

Figure 10.17 The inverse Weibull distribution of Eq. (10.117) with $\alpha = 1$ a.u. and $\beta = 2$.

enough to calculate Weibull parameter estimators a and b. Figure 10.18 shows the values of the estimators at each sample dot, which can be compared with the true Weibull parameters $\alpha = 1$ a.u. and $\beta = 2$. Dots near the diagonal $F_1 = F_2$ yield high values for b. The lower subspace means that F_1 and F_2 are both low, which yields small t_1 and t_2. Consequently, a low value for the estimated location parameter a is found. Vice versa, in the top right-hand corner F_1 and F_2 are both high and therefore also a high a value is found. This representation gives insight into what estimators can be expected from which part of the ranked sample space.

Taking this approach one step further, another property may be mapped, namely the survival after a certain amount of time. Section 2.6.2 described how accelerated ageing can be used to take out child mortality failure, preventing early failures. Two types of

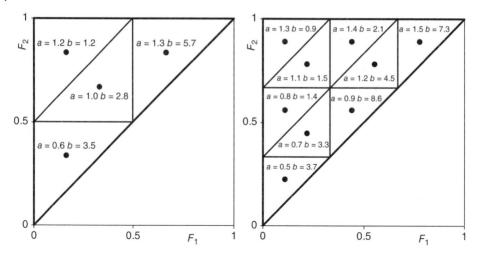

Figure 10.18 Same sampling as in Figure 10.10. Again, the dots are the means that represent the distribution in each subspace. At each dot the estimated Weibull parameters are shown.

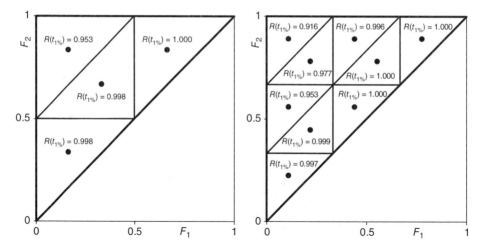

Figure 10.19 Surviving fractions per subspace after an accelerated ageing test that corresponds to sacrificing 1% of the total population. The reliabilities are determined for the representative means of the subspaces, which are marked with black dots.

child mortality process were discussed: the competing process (Section 2.4.5) and the fast ageing subpopulation (Section 2.5.3). The idea is to apply a high stress B_2 during a fixed period τ_2, after which a possible child mortality process is largely taken out (see Sections 2.6.1 and 2.6.2). With known Weibull parameters this period τ_2 at high stress B_2 can be translated into a time τ_1 at (e.g. nominal) stress B_1. One may wonder what this test does to a regular Weibull-distributed population (without a ruling child mortality process) and particularly to the probabilities of the subspaces.

As an example, it is assumed that the accelerating ageing corresponds to the time τ_1 at which 1% of the Weibull-distributed population would fail: $\tau_1 = t_{1\%}$. For each subspace the Weibull parameters were estimated and therefore the fractions of those subpopulations surviving the accelerated ageing test can be calculated. Figure 10.19 shows the surviving fraction of that subspace (represented by the local mean set) after an equivalent time $t_{1\%}$ (i.e. the time in which 1% of the total population would fail).

Figure 10.19 shows that particularly the upper-left part of the distribution space is reduced (i.e. with low F_1 and high F_2), which leads to distributions with low shape parameters b. Such tests and other quality tests are able to reduce fractions of populations. The consequence is that the confidence limits of the distribution after the endurance test can be influenced. Not all subspaces have equal probability after this test anymore.

The following subsection will concentrate on the confidence limits, assuming the populations are regular and not disturbed, for example, by accelerated ageing tests.

10.4.3.2 Confidence Limits for Population Fractions

In the following, the confidence limits based on the beta distribution (Section 4.1.2.1) and regression (Section 10.3) are related to distribution subspaces and compared with each other. The $n = 2$ case as discussed in the previous subsection is illustrative for mapping estimators, but the sample size is too small to calculate the variances in parameter estimation. For $n > 2$ the residues can be determined and regression-based confidence intervals can be compared with the beta distribution-based confidence intervals. In order to have a quantifiable variance, this will be elaborated for $n = 4$ in the following.

The beta $A\%$ confidence limits can be calculated as in Section 5.2.4, Eq. (5.14):

$$F_{i,n,A\%} = B^{inv}(A\%; i, n + 1 - i) \qquad (10.118)$$

A way to produce cumulative distribution samples was discussed in Section 10.4.1. A set for $n = 4$ and $m = 2$ is listed in Table 10.5. That particular case is used to illustrate the effect of sampling from a population. The cumulative distribution values for $n = 4$ and $m = 2$ are used to produce times to failure with Eq. (10.117). The samples from the population of failure times are Weibull-distributed with parameters $\alpha = 1$ a.u. and $\beta = 2$. These sets are shown in Table 10.7.

Figure 10.20 shows the Weibull distribution with parameters $\alpha = 1$ a.u. and $\beta = 2$ as a fat line together with its confidence limits 5% and 95%. The figure also shows the LS best fits of the 16 failure time data sets from Table 10.7. The confidence limits form an envelope reasonably about these best fits. This is essentially what the beta distribution describes. All these lines are the result of limited sampling (here, sample size $n = 4$) from the large population. Overall, the generated best fits and the confidence limits due to the beta distribution are in good agreement.

Next, the effect of the regression error analysis is discussed. Each of the best fits is determined with linear regression. Figure 10.21 shows 4 of the 16 data sets together with their 5% and 95% confidence limit based on the regression analysis. These intervals are much smaller than the confidence intervals based on the beta distribution.

As mentioned in Section 5.2.6, the beta distribution describes the random selection of data from the population. The scatter in data is reflected rather in the wide scatter of the best-fit lines than in the scatter about the individual best fits. The regression error is based on the residues with respect to the best fit and not their deviation from the true line. A mechanism that would influence the scatter of lines is formed by quality control tests like the accelerated ageing test discussed at the end of Section 10.4.3.1.

Table 10.7 The sampled cumulative distribution values from Table 10.5 and corresponding times as calculated with Eq. (10.117).

F_1	F_2	F_3	F_4	t_1 (a.u.)	t_2 (a.u.)	t_3 (a.u.)	t_4 (a.u.)	a (a.u.)	b	$s^2 (\times 10^{-3})$
0.1	0.2	0.3	0.4	0.325	0.472	0.597	0.715	0.6	3.2	0.03
0.1	0.2	0.3	0.9	0.325	0.472	0.597	1.517	0.8	1.8	9.1
0.1	0.2	0.4	0.8	0.325	0.472	0.715	1.269	0.8	1.9	2.5
0.1	0.2	0.8	0.9	0.325	0.472	1.269	1.517	1.1	1.5	7.2
0.1	0.3	0.4	0.7	0.325	0.597	0.715	1.097	0.8	2.2	0.6
0.1	0.3	0.7	0.9	0.325	0.597	1.097	1.517	1.1	1.6	0.8
0.1	0.4	0.7	0.8	0.325	0.715	1.097	1.269	1.0	1.8	1.7
0.1	0.7	0.8	0.9	0.325	1.097	1.269	1.517	1.2	1.7	8.3
0.2	0.3	0.4	0.6	0.472	0.597	0.715	0.957	0.8	3.7	0.4
0.2	0.3	0.6	0.9	0.472	0.597	0.957	1.517	1.0	2.2	3.0
0.2	0.4	0.6	0.8	0.472	0.715	0.957	1.269	1.0	2.6	0.1
0.2	0.6	0.8	0.9	0.472	0.957	1.269	1.517	1.3	2.1	1.1
0.3	0.4	0.6	0.7	0.597	0.715	0.957	1.097	1.0	4.0	0.4
0.3	0.6	0.7	0.9	0.597	0.957	1.097	1.517	1.2	2.8	0.3
0.4	0.6	0.7	0.8	0.715	0.957	1.097	1.269	1.1	4.4	0.03
0.6	0.7	0.8	0.9	0.957	1.097	1.269	1.517	1.3	5.6	0.2

Note: The times are used to produce best fits that are plotted in Figure 10.20. The estimated Weibull parameters a and b are shown (the true parameters are $\alpha = 1$ a.u. and $\beta = 2$). The last column gives the mean squared error s^2 (cf. Section 10.3, Eq. (10.92)) as a measure of data scatter about the best fit. The underlined sets are plotted in Figures 10.21 and 10.22.

Particularly, subspaces that lead to low shape parameter estimators b will be reduced more. The discussed example is not supposed to have undergone such interventions.

The actual distribution is unknown in practice, which makes it impossible to evaluate the residues with respect to the true line (marked in the figures with $\alpha = 1$ and $\beta = 2$). The confidence limits based on the regression error give an impression of the accuracy of the fit and the scatter of data about the best fit. However, they are insufficient to describe the population. It is noteworthy that, despite the fact that their underlying data are drawn from the same population and that 5%–95% confidence intervals are used, the lines hardly resemble each other, whereas there should be an indication that they belong to the same population. Therefore, however inconvenient the wide beta confidence intervals may be, this is nevertheless the route that indicates the resemblance of the best fits by largely overlapping confidence intervals of these lines (as should be). The underlying assumption is that all samples are drawn from the same distribution. Section 5.2.6 also discusses the possibility of an inhomogeneous population where this assumption might not apply.

Figure 10.22 shows the same data as Figure 10.21, but the regression-based confidence intervals about the four best fits are now replaced by beta distribution-based confidence intervals. These intervals are typical of a comparable width as those about the true Weibull distribution (the lower the shape parameter β, the less steep the graph and

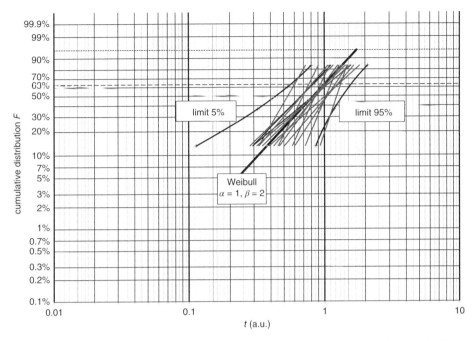

Figure 10.20 Weibull distribution with parameters $\alpha = 1$ a.u. and $\beta = 2$ with confidence limits 5% and 95%. The remaining solid lines are produced as best fits for all time data sets of Table 10.7.

the wider the confidence interval). The mutual overlap is larger and is an indication that these data all come from the same population.

As mentioned in Section 5.2.6, the regression-based confidence intervals are a measure of the suitability of the fit for the particular data set. The beta distribution-based confidence intervals are a measure for the scatter of a homogeneous population given the distribution.

10.4.4 Monte Carlo Experiments and Random Number Generators

In the previous sections samples were drawn from a population by selecting centres of mass in subspaces (cf. Figure 10.10). This enables us to demonstrate how parts of the cumulative distribution space behave. For many simulations this approach does not appear efficient. For example, Table 10.7 lists 16 drawn sets of 4 cumulative distribution values (i.e. 64 values in total). The lowest F_1 is 1/10 in that approach. However, if 64 cumulative distribution values were drawn at once before subdividing them into sets of 4, then the lowest F_1 would be expected to be $<F_{1,64}> = 1/65$ instead of 1/10. The sampling of subspaces by their centres of mass does not go to the extremes that a true random sampling of $n \cdot m$ would (n sample size and m subdivision factor).

As an alternative, generally (semi)-random number generators are used in simulations. These generators produce uniformly distributed numbers between 0 and 1 as cumulative distribution samples. Subsequently these are grouped into n-sized ranked sets $\{F_1, ..., F_n\}$ and converted to, for example, time to failure sets $\{t_1, ..., t_n\}$. Among these sets the extremes are typically covered if the generator is designed well.

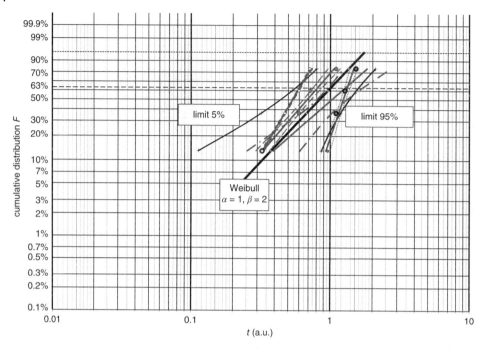

Figure 10.21 Weibull distribution with parameters $\alpha = 1$ a.u. and $\beta = 2$ with confidence limits 5% and 95%. From Table 10.7 the data sets {0.325, 0.472, 0.597, 0.715}, {0.325, 0.597, 0.715, 1.097}, {0.325, 1.097, 1.269, 1.517} and {0.957, 1.097, 1.269, 1.517} are shown with their 5% and 95% confidence limits based on the regression error. The third data set has the largest scatter and the widest confidence intervals. However, generally these intervals are still much smaller than the limits based on the beta distribution.

This method of producing samples is called the Monte Carlo method. A range of generators exist. Here an introduction to traditional generators is discussed. More advanced generators and approaches can be found in [59]. If a mathematical formula is used, the generators are normally referred to as semi- or pseudo-random number generators (see e.g. [68], pp. 949–950). Widely used random number generators are the so-called linear congruent random number generators. These use a seed X_0 (start value) and a recurrent relation to produce the pseudo-random numbers X_k ($k \geq 0$):

$$X_{k+1} = (A \cdot X_k + B) \bmod M \tag{10.119}$$

where A is the multiplier, B the increment and M the modulus. A, B and M are (often large) integer coefficients of the recurrent expression. Dividing X_k by M produces numbers between 0 and 1 as pseudo-random cumulative distribution values F_k.

If $B = 0$ the random number generator is called linear congruential, if $B \neq 0$ mixed congruential. There are some basic requirements for a random number generator:

1. The sequence of random numbers X_k must be (very) large before it starts to repeat itself.
2. The numbers X_k must appear random. A measure of randomness can be the correlation between X_k and X_{k+s} with s an arbitrary integer.

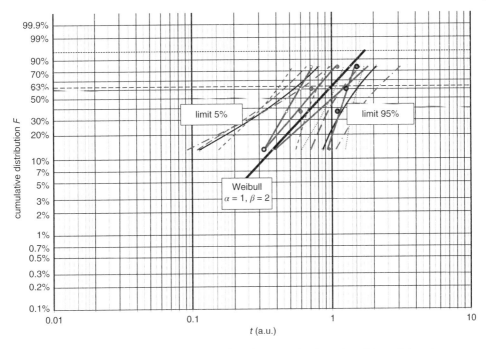

Figure 10.22 The same data as in Figure 10.21, but the regression-based confidence intervals are replaced by beta distribution-based confidence intervals. The overlap is larger (but naturally not complete). The smaller the shape parameter, the wider the confidence intervals.

3. Also, a limited sequence must appear random (this requirement is usually not mentioned, but in case of small simulations it should be tested).

From these requirements, relationships between the A, B and M follow. Also, X_0 may depend on them. The correlation ρ_s between X_k and X_{k+s} is approximated by Greenberger [60] within a margin ε as:

$$\rho_s = \frac{1 - 6 \cdot \dfrac{B_s}{M} \cdot \left(1 - \dfrac{B_s}{M}\right)}{A_s} + \varepsilon \qquad (10.120)$$

where:

$$A_s = (A^s) \bmod M$$

$$B_s = \left(B \cdot \sum_{k=0}^{s-1} A^k\right) \bmod M$$

$$|\varepsilon| < \frac{A_s}{M} \qquad (10.121)$$

If $M \gg B$ and $A \approx \sqrt{M}$ then the correlation $\rho_s \approx \sqrt{M}^{-1}$ and if $M \gg 1$ the correlation approaches 0 which fulfils the above-mentioned second requirement. If the parameters are chosen poorly, the sequence of generated numbers X_k may repeat itself with a relatively short cycle or period compared to M, $X_{k+s} = X_k$ with $S \ll M$, which should be

Table 10.8 Linear and mixed congruential random number generators.

Source	A	B	M	X_0	Remark
Lehmer [62]	23	0	$10^8 + 1$	47 594 118	for 8-bit machine
Matlab v4	$7^5 = 16\,807$	0	$2^{31} - 1$		
Leemis and Park [64]	48 271	0	$2^{31} - 1$	123 456 789	
Jain [65]	$2^{18} + 1$	1	2^{35}		$p_s < 2^{-18}$
van Asselt *et al.* [66]	25 173	13 849	65 536	$X_0 \in \mathbb{N}\backslash\{0\}$	

avoided. With properly chosen parameters it is possible to have a period that equals M. Hull and Dobell [61] proved that this is achieved if and only if:

1. B is relatively prime to M. This means that the only positive integer factor that divides B and M is 1.
2. $A \equiv 1 \pmod{p}$ if p is a prime factor of M. This means applying every prime factor p of M.
3. $A \equiv 1 \pmod{4}$ if 4 is a factor of M.

These requirements are boundary conditions for well-designed random number generators. The first linear congruential generator was presented by Lehmer [62] and used a modulus of 10 power plus 1. As most computers perform binary calculations, a modulus of $M = 2^Q$ (with Q a positive integer) has advantages in calculation speed and truncation, which is one of the popular choices.

B is either 0 or, being relatively prime to M, B must be an odd number if $M = 2^Q$. Often B is chosen as 1 and if not, series produced with other B values can be rewritten as a simple function of the series when $B = 1$ according to Knuth [63]. If $Q > 1$ (which it should be in order to have a large M and large period), then M is a multiple of 4 and so $A - 1$ is a power of 2. Therefore, $A = 4C + 1$ with C a positive integer.

Other choices are to take M a prime number as Lehmer did. Some choices of random number generators are shown in Table 10.8. More advanced random number generators have been developed which are beyond the scope of this book. Recommended for further reading is [59].

A test on random number generators firstly concerns the period after which the sequence repeats itself. Another test is a Kaplan–Meier plot (see Section 5.2.3) to evaluate whether the numbers are uniformly distributed. A spectral test on uniformity [63, 67] tests numbers $(X_0, ..., X_{q-1})$, $(X_1, ..., X_q)$, $(X_2, ..., X_{q+1})$ to show a lattice structure in a q-dimensional space. For $q = 2$ this can be conveniently represented by a two-dimensional plot. This test can also be done with a limited number of samples (e.g. X_k with $k = 400$) and can evaluate whether the specific choice of the seed X_0 matters much.

Other tests can involve typical functions that are related to the subjects under investigation. For instance, with Weibull-related subjects the expected values of the uniform distribution of F and of some functions related to the Weibull-distributed variable t can be tested. In these tests it can be checked whether theoretically expected values

Table 10.9 Some functions to test random number generators.

Function	Mean	Variance
Standard uniformly distributed F		
$\langle F \rangle$	$\dfrac{1}{2}$	$\dfrac{1}{12n}$
$\langle F^2 \rangle$	$\dfrac{1}{3}$	$\dfrac{4}{45n}$
$\langle F_1 \rangle$	$\dfrac{1}{n+1}$	$\dfrac{n}{n^3 + 4n^2 + 5n + 2}$
$\langle F_1^2 \rangle$	$\dfrac{2}{n^2 + 3n + 2}$	$\dfrac{24}{(n+1)(n+2)(n+3)(n+4)} - \langle F_1^2 \rangle^2$
$\langle F_n \rangle$	$\dfrac{n}{n+1}$	$\dfrac{n}{n^3 + 4n^2 + 5n + 2}$
$\langle F_n^2 \rangle$	$\dfrac{n}{n+2}$	$\left(\dfrac{n}{n+2}\right)^2 \cdot \dfrac{n}{n+4}$
Weibull distributed variable x with $\alpha = \beta = 1$		
$\langle x \rangle$	1	$\dfrac{1}{n}$
$\langle x^2 \rangle$	2	$\dfrac{20}{n}$
$\langle \ln x \rangle$	$-\gamma$	$\dfrac{\pi^2}{6}$
$\langle (\ln x)^2 \rangle$	$\gamma^2 + \dfrac{\pi^2}{6}$	
$\langle \ln x_1 \rangle$	$-\gamma - \ln n$	$\dfrac{\pi^2}{6}$
$\langle (\ln x_1)^2 \rangle$	$(\gamma + \ln n)^2 + \dfrac{\pi^2}{6}$	

Note: The variance and expected values of the square can be used to check whether means fall within confidence intervals set by the standard deviations.

and observed means correspond reasonably within the confidence limits based on the calculated variance. A set of test functions is shown in Table 10.9 [27].

10.4.5 Alternative Sampling and Fractals

Monte Carlo number generation is more efficient than numerical integration with the aid of the subspaces and representative means as in Section 10.4.3. An advantage of mapping the distribution space is to be able to secure which parts are covered and with what density. This can be relevant to properties that have to be explored and tested if Monte Carlo simulations are used.

An alternative approach to random number generation is to not uniformly and randomly sample the distribution space, but increase the density in areas where the explored property varies most and decrease the density in areas where the property

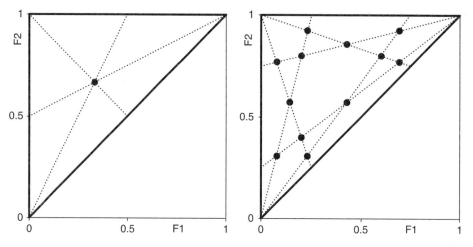

Figure 10.23 A geometrical construction to define samples from a two-dimensional distribution space. The dotted lines are median lines of either the complete space (a) or of half-spaces (b). The intersections of the lines form the samples. The coordinates of the samples are shown in Table 10.10.

is fairly stable. Fractal or geometric constructions can be developed. An example of geometric constructions for the two-dimensional case is shown in Figure 10.23.

The left construction consists of the three medians that bisect the two-dimensional space. These so-called bisectors intersect at the centre of mass (i.e. the mean set of the distributions): $(F_1,F_2) = (1/3,2/3)$.

The right construction consists of the bisectors of the half spaces from the left construction. These generate 12 (F_1,F_2) sets, of which two sets have $F_1 = 1/13$ and two have $F_2 = 12/13$. The sampling may therefore put more weight on the extremes. This might be compensated by a weighted mean with the left construction, (which involves 13 sets). The sets and tests are listed in Table 10.10.

Table 10.10 The coordinates of the distribution sets (F_1,F_2) in Figure 10.23 with test results.

	Left	Right construction												Theory
F_1	1/3	1/13	1/13	1/7	1/5	1/5	3/13	3/13	3/7	3/7	3/5	9/13	9/13	
F_2	2/3	4/13	10/13	4/7	2/5	4/5	4/13	12/13	4/7	6/7	4/5	10/13	12/13	
$<F>$	1/2	1/2												1/2
$<F_1>$	1/3	1/3												1/3
$<F_2>$	2/3	2/3												2/3
$<F^2>$	5/9	0.3254												1/3
$<F_1^2>$	1/9	0.1587												1/6
$<F_2^2>$	4/9	0.4921												1/2

Note: The sequence for the right construction is first ranking F_1 from small to large and second ranking F_2 also from small to large. In Figure 10.23 these show firstly from left to right and secondly from low to high. The last six rows are calculated test functions with $n = 2$ (cf. Table 10.9).

At higher dimensions n there will be $n + 1$ corners from which medians or other quantile lines can be drawn and from which points can be constructed. Various fractal structures can be developed as well.

10.5 Hypothesis Testing

Asset management involves decision-making about actions like maintenance, repair, revision, replacement, and so on. The ground for decision-making is the perceived conception of the situation. Often this involves assumptions such as the likelihood that more failures will follow some event and whether or not specified asset properties are met. Hypotheses are statements about a situation characterized by a measure such as a parameter value or distribution.

An example of the first is how a measured property (e.g. breakdown strength, loss factor or humidity) compares to a reference or specification. Statements could be: the samples meet a certain strength; they are stronger than a given group of other samples; and so on. An example of the second type is how the distribution of a set of observations compares to a specified distribution. Statements could then be: the failure behaviour is exponentially distributed; the distribution of samples appears mixed; and so on.

With hypothesis testing generally, two complementary hypotheses are formulated: the null hypothesis H_0 and the alternative hypothesis H_1. H_0 is tested and if the findings support H_0 to pass the test, H_0 is confirmed to be true. If H_0 does not pass the test, then H_0 is rejected and the alternative H_1 is accepted as the truth. A test is designed that produces a quantitative result that falls in a defined interval if H_0 is to be justified and outside this interval if H_0 is to be rejected. Boundaries of such intervals are called 'critical values'.

As an example, the null hypothesis H_0 can be that a certain installed cable meets the breakdown strength requirements. The alternative hypothesis H_1 is simply that it fails to meet the requirements. The breakdown strength is normally specified in terms of being able to withstand a specified enhanced voltage during a specific test time t_{test} without breaking down. If a withstand test is carried out and the cable does not fail, then apparently $t_{bd} > t_{test}$ and H_0 is accepted. However, if it fails during the test then obviously $t_{bd} \leq t_{test}$ and H_0 is rejected. Consequently, H_1 is then accepted.

Another example is the hypothesis H_0 that an insulator has an adequate length L. It may not be too short, but also not too long. In this case the test result has to fall in a specified length interval $[L_{min}, L_{max}]$. The alternative hypothesis H_1 is that the insulator does not have an adequate length. The test can be a length measurement and comparing the length to the interval. If the measured length L falls in the interval H_0 applies, otherwise H_1. It is worth mentioning that accepting or rejecting hypotheses involves boundaries. Setting these boundaries can be a difficult task.

With hypothesis testing there is a risk of an incorrect outcome. The two errors that can be made are (see also Table 10.11):

1. Type I error: the hypothesis H_0 is true but is rejected in the test because the test result falls outside the interval. The probability α that this happens is called the significance level of the test.

Table 10.11 Overview of factual and perceived truth-related decisions.

Perceived truth	Factual truth	
	H_0 true, so H_1 false	H_0 false, so H_1 true
Acceptance based on test	$P(\text{right decision}) = 1 - \alpha$	Type II error $P(\text{wrong decision}) = \beta$
Rejection based on test	Type I error $P(\text{wrong decision}) = \alpha$	$P(\text{right decision}) = 1 - \beta$

Note: A Type I error occurs when H_0 is true, but is rejected. A Type II error occurs when H_0 is false, but is accepted anyway.

2. Type II error: the hypothesis H_0 is false but is accepted in the test because the test result falls inside the interval. If the probability that H_0 is false (i.e. that a Type II error occurs) equals β, then $\eta = 1 - \beta$ is called the power of the test.

An example of a Type I error is the following. Assume a certain material A is known to have electric strength E_{bd} that shows a certain distribution. It may be known that in 99% of cases a randomly selected sample will have $E_{bd} > E_0$. This electric strength may be taken as a measure to distinguish material A from another material B with a considerably lower electric strength. The null hypothesis H_0 is that a certain sample consists of material A and the alternative hypothesis is that it is made of B (it is assumed it must be either A or B). During a prescribed test period an electric field is applied to a material sample in order to test H_0. If the sample breaks down it is regarded as not consisting of A, but of B. However, there is a probability $\alpha = 1\%$ that the material sample is made of A while it still has $E_{bd} \leq E_0$. For that sample H_0 would be rejected erroneously, which is a Type I error.

Similarly, an example of a Type II error would be the following. Imagine, as in the example above, it is known that 0.5% of samples made of material B would have $E_{bd} > E_0$. If the same H_0 is tested (stating that a sample is made of A), then a probability $\beta = 0.5\%$ exists that a sample made of B does withstand the electric test. For that sample H_0 would then be accepted erroneously, which is a Type II error.

With hypothesis testing a test should be developed with error probabilities α and β as small as possible. Usually, both α and β depend on the critical value(s) at the expense of each other. The significance level α and the power $\eta = 1 - \beta$ of the test depend on the critical value(s) c, which basically are chosen variables. Therefore, α, β and η are also viewed as functions of the critical value(s): $\alpha(c)$, $\beta(c)$ and $\eta(c)$.

10.6 Approximations for the Normal Distribution

The cumulative distribution $F(x)$ is the integral of $f(x)$:

$$F(x) = \int_{-\infty}^{x} f(x)dx = \int_{-\infty}^{x} \frac{1}{\sigma\sqrt{2\pi}} \cdot e^{-\frac{1}{2}\left(\frac{x-\mu}{\sigma}\right)^2} dx \qquad (10.122)$$

As mentioned before, there is no analytical expression for the cumulative normal distribution, nor for the reliability function or the hazard rate. In practice, most mathematical software and spreadsheets readily produce numerical values for $F(x)$. From these and the analytical expression for $f(x)$, the other fundamental functions R, h and H can also be calculated.

Before such software became available, various methods were developed to approximate the cumulative distribution $F(x)$. Such approximations show that $F(x)$ can be calculated with great precision and are illustrations of the various techniques shown in Section 10.2 and following. Apart from the illustrative purpose of approximation with power series and power functions, even with the present numerical facilities there is virtue in discussing the approximation methods and their results. For instance, it may be desirable to know with what power of x these functions vary.

This section describes a few of those expressions. In order to allow more compact equations, approximations are given for the normalized normal distribution. These can be converted to any normal distribution using Eqs (4.53)–(4.56).

10.6.1 Power Series

One approach is to write $F(y)$ as a Taylor (or Maclaurin, see Section 10.2.1) series. The series is infinite and therefore a finite power series remains an approximation. On the other hand, the power series can be taken as long as necessary to meet any desired resolution in $F(y)$. The power expansion of $F(y)$ for $y \geq 0$ becomes (cf. Eq. (26.2.10) in [68] and Example 10.1 in Section 10.2.1):

$$y \geq 0 : F(y) = \frac{1}{2} + \frac{1}{\sqrt{2\pi}} \sum_{n=0}^{\infty} \frac{(-1)^n y^{2n+1}}{n! 2^n (2n+1)} \tag{10.123}$$

For $y \leq 0$:

$$y \leq 0 : F(y) = 1 - F(|y|) = \frac{1}{2} - \frac{1}{\sqrt{2\pi}} \sum_{n=0}^{\infty} \frac{(-1)^n (-y)^{2n+1}}{n! 2^n (2n+1)} \tag{10.124}$$

From these expressions $R(y)$ follows as:

$$y \geq 0 : R(y) = 1 - F(y) = \frac{1}{2} - \frac{1}{\sqrt{2\pi}} \sum_{n=0}^{\infty} \frac{(-1)^n y^{2n+1}}{n! 2^n (2n+1)} \tag{10.125}$$

For $y \leq 0$:

$$y \leq 0 : R(y) = 1 - F(y) = F(|y|) = \frac{1}{2} + \frac{1}{\sqrt{2\pi}} \sum_{n=0}^{\infty} \frac{(-1)^n (-y)^{2n+1}}{n! 2^n (2n+1)} \tag{10.126}$$

10.6.2 Power Series Times Density $f(y)$

The distribution density f contains an exponential function $\exp(-\frac{1}{2}y^2)$. The derivative of that function contains this function again:

$$\frac{d}{dy} f(y) = \frac{1}{\sqrt{2\pi}} \frac{d}{dy} e^{-\frac{1}{2}y^2} = -y \cdot \frac{1}{\sqrt{2\pi}} \cdot e^{-\frac{1}{2}y^2} = -y \cdot f(y) \tag{10.127}$$

In all further derivatives this term appears multiplied by a power series of y. Various approximations are based on this term, which also appears in $R(y)$ and $F(y)$. An expression for $F(y)$ with $y \geq 0$ is (cf. Eq. (26.2.11) in [68]):

$$y \geq 0 : F(y) = \frac{1}{2} + f(y) \cdot \sum_{n=0}^{\infty} \frac{y^{2n+1}}{1 \cdot 3 \cdot 5 \ldots (2n+1)} \tag{10.128}$$

The expression for $y \leq 0$ can be found in a similar way as in Eq. (10.124).

10.6.3 Inequalities for Boxing R(y) and h(y) for Large y

For large values of y (i.e. for $y > 1$, but with better than 5% approximation for $y > 2.5$) the following relation boxes in the value of $R(y)$ (cf. p. 333 in [69]):

$$f(y) \cdot \left(\frac{1}{y} - \frac{1}{y^3} \right) < R(y) < f(y) \cdot \frac{1}{y} \tag{10.129}$$

This can be rewritten in terms of the hazard rate:

$$y \cdot \left(1 + \frac{1}{y^2 - 1} \right) > h(y) > y \tag{10.130}$$

10.6.4 Polynomial Expression for F(y)

A polynomial expression is given by Zelen and Severo with a finite number of terms and a high accuracy of $|\varepsilon(y)| < 7.5 \times 10^{-8}$ (cf. Eq. (26.2.17) in [68]):

$$y \geq 0 : F(y) = 1 - f(y) \cdot (b_1 t + b_2 t^2 + b_3 t^3 + b_4 t^4 + b_5 t^5) + \varepsilon(y),$$

$$t = \frac{1}{1 + py} \tag{10.131}$$

with $p = 0.2316419$, $b_1 = 0.319381530$, $b_2 = -0.356563782$, $b_3 = 1.781477937$, $b_4 = -1.821255978$, $b_5 = 1.330274429$.

10.6.5 Power Function for the Reliability Function R(y)

Finally, three functions are presented with two parameters. These are derived with methods described in Section 10.2.3. Rather than a polynomial or a power expansion, a power function is obtained that fits the asymptotic difference between $R(y)$ and $f(y)$. To be more precise, a power function $D(y;P,Q,R) = D(y;-B,A^B,-A)$ (cf Eq. 10.47) with only two parameters A and B is used:

$$y \geq 0 : \frac{R(y)}{f(y)} \cdot \frac{f(0)}{R(0)} \approx \left(\frac{y}{A} + 1 \right)^{-B} \tag{10.132}$$

where $R(0) = 1/2$, $f(0) = 1/\sqrt{(2\pi)}$ and $f(0)/R(0) = \sqrt{(2/\pi)}$. The three-parameter power function of Section 10.2.3 can be recognized, if a term $(1/A)^{-B} = A^B$ is put outside the brackets. Under the assumption that the approximation model holds, it follows that:

$$f(y) = -\frac{dR}{dy} = -\frac{R(0)}{f(0)} \cdot \frac{d}{dy} \left(\left(\frac{y}{A} + 1 \right)^{-B} \cdot f(y) \right)$$

$$= \frac{R(0)}{f(0)} \left(\frac{B}{A} \cdot \left(\frac{y}{A} + 1 \right)^{-1} + y \right) \cdot \left(\frac{y}{A} + 1 \right)^{-B} \cdot f(y) \tag{10.133}$$

Table 10.12 Two parameter power functions as fits for the asymptotic behaviour of Eq. (10.132).

	R(0) and R(A) correct	Low relative error	Simple
Parameters			
A	1.96018486234016	1.53	1
B	1.5640012379807	1.298	1
Expression for $y \geq 0$			
$R(y) \approx \sqrt{\dfrac{\pi}{2}} \cdot \dfrac{1}{2} e^{-\frac{1}{2}y^2} * \ldots$	$\left(\dfrac{y}{1.96\ldots} + 1\right)^{-1.564\ldots}$	$\left(\dfrac{y}{1.53} + 1\right)^{-1.298}$	$\dfrac{1}{y+1}$
Deviations over interval [0,5]			
Maximum ΔR	0.0008	0.000128	7×10^{-5}
Minimum ΔR	−0.00022	−0.0032	−0.015
Maximum $\Delta R/R$	0.003	0.0064	0.08
Minimum $\Delta R/R$	−0.10	−0.012	−0.05
Deviations over interval [0,3]			
Maximum ΔR	0.0008	0.000128	7×10^{-5}
Minimum ΔR	−0.00022	−0.0032	−0.015
Maximum $\Delta R/R$	0.003	0.0064	0.029
Minimum $\Delta R/R$	−0.04	−0.009	−0.05

Note: The deviations are investigated for y values over the intervals [0,5] and [0,3] in steps of 0.001 ($y \geq 0$). For negative y and for F, similar manipulations as in Eqs (10.125) and (10.126) can be used.

Therefore, the model of Eq. (10.132) yields:

$$1 = \frac{R(0)}{f(0)} \left(\frac{B}{A} \cdot \left(\frac{y}{A} + 1 \right)^{-1} + y \right) \cdot \left(\frac{y}{A} + 1 \right)^{-B} \tag{10.134}$$

The special case of $y = 0$ yields:

$$1 = \frac{R(0)}{f(0)} \cdot \left(\frac{B}{A} \right) \Leftrightarrow \frac{A}{B} = \frac{R(0)}{f(0)} \tag{10.135}$$

Another special case is when $y = A$. This yields:

$$1 = \frac{R(0)}{f(0)} \left(\frac{B}{A} \cdot \frac{1}{2^{B+1}} + A \frac{1}{2^B} \right) = \frac{1}{2^B} \left(\frac{1}{2} + \frac{A^2}{B} \right) \Leftrightarrow B \cdot \ln 2 = \ln \left(\frac{1}{2} + \frac{A^2}{B} \right) \tag{10.136}$$

This yields two sets of solutions for A and B. The set with the smallest error is shown in the second column of Table 10.12. This fit is exact in two points.

If the model of Eq. (10.132) is used without the constraints of the special cases, a best fit in terms of a low relative error can be searched for. This exercise leads to the 'low relative error' solution in the third column of Table 10.12. For $0 < y < 3\sigma$ (i.e. 99.73% of the population with $y > 0$, see Table 4.15), this approximation estimates the reliability better than 1% and the maximum absolute error in R and F is about 0.0025.

A third case is shown where $A = 1$ and $B = 1$, which means:

$$y \geq 0 : R(y) \approx \sqrt{\frac{\pi}{2} \cdot \frac{1}{y+1}} \cdot f(y) \tag{10.137}$$

This solution exactly meets the special case of $y = 0$ but not of $y = A$, and it does not give the smallest error, but its advantage is that it is very simple.

With this expression the maximum absolute error in R is 0.015 for $0 \leq y \leq 5$ (y increment of 0.001) and the maximum absolute relative error in R is 0.08 according to Table 10.12. For $0 \leq y \leq 3$ the maximum absolute error remains the same, but the maximum absolute relative error in R now drops to 0.05.

10.6.6 Wrap-up of Approximations

Various approximations of $F(y)$ and $R(y)$ have been shown. The power-series approaches can reach any precision by expanding the power series until infinity. With six parameters the approach of Zelen and Severo ([68], p. 932; see Eq. (10.131)) offers high accuracy with maximum absolute error in $F(y)$ of 7.5×10^{-8}.

Simple expressions with two parameters are achieved under various boundary conditions with power functions that may be optimized further. The accuracy is typically between 0.3% and 5%, or better for $0 \leq y \leq 3$.

Appendix A

Weibull Plot

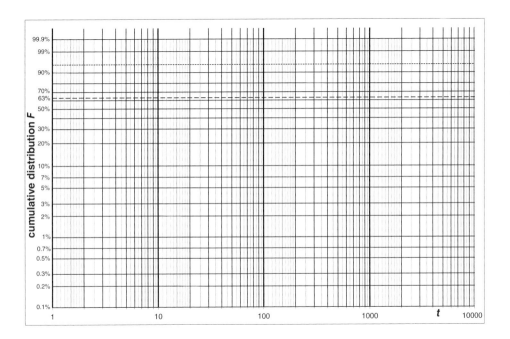

Reliability Analysis for Asset Management of Electric Power Grids, First Edition. Robert Ross.
© 2019 John Wiley & Sons Ltd. Published 2019 by John Wiley & Sons Ltd.
Companion website: www.wiley.com/go/ross/reliabilityanalysis

Appendix B

Laplace Transforms

The Laplace transform $L(P(t))$ of a function $P(t)$ is defined as:

$$L(P(t)) = \int_0^\infty e^{-st} P(t)dt = P(s) \tag{B.1}$$

Here, s is a complex number.

The inverse Laplace transform $L^{-1}(P(s))$ is a complex integral:

$$L^{-1}(P(s)) = \int_{\gamma-i\infty}^{\gamma+i\infty} e^{st} P(s)ds = P(t) \tag{B.2}$$

Here, $\gamma = \mathrm{Re}(s)$ is the real part of s.

The table below shows various relevant transforms for the Markov–Laplace method (cf. Chapter 8).

Time function $P(t)$	Laplace-transformed function $P(s)$
1	$\dfrac{1}{s}$
$e^{-\lambda t}$	$\dfrac{1}{s+\lambda}$
$\dfrac{t^n}{n!} e^{-\lambda t}$	$\dfrac{1}{(s+\lambda)^{n+1}}$
$\dfrac{d}{dt} P(t)$	$s \cdot P(s) - P(t = 0)$
$\int_0^t P(t)\ dt$	$\dfrac{1}{s}\ P(s)$
$\lim\limits_{t \to \infty} P(t)$	$\lim\limits_{s \downarrow 0}\ s\ P(s)$
$\lim\limits_{t \downarrow 0}\ t\ P(t)$	$\lim\limits_{s \to \infty}\ s\ P(s)$

Reliability Analysis for Asset Management of Electric Power Grids, First Edition. Robert Ross.
© 2019 John Wiley & Sons Ltd. Published 2019 by John Wiley & Sons Ltd.
Companion website: www.wiley.com/go/ross/reliabilityanalysis

Appendix C

Taylor Series

The Taylor series is discussed in Section 10.2.1. Below is a table of expansions of functions. The binomial coefficient for any $A \in \mathbb{R}$ is defined as (see Section 10.1.1):

$$\binom{A}{n} = \frac{A \cdot (A-1) \cdot \ldots \cdot (A+1-n)}{n!} \qquad (C.1)$$

Function	Series expansion	Interval
$g(x)$	$g(x) \approx \sum_{n=0}^{k} \frac{g^{(n)}(a)}{n!} \cdot (x-a)^n + o((x-a)^n)$	Depends on $g(x)$
$(1+x)^A$	$1 + A \cdot x + \binom{A}{2} \cdot x^2 + \binom{A}{3} \cdot x^3 + \ldots + \binom{A}{n} \cdot x^n + \ldots$	$-1 < x < 1$
$(1-x)^A$	$1 - A \cdot x + \binom{A}{2} \cdot x^2 - \binom{A}{3} \cdot x^3 + \ldots + (-1)^{n+1} \cdot \binom{A}{n} \cdot x^n + \ldots$	$-1 < x < 1$
$\dfrac{1}{1+x}$	$1 - x + x^2 - x^3 + \ldots + (-1)^n \cdot x^n + \ldots$	$-1 < x < 1$
$\dfrac{1}{1-x}$	$1 + x + x^2 + x^3 + \ldots + x^n + \ldots$	$-1 < x < 1$
$\dfrac{1}{(1-x)^2}$	$1 + 2x + 3x^2 + \ldots + (n+1)x^n + \ldots$	$-1 < x < 1$
$\sqrt{1+x}$	$1 + \dfrac{x}{2} - \dfrac{x^2}{8} + \dfrac{x^3}{16} - \dfrac{5 \cdot x^4}{128} + \dfrac{7 \cdot x^5}{256} - \dfrac{21 \cdot x^6}{1024} - \ldots + (-1)^{n+1} \cdot \binom{\frac{1}{2}}{n} \cdot x^n + \ldots$	$-1 < x < 1$
$\dfrac{1}{\sqrt{1+x}}$	$1 - \dfrac{x}{2} + \dfrac{3 \cdot x^2}{8} - \dfrac{5 \cdot x^3}{16} + \dfrac{35 \cdot x^4}{128} - \dfrac{63 \cdot x^5}{256} + \dfrac{63 \cdot x^6}{1024} - \ldots + (-1)^n \cdot \binom{\frac{1}{2}}{n} \cdot x^n + \ldots$	$-1 < x < 1$
$\dfrac{1}{A-B \cdot x}$	$\dfrac{1}{A}\left[1 + \dfrac{B \cdot x}{A} + \left(\dfrac{B \cdot x}{A}\right)^2 + \left(\dfrac{B \cdot x}{A}\right)^3 + \ldots + \left(\dfrac{B \cdot x}{A}\right)^n \right] + \ldots$	$\|x\| < \left\|\dfrac{A}{B}\right\|$
$\dfrac{1}{A-B \cdot x}$	$\dfrac{-1}{B \cdot x}\left[1 + \dfrac{A}{B \cdot x} + \left(\dfrac{A}{B \cdot x}\right)^2 + \left(\dfrac{A}{B \cdot x}\right)^3 + \ldots + \left(\dfrac{A}{B \cdot x}\right)^n \right] + \ldots$	$\|x\| > \left\|\dfrac{A}{B}\right\|$
e^x	$1 + x + \dfrac{x^2}{2!} + \dfrac{x^3}{3!} + \ldots + \dfrac{x^n}{n!} + \ldots$	$-\infty < x < \infty$
e^{-x}	$1 - x + \dfrac{x^2}{2!} - \dfrac{x^3}{3!} + \ldots + (-1)^{n-1} \cdot \dfrac{x^n}{n!} + \ldots$	$-\infty < x < \infty$

Reliability Analysis for Asset Management of Electric Power Grids, First Edition. Robert Ross.
© 2019 John Wiley & Sons Ltd. Published 2019 by John Wiley & Sons Ltd.
Companion website: www.wiley.com/go/ross/reliabilityanalysis

Function	Series expansion	Interval
a^x	$1 + x \cdot \ln a + \dfrac{(x \cdot \ln a)^2}{2!} + \dfrac{(x \cdot \ln a)^3}{3!} + \ldots + \dfrac{(x \cdot \ln n)^n}{n!} + \ldots$	$-\infty < x < \infty$
$\ln(1+x)$	$x - \dfrac{x^2}{2} + \dfrac{x^3}{3} - \dfrac{x^4}{4} + \ldots + (-1)^{n-1} \cdot \dfrac{x^n}{n} + \ldots$	$-1 < x \leq 1$
$\ln(a+x)$	$\ln a + \dfrac{x}{a} - \dfrac{1}{2} \cdot \left(\dfrac{x}{a}\right)^2 + \dfrac{1}{3} \cdot \left(\dfrac{x}{a}\right)^3 - \dfrac{1}{4} \cdot \left(\dfrac{x}{a}\right)^4 + \ldots + \dfrac{(-1)^{n-1}}{n} \cdot \left(\dfrac{x}{a}\right)^n + \ldots$	$-a < x \leq a$

Appendix D

SI Prefixes

Factor	Prefix name	Prefix symbol
10^{24}	Yotta	Y
10^{21}	Zetta	Z
10^{18}	Exa	E
10^{15}	Peta	P
10^{12}	Tera	T
10^{9}	Giga	G
10^{6}	Mega	M
10^{3}	Kilo	k
10^{2}	Hecto	h
10^{1}	Deca	da
10^{-1}	Deci	d
10^{-2}	Centi	c
10^{-3}	Milli	m
10^{-6}	Micro	μ
10^{-9}	Nano	n
10^{-12}	Pico	p
10^{-15}	Femto	f
10^{-18}	Atto	a
10^{-21}	Zepto	z
10^{-24}	Yocto	y

Reliability Analysis for Asset Management of Electric Power Grids, First Edition. Robert Ross.
© 2019 John Wiley & Sons Ltd. Published 2019 by John Wiley & Sons Ltd.
Companion website: www.wiley.com/go/ross/reliabilityanalysis

Appendix E

Greek Characters

Name	Greek uppercase (roman)	Greek lowercase (roman)	Greek uppercase (italic)	Greek lowercase (italic)
Alpha	A	α	A	α
Beta	B	β	B	β
Gamma	Γ	γ	Γ	γ
Delta	Δ	δ	Δ	δ
Epsilon	E	ϵ	E	ϵ
Zeta	Z	ζ	Z	ζ
Eta	H	η	H	η
Theta	Θ	θ	Θ	θ
Iota	I	ι	I	ι
Kappa	K	κ	K	κ
Lambda	Λ	λ	Λ	λ
Mu	M	μ	M	μ
Nu	N	ν	N	ν
Xi	Ξ	ξ	Ξ	ξ
Omicron	O	o	O	o
Pi	Π	π	Π	π
Rho	P	ρ	P	ρ
Sigma	Σ	σ	Σ	σ
Tau	T	τ	T	τ
Upsilon	Υ	υ	Y	υ
Phi	Φ	φ, ϕ	Φ	φ, ϕ
Chi	X	χ	X	χ
Psi	Ψ	ψ	Ψ	ψ
Omega	Ω	ω	Ω	ω

Reliability Analysis for Asset Management of Electric Power Grids, First Edition. Robert Ross.
© 2019 John Wiley & Sons Ltd. Published 2019 by John Wiley & Sons Ltd.
Companion website: www.wiley.com/go/ross/reliabilityanalysis

Appendix F

Standard Weibull and Exponential Distribution

The table below gives numerical values for the standardized Weibull cumulative distribution (i.e. $\alpha = 1$ and $\beta = 1$), which equals the standardized cumulative exponential distribution (i.e. with $\theta = 1$). The conversion to a Weibull distribution as a function of x with scale parameter α and shape parameter β can be found with $y = (x/\alpha)^\beta$ and $x = \alpha \cdot y^{1/\beta}$ (cf. Section 4.2.1). The conversion to an exponential distribution as a function of x with characteristic parameter α can be found with $y = x/\theta$ and $x = y \cdot \theta$ (cf. Section 4.2.4).

Reliability Analysis for Asset Management of Electric Power Grids, First Edition. Robert Ross.
© 2019 John Wiley & Sons Ltd. Published 2019 by John Wiley & Sons Ltd.
Companion website: www.wiley.com/go/ross/reliabilityanalysis

Standardized Weibull and exponential cumulative distribution for $y \in [0.000, 2.000]$:

y	F(y)	y	F(y)	y	F(y)	y	F(y)	y	F(y)
0.000	0.0000000	0.400	0.3296800	0.800	0.5506710	1.200	0.6988058	1.600	0.7981035
0.010	0.0099502	0.410	0.3363497	0.810	0.5551419	1.210	0.7018027	1.610	0.8001124
0.020	0.0198013	0.420	0.3429532	0.820	0.5595683	1.220	0.7047698	1.620	0.8021013
0.030	0.0295545	0.430	0.3494909	0.830	0.5639507	1.230	0.7077074	1.630	0.8040704
0.040	0.0392106	0.440	0.3559636	0.840	0.5682895	1.240	0.7106158	1.640	0.8060200
0.050	0.0487706	0.450	0.3623718	0.850	0.5725851	1.250	0.7134952	1.650	0.8079501
0.060	0.0582355	0.460	0.3687164	0.860	0.5768379	1.260	0.7163460	1.660	0.8098610
0.070	0.0676062	0.470	0.3749977	0.870	0.5810485	1.270	0.7191684	1.670	0.8117529
0.080	0.0768837	0.480	0.3812166	0.880	0.5852171	1.280	0.7219627	1.680	0.8136260
0.090	0.0860688	0.490	0.3873736	0.890	0.5893442	1.290	0.7247292	1.690	0.8154805
0.100	0.0951626	0.500	0.3934693	0.900	0.5934303	1.300	0.7274682	1.700	0.8173165
0.110	0.1041659	0.510	0.3995044	0.910	0.5974758	1.310	0.7301799	1.710	0.8191342
0.120	0.1130796	0.520	0.4054795	0.920	0.6014810	1.320	0.7328647	1.720	0.8209339
0.130	0.1219046	0.530	0.4113950	0.930	0.6054463	1.330	0.7355227	1.730	0.8227156
0.140	0.1306418	0.540	0.4172517	0.940	0.6093722	1.340	0.7381543	1.740	0.8244796
0.150	0.1392920	0.550	0.4230502	0.950	0.6132590	1.350	0.7407597	1.750	0.8262261
0.160	0.1478562	0.560	0.4287909	0.960	0.6171071	1.360	0.7433392	1.760	0.8279551
0.170	0.1563352	0.570	0.4344746	0.970	0.6209170	1.370	0.7458930	1.770	0.8296670
0.180	0.1647298	0.580	0.4401016	0.980	0.6246889	1.380	0.7484214	1.780	0.8313619
0.190	0.1730409	0.590	0.4456727	0.990	0.6284233	1.390	0.7509247	1.790	0.8330398
0.200	0.1812692	0.600	0.4511884	1.000	0.6321206	1.400	0.7534030	1.800	0.8347011
0.210	0.1894158	0.610	0.4566491	1.010	0.6357810	1.410	0.7558567	1.810	0.8363459
0.220	0.1974812	0.620	0.4620556	1.020	0.6394051	1.420	0.7582860	1.820	0.8379742
0.230	0.2054664	0.630	0.4674082	1.030	0.6429930	1.430	0.7606911	1.830	0.8395864
0.240	0.2133721	0.640	0.4727076	1.040	0.6465453	1.440	0.7630722	1.840	0.8411826
0.250	0.2211992	0.650	0.4779542	1.050	0.6500623	1.450	0.7654297	1.850	0.8427628
0.260	0.2289484	0.660	0.4831487	1.060	0.6535442	1.460	0.7677637	1.860	0.8443274
0.270	0.2366205	0.670	0.4882914	1.070	0.6569915	1.470	0.7700745	1.870	0.8458763
0.280	0.2442163	0.680	0.4933830	1.080	0.6604045	1.480	0.7723623	1.880	0.8474099
0.290	0.2517364	0.690	0.4984239	1.090	0.6637835	1.490	0.7746273	1.890	0.8489282
0.300	0.2591818	0.700	0.5034147	1.100	0.6671289	1.500	0.7768698	1.900	0.8504314
0.310	0.2665530	0.710	0.5083558	1.110	0.6704410	1.510	0.7790900	1.910	0.8519196
0.320	0.2738510	0.720	0.5132477	1.120	0.6737202	1.520	0.7812881	1.920	0.8533930
0.330	0.2810763	0.730	0.5180910	1.130	0.6769667	1.530	0.7834643	1.930	0.8548518
0.340	0.2882297	0.740	0.5228861	1.140	0.6801810	1.540	0.7856189	1.940	0.8562961
0.350	0.2953119	0.750	0.5276334	1.150	0.6833632	1.550	0.7877520	1.950	0.8577259
0.360	0.3023237	0.760	0.5323336	1.160	0.6865138	1.560	0.7898639	1.960	0.8591416
0.370	0.3092657	0.770	0.5369869	1.170	0.6896331	1.570	0.7919548	1.970	0.8605431
0.380	0.3161386	0.780	0.5415940	1.180	0.6927213	1.580	0.7940249	1.980	0.8619308
0.390	0.3229431	0.790	0.5461552	1.190	0.6957787	1.590	0.7960744	1.990	0.8633046
0.400	0.3296800	0.800	0.5506710	1.200	0.6988058	1.600	0.7981035	2.000	0.8646647

Standardized Weibull and exponential cumulative distribution for $y \in [2.000, 4.000]$:

y	F(y)	y	F(y)	y	F(y)	y	F(y)	y	F(y)
2.000	0 8646647	2.400	0.9092820	2.800	0.9391899	3.200	0.9592378	3.600	0.9726763
2.010	0.8660113	2.410	0.9101847	2.810	0.9397950	3.210	0.9596434	3.610	0.9729482
2.020	0.8673445	2.420	0.9110784	2.820	0.9403941	3.220	0.9600449	3.620	0.9732173
2.030	0.8686645	2.430	0.9119632	2.830	0.9409871	3.230	0.9604425	3.630	0.9734838
2.040	0.8699713	2.440	0.9128391	2.840	0.9415743	3.240	0.9608361	3.640	0.9737477
2.050	0.8712651	2.450	0.9137064	2.850	0.9421557	3.250	0.9612258	3.650	0.9740089
2.060	0.8725460	2.460	0.9145650	2.860	0.9427312	3.260	0.9616116	3.660	0.9742675
2.070	0.8738142	2.470	0.9154151	2.870	0.9433011	3.270	0.9619936	3.670	0.9745235
2.080	0.8750698	2.480	0.9162568	2.880	0.9438652	3.280	0.9623717	3.680	0.9747770
2.090	0.8763129	2.490	0.9170900	2.890	0.9444238	3.290	0.9627462	3.690	0.9750280
2.100	0.8775436	2.500	0.9179150	2.900	0.9449768	3.300	0.9631168	3.700	0.9752765
2.110	0.8787620	2.510	0.9187318	2.910	0.9455243	3.310	0.9634838	3.710	0.9755225
2.120	0.8799684	2.520	0.9195404	2.920	0.9460663	3.320	0.9638472	3.720	0.9757660
2.130	0.8811627	2.530	0.9203410	2.930	0.9466030	3.330	0.9642069	3.730	0.9760072
2.140	0.8823452	2.540	0.9211336	2.940	0.9471343	3.340	0.9645630	3.740	0.9762459
2.150	0.8835158	2.550	0.9219183	2.950	0.9476603	3.350	0.9649156	3.750	0.9764823
2.160	0.8846749	2.560	0.9226953	2.960	0.9481811	3.360	0.9652647	3.760	0.9767163
2.170	0.8858224	2.570	0.9234645	2.970	0.9486967	3.370	0.9656104	3.770	0.9769479
2.180	0.8869585	2.580	0.9242260	2.980	0.9492072	3.380	0.9659525	3.780	0.9771773
2.190	0.8880833	2.590	0.9249800	2.990	0.9497126	3.390	0.9662913	3.790	0.9774044
2.200	0.8891968	2.600	0.9257264	3.000	0.9502129	3.400	0.9666267	3.800	0.9776292
2.210	0.8902994	2.610	0.9264655	3.010	0.9507083	3.410	0.9669588	3.810	0.9778518
2.220	0.8913909	2.620	0.9271971	3.020	0.9511988	3.420	0.9672876	3.820	0.9780722
2.230	0.8924716	2.630	0.9279215	3.030	0.9516844	3.430	0.9676131	3.830	0.9782904
2.240	0.8935415	2.640	0.9286387	3.040	0.9521651	3.440	0.9679353	3.840	0.9785064
2.250	0.8946008	2.650	0.9293488	3.050	0.9526411	3.450	0.9682544	3.850	0.9787203
2.260	0.8956495	2.660	0.9300518	3.060	0.9531123	3.460	0.9685702	3.860	0.9789320
2.270	0.8966878	2.670	0.9307478	3.070	0.9535788	3.470	0.9688830	3.870	0.9791416
2.280	0.8977158	2.680	0.9314368	3.080	0.9540407	3.480	0.9691926	3.880	0.9793492
2.290	0.8987335	2.690	0.9321191	3.090	0.9544980	3.490	0.9694991	3.890	0.9795547
2.300	0.8997412	2.700	0.9327945	3.100	0.9549508	3.500	0.9698026	3.900	0.9797581
2.310	0.9007387	2.710	0.9334632	3.110	0.9553990	3.510	0.9701031	3.910	0.9799595
2.320	0.9017264	2.720	0.9341252	3.120	0.9558428	3.520	0.9704006	3.920	0.9801589
2.330	0.9027043	2.730	0.9347807	3.130	0.9562822	3.530	0.9706951	3.930	0.9803563
2.340	0.9036724	2.740	0.9354297	3.140	0.9567172	3.540	0.9709867	3.940	0.9805518
2.350	0.9046308	2.750	0.9360721	3.150	0.9571479	3.550	0.9712754	3.950	0.9807453
2.360	0.9055798	2.760	0.9367082	3.160	0.9575743	3.560	0.9715612	3.960	0.9809369
2.370	0.9065193	2.770	0.9373380	3.170	0.9579964	3.570	0.9718441	3.970	0.9811266
2.380	0.9074494	2.780	0.9379615	3.180	0.9584143	3.580	0.9721243	3.980	0.9813144
2.390	0.9083703	2.790	0.9385788	3.190	0.9588281	3.590	0.9724017	3.990	0.9815003
2.400	0.9092820	2.800	0.9391899	3.200	0.9592378	3.600	0.9726763	4.000	0.9816844

Standardized Weibull and exponential cumulative distribution for $y \in [4.000, 6.000]$:

y	F(y)	y	F(y)	y	F(y)	y	F(y)	y	F(y)
4.000	0.9816844	4.400	0.9877227	4.800	0.9917703	5.200	0.9944834	5.600	0.9963021
4.010	0.9818666	4.410	0.9878448	4.810	0.9918521	5.210	0.9945383	5.610	0.9963389
4.020	0.9820470	4.420	0.9879658	4.820	0.9919332	5.220	0.9945927	5.620	0.9963754
4.030	0.9822257	4.430	0.9880855	4.830	0.9920135	5.230	0.9946465	5.630	0.9964114
4.040	0.9824025	4.440	0.9882041	4.840	0.9920929	5.240	0.9946997	5.640	0.9964471
4.050	0.9825776	4.450	0.9883214	4.850	0.9921716	5.250	0.9947525	5.650	0.9964825
4.060	0.9827510	4.460	0.9884376	4.860	0.9922495	5.260	0.9948047	5.660	0.9965175
4.070	0.9829226	4.470	0.9885527	4.870	0.9923266	5.270	0.9948564	5.670	0.9965521
4.080	0.9830925	4.480	0.9886666	4.880	0.9924030	5.280	0.9949076	5.680	0.9965864
4.090	0.9832608	4.490	0.9887794	4.890	0.9924786	5.290	0.9949582	5.690	0.9966204
4.100	0.9834273	4.500	0.9888910	4.900	0.9925534	5.300	0.9950084	5.700	0.9966540
4.110	0.9835922	4.510	0.9890015	4.910	0.9926275	5.310	0.9950581	5.710	0.9966873
4.120	0.9837555	4.520	0.9891110	4.920	0.9927009	5.320	0.9951072	5.720	0.9967203
4.130	0.9839171	4.530	0.9892193	4.930	0.9927735	5.330	0.9951559	5.730	0.9967529
4.140	0.9840771	4.540	0.9893266	4.940	0.9928454	5.340	0.9952041	5.740	0.9967852
4.150	0.9842356	4.550	0.9894328	4.950	0.9929166	5.350	0.9952518	5.750	0.9968172
4.160	0.9843924	4.560	0.9895379	4.960	0.9929871	5.360	0.9952991	5.760	0.9968489
4.170	0.9845477	4.570	0.9896420	4.970	0.9930569	5.370	0.9953459	5.770	0.9968802
4.180	0.9847015	4.580	0.9897451	4.980	0.9931259	5.380	0.9953922	5.780	0.9969113
4.190	0.9848537	4.590	0.9898471	4.990	0.9931943	5.390	0.9954380	5.790	0.9969420
4.200	0.9850044	4.600	0.9899482	5.000	0.9932621	5.400	0.9954834	5.800	0.9969724
4.210	0.9851536	4.610	0.9900482	5.010	0.9933291	5.410	0.9955284	5.810	0.9970026
4.220	0.9853014	4.620	0.9901472	5.020	0.9933955	5.420	0.9955729	5.820	0.9970324
4.230	0.9854476	4.630	0.9902452	5.030	0.9934612	5.430	0.9956169	5.830	0.9970619
4.240	0.9855924	4.640	0.9903423	5.040	0.9935263	5.440	0.9956605	5.840	0.9970912
4.250	0.9857358	4.650	0.9904384	5.050	0.9935907	5.450	0.9957037	5.850	0.9971201
4.260	0.9858777	4.660	0.9905335	5.060	0.9936544	5.460	0.9957464	5.860	0.9971488
4.270	0.9860182	4.670	0.9906277	5.070	0.9937176	5.470	0.9957888	5.870	0.9971771
4.280	0.9861573	4.680	0.9907210	5.080	0.9937801	5.480	0.9958307	5.880	0.9972052
4.290	0.9862951	4.690	0.9908133	5.090	0.9938420	5.490	0.9958722	5.890	0.9972330
4.300	0.9864314	4.700	0.9909047	5.100	0.9939033	5.500	0.9959132	5.900	0.9972606
4.310	0.9865665	4.710	0.9909952	5.110	0.9939639	5.510	0.9959539	5.910	0.9972878
4.320	0.9867001	4.720	0.9910848	5.120	0.9940240	5.520	0.9959942	5.920	0.9973148
4.330	0.9868325	4.730	0.9911735	5.130	0.9940834	5.530	0.9960340	5.930	0.9973415
4.340	0.9869635	4.740	0.9912614	5.140	0.9941423	5.540	0.9960735	5.940	0.9973680
4.350	0.9870932	4.750	0.9913483	5.150	0.9942006	5.550	0.9961125	5.950	0.9973942
4.360	0.9872216	4.760	0.9914344	5.160	0.9942583	5.560	0.9961512	5.960	0.9974201
4.370	0.9873488	4.770	0.9915196	5.170	0.9943154	5.570	0.9961895	5.970	0.9974458
4.380	0.9874746	4.780	0.9916040	5.180	0.9943720	5.580	0.9962274	5.980	0.9974712
4.390	0.9875993	4.790	0.9916875	5.190	0.9944280	5.590	0.9962650	5.990	0.9974963
4.400	0.9877227	4.800	0.9917703	5.200	0.9944834	5.600	0.9963021	6.000	0.9975212

Standardized Weibull and exponential cumulative distribution for $y \in [6.000, 8.000]$:

y	F(y)	y	F(y)	y	F(y)	y	F(y)	y	F(y)
6.000	0.9975212	6.400	0.9983384	6.800	0.9988862	7.200	0.9992534	7.600	0.9994995
6.010	0.9975459	6.410	0.9983550	6.810	0.9988973	7.210	0.9992608	7.610	0.9995045
6.020	0.9975703	6.420	0.9983713	6.820	0.9989083	7.220	0.9992682	7.620	0.9995095
6.030	0.9975945	6.430	0.9983875	6.830	0.9989191	7.230	0.9992755	7.630	0.9995143
6.040	0.9976184	6.440	0.9984036	6.840	0.9989299	7.240	0.9992827	7.640	0.9995192
6.050	0.9976421	6.450	0.9984195	6.850	0.9989405	7.250	0.9992898	7.650	0.9995240
6.060	0.9976656	6.460	0.9984352	6.860	0.9989511	7.260	0.9992969	7.660	0.9995287
6.070	0.9976888	6.470	0.9984508	6.870	0.9989615	7.270	0.9993039	7.670	0.9995334
6.080	0.9977118	6.480	0.9984662	6.880	0.9989719	7.280	0.9993108	7.680	0.9995380
6.090	0.9977346	6.490	0.9984815	6.890	0.9989821	7.290	0.9993177	7.690	0.9995426
6.100	0.9977571	6.500	0.9984966	6.900	0.9989922	7.300	0.9993245	7.700	0.9995472
6.110	0.9977794	6.510	0.9985115	6.910	0.9990022	7.310	0.9993312	7.710	0.9995517
6.120	0.9978015	6.520	0.9985263	6.920	0.9990122	7.320	0.9993378	7.720	0.9995561
6.130	0.9978234	6.530	0.9985410	6.930	0.9990220	7.330	0.9993444	7.730	0.9995606
6.140	0.9978451	6.540	0.9985555	6.940	0.9990317	7.340	0.9993509	7.740	0.9995649
6.150	0.9978665	6.550	0.9985699	6.950	0.9990414	7.350	0.9993574	7.750	0.9995693
6.160	0.9978877	6.560	0.9985841	6.960	0.9990509	7.360	0.9993638	7.760	0.9995735
6.170	0.9979088	6.570	0.9985982	6.970	0.9990603	7.370	0.9993701	7.770	0.9995778
6.180	0.9979296	6.580	0.9986122	6.980	0.9990697	7.380	0.9993764	7.780	0.9995820
6.190	0.9979502	6.590	0.9986260	6.990	0.9990790	7.390	0.9993826	7.790	0.9995861
6.200	0.9979706	6.600	0.9986396	7.000	0.9990881	7.400	0.9993887	7.800	0.9995903
6.210	0.9979908	6.610	0.9986532	7.010	0.9990972	7.410	0.9993948	7.810	0.9995943
6.220	0.9980108	6.620	0.9986666	7.020	0.9991062	7.420	0.9994009	7.820	0.9995984
6.230	0.9980305	6.630	0.9986798	7.030	0.9991151	7.430	0.9994068	7.830	0.9996024
6.240	0.9980501	6.640	0.9986930	7.040	0.9991239	7.440	0.9994127	7.840	0.9996063
6.250	0.9980695	6.650	0.9987060	7.050	0.9991326	7.450	0.9994186	7.850	0.9996102
6.260	0.9980888	6.660	0.9987189	7.060	0.9991412	7.460	0.9994243	7.860	0.9996141
6.270	0.9981078	6.670	0.9987316	7.070	0.9991498	7.470	0.9994301	7.870	0.9996180
6.280	0.9981266	6.680	0.9987442	7.080	0.9991582	7.480	0.9994357	7.880	0.9996218
6.290	0.9981452	6.690	0.9987567	7.090	0.9991666	7.490	0.9994414	7.890	0.9996255
6.300	0.9981637	6.700	0.9987691	7.100	0.9991749	7.500	0.9994469	7.900	0.9996293
6.310	0.9981820	6.710	0.9987813	7.110	0.9991831	7.510	0.9994524	7.910	0.9996329
6.320	0.9982001	6.720	0.9987935	7.120	0.9991912	7.520	0.9994579	7.920	0.9996366
6.330	0.9982180	6.730	0.9988055	7.130	0.9991993	7.530	0.9994633	7.930	0.9996402
6.340	0.9982357	6.740	0.9988174	7.140	0.9992072	7.540	0.9994686	7.940	0.9996438
6.350	0.9982533	6.750	0.9988291	7.150	0.9992151	7.550	0.9994739	7.950	0.9996473
6.360	0.9982706	6.760	0.9988408	7.160	0.9992229	7.560	0.9994791	7.960	0.9996508
6.370	0.9982878	6.770	0.9988523	7.170	0.9992307	7.570	0.9994843	7.970	0.9996543
6.380	0.9983049	6.780	0.9988637	7.180	0.9992383	7.580	0.9994894	7.980	0.9996578
6.390	0.9983217	6.790	0.9988750	7.190	0.9992459	7.590	0.9994945	7.990	0.9996612
6.400	0.9983384	6.800	0.9988862	7.200	0.9992534	7.600	0.9994995	8.000	0.9996645

Standardized Weibull and exponential cumulative distribution for $y \in [8.000, 10.000]$:

y	F(y)	y	F(y)	y	F(y)	y	F(y)	y	F(y)
8.000	0.9996645	8.400	0.9997751	8.800	0.9998493	9.200	0.9998990	9.600	0.9999323
8.010	0.9996679	8.410	0.9997774	8.810	0.9998508	9.210	0.9999000	9.610	0.9999329
8.020	0.9996712	8.420	0.9997796	8.820	0.9998523	9.220	0.9999010	9.620	0.9999336
8.030	0.9996745	8.430	0.9997818	8.830	0.9998537	9.230	0.9999019	9.630	0.9999343
8.040	0.9996777	8.440	0.9997839	8.840	0.9998552	9.240	0.9999029	9.640	0.9999349
8.050	0.9996809	8.450	0.9997861	8.850	0.9998566	9.250	0.9999039	9.650	0.9999356
8.060	0.9996841	8.460	0.9997882	8.860	0.9998580	9.260	0.9999048	9.660	0.9999362
8.070	0.9996872	8.470	0.9997903	8.870	0.9998595	9.270	0.9999058	9.670	0.9999369
8.080	0.9996903	8.480	0.9997924	8.880	0.9998609	9.280	0.9999067	9.680	0.9999375
8.090	0.9996934	8.490	0.9997945	8.890	0.9998622	9.290	0.9999077	9.690	0.9999381
8.100	0.9996965	8.500	0.9997965	8.900	0.9998636	9.300	0.9999086	9.700	0.9999387
8.110	0.9996995	8.510	0.9997986	8.910	0.9998650	9.310	0.9999095	9.710	0.9999393
8.120	0.9997025	8.520	0.9998006	8.920	0.9998663	9.320	0.9999104	9.720	0.9999399
8.130	0.9997054	8.530	0.9998025	8.930	0.9998676	9.330	0.9999113	9.730	0.9999405
8.140	0.9997084	8.540	0.9998045	8.940	0.9998690	9.340	0.9999122	9.740	0.9999411
8.150	0.9997113	8.550	0.9998065	8.950	0.9998703	9.350	0.9999130	9.750	0.9999417
8.160	0.9997141	8.560	0.9998084	8.960	0.9998716	9.360	0.9999139	9.760	0.9999423
8.170	0.9997170	8.570	0.9998103	8.970	0.9998728	9.370	0.9999148	9.770	0.9999429
8.180	0.9997198	8.580	0.9998122	8.980	0.9998741	9.380	0.9999156	9.780	0.9999434
8.190	0.9997226	8.590	0.9998140	8.990	0.9998753	9.390	0.9999164	9.790	0.9999440
8.200	0.9997253	8.600	0.9998159	9.000	0.9998766	9.400	0.9999173	9.800	0.9999445
8.210	0.9997281	8.610	0.9998177	9.010	0.9998778	9.410	0.9999181	9.810	0.9999451
8.220	0.9997308	8.620	0.9998195	9.020	0.9998790	9.420	0.9999189	9.820	0.9999456
8.230	0.9997335	8.630	0.9998213	9.030	0.9998802	9.430	0.9999197	9.830	0.9999462
8.240	0.9997361	8.640	0.9998231	9.040	0.9998814	9.440	0.9999205	9.840	0.9999467
8.250	0.9997387	8.650	0.9998249	9.050	0.9998826	9.450	0.9999213	9.850	0.9999473
8.260	0.9997413	8.660	0.9998266	9.060	0.9998838	9.460	0.9999221	9.860	0.9999478
8.270	0.9997439	8.670	0.9998283	9.070	0.9998849	9.470	0.9999229	9.870	0.9999483
8.280	0.9997465	8.680	0.9998300	9.080	0.9998861	9.480	0.9999236	9.880	0.9999488
8.290	0.9997490	8.690	0.9998317	9.090	0.9998872	9.490	0.9999244	9.890	0.9999493
8.300	0.9997515	8.700	0.9998334	9.100	0.9998883	9.500	0.9999251	9.900	0.9999498
8.310	0.9997540	8.710	0.9998351	9.110	0.9998894	9.510	0.9999259	9.910	0.9999503
8.320	0.9997564	8.720	0.9998367	9.120	0.9998905	9.520	0.9999266	9.920	0.9999508
8.330	0.9997588	8.730	0.9998383	9.130	0.9998916	9.530	0.9999274	9.930	0.9999513
8.340	0.9997612	8.740	0.9998399	9.140	0.9998927	9.540	0.9999281	9.940	0.9999518
8.350	0.9997636	8.750	0.9998415	9.150	0.9998938	9.550	0.9999288	9.950	0.9999523
8.360	0.9997660	8.760	0.9998431	9.160	0.9998948	9.560	0.9999295	9.960	0.9999527
8.370	0.9997683	8.770	0.9998447	9.170	0.9998959	9.570	0.9999302	9.970	0.9999532
8.380	0.9997706	8.780	0.9998462	9.180	0.9998969	9.580	0.9999309	9.980	0.9999537
8.390	0.9997729	8.790	0.9998478	9.190	0.9998979	9.590	0.9999316	9.990	0.9999541
8.400	0.9997751	8.800	0.9998493	9.200	0.9998990	9.600	0.9999323	10.000	0.9999546

Appendix G

Standardized Normal Distribution

The table below gives numerical values for the standardized normal cumulative distribution (i.e. $\mu = 0$ and $\sigma = 1$). The conversion to normal distributions as a function of x with mean μ and standard deviation σ can be found with $y = (x - \mu)/\sigma$ and $x = \sigma \cdot y + \mu$ (cf. Section 4.3.1.1).

Reliability Analysis for Asset Management of Electric Power Grids, First Edition. Robert Ross.
© 2019 John Wiley & Sons Ltd. Published 2019 by John Wiley & Sons Ltd.
Companion website: www.wiley.com/go/ross/reliabilityanalysis

The standardized normal function for $y \in [0.000, 1.000]$. Note: $F(-y) = 1 - F(y)$:

y	F(y)	y	F(y)	y	F(y)	y	F(y)	y	F(y)
0.000	0.5000000	0.200	0.5792597	0.400	0.6554217	0.600	0.7257469	0.800	0.7881446
0.005	0.5019947	0.205	0.5812139	0.405	0.6572612	0.605	0.7274105	0.805	0.7895902
0.010	0.5039894	0.210	0.5831662	0.410	0.6590970	0.610	0.7290691	0.810	0.7910299
0.015	0.5059839	0.215	0.5851163	0.415	0.6609290	0.615	0.7307226	0.815	0.7924638
0.020	0.5079783	0.220	0.5870644	0.420	0.6627573	0.620	0.7323711	0.820	0.7938919
0.025	0.5099725	0.225	0.5890104	0.425	0.6645817	0.625	0.7340145	0.825	0.7953142
0.030	0.5119665	0.230	0.5909541	0.430	0.6664022	0.630	0.7356527	0.830	0.7967306
0.035	0.5139601	0.235	0.5928956	0.435	0.6682188	0.635	0.7372858	0.835	0.7981411
0.040	0.5159534	0.240	0.5948349	0.440	0.6700314	0.640	0.7389137	0.840	0.7995458
0.045	0.5179463	0.245	0.5967718	0.445	0.6718401	0.645	0.7405364	0.845	0.8009446
0.050	0.5199388	0.250	0.5987063	0.450	0.6736448	0.650	0.7421539	0.850	0.8023375
0.055	0.5219308	0.255	0.6006385	0.455	0.6754454	0.655	0.7437661	0.855	0.8037244
0.060	0.5239222	0.260	0.6025681	0.460	0.6772419	0.660	0.7453731	0.860	0.8051055
0.065	0.5259130	0.265	0.6044953	0.465	0.6790343	0.665	0.7469748	0.865	0.8064806
0.070	0.5279032	0.270	0.6064199	0.470	0.6808225	0.670	0.7485711	0.870	0.8078498
0.075	0.5298926	0.275	0.6083419	0.475	0.6826065	0.675	0.7501621	0.875	0.8092130
0.080	0.5318814	0.280	0.6102612	0.480	0.6843863	0.680	0.7517478	0.880	0.8105703
0.085	0.5338693	0.285	0.6121779	0.485	0.6861618	0.685	0.7533280	0.885	0.8119217
0.090	0.5358564	0.290	0.6140919	0.490	0.6879331	0.690	0.7549029	0.890	0.8132671
0.095	0.5378426	0.295	0.6160031	0.495	0.6896999	0.695	0.7564723	0.895	0.8146065
0.100	0.5398278	0.300	0.6179114	0.500	0.6914625	0.700	0.7580363	0.900	0.8159399
0.105	0.5418121	0.305	0.6198169	0.505	0.6932206	0.705	0.7595949	0.905	0.8172673
0.110	0.5437953	0.310	0.6217195	0.510	0.6949743	0.710	0.7611479	0.910	0.8185887
0.115	0.5457774	0.315	0.6236192	0.515	0.6967235	0.715	0.7626955	0.915	0.8199042
0.120	0.5477584	0.320	0.6255158	0.520	0.6984682	0.720	0.7642375	0.920	0.8212136
0.125	0.5497382	0.325	0.6274095	0.525	0.7002084	0.725	0.7657740	0.925	0.8225170
0.130	0.5517168	0.330	0.6293000	0.530	0.7019440	0.730	0.7673049	0.930	0.8238145
0.135	0.5536941	0.335	0.6311875	0.535	0.7036751	0.735	0.7688303	0.935	0.8251059
0.140	0.5556700	0.340	0.6330717	0.540	0.7054015	0.740	0.7703500	0.940	0.8263912
0.145	0.5576446	0.345	0.6349528	0.545	0.7071232	0.745	0.7718641	0.945	0.8276706
0.150	0.5596177	0.350	0.6368307	0.550	0.7088403	0.750	0.7733726	0.950	0.8289439
0.155	0.5615893	0.355	0.6387052	0.555	0.7105527	0.755	0.7748755	0.955	0.8302112
0.160	0.5635595	0.360	0.6405764	0.560	0.7122603	0.760	0.7763727	0.960	0.8314724
0.165	0.5655280	0.365	0.6424443	0.565	0.7139631	0.765	0.7778642	0.965	0.8327276
0.170	0.5674949	0.370	0.6443088	0.570	0.7156612	0.770	0.7793501	0.970	0.8339768
0.175	0.5694602	0.375	0.6461698	0.575	0.7173544	0.775	0.7808302	0.975	0.8352199
0.180	0.5714237	0.380	0.6480273	0.580	0.7190427	0.780	0.7823046	0.980	0.8364569
0.185	0.5733855	0.385	0.6498813	0.585	0.7207261	0.785	0.7837732	0.985	0.8376880
0.190	0.5753454	0.390	0.6517317	0.590	0.7224047	0.790	0.7852361	0.990	0.8389129
0.195	0.5773035	0.395	0.6535786	0.595	0.7240783	0.795	0.7866932	0.995	0.8401319
0.200	0.5792597	0.400	0.6554217	0.600	0.7257469	0.800	0.7881446	1.000	0.8413447

The standardized normal function for $y \in [1.000, 2.000]$. Note: $F(-y) = 1 - F(y)$:

y	F(y)	y	F(y)	y	F(y)	y	F(y)	y	F(y)
1.000	0.8413447	1.200	0.8849303	1.400	0.9192433	1.600	0.9452007	1.800	0.9640697
1.005	0.8425516	1.205	0.8858983	1.405	0.9199894	1.605	0.9457531	1.805	0.9644627
1.010	0.8437524	1.210	0.8868606	1.410	0.9207302	1.610	0.9463011	1.810	0.9648521
1.015	0.8449471	1.215	0.8878170	1.415	0.9214658	1.615	0.9468447	1.815	0.9652380
1.020	0.8461358	1.220	0.8887676	1.420	0.9221962	1.620	0.9473839	1.820	0.9656205
1.025	0.8473184	1.225	0.8897124	1.425	0.9229214	1.625	0.9479187	1.825	0.9659995
1.030	0.8484950	1.230	0.8906514	1.430	0.9236415	1.630	0.9484493	1.830	0.9663750
1.035	0.8496655	1.235	0.8915847	1.435	0.9243565	1.635	0.9489755	1.835	0.9667472
1.040	0.8508300	1.240	0.8925123	1.440	0.9250663	1.640	0.9494974	1.840	0.9671159
1.045	0.8519885	1.245	0.8934341	1.445	0.9257711	1.645	0.9500151	1.845	0.9674812
1.050	0.8531409	1.250	0.8943502	1.450	0.9264707	1.650	0.9505285	1.850	0.9678432
1.055	0.8542873	1.255	0.8952606	1.455	0.9271654	1.655	0.9510378	1.855	0.9682019
1.060	0.8554277	1.260	0.8961653	1.460	0.9278550	1.660	0.9515428	1.860	0.9685572
1.065	0.8565620	1.265	0.8970643	1.465	0.9285395	1.665	0.9520436	1.865	0.9689093
1.070	0.8576903	1.270	0.8979577	1.470	0.9292191	1.670	0.9525403	1.870	0.9692581
1.075	0.8588126	1.275	0.8988454	1.475	0.9298937	1.675	0.9530329	1.875	0.9696036
1.080	0.8599289	1.280	0.8997274	1.480	0.9305634	1.680	0.9535213	1.880	0.9699460
1.085	0.8610392	1.285	0.9006039	1.485	0.9312281	1.685	0.9540057	1.885	0.9702851
1.090	0.8621434	1.290	0.9014747	1.490	0.9318879	1.690	0.9544860	1.890	0.9706210
1.095	0.8632417	1.295	0.9023399	1.495	0.9325428	1.695	0.9549623	1.895	0.9709538
1.100	0.8643339	1.300	0.9031995	1.500	0.9331928	1.700	0.9554345	1.900	0.9712834
1.105	0.8654202	1.305	0.9040536	1.505	0.9338380	1.705	0.9559028	1.905	0.9716100
1.110	0.8665005	1.310	0.9049021	1.510	0.9344783	1.710	0.9563671	1.910	0.9719334
1.115	0.8675748	1.315	0.9057450	1.515	0.9351138	1.715	0.9568274	1.915	0.9722537
1.120	0.8686431	1.320	0.9065825	1.520	0.9357445	1.720	0.9572838	1.920	0.9725711
1.125	0.8697055	1.325	0.9074144	1.525	0.9363705	1.725	0.9577363	1.925	0.9728853
1.130	0.8707619	1.330	0.9082409	1.530	0.9369916	1.730	0.9581849	1.930	0.9731966
1.135	0.8718123	1.335	0.9090618	1.535	0.9376081	1.735	0.9586296	1.935	0.9735049
1.140	0.8728568	1.340	0.9098773	1.540	0.9382198	1.740	0.9590705	1.940	0.9738102
1.145	0.8738954	1.345	0.9106874	1.545	0.9388269	1.745	0.9595076	1.945	0.9741125
1.150	0.8749281	1.350	0.9114920	1.550	0.9394292	1.750	0.9599408	1.950	0.9744119
1.155	0.8759548	1.355	0.9122912	1.555	0.9400270	1.755	0.9603703	1.955	0.9747085
1.160	0.8769756	1.360	0.9130850	1.560	0.9406201	1.760	0.9607961	1.960	0.9750021
1.165	0.8779905	1.365	0.9138735	1.565	0.9412085	1.765	0.9612181	1.965	0.9752929
1.170	0.8789995	1.370	0.9146565	1.570	0.9417924	1.770	0.9616364	1.970	0.9755808
1.175	0.8800026	1.375	0.9154343	1.575	0.9423718	1.775	0.9620511	1.975	0.9758659
1.180	0.8809999	1.380	0.9162067	1.580	0.9429466	1.780	0.9624620	1.980	0.9761482
1.185	0.8819913	1.385	0.9169738	1.585	0.9435168	1.785	0.9628693	1.985	0.9764278
1.190	0.8829768	1.390	0.9177356	1.590	0.9440826	1.790	0.9632730	1.990	0.9767045
1.195	0.8839565	1.395	0.9184921	1.595	0.9446439	1.795	0.9636732	1.995	0.9769786
1.200	0.8849303	1.400	0.9192433	1.600	0.9452007	1.800	0.9640697	2.000	0.9772499

The standardized normal function for $y \in [2.000, 3.000]$. Note: $F(-y) = 1 - F(y)$:

y	F(y)	y	F(y)	y	F(y)	y	F(y)	y	F(y)
2.000	0.9772499	2.200	0.9860966	2.400	0.9918025	2.600	0.9953388	2.800	0.9974449
2.005	0.9775185	2.205	0.9862730	2.405	0.9919138	2.605	0.9954063	2.805	0.9974842
2.010	0.9777844	2.210	0.9864474	2.410	0.9920237	2.610	0.9954729	2.810	0.9975229
2.015	0.9780477	2.215	0.9866200	2.415	0.9921324	2.615	0.9955386	2.815	0.9975611
2.020	0.9783083	2.220	0.9867906	2.420	0.9922397	2.620	0.9956035	2.820	0.9975988
2.025	0.9785663	2.225	0.9869594	2.425	0.9923458	2.625	0.9956676	2.825	0.9976360
2.030	0.9788217	2.230	0.9871263	2.430	0.9924506	2.630	0.9957308	2.830	0.9976726
2.035	0.9790746	2.235	0.9872913	2.435	0.9925541	2.635	0.9957931	2.835	0.9977087
2.040	0.9793248	2.240	0.9874545	2.440	0.9926564	2.640	0.9958547	2.840	0.9977443
2.045	0.9795726	2.245	0.9876159	2.445	0.9927574	2.645	0.9959155	2.845	0.9977794
2.050	0.9798178	2.250	0.9877755	2.450	0.9928572	2.650	0.9959754	2.850	0.9978140
2.055	0.9800605	2.255	0.9879333	2.455	0.9929558	2.655	0.9960346	2.855	0.9978482
2.060	0.9803007	2.260	0.9880894	2.460	0.9930531	2.660	0.9960930	2.860	0.9978818
2.065	0.9805385	2.265	0.9882437	2.465	0.9931493	2.665	0.9961506	2.865	0.9979150
2.070	0.9807738	2.270	0.9883962	2.470	0.9932443	2.670	0.9962074	2.870	0.9979476
2.075	0.9810067	2.275	0.9885470	2.475	0.9933382	2.675	0.9962635	2.875	0.9979799
2.080	0.9812372	2.280	0.9886962	2.480	0.9934309	2.680	0.9963189	2.880	0.9980116
2.085	0.9814653	2.285	0.9888436	2.485	0.9935224	2.685	0.9963735	2.885	0.9980429
2.090	0.9816911	2.290	0.9889893	2.490	0.9936128	2.690	0.9964274	2.890	0.9980738
2.095	0.9819145	2.295	0.9891334	2.495	0.9937021	2.695	0.9964806	2.895	0.9981042
2.100	0.9821356	2.300	0.9892759	2.500	0.9937903	2.700	0.9965330	2.900	0.9981342
2.105	0.9823543	2.305	0.9894167	2.505	0.9938774	2.705	0.9965848	2.905	0.9981637
2.110	0.9825708	2.310	0.9895559	2.510	0.9939634	2.710	0.9966358	2.910	0.9981929
2.115	0.9827850	2.315	0.9896935	2.515	0.9940484	2.715	0.9966862	2.915	0.9982216
2.120	0.9829970	2.320	0.9898296	2.520	0.9941323	2.720	0.9967359	2.920	0.9982498
2.125	0.9832067	2.325	0.9899640	2.525	0.9942151	2.725	0.9967849	2.925	0.9982777
2.130	0.9834142	2.330	0.9900969	2.530	0.9942969	2.730	0.9968333	2.930	0.9983052
2.135	0.9836195	2.335	0.9902283	2.535	0.9943776	2.735	0.9968810	2.935	0.9983323
2.140	0.9838226	2.340	0.9903581	2.540	0.9944574	2.740	0.9969280	2.940	0.9983589
2.145	0.9840236	2.345	0.9904865	2.545	0.9945361	2.745	0.9969745	2.945	0.9983852
2.150	0.9842224	2.350	0.9906133	2.550	0.9946139	2.750	0.9970202	2.950	0.9984111
2.155	0.9844191	2.355	0.9907386	2.555	0.9946906	2.755	0.9970654	2.955	0.9984367
2.160	0.9846137	2.360	0.9908625	2.560	0.9947664	2.760	0.9971099	2.960	0.9984618
2.165	0.9848062	2.365	0.9909850	2.565	0.9948412	2.765	0.9971539	2.965	0.9984866
2.170	0.9849966	2.370	0.9911060	2.570	0.9949151	2.770	0.9971972	2.970	0.9985110
2.175	0.9851849	2.375	0.9912255	2.575	0.9949880	2.775	0.9972399	2.975	0.9985351
2.180	0.9853713	2.380	0.9913437	2.580	0.9950600	2.780	0.9972821	2.980	0.9985588
2.185	0.9855556	2.385	0.9914604	2.585	0.9951311	2.785	0.9973236	2.985	0.9985821
2.190	0.9857379	2.390	0.9915758	2.590	0.9952012	2.790	0.9973646	2.990	0.9986051
2.195	0.9859182	2.395	0.9916898	2.595	0.9952705	2.795	0.9974050	2.995	0.9986278
2.200	0.9860966	2.400	0.9918025	2.600	0.9953388	2.800	0.9974449	3.000	0.9986501

The standardized normal function for $y \in [3.000, 4.000]$. Note: $F(-y) = 1 - F(y)$:

y	F(y)	y	F(y)	y	F(y)	y	F(y)	y	F(y)
3.000	0.9986501	3.200	0.9993129	3.400	0.9996631	3.600	0.9998409	3.800	0.9999277
3.005	0.9986721	3.205	0.9993247	3.405	0.9996692	3.605	0.9998439	3.805	0.9999291
3.010	0.9986938	3.210	0.9993363	3.410	0.9996752	3.610	0.9998469	3.810	0.9999305
3.015	0.9987151	3.215	0.9993478	3.415	0.9996811	3.615	0.9998498	3.815	0.9999319
3.020	0.9987361	3.220	0.9993590	3.420	0.9996869	3.620	0.9998527	3.820	0.9999333
3.025	0.9987568	3.225	0.9993701	3.425	0.9996926	3.625	0.9998555	3.825	0.9999346
3.030	0.9987772	3.230	0.9993810	3.430	0.9996982	3.630	0.9998583	3.830	0.9999359
3.035	0.9987973	3.235	0.9993918	3.435	0.9997037	3.635	0.9998610	3.835	0.9999372
3.040	0.9988171	3.240	0.9994024	3.440	0.9997091	3.640	0.9998637	3.840	0.9999385
3.045	0.9988366	3.245	0.9994127	3.445	0.9997145	3.645	0.9998663	3.845	0.9999397
3.050	0.9988558	3.250	0.9994230	3.450	0.9997197	3.650	0.9998689	3.850	0.9999409
3.055	0.9988747	3.255	0.9994330	3.455	0.9997249	3.655	0.9998714	3.855	0.9999421
3.060	0.9988933	3.260	0.9994429	3.460	0.9997299	3.660	0.9998739	3.860	0.9999433
3.065	0.9989117	3.265	0.9994527	3.465	0.9997349	3.665	0.9998763	3.865	0.9999445
3.070	0.9989297	3.270	0.9994623	3.470	0.9997398	3.670	0.9998787	3.870	0.9999456
3.075	0.9989475	3.275	0.9994717	3.475	0.9997446	3.675	0.9998811	3.875	0.9999467
3.080	0.9989650	3.280	0.9994810	3.480	0.9997493	3.680	0.9998834	3.880	0.9999478
3.085	0.9989822	3.285	0.9994901	3.485	0.9997539	3.685	0.9998856	3.885	0.9999488
3.090	0.9989992	3.290	0.9994991	3.490	0.9997585	3.690	0.9998879	3.890	0.9999499
3.095	0.9990159	3.295	0.9995079	3.495	0.9997630	3.695	0.9998901	3.895	0.9999509
3.100	0.9990324	3.300	0.9995166	3.500	0.9997674	3.700	0.9998922	3.900	0.9999519
3.105	0.9990486	3.305	0.9995251	3.505	0.9997717	3.705	0.9998943	3.905	0.9999529
3.110	0.9990646	3.310	0.9995335	3.510	0.9997759	3.710	0.9998964	3.910	0.9999539
3.115	0.9990803	3.315	0.9995418	3.515	0.9997801	3.715	0.9998984	3.915	0.9999548
3.120	0.9990957	3.320	0.9995499	3.520	0.9997842	3.720	0.9999004	3.920	0.9999557
3.125	0.9991110	3.325	0.9995579	3.525	0.9997883	3.725	0.9999023	3.925	0.9999566
3.130	0.9991260	3.330	0.9995658	3.530	0.9997922	3.730	0.9999043	3.930	0.9999575
3.135	0.9991407	3.335	0.9995735	3.535	0.9997961	3.735	0.9999061	3.935	0.9999584
3.140	0.9991553	3.340	0.9995811	3.540	0.9997999	3.740	0.9999080	3.940	0.9999593
3.145	0.9991696	3.345	0.9995886	3.545	0.9998037	3.745	0.9999098	3.945	0.9999601
3.150	0.9991836	3.350	0.9995959	3.550	0.9998074	3.750	0.9999116	3.950	0.9999609
3.155	0.9991975	3.355	0.9996032	3.555	0.9998110	3.755	0.9999133	3.955	0.9999617
3.160	0.9992112	3.360	0.9996103	3.560	0.9998146	3.760	0.9999150	3.960	0.9999625
3.165	0.9992246	3.365	0.9996173	3.565	0.9998181	3.765	0.9999167	3.965	0.9999633
3.170	0.9992378	3.370	0.9996242	3.570	0.9998215	3.770	0.9999184	3.970	0.9999641
3.175	0.9992508	3.375	0.9996309	3.575	0.9998249	3.775	0.9999200	3.975	0.9999648
3.180	0.9992636	3.380	0.9996376	3.580	0.9998282	3.780	0.9999216	3.980	0.9999655
3.185	0.9992762	3.385	0.9996441	3.585	0.9998315	3.785	0.9999231	3.985	0.9999663
3.190	0.9992886	3.390	0.9996505	3.590	0.9998347	3.790	0.9999247	3.990	0.9999670
3.195	0.9993008	3.395	0.9996569	3.595	0.9998378	3.795	0.9999262	3.995	0.9999677
3.200	0.9993129	3.400	0.9996631	3.600	0.9998409	3.800	0.9999277	4.000	0.9999683

The standardized normal function for $y \in [4.000, 5.000]$. Note: $F(-y) = 1 - F(y)$:

y	F(y)	y	F(y)	y	F(y)	y	F(y)	y	F(y)
4.000	0.9999683	4.200	0.9999867	4.400	0.9999946	4.600	0.9999979	4.800	0.9999992
4.005	0.9999690	4.205	0.9999869	4.405	0.9999947	4.605	0.9999979	4.805	0.9999992
4.010	0.9999696	4.210	0.9999872	4.410	0.9999948	4.610	0.9999980	4.810	0.9999992
4.015	0.9999703	4.215	0.9999875	4.415	0.9999949	4.615	0.9999980	4.815	0.9999993
4.020	0.9999709	4.220	0.9999878	4.420	0.9999951	4.620	0.9999981	4.820	0.9999993
4.025	0.9999715	4.225	0.9999881	4.425	0.9999952	4.625	0.9999981	4.825	0.9999993
4.030	0.9999721	4.230	0.9999883	4.430	0.9999953	4.630	0.9999982	4.830	0.9999993
4.035	0.9999727	4.235	0.9999886	4.435	0.9999954	4.635	0.9999982	4.835	0.9999993
4.040	0.9999733	4.240	0.9999888	4.440	0.9999955	4.640	0.9999983	4.840	0.9999994
4.045	0.9999738	4.245	0.9999891	4.445	0.9999956	4.645	0.9999983	4.845	0.9999994
4.050	0.9999744	4.250	0.9999893	4.450	0.9999957	4.650	0.9999983	4.850	0.9999994
4.055	0.9999749	4.255	0.9999895	4.455	0.9999958	4.655	0.9999984	4.855	0.9999994
4.060	0.9999755	4.260	0.9999898	4.460	0.9999959	4.660	0.9999984	4.860	0.9999994
4.065	0.9999760	4.265	0.9999900	4.465	0.9999960	4.665	0.9999985	4.865	0.9999994
4.070	0.9999765	4.270	0.9999902	4.470	0.9999961	4.670	0.9999985	4.870	0.9999994
4.075	0.9999770	4.275	0.9999904	4.475	0.9999962	4.675	0.9999985	4.875	0.9999995
4.080	0.9999775	4.280	0.9999907	4.480	0.9999963	4.680	0.9999986	4.880	0.9999995
4.085	0.9999780	4.285	0.9999909	4.485	0.9999964	4.685	0.9999986	4.885	0.9999995
4.090	0.9999784	4.290	0.9999911	4.490	0.9999964	4.690	0.9999986	4.890	0.9999995
4.095	0.9999789	4.295	0.9999913	4.495	0.9999965	4.695	0.9999987	4.895	0.9999995
4.100	0.9999793	4.300	0.9999915	4.500	0.9999966	4.700	0.9999987	4.900	0.9999995
4.105	0.9999798	4.305	0.9999917	4.505	0.9999967	4.705	0.9999987	4.905	0.9999995
4.110	0.9999802	4.310	0.9999918	4.510	0.9999968	4.710	0.9999988	4.910	0.9999995
4.115	0.9999806	4.315	0.9999920	4.515	0.9999968	4.715	0.9999988	4.915	0.9999996
4.120	0.9999811	4.320	0.9999922	4.520	0.9999969	4.720	0.9999988	4.920	0.9999996
4.125	0.9999815	4.325	0.9999924	4.525	0.9999970	4.725	0.9999988	4.925	0.9999996
4.130	0.9999819	4.330	0.9999925	4.530	0.9999971	4.730	0.9999989	4.930	0.9999996
4.135	0.9999823	4.335	0.9999927	4.535	0.9999971	4.735	0.9999989	4.935	0.9999996
4.140	0.9999826	4.340	0.9999929	4.540	0.9999972	4.740	0.9999989	4.940	0.9999996
4.145	0.9999830	4.345	0.9999930	4.545	0.9999973	4.745	0.9999990	4.945	0.9999996
4.150	0.9999834	4.350	0.9999932	4.550	0.9999973	4.750	0.9999990	4.950	0.9999996
4.155	0.9999837	4.355	0.9999933	4.555	0.9999974	4.755	0.9999990	4.955	0.9999996
4.160	0.9999841	4.360	0.9999935	4.560	0.9999974	4.760	0.9999990	4.960	0.9999996
4.165	0.9999844	4.365	0.9999936	4.565	0.9999975	4.765	0.9999991	4.965	0.9999997
4.170	0.9999848	4.370	0.9999938	4.570	0.9999976	4.770	0.9999991	4.970	0.9999997
4.175	0.9999851	4.375	0.9999939	4.575	0.9999976	4.775	0.9999991	4.975	0.9999997
4.180	0.9999854	4.380	0.9999941	4.580	0.9999977	4.780	0.9999991	4.980	0.9999997
4.185	0.9999857	4.385	0.9999942	4.585	0.9999977	4.785	0.9999991	4.985	0.9999997
4.190	0.9999861	4.390	0.9999943	4.590	0.9999978	4.790	0.9999992	4.990	0.9999997
4.195	0.9999864	4.395	0.9999945	4.595	0.9999978	4.795	0.9999992	4.995	0.9999997
4.200	0.9999867	4.400	0.9999946	4.600	0.9999979	4.800	0.9999992	5.000	0.9999997

Appendix H

Standardized Lognormal Distribution

The table below gives numerical values for the standardized lognormal cumulative distribution (i.e. $\ln(y)$ is normally distributed and $\mu = <\ln(y)> = 0$, $\sigma = \sqrt{\text{var}(\ln(y))} = 1$). The conversion to lognormal distributions as a function of x with mean μ and standard deviation σ can be found with $\ln(y) = (\ln(x) - \mu)/\sigma$ and $\ln(x) = \sigma \cdot \ln(y) + \mu$ (cf. Section 4.3.2).

Reliability Analysis for Asset Management of Electric Power Grids, First Edition. Robert Ross.
© 2019 John Wiley & Sons Ltd. Published 2019 by John Wiley & Sons Ltd.
Companion website: www.wiley.com/go/ross/reliabilityanalysis

The standardized lognormal function for $y \in [0.000, 2.000]$:

y	F(y)	y	F(y)	y	F(y)	y	F(y)	y	F(y)
0.000	0.0000000	0.400	0.1797572	0.800	0.4117119	1.200	0.5723348	1.600	0.6808238
0.010	0.0000021	0.410	0.1863042	0.810	0.4165525	1.210	0.5755885	1.610	0.6830462
0.020	0.0000458	0.420	0.1928339	0.820	0.4213461	1.220	0.5788103	1.620	0.6852484
0.030	0.0002270	0.430	0.1993431	0.830	0.4260932	1.230	0.5820006	1.630	0.6874305
0.040	0.0006435	0.440	0.2058287	0.840	0.4307939	1.240	0.5851598	1.640	0.6895927
0.050	0.0013689	0.450	0.2122880	0.850	0.4354486	1.250	0.5882881	1.650	0.6917354
0.060	0.0024509	0.460	0.2187184	0.860	0.4400577	1.260	0.5913860	1.660	0.6938586
0.070	0.0039156	0.470	0.2251177	0.870	0.4446215	1.270	0.5944538	1.670	0.6959626
0.080	0.0057729	0.480	0.2314838	0.880	0.4491404	1.280	0.5974917	1.680	0.6980477
0.090	0.0080213	0.490	0.2378146	0.890	0.4536147	1.290	0.6005003	1.690	0.7001140
0.100	0.0106511	0.500	0.2441086	0.900	0.4580449	1.300	0.6034797	1.700	0.7021618
0.110	0.0136474	0.510	0.2503641	0.910	0.4624312	1.310	0.6064303	1.710	0.7041912
0.120	0.0169919	0.520	0.2565796	0.920	0.4667741	1.320	0.6093525	1.720	0.7062024
0.130	0.0206642	0.530	0.2627539	0.930	0.4710739	1.330	0.6122465	1.730	0.7081957
0.140	0.0246428	0.540	0.2688858	0.940	0.4753310	1.340	0.6151127	1.740	0.7101713
0.150	0.0289061	0.550	0.2749743	0.950	0.4795459	1.350	0.6179513	1.750	0.7121292
0.160	0.0334324	0.560	0.2810185	0.960	0.4837189	1.360	0.6207628	1.760	0.7140698
0.170	0.0382009	0.570	0.2870175	0.970	0.4878504	1.370	0.6235473	1.770	0.7159932
0.180	0.0431911	0.580	0.2929706	0.980	0.4919408	1.380	0.6263053	1.780	0.7178997
0.190	0.0483837	0.590	0.2988771	0.990	0.4959906	1.390	0.6290369	1.790	0.7197893
0.200	0.0537603	0.600	0.3047366	1.000	0.5000000	1.400	0.6317426	1.800	0.7216623
0.210	0.0593034	0.610	0.3105485	1.010	0.5039695	1.410	0.6344226	1.810	0.7235188
0.220	0.0649967	0.620	0.3163124	1.020	0.5078996	1.420	0.6370771	1.820	0.7253590
0.230	0.0708248	0.630	0.3220279	1.030	0.5117905	1.430	0.6397065	1.830	0.7271832
0.240	0.0767732	0.640	0.3276949	1.040	0.5156428	1.440	0.6423111	1.840	0.7289914
0.250	0.0828285	0.650	0.3333131	1.050	0.5194567	1.450	0.6448911	1.850	0.7307839
0.260	0.0889783	0.660	0.3388823	1.060	0.5232328	1.460	0.6474468	1.860	0.7325608
0.270	0.0952107	0.670	0.3444024	1.070	0.5269713	1.470	0.6499785	1.870	0.7343223
0.280	0.1015151	0.680	0.3498733	1.080	0.5306727	1.480	0.6524864	1.880	0.7360686
0.290	0.1078813	0.690	0.3552951	1.090	0.5343374	1.490	0.6549709	1.890	0.7377997
0.300	0.1143000	0.700	0.3606676	1.100	0.5379658	1.500	0.6574322	1.900	0.7395160
0.310	0.1207626	0.710	0.3659910	1.110	0.5415582	1.510	0.6598705	1.910	0.7412174
0.320	0.1272610	0.720	0.3712653	1.120	0.5451150	1.520	0.6622861	1.920	0.7429043
0.330	0.1337879	0.730	0.3764906	1.130	0.5486367	1.530	0.6646792	1.930	0.7445767
0.340	0.1403363	0.740	0.3816672	1.140	0.5521235	1.540	0.6670502	1.940	0.7462348
0.350	0.1469000	0.750	0.3867951	1.150	0.5555760	1.550	0.6693993	1.950	0.7478787
0.360	0.1534730	0.760	0.3918745	1.160	0.5589943	1.560	0.6717266	1.960	0.7495087
0.370	0.1600500	0.770	0.3969056	1.170	0.5623791	1.570	0.6740325	1.970	0.7511248
0.380	0.1666261	0.780	0.4018887	1.180	0.5657305	1.580	0.6763171	1.980	0.7527272
0.390	0.1731965	0.790	0.4068241	1.190	0.5690489	1.590	0.6785808	1.990	0.7543160
0.400	0.1797572	0.800	0.4117119	1.200	0.5723348	1.600	0.6808238	2.000	0.7558914

The standardized lognormal function for $y \in [2.000, 4.000]$:

y	F(y)	y	F(y)	y	F(y)	y	F(y)	y	F(y)
2.000	0.7558914	2.400	0.8093405	2.800	0.8484057	3.200	0.8776158	3.600	0.8998915
2.010	0.7574535	2.410	0.8104692	2.810	0.8492412	3.210	0.8782475	3.610	0.9003779
2.020	0.7590025	2.420	0.8115891	2.820	0.8500708	3.220	0.8788750	3.620	0.9008612
2.030	0.7605385	2.430	0.8127004	2.830	0.8508943	3.230	0.8794982	3.630	0.9013414
2.040	0.7620616	2.440	0.8138030	2.840	0.8517120	3.240	0.8801173	3.640	0.9018186
2.050	0.7635720	2.450	0.8148971	2.850	0.8525238	3.250	0.8807322	3.650	0.9022928
2.060	0.7650697	2.460	0.8159828	2.860	0.8533297	3.260	0.8813430	3.660	0.9027641
2.070	0.7665550	2.470	0.8170601	2.870	0.8541300	3.270	0.8819498	3.670	0.9032324
2.080	0.7680280	2.480	0.8181292	2.880	0.8549245	3.280	0.8825524	3.680	0.9036977
2.090	0.7694887	2.490	0.8191900	2.890	0.8557133	3.290	0.8831511	3.690	0.9041602
2.100	0.7709374	2.500	0.8202428	2.900	0.8564966	3.300	0.8837459	3.700	0.9046198
2.110	0.7723740	2.510	0.8212875	2.910	0.8572743	3.310	0.8843366	3.710	0.9050765
2.120	0.7737989	2.520	0.8223242	2.920	0.8580465	3.320	0.8849235	3.720	0.9055304
2.130	0.7752120	2.530	0.8233531	2.930	0.8588132	3.330	0.8855065	3.730	0.9059815
2.140	0.7766135	2.540	0.8243741	2.940	0.8595745	3.340	0.8860856	3.740	0.9064298
2.150	0.7780035	2.550	0.8253874	2.950	0.8603304	3.350	0.8866610	3.750	0.9068753
2.160	0.7793822	2.560	0.8263931	2.960	0.8610811	3.360	0.8872325	3.760	0.9073180
2.170	0.7807496	2.570	0.8273911	2.970	0.8618264	3.370	0.8878003	3.770	0.9077581
2.180	0.7821058	2.580	0.8283817	2.980	0.8625666	3.380	0.8883644	3.780	0.9081954
2.190	0.7834511	2.590	0.8293647	2.990	0.8633016	3.390	0.8889248	3.790	0.9086300
2.200	0.7847854	2.600	0.8303404	3.000	0.8640314	3.400	0.8894815	3.800	0.9090620
2.210	0.7861089	2.610	0.8313088	3.010	0.8647561	3.410	0.8900346	3.810	0.9094913
2.220	0.7874217	2.620	0.8322700	3.020	0.8654758	3.420	0.8905841	3.820	0.9099180
2.230	0.7887240	2.630	0.8332239	3.030	0.8661905	3.430	0.8911301	3.830	0.9103421
2.240	0.7900157	2.640	0.8341708	3.040	0.8669003	3.440	0.8916725	3.840	0.9107636
2.250	0.7912971	2.650	0.8351106	3.050	0.8676051	3.450	0.8922113	3.850	0.9111826
2.260	0.7925682	2.660	0.8360435	3.060	0.8683051	3.460	0.8927467	3.860	0.9115990
2.270	0.7938292	2.670	0.8369694	3.070	0.8690002	3.470	0.8932787	3.870	0.9120128
2.280	0.7950801	2.680	0.8378885	3.080	0.8696906	3.480	0.8938072	3.880	0.9124242
2.290	0.7963210	2.690	0.8388008	3.090	0.8703762	3.490	0.8943323	3.890	0.9128331
2.300	0.7975520	2.700	0.8397064	3.100	0.8710571	3.500	0.8948540	3.900	0.9132395
2.310	0.7987733	2.710	0.8406053	3.110	0.8717333	3.510	0.8953724	3.910	0.9136434
2.320	0.7999849	2.720	0.8414976	3.120	0.8724049	3.520	0.8958875	3.920	0.9140449
2.330	0.8011869	2.730	0.8423834	3.130	0.8730719	3.530	0.8963992	3.930	0.9144440
2.340	0.8023794	2.740	0.8432627	3.140	0.8737344	3.540	0.8969077	3.940	0.9148407
2.350	0.8035625	2.750	0.8441355	3.150	0.8743923	3.550	0.8974130	3.950	0.9152350
2.360	0.8047364	2.760	0.8450021	3.160	0.8750458	3.560	0.8979150	3.960	0.9156270
2.370	0.8059010	2.770	0.8458623	3.170	0.8756949	3.570	0.8984139	3.970	0.9160166
2.380	0.8070565	2.780	0.8467162	3.180	0.8763395	3.580	0.8989096	3.980	0.9164038
2.390	0.8082030	2.790	0.8475640	3.190	0.8769799	3.590	0.8994021	3.990	0.9167888
2.400	0.8093405	2.800	0.8484057	3.200	0.8776158	3.600	0.8998915	4.000	0.9171715

The standardized lognormal function for $y \in [4.000, 6.000]$:

y	F(y)	y	F(y)	y	F(y)	y	F(y)	y	F(y)
4.000	0.9171715	4.400	0.9307772	4.800	0.9416313	5.200	0.9503912	5.600	0.9575346
4.010	0.9175519	4.410	0.9310789	4.810	0.9418735	5.210	0.9505878	5.610	0.9576958
4.020	0.9179300	4.420	0.9313789	4.820	0.9421144	5.220	0.9507834	5.620	0.9578561
4.030	0.9183059	4.430	0.9316772	4.830	0.9423541	5.230	0.9509780	5.630	0.9580157
4.040	0.9186796	4.440	0.9319739	4.840	0.9425924	5.240	0.9511716	5.640	0.9581745
4.050	0.9190511	4.450	0.9322689	4.850	0.9428295	5.250	0.9513643	5.650	0.9583325
4.060	0.9194203	4.460	0.9325622	4.860	0.9430654	5.260	0.9515560	5.660	0.9584898
4.070	0.9197874	4.470	0.9328539	4.870	0.9433000	5.270	0.9517467	5.670	0.9586463
4.080	0.9201523	4.480	0.9331440	4.880	0.9435333	5.280	0.9519364	5.680	0.9588021
4.090	0.9205151	4.490	0.9334324	4.890	0.9437655	5.290	0.9521252	5.690	0.9589571
4.100	0.9208758	4.500	0.9337193	4.900	0.9439963	5.300	0.9523130	5.700	0.9591114
4.110	0.9212343	4.510	0.9340045	4.910	0.9442260	5.310	0.9524999	5.710	0.9592649
4.120	0.9215907	4.520	0.9342882	4.920	0.9444545	5.320	0.9526859	5.720	0.9594177
4.130	0.9219451	4.530	0.9345703	4.930	0.9446818	5.330	0.9528709	5.730	0.9595698
4.140	0.9222974	4.540	0.9348509	4.940	0.9449079	5.340	0.9530550	5.740	0.9597211
4.150	0.9226476	4.550	0.9351299	4.950	0.9451327	5.350	0.9532382	5.750	0.9598718
4.160	0.9229958	4.560	0.9354074	4.960	0.9453565	5.360	0.9534205	5.760	0.9600217
4.170	0.9233420	4.570	0.9356833	4.970	0.9455790	5.370	0.9536018	5.770	0.9601708
4.180	0.9236861	4.580	0.9359577	4.980	0.9458004	5.380	0.9537823	5.780	0.9603193
4.190	0.9240283	4.590	0.9362307	4.990	0.9460206	5.390	0.9539618	5.790	0.9604671
4.200	0.9243685	4.600	0.9365021	5.000	0.9462397	5.400	0.9541405	5.800	0.9606142
4.210	0.9247067	4.610	0.9367720	5.010	0.9464576	5.410	0.9543183	5.810	0.9607605
4.220	0.9250430	4.620	0.9370405	5.020	0.9466744	5.420	0.9544952	5.820	0.9609062
4.230	0.9253773	4.630	0.9373075	5.030	0.9468901	5.430	0.9546712	5.830	0.9610512
4.240	0.9257097	4.640	0.9375730	5.040	0.9471047	5.440	0.9548464	5.840	0.9611955
4.250	0.9260402	4.650	0.9378371	5.050	0.9473181	5.450	0.9550207	5.850	0.9613391
4.260	0.9263688	4.660	0.9380997	5.060	0.9475305	5.460	0.9551941	5.860	0.9614821
4.270	0.9266955	4.670	0.9383610	5.070	0.9477417	5.470	0.9553667	5.870	0.9616243
4.280	0.9270203	4.680	0.9386208	5.080	0.9479519	5.480	0.9555384	5.880	0.9617659
4.290	0.9273433	4.690	0.9388792	5.090	0.9481609	5.490	0.9557093	5.890	0.9619069
4.300	0.9276645	4.700	0.9391362	5.100	0.9483689	5.500	0.9558793	5.900	0.9620471
4.310	0.9279838	4.710	0.9393918	5.110	0.9485759	5.510	0.9560485	5.910	0.9621867
4.320	0.9283012	4.720	0.9396460	5.120	0.9487817	5.520	0.9562169	5.920	0.9623257
4.330	0.9286169	4.730	0.9398989	5.130	0.9489865	5.530	0.9563844	5.930	0.9624640
4.340	0.9289308	4.740	0.9401504	5.140	0.9491903	5.540	0.9565512	5.940	0.9626017
4.350	0.9292429	4.750	0.9404005	5.150	0.9493930	5.550	0.9567171	5.950	0.9627387
4.360	0.9295533	4.760	0.9406493	5.160	0.9495947	5.560	0.9568822	5.960	0.9628751
4.370	0.9298619	4.770	0.9408968	5.170	0.9497954	5.570	0.9570465	5.970	0.9630108
4.380	0.9301687	4.780	0.9411429	5.180	0.9499950	5.580	0.9572100	5.980	0.9631459
4.390	0.9304738	4.790	0.9413878	5.190	0.9501936	5.590	0.9573727	5.990	0.9632804
4.400	0.9307772	4.800	0.9416313	5.200	0.9503912	5.600	0.9575346	6.000	0.9634142

The standardized lognormal function for $y \in [6.000, 8.000]$

y	F(y)	y	F(y)	y	F(y)	y	F(y)	y	F(y)
6.000	0.9634142	6.400	0.9682945	6.800	0.9723761	7.200	0.9758137	7.600	0.9787274
6.010	0.9635475	6.410	0.9684055	6.810	0.9724693	7.210	0.9758925	7.610	0.9787944
6.020	0.9636801	6.420	0.9685161	6.820	0.9725622	7.220	0.9759710	7.620	0.9788612
6.030	0.9638121	6.430	0.9686261	6.830	0.9726546	7.230	0.9760491	7.630	0.9789276
6.040	0.9639435	6.440	0.9687357	6.840	0.9727466	7.240	0.9761269	7.640	0.9789938
6.050	0.9640743	6.450	0.9688448	6.850	0.9728383	7.250	0.9762044	7.650	0.9790598
6.060	0.9642045	6.460	0.9689534	6.860	0.9729295	7.260	0.9762816	7.660	0.9791254
6.070	0.9643340	6.470	0.9690615	6.870	0.9730204	7.270	0.9763585	7.670	0.9791908
6.080	0.9644630	6.480	0.9691691	6.880	0.9731109	7.280	0.9764350	7.680	0.9792560
6.090	0.9645914	6.490	0.9692763	6.890	0.9732010	7.290	0.9765113	7.690	0.9793209
6.100	0.9647192	6.500	0.9693830	6.900	0.9732907	7.300	0.9765872	7.700	0.9793855
6.110	0.9648464	6.510	0.9694892	6.910	0.9733800	7.310	0.9766628	7.710	0.9794499
6.120	0.9649731	6.520	0.9695950	6.920	0.9734690	7.320	0.9767381	7.720	0.9795141
6.130	0.9650991	6.530	0.9697003	6.930	0.9735576	7.330	0.9768131	7.730	0.9795779
6.140	0.9652246	6.540	0.9698051	6.940	0.9736458	7.340	0.9768878	7.740	0.9796416
6.150	0.9653495	6.550	0.9699095	6.950	0.9737336	7.350	0.9769622	7.750	0.9797049
6.160	0.9654738	6.560	0.9700134	6.960	0.9738211	7.360	0.9770363	7.760	0.9797681
6.170	0.9655976	6.570	0.9701169	6.970	0.9739082	7.370	0.9771101	7.770	0.9798310
6.180	0.9657208	6.580	0.9702199	6.980	0.9739949	7.380	0.9771836	7.780	0.9798936
6.190	0.9658434	6.590	0.9703224	6.990	0.9740812	7.390	0.9772568	7.790	0.9799560
6.200	0.9659655	6.600	0.9704245	7.000	0.9741672	7.400	0.9773297	7.800	0.9800181
6.210	0.9660870	6.610	0.9705262	7.010	0.9742529	7.410	0.9774023	7.810	0.9800800
6.220	0.9662080	6.620	0.9706274	7.020	0.9743381	7.420	0.9774746	7.820	0.9801417
6.230	0.9663284	6.630	0.9707282	7.030	0.9744231	7.430	0.9775466	7.830	0.9802031
6.240	0.9664482	6.640	0.9708285	7.040	0.9745076	7.440	0.9776183	7.840	0.9802643
6.250	0.9665676	6.650	0.9709284	7.050	0.9745918	7.450	0.9776897	7.850	0.9803253
6.260	0.9666864	6.660	0.9710279	7.060	0.9746757	7.460	0.9777609	7.860	0.9803860
6.270	0.9668046	6.670	0.9711269	7.070	0.9747592	7.470	0.9778317	7.870	0.9804464
6.280	0.9669224	6.680	0.9712255	7.080	0.9748424	7.480	0.9779023	7.880	0.9805067
6.290	0.9670395	6.690	0.9713237	7.090	0.9749252	7.490	0.9779726	7.890	0.9805667
6.300	0.9671562	6.700	0.9714215	7.100	0.9750076	7.500	0.9780426	7.900	0.9806265
6.310	0.9672723	6.710	0.9715188	7.110	0.9750898	7.510	0.9781123	7.910	0.9806860
6.320	0.9673880	6.720	0.9716157	7.120	0.9751716	7.520	0.9781818	7.920	0.9807453
6.330	0.9675031	6.730	0.9717122	7.130	0.9752530	7.530	0.9782509	7.930	0.9808044
6.340	0.9676176	6.740	0.9718083	7.140	0.9753341	7.540	0.9783198	7.940	0.9808632
6.350	0.9677317	6.750	0.9719039	7.150	0.9754149	7.550	0.9783884	7.950	0.9809219
6.360	0.9678453	6.760	0.9719992	7.160	0.9754953	7.560	0.9784568	7.960	0.9809803
6.370	0.9679583	6.770	0.9720940	7.170	0.9755754	7.570	0.9785249	7.970	0.9810384
6.380	0.9680709	6.780	0.9721885	7.180	0.9756552	7.580	0.9785927	7.980	0.9810964
6.390	0.9681829	6.790	0.9722825	7.190	0.9757346	7.590	0.9786602	7.990	0.9811541
6.400	0.9682945	6.800	0.9723761	7.200	0.9758137	7.600	0.9787274	8.000	0.9812116

The standardized lognormal function for $y \in [8.000, 10.000]$:

y	F(y)	y	F(y)	y	F(y)	y	F(y)	y	F(y)
8.000	0.9812116	8.400	0.9833411	8.800	0.9851756	9.200	0.9867636	9.600	0.9881440
8.010	0.9812689	8.410	0.9833903	8.810	0.9852182	9.210	0.9868005	9.610	0.9881761
8.020	0.9813260	8.420	0.9834394	8.820	0.9852605	9.220	0.9868372	9.620	0.9882082
8.030	0.9813828	8.430	0.9834882	8.830	0.9853028	9.230	0.9868739	9.630	0.9882401
8.040	0.9814394	8.440	0.9835369	8.840	0.9853448	9.240	0.9869104	9.640	0.9882719
8.050	0.9814958	8.450	0.9835854	8.850	0.9853867	9.250	0.9869467	9.650	0.9883036
8.060	0.9815520	8.460	0.9836338	8.860	0.9854285	9.260	0.9869830	9.660	0.9883352
8.070	0.9816080	8.470	0.9836819	8.870	0.9854701	9.270	0.9870191	9.670	0.9883667
8.080	0.9816638	8.480	0.9837299	8.880	0.9855116	9.280	0.9870551	9.680	0.9883981
8.090	0.9817193	8.490	0.9837777	8.890	0.9855529	9.290	0.9870910	9.690	0.9884294
8.100	0.9817747	8.500	0.9838253	8.900	0.9855941	9.300	0.9871268	9.700	0.9884605
8.110	0.9818298	8.510	0.9838727	8.910	0.9856351	9.310	0.9871624	9.710	0.9884916
8.120	0.9818847	8.520	0.9839200	8.920	0.9856760	9.320	0.9871979	9.720	0.9885226
8.130	0.9819394	8.530	0.9839671	8.930	0.9857167	9.330	0.9872333	9.730	0.9885534
8.140	0.9819939	8.540	0.9840140	8.940	0.9857573	9.340	0.9872685	9.740	0.9885842
8.150	0.9820482	8.550	0.9840608	8.950	0.9857977	9.350	0.9873037	9.750	0.9886149
8.160	0.9821023	8.560	0.9841073	8.960	0.9858380	9.360	0.9873387	9.760	0.9886454
8.170	0.9821562	8.570	0.9841538	8.970	0.9858782	9.370	0.9873736	9.770	0.9886759
8.180	0.9822099	8.580	0.9842000	8.980	0.9859182	9.380	0.9874084	9.780	0.9887062
8.190	0.9822634	8.590	0.9842461	8.990	0.9859581	9.390	0.9874430	9.790	0.9887365
8.200	0.9823166	8.600	0.9842920	9.000	0.9859978	9.400	0.9874776	9.800	0.9887666
8.210	0.9823697	8.610	0.9843377	9.010	0.9860374	9.410	0.9875120	9.810	0.9887967
8.220	0.9824226	8.620	0.9843832	9.020	0.9860768	9.420	0.9875463	9.820	0.9888266
8.230	0.9824752	8.630	0.9844286	9.030	0.9861161	9.430	0.9875805	9.830	0.9888564
8.240	0.9825277	8.640	0.9844739	9.040	0.9861553	9.440	0.9876145	9.840	0.9888862
8.250	0.9825800	8.650	0.9845189	9.050	0.9861943	9.450	0.9876485	9.850	0.9889158
8.260	0.9826321	8.660	0.9845638	9.060	0.9862332	9.460	0.9876823	9.860	0.9889454
8.270	0.9826840	8.670	0.9846086	9.070	0.9862720	9.470	0.9877160	9.870	0.9889748
8.280	0.9827357	8.680	0.9846532	9.080	0.9863106	9.480	0.9877496	9.880	0.9890042
8.290	0.9827872	8.690	0.9846976	9.090	0.9863491	9.490	0.9877831	9.890	0.9890334
8.300	0.9828385	8.700	0.9847418	9.100	0.9863874	9.500	0.9878165	9.900	0.9890626
8.310	0.9828896	8.710	0.9847859	9.110	0.9864256	9.510	0.9878497	9.910	0.9890917
8.320	0.9829405	8.720	0.9848299	9.120	0.9864637	9.520	0.9878829	9.920	0.9891206
8.330	0.9829912	8.730	0.9848736	9.130	0.9865017	9.530	0.9879159	9.930	0.9891495
8.340	0.9830418	8.740	0.9849172	9.140	0.9865395	9.540	0.9879488	9.940	0.9891783
8.350	0.9830921	8.750	0.9849607	9.150	0.9865771	9.550	0.9879816	9.950	0.9892069
8.360	0.9831423	8.760	0.9850040	9.160	0.9866147	9.560	0.9880143	9.960	0.9892355
8.370	0.9831922	8.770	0.9850471	9.170	0.9866521	9.570	0.9880469	9.970	0.9892640
8.380	0.9832420	8.780	0.9850901	9.180	0.9866894	9.580	0.9880794	9.980	0.9892924
8.390	0.9832916	8.790	0.9851330	9.190	0.9867265	9.590	0.9881117	9.990	0.9893207
8.400	0.9833411	8.800	0.9851756	9.200	0.9867636	9.600	0.9881440	10.000	0.9893489

Appendix I

Gamma Function

Reliability Analysis for Asset Management of Electric Power Grids, First Edition. Robert Ross.
© 2019 John Wiley & Sons Ltd. Published 2019 by John Wiley & Sons Ltd.
Companion website: www.wiley.com/go/ross/reliabilityanalysis

The gamma function for $x \in [1.000, 1.200]$ (range extendable by $\Gamma(x + 1) = x \cdot \Gamma(x)$):

x	$\Gamma(x)$	x	$\Gamma(x)$	x	$\Gamma(x)$	x	$\Gamma(x)$	x	$\Gamma(x)$
1.000	1.000000	1.040	0.978438	1.080	0.959725	1.120	0.943590	1.160	0.929803
1.001	0.999424	1.041	0.977937	1.081	0.959292	1.121	0.943218	1.161	0.929487
1.002	0.998850	1.042	0.977437	1.082	0.958859	1.122	0.942847	1.162	0.929172
1.003	0.998277	1.043	0.976940	1.083	0.958429	1.123	0.942477	1.163	0.928858
1.004	0.997707	1.044	0.976444	1.084	0.958000	1.124	0.942109	1.164	0.928546
1.005	0.997139	1.045	0.975949	1.085	0.957573	1.125	0.941743	1.165	0.928235
1.006	0.996572	1.046	0.975457	1.086	0.957147	1.126	0.941378	1.166	0.927925
1.007	0.996008	1.047	0.974966	1.087	0.956723	1.127	0.941014	1.167	0.927617
1.008	0.995445	1.048	0.974477	1.088	0.956300	1.128	0.940652	1.168	0.927310
1.009	0.994885	1.049	0.973990	1.089	0.955879	1.129	0.940291	1.169	0.927004
1.010	0.994326	1.050	0.973504	1.090	0.955459	1.130	0.939931	1.170	0.926700
1.011	0.993769	1.051	0.973020	1.091	0.955042	1.131	0.939574	1.171	0.926397
1.012	0.993214	1.052	0.972538	1.092	0.954625	1.132	0.939217	1.172	0.926095
1.013	0.992661	1.053	0.972058	1.093	0.954211	1.133	0.938862	1.173	0.925794
1.014	0.992110	1.054	0.971579	1.094	0.953797	1.134	0.938508	1.174	0.925495
1.015	0.991561	1.055	0.971103	1.095	0.953386	1.135	0.938156	1.175	0.925197
1.016	0.991014	1.056	0.970627	1.096	0.952976	1.136	0.937805	1.176	0.924901
1.017	0.990469	1.057	0.970154	1.097	0.952567	1.137	0.937456	1.177	0.924606
1.018	0.989925	1.058	0.969682	1.098	0.952160	1.138	0.937108	1.178	0.924312
1.019	0.989384	1.059	0.969212	1.099	0.951755	1.139	0.936761	1.179	0.924019
1.020	0.988844	1.060	0.968744	1.100	0.951351	1.140	0.936416	1.180	0.923728
1.021	0.988306	1.061	0.968277	1.101	0.950948	1.141	0.936072	1.181	0.923438
1.022	0.987771	1.062	0.967812	1.102	0.950548	1.142	0.935730	1.182	0.923149
1.023	0.987236	1.063	0.967349	1.103	0.950148	1.143	0.935389	1.183	0.922862
1.024	0.986704	1.064	0.966887	1.104	0.949750	1.144	0.935049	1.184	0.922575
1.025	0.986174	1.065	0.966427	1.105	0.949354	1.145	0.934711	1.185	0.922290
1.026	0.985645	1.066	0.965969	1.106	0.948959	1.146	0.934374	1.186	0.922007
1.027	0.985119	1.067	0.965512	1.107	0.948566	1.147	0.934039	1.187	0.921724
1.028	0.984594	1.068	0.965057	1.108	0.948174	1.148	0.933705	1.188	0.921443
1.029	0.984071	1.069	0.964604	1.109	0.947784	1.149	0.933372	1.189	0.921164
1.030	0.983550	1.070	0.964152	1.110	0.947396	1.150	0.933041	1.190	0.920885
1.031	0.983031	1.071	0.963702	1.111	0.947008	1.151	0.932711	1.191	0.920608
1.032	0.982513	1.072	0.963254	1.112	0.946623	1.152	0.932382	1.192	0.920332
1.033	0.981997	1.073	0.962807	1.113	0.946238	1.153	0.932055	1.193	0.920057
1.034	0.981484	1.074	0.962362	1.114	0.945856	1.154	0.931729	1.194	0.919783
1.035	0.980972	1.075	0.961918	1.115	0.945474	1.155	0.931405	1.195	0.919511
1.036	0.980461	1.076	0.961476	1.116	0.945095	1.156	0.931082	1.196	0.919240
1.037	0.979953	1.077	0.961036	1.117	0.944716	1.157	0.930760	1.197	0.918970
1.038	0.979446	1.078	0.960598	1.118	0.944339	1.158	0.930440	1.198	0.918702
1.039	0.978941	1.079	0.960161	1.119	0.943964	1.159	0.930121	1.199	0.918435
1.040	0.978438	1.080	0.959725	1.120	0.943590	1.160	0.929803	1.200	0.918169

The gamma function for $x \in [1.200, 1.400]$ (range extendable by $\Gamma(x + 1) = x \cdot \Gamma(x)$):

x	$\Gamma(x)$	x	$\Gamma(x)$	x	$\Gamma(x)$	x	$\Gamma(x)$	x	$\Gamma(x)$
1.200	0.918169	1.240	0.908521	1.280	0.900718	1.320	0.894640	1.360	0.890185
1.201	0.917904	1.241	0.908304	1.281	0.900546	1.321	0.894510	1.361	0.890093
1.202	0.917640	1.242	0.908088	1.282	0.900375	1.322	0.894380	1.362	0.890003
1.203	0.917378	1.243	0.907873	1.283	0.900204	1.323	0.894251	1.363	0.889913
1.204	0.917117	1.244	0.907660	1.284	0.900035	1.324	0.894123	1.364	0.889825
1.205	0.916857	1.245	0.907447	1.285	0.899867	1.325	0.893997	1.365	0.889737
1.206	0.916599	1.246	0.907236	1.286	0.899700	1.326	0.893871	1.366	0.889650
1.207	0.916341	1.247	0.907026	1.287	0.899533	1.327	0.893746	1.367	0.889565
1.208	0.916085	1.248	0.906817	1.288	0.899368	1.328	0.893623	1.368	0.889480
1.209	0.915830	1.249	0.906609	1.289	0.899204	1.329	0.893500	1.369	0.889396
1.210	0.915576	1.250	0.906402	1.290	0.899042	1.330	0.893378	1.370	0.889314
1.211	0.915324	1.251	0.906197	1.291	0.898880	1.331	0.893257	1.371	0.889232
1.212	0.915073	1.252	0.905992	1.292	0.898719	1.332	0.893138	1.372	0.889151
1.213	0.914823	1.253	0.905789	1.293	0.898559	1.333	0.893019	1.373	0.889071
1.214	0.914574	1.254	0.905587	1.294	0.898401	1.334	0.892901	1.374	0.888992
1.215	0.914326	1.255	0.905386	1.295	0.898243	1.335	0.892784	1.375	0.888914
1.216	0.914080	1.256	0.905186	1.296	0.898086	1.336	0.892669	1.376	0.888836
1.217	0.913834	1.257	0.904987	1.297	0.897931	1.337	0.892554	1.377	0.888760
1.218	0.913590	1.258	0.904789	1.298	0.897776	1.338	0.892440	1.378	0.888685
1.219	0.913348	1.259	0.904593	1.299	0.897623	1.339	0.892327	1.379	0.888611
1.220	0.913106	1.260	0.904397	1.300	0.897471	1.340	0.892216	1.380	0.888537
1.221	0.912865	1.261	0.904203	1.301	0.897319	1.341	0.892105	1.381	0.888465
1.222	0.912626	1.262	0.904009	1.302	0.897169	1.342	0.891995	1.382	0.888393
1.223	0.912388	1.263	0.903817	1.303	0.897020	1.343	0.891886	1.383	0.888323
1.224	0.912151	1.264	0.903626	1.304	0.896872	1.344	0.891778	1.384	0.888253
1.225	0.911916	1.265	0.903436	1.305	0.896724	1.345	0.891671	1.385	0.888184
1.226	0.911681	1.266	0.903247	1.306	0.896578	1.346	0.891565	1.386	0.888116
1.227	0.911448	1.267	0.903060	1.307	0.896433	1.347	0.891460	1.387	0.888049
1.228	0.911216	1.268	0.902873	1.308	0.896289	1.348	0.891356	1.388	0.887983
1.229	0.910985	1.269	0.902688	1.309	0.896146	1.349	0.891253	1.389	0.887918
1.230	0.910755	1.270	0.902503	1.310	0.896004	1.350	0.891151	1.390	0.887854
1.231	0.910526	1.271	0.902320	1.311	0.895863	1.351	0.891050	1.391	0.887791
1.232	0.910299	1.272	0.902137	1.312	0.895723	1.352	0.890950	1.392	0.887729
1.233	0.910072	1.273	0.901956	1.313	0.895584	1.353	0.890851	1.393	0.887667
1.234	0.909847	1.274	0.901776	1.314	0.895446	1.354	0.890753	1.394	0.887607
1.235	0.909623	1.275	0.901597	1.315	0.895310	1.355	0.890656	1.395	0.887548
1.236	0.909401	1.276	0.901419	1.316	0.895174	1.356	0.890560	1.396	0.887489
1.237	0.909179	1.277	0.901242	1.317	0.895039	1.357	0.890464	1.397	0.887431
1.238	0.908959	1.278	0.901067	1.318	0.894905	1.358	0.890370	1.398	0.887375
1.239	0.908739	1.279	0.900892	1.319	0.894772	1.359	0.890277	1.399	0.887319
1.240	0.908521	1.280	0.900718	1.320	0.894640	1.360	0.890185	1.400	0.887264

The gamma function for $x \in [1.400, 1.600]$ (range extendable by $\Gamma(x+1) = x \cdot \Gamma(x)$):

x	$\Gamma(x)$	x	$\Gamma(x)$	x	$\Gamma(x)$	x	$\Gamma(x)$	x	$\Gamma(x)$
1.400	0.887264	1.440	0.885805	1.480	0.885747	1.520	0.887039	1.560	0.889639
1.401	0.887210	1.441	0.885787	1.481	0.885763	1.521	0.887088	1.561	0.889721
1.402	0.887157	1.442	0.885769	1.482	0.885780	1.522	0.887138	1.562	0.889803
1.403	0.887105	1.443	0.885753	1.483	0.885798	1.523	0.887189	1.563	0.889886
1.404	0.887053	1.444	0.885737	1.484	0.885816	1.524	0.887241	1.564	0.889970
1.405	0.887003	1.445	0.885722	1.485	0.885836	1.525	0.887293	1.565	0.890055
1.406	0.886953	1.446	0.885708	1.486	0.885856	1.526	0.887346	1.566	0.890140
1.407	0.886905	1.447	0.885695	1.487	0.885877	1.527	0.887400	1.567	0.890226
1.408	0.886857	1.448	0.885683	1.488	0.885899	1.528	0.887455	1.568	0.890313
1.409	0.886810	1.449	0.885672	1.489	0.885922	1.529	0.887511	1.569	0.890401
1.410	0.886765	1.450	0.885661	1.490	0.885945	1.530	0.887568	1.570	0.890490
1.411	0.886720	1.451	0.885652	1.491	0.885970	1.531	0.887625	1.571	0.890579
1.412	0.886676	1.452	0.885643	1.492	0.885995	1.532	0.887683	1.572	0.890669
1.413	0.886633	1.453	0.885635	1.493	0.886021	1.533	0.887742	1.573	0.890760
1.414	0.886590	1.454	0.885628	1.494	0.886048	1.534	0.887802	1.574	0.890852
1.415	0.886549	1.455	0.885622	1.495	0.886076	1.535	0.887863	1.575	0.890945
1.416	0.886509	1.456	0.885617	1.496	0.886104	1.536	0.887924	1.576	0.891038
1.417	0.886469	1.457	0.885612	1.497	0.886134	1.537	0.887986	1.577	0.891132
1.418	0.886430	1.458	0.885609	1.498	0.886164	1.538	0.888049	1.578	0.891227
1.419	0.886393	1.459	0.885606	1.499	0.886195	1.539	0.888113	1.579	0.891323
1.420	0.886356	1.460	0.885604	1.500	0.886227	1.540	0.888178	1.580	0.891420
1.421	0.886320	1.461	0.885603	1.501	0.886260	1.541	0.888243	1.581	0.891517
1.422	0.886285	1.462	0.885603	1.502	0.886293	1.542	0.888309	1.582	0.891615
1.423	0.886251	1.463	0.885604	1.503	0.886328	1.543	0.888376	1.583	0.891714
1.424	0.886217	1.464	0.885606	1.504	0.886363	1.544	0.888444	1.584	0.891814
1.425	0.886185	1.465	0.885608	1.505	0.886399	1.545	0.888513	1.585	0.891914
1.426	0.886153	1.466	0.885611	1.506	0.886436	1.546	0.888582	1.586	0.892015
1.427	0.886123	1.467	0.885616	1.507	0.886474	1.547	0.888653	1.587	0.892117
1.428	0.886093	1.468	0.885621	1.508	0.886512	1.548	0.888724	1.588	0.892220
1.429	0.886064	1.469	0.885626	1.509	0.886551	1.549	0.888796	1.589	0.892324
1.430	0.886036	1.470	0.885633	1.510	0.886592	1.550	0.888868	1.590	0.892428
1.431	0.886009	1.471	0.885641	1.511	0.886633	1.551	0.888942	1.591	0.892533
1.432	0.885983	1.472	0.885649	1.512	0.886675	1.552	0.889016	1.592	0.892639
1.433	0.885958	1.473	0.885658	1.513	0.886717	1.553	0.889091	1.593	0.892746
1.434	0.885933	1.474	0.885668	1.514	0.886761	1.554	0.889167	1.594	0.892854
1.435	0.885910	1.475	0.885679	1.515	0.886805	1.555	0.889244	1.595	0.892962
1.436	0.885887	1.476	0.885691	1.516	0.886850	1.556	0.889321	1.596	0.893071
1.437	0.885865	1.477	0.885704	1.517	0.886896	1.557	0.889400	1.597	0.893181
1.438	0.885844	1.478	0.885717	1.518	0.886943	1.558	0.889479	1.598	0.893292
1.439	0.885824	1.479	0.885732	1.519	0.886990	1.559	0.889559	1.599	0.893403
1.440	0.885805	1.480	0.885747	1.520	0.887039	1.560	0.889639	1.600	0.893515

The gamma function for $x \in [1.600, 1.800]$ (range extendable by $\Gamma(x + 1) = x \cdot \Gamma(x)$):

x	$\Gamma(x)$	x	$\Gamma(x)$	x	$\Gamma(x)$	x	$\Gamma(x)$	x	$\Gamma(x)$
1.600	0.893515	1.640	0.898642	1.680	0.905001	1.720	0.912581	1.760	0.921375
1.601	0.893628	1.641	0.898786	1.681	0.905176	1.721	0.912786	1.761	0.921610
1.602	0.893742	1.642	0.898931	1.682	0.905351	1.722	0.912991	1.762	0.921846
1.603	0.893857	1.643	0.899076	1.683	0.905527	1.723	0.913198	1.763	0.922083
1.604	0.893972	1.644	0.899223	1.684	0.905704	1.724	0.913405	1.764	0.922321
1.605	0.894088	1.645	0.899370	1.685	0.905882	1.725	0.913613	1.765	0.922560
1.606	0.894205	1.646	0.899518	1.686	0.906060	1.726	0.913822	1.766	0.922799
1.607	0.894323	1.647	0.899666	1.687	0.906240	1.727	0.914032	1.767	0.923039
1.608	0.894441	1.648	0.899816	1.688	0.906420	1.728	0.914242	1.768	0.923279
1.609	0.894561	1.649	0.899966	1.689	0.906600	1.729	0.914453	1.769	0.923521
1.610	0.894681	1.650	0.900117	1.690	0.906782	1.730	0.914665	1.770	0.923763
1.611	0.894801	1.651	0.900269	1.691	0.906964	1.731	0.914878	1.771	0.924006
1.612	0.894923	1.652	0.900421	1.692	0.907147	1.732	0.915091	1.772	0.924250
1.613	0.895045	1.653	0.900574	1.693	0.907331	1.733	0.915306	1.773	0.924494
1.614	0.895169	1.654	0.900728	1.694	0.907515	1.734	0.915521	1.774	0.924740
1.615	0.895292	1.655	0.900883	1.695	0.907701	1.735	0.915736	1.775	0.924986
1.616	0.895417	1.656	0.901039	1.696	0.907887	1.736	0.915953	1.776	0.925233
1.617	0.895543	1.657	0.901195	1.697	0.908074	1.737	0.916170	1.777	0.925480
1.618	0.895669	1.658	0.901352	1.698	0.908261	1.738	0.916388	1.778	0.925728
1.619	0.895796	1.659	0.901510	1.699	0.908450	1.739	0.916607	1.779	0.925977
1.620	0.895924	1.660	0.901668	1.700	0.908639	1.740	0.916826	1.780	0.926227
1.621	0.896052	1.661	0.901828	1.701	0.908829	1.741	0.917046	1.781	0.926478
1.622	0.896182	1.662	0.901988	1.702	0.909019	1.742	0.917267	1.782	0.926729
1.623	0.896312	1.663	0.902149	1.703	0.909211	1.743	0.917489	1.783	0.926981
1.624	0.896443	1.664	0.902310	1.704	0.909403	1.744	0.917712	1.784	0.927234
1.625	0.896574	1.665	0.902473	1.705	0.909596	1.745	0.917935	1.785	0.927488
1.626	0.896707	1.666	0.902636	1.706	0.909789	1.746	0.918159	1.786	0.927742
1.627	0.896840	1.667	0.902800	1.707	0.909984	1.747	0.918384	1.787	0.927997
1.628	0.896974	1.668	0.902965	1.708	0.910179	1.748	0.918609	1.788	0.928253
1.629	0.897109	1.669	0.903130	1.709	0.910375	1.749	0.918835	1.789	0.928510
1.630	0.897244	1.670	0.903296	1.710	0.910572	1.750	0.919063	1.790	0.928767
1.631	0.897381	1.671	0.903464	1.711	0.910769	1.751	0.919290	1.791	0.929026
1.632	0.897518	1.672	0.903631	1.712	0.910967	1.752	0.919519	1.792	0.929285
1.633	0.897655	1.673	0.903800	1.713	0.911166	1.753	0.919748	1.793	0.929544
1.634	0.897794	1.674	0.903969	1.714	0.911366	1.754	0.919978	1.794	0.929805
1.635	0.897933	1.675	0.904139	1.715	0.911567	1.755	0.920209	1.795	0.930066
1.636	0.898074	1.676	0.904310	1.716	0.911768	1.756	0.920441	1.796	0.930328
1.637	0.898215	1.677	0.904482	1.717	0.911970	1.757	0.920673	1.797	0.930591
1.638	0.898356	1.678	0.904654	1.718	0.912173	1.758	0.920906	1.798	0.930854
1.639	0.898499	1.679	0.904827	1.719	0.912376	1.759	0.921140	1.799	0.931119
1.640	0.898642	1.680	0.905001	1.720	0.912581	1.760	0.921375	1.800	0.931384

The gamma function for $x \in [1.800, 2.000]$ (range extendable by $\Gamma(x + 1) = x \cdot \Gamma(x)$):

x	Γ(x)	x	Γ(x)	x	Γ(x)	x	Γ(x)	x	Γ(x)
1.800	0.931384	1.840	0.942612	1.880	0.955071	1.920	0.968774	1.960	0.983743
1.801	0.931650	1.841	0.942909	1.881	0.955398	1.921	0.969133	1.961	0.984133
1.802	0.931916	1.842	0.943206	1.882	0.955726	1.922	0.969492	1.962	0.984525
1.803	0.932184	1.843	0.943504	1.883	0.956055	1.923	0.969853	1.963	0.984917
1.804	0.932452	1.844	0.943803	1.884	0.956385	1.924	0.970214	1.964	0.985310
1.805	0.932720	1.845	0.944102	1.885	0.956715	1.925	0.970576	1.965	0.985704
1.806	0.932990	1.846	0.944402	1.886	0.957047	1.926	0.970938	1.966	0.986098
1.807	0.933261	1.847	0.944703	1.887	0.957379	1.927	0.971302	1.967	0.986494
1.808	0.933532	1.848	0.945005	1.888	0.957711	1.928	0.971666	1.968	0.986890
1.809	0.933804	1.849	0.945308	1.889	0.958045	1.929	0.972031	1.969	0.987287
1.810	0.934076	1.850	0.945611	1.890	0.958379	1.930	0.972397	1.970	0.987685
1.811	0.934350	1.851	0.945915	1.891	0.958714	1.931	0.972764	1.971	0.988084
1.812	0.934624	1.852	0.946220	1.892	0.959050	1.932	0.973131	1.972	0.988483
1.813	0.934899	1.853	0.946526	1.893	0.959387	1.933	0.973499	1.973	0.988883
1.814	0.935175	1.854	0.946832	1.894	0.959725	1.934	0.973868	1.974	0.989285
1.815	0.935451	1.855	0.947139	1.895	0.960063	1.935	0.974238	1.975	0.989687
1.816	0.935728	1.856	0.947447	1.896	0.960402	1.936	0.974609	1.976	0.990089
1.817	0.936006	1.857	0.947756	1.897	0.960742	1.937	0.974980	1.977	0.990493
1.818	0.936285	1.858	0.948066	1.898	0.961082	1.938	0.975352	1.978	0.990897
1.819	0.936565	1.859	0.948376	1.899	0.961424	1.939	0.975725	1.979	0.991302
1.820	0.936845	1.860	0.948687	1.900	0.961766	1.940	0.976099	1.980	0.991708
1.821	0.937126	1.861	0.948999	1.901	0.962109	1.941	0.976473	1.981	0.992115
1.822	0.937408	1.862	0.949311	1.902	0.962453	1.942	0.976849	1.982	0.992523
1.823	0.937691	1.863	0.949625	1.903	0.962797	1.943	0.977225	1.983	0.992931
1.824	0.937974	1.864	0.949939	1.904	0.963142	1.944	0.977602	1.984	0.993341
1.825	0.938258	1.865	0.950254	1.905	0.963488	1.945	0.977980	1.985	0.993751
1.826	0.938543	1.866	0.950570	1.906	0.963835	1.946	0.978358	1.986	0.994162
1.827	0.938829	1.867	0.950886	1.907	0.964183	1.947	0.978738	1.987	0.994573
1.828	0.939115	1.868	0.951203	1.908	0.964531	1.948	0.979118	1.988	0.994986
1.829	0.939402	1.869	0.951521	1.909	0.964881	1.949	0.979499	1.989	0.995399
1.830	0.939690	1.870	0.951840	1.910	0.965231	1.950	0.979881	1.990	0.995813
1.831	0.939979	1.871	0.952160	1.911	0.965582	1.951	0.980263	1.991	0.996228
1.832	0.940269	1.872	0.952480	1.912	0.965933	1.952	0.980647	1.992	0.996644
1.833	0.940559	1.873	0.952801	1.913	0.966286	1.953	0.981031	1.993	0.997061
1.834	0.940850	1.874	0.953123	1.914	0.966639	1.954	0.981416	1.994	0.997478
1.835	0.941142	1.875	0.953446	1.915	0.966993	1.955	0.981802	1.995	0.997896
1.836	0.941434	1.876	0.953769	1.916	0.967347	1.956	0.982188	1.996	0.998315
1.837	0.941728	1.877	0.954093	1.917	0.967703	1.957	0.982576	1.997	0.998735
1.838	0.942022	1.878	0.954419	1.918	0.968059	1.958	0.982964	1.998	0.999156
1.839	0.942317	1.879	0.954744	1.919	0.968416	1.959	0.983353	1.999	0.999578
1.840	0.942612	1.880	0.955071	1.920	0.968774	1.960	0.983743	2.000	1.000000

Appendix J

Plotting Positions

Reliability Analysis for Asset Management of Electric Power Grids, First Edition. Robert Ross.
© 2019 John Wiley & Sons Ltd. Published 2019 by John Wiley & Sons Ltd.
Companion website: www.wiley.com/go/ross/reliabilityanalysis

J.1 Expected Ranked Probability $<F_{i,n}>$ for $n = 1, \dots, 30$

The equation for $<F_{i,n}>$ is (cf. Section 4.1.2.1, Eq. (4.15)):

$$< F_{i,n} >= \frac{i}{n+1}$$

							n								
i	1	2	3	4	5	6	7	8	9	10	11	12	13	14	15
1	0.5000	0.3333	0.2500	0.2000	0.1667	0.1429	0.1250	0.1111	0.1000	0.0909	0.0833	0.0769	0.0714	0.0667	0.0625
2		0.6667	0.5000	0.4000	0.3333	0.2857	0.2500	0.2222	0.2000	0.1818	0.1667	0.1538	0.1429	0.1333	0.1250
3			0.7500	0.6000	0.5000	0.4286	0.3750	0.3333	0.3000	0.2727	0.2500	0.2308	0.2143	0.2000	0.1875
4				0.8000	0.6667	0.5714	0.5000	0.4444	0.4000	0.3636	0.3333	0.3077	0.2857	0.2667	0.2500
5					0.8333	0.7143	0.6250	0.5556	0.5000	0.4545	0.4167	0.3846	0.3571	0.3333	0.3125
6						0.8571	0.7500	0.6667	0.6000	0.5455	0.5000	0.4615	0.4286	0.4000	0.3750
7							0.8750	0.7778	0.7000	0.6364	0.5833	0.5385	0.5000	0.4667	0.4375
8								0.8889	0.8000	0.7273	0.6667	0.6154	0.5714	0.5333	0.5000
9									0.9000	0.8182	0.7500	0.6923	0.6429	0.6000	0.5625
10										0.9091	0.8333	0.7692	0.7143	0.6667	0.6250
11											0.9167	0.8462	0.7857	0.7333	0.6875
12												0.9231	0.8571	0.8000	0.7500
13													0.9286	0.8667	0.8125
14														0.9333	0.8750
15															0.9375

							n								
i	16	17	18	19	20	21	22	23	24	25	26	27	28	29	30
1	0.0588	0.0556	0.0526	0.0500	0.0476	0.0455	0.0435	0.0417	0.0400	0.0385	0.0370	0.0357	0.0345	0.0333	0.0323
2	0.1176	0.1111	0.1053	0.1000	0.0952	0.0909	0.0870	0.0833	0.0800	0.0769	0.0741	0.0714	0.0690	0.0667	0.0645
3	0.1765	0.1667	0.1579	0.1500	0.1429	0.1364	0.1304	0.1250	0.1200	0.1154	0.1111	0.1071	0.1034	0.1000	0.0968
4	0.2353	0.2222	0.2105	0.2000	0.1905	0.1818	0.1739	0.1667	0.1600	0.1538	0.1481	0.1429	0.1379	0.1333	0.1290
5	0.2941	0.2778	0.2632	0.2500	0.2381	0.2273	0.2174	0.2083	0.2000	0.1923	0.1852	0.1786	0.1724	0.1667	0.1613
6	0.3529	0.3333	0.3158	0.3000	0.2857	0.2727	0.2609	0.2500	0.2400	0.2308	0.2222	0.2143	0.2069	0.2000	0.1935
7	0.4118	0.3889	0.3684	0.3500	0.3333	0.3182	0.3043	0.2917	0.2800	0.2692	0.2593	0.2500	0.2414	0.2333	0.2258
8	0.4706	0.4444	0.4211	0.4000	0.3810	0.3636	0.3478	0.3333	0.3200	0.3077	0.2963	0.2857	0.2759	0.2667	0.2581
9	0.5294	0.5000	0.4737	0.4500	0.4286	0.4091	0.3913	0.3750	0.3600	0.3462	0.3333	0.3214	0.3103	0.3000	0.2903
10	0.5882	0.5556	0.5263	0.5000	0.4762	0.4545	0.4348	0.4167	0.4000	0.3846	0.3704	0.3571	0.3448	0.3333	0.3226
11	0.6471	0.6111	0.5789	0.5500	0.5238	0.5000	0.4783	0.4583	0.4400	0.4231	0.4074	0.3929	0.3793	0.3667	0.3548
12	0.7059	0.6667	0.6316	0.6000	0.5714	0.5455	0.5217	0.5000	0.4800	0.4615	0.4444	0.4286	0.4138	0.4000	0.3871
13	0.7647	0.7222	0.6842	0.6500	0.6190	0.5909	0.5652	0.5417	0.5200	0.5000	0.4815	0.4643	0.4483	0.4333	0.4194
14	0.8235	0.7778	0.7368	0.7000	0.6667	0.6364	0.6087	0.5833	0.5600	0.5385	0.5185	0.5000	0.4828	0.4667	0.4516
15	0.8824	0.8333	0.7895	0.7500	0.7143	0.6818	0.6522	0.6250	0.6000	0.5769	0.5556	0.5357	0.5172	0.5000	0.4839
16	0.9412	0.8889	0.8421	0.8000	0.7619	0.7273	0.6957	0.6667	0.6400	0.6154	0.5926	0.5714	0.5517	0.5333	0.5161
17		0.9444	0.8947	0.8500	0.8095	0.7727	0.7391	0.7083	0.6800	0.6538	0.6296	0.6071	0.5862	0.5667	0.5484
18			0.9474	0.9000	0.8571	0.8182	0.7826	0.7500	0.7200	0.6923	0.6667	0.6429	0.6207	0.6000	0.5806
19				0.9500	0.9048	0.8636	0.8261	0.7917	0.7600	0.7308	0.7037	0.6786	0.6552	0.6333	0.6129
20					0.9524	0.9091	0.8696	0.8333	0.8000	0.7692	0.7407	0.7143	0.6897	0.6667	0.6452
21						0.9545	0.9130	0.8750	0.8400	0.8077	0.7778	0.7500	0.7241	0.7000	0.6774
22							0.9565	0.9167	0.8800	0.8462	0.8148	0.7857	0.7586	0.7333	0.7097
23								0.9583	0.9200	0.8846	0.8519	0.8214	0.7931	0.7667	0.7419
24									0.9600	0.9231	0.8889	0.8571	0.8276	0.8000	0.7742
25										0.9615	0.9259	0.8929	0.8621	0.8333	0.8065
26											0.9630	0.9286	0.8966	0.8667	0.8387
27												0.9643	0.9310	0.9000	0.8710
28													0.9655	0.9333	0.9032
29														0.9667	0.9355
30															0.9677

J.2 Expected Ranked Probability $<F_{i,n}>$ for $n = 31, \ldots, 45$

The equation for $<F_{i,n}>$ is (cf. Section 4.1.2.1, Eq. (4.15)):

$$<F_{i,n}> = \frac{i}{n+1}$$

								n							
i	31	32	33	34	35	36	37	38	39	40	41	42	43	44	45
1	0.0313	0.0303	0.0294	0.0286	0.0278	0.0270	0.0263	0.0256	0.0250	0.0244	0.0238	0.0233	0.0227	0.0222	0.0217
2	0.0625	0.0606	0.0588	0.0571	0.0556	0.0541	0.0526	0.0513	0.0500	0.0488	0.0476	0.0465	0.0455	0.0444	0.0435
3	0.0938	0.0909	0.0882	0.0857	0.0833	0.0811	0.0789	0.0769	0.0750	0.0732	0.0714	0.0698	0.0682	0.0667	0.0652
4	0.1250	0.1212	0.1176	0.1143	0.1111	0.1081	0.1053	0.1026	0.1000	0.0976	0.0952	0.0930	0.0909	0.0889	0.0870
5	0.1563	0.1515	0.1471	0.1429	0.1389	0.1351	0.1316	0.1282	0.1250	0.1220	0.1190	0.1163	0.1136	0.1111	0.1087
6	0.1875	0.1818	0.1765	0.1714	0.1667	0.1622	0.1579	0.1538	0.1500	0.1463	0.1429	0.1395	0.1364	0.1333	0.1304
7	0.2188	0.2121	0.2059	0.2000	0.1944	0.1892	0.1842	0.1795	0.1750	0.1707	0.1667	0.1628	0.1591	0.1556	0.1522
8	0.2500	0.2424	0.2353	0.2286	0.2222	0.2162	0.2105	0.2051	0.2000	0.1951	0.1905	0.1860	0.1818	0.1778	0.1739
9	0.2813	0.2727	0.2647	0.2571	0.2500	0.2432	0.2368	0.2308	0.2250	0.2195	0.2143	0.2093	0.2045	0.2000	0.1957
10	0.3125	0.3030	0.2941	0.2857	0.2778	0.2703	0.2632	0.2564	0.2500	0.2439	0.2381	0.2326	0.2273	0.2222	0.2174
11	0.3438	0.3333	0.3235	0.3143	0.3056	0.2973	0.2895	0.2821	0.2750	0.2683	0.2619	0.2558	0.2500	0.2444	0.2391
12	0.3750	0.3636	0.3529	0.3429	0.3333	0.3243	0.3158	0.3077	0.3000	0.2927	0.2857	0.2791	0.2727	0.2667	0.2609
13	0.4063	0.3939	0.3824	0.3714	0.3611	0.3514	0.3421	0.3333	0.3250	0.3171	0.3095	0.3023	0.2955	0.2889	0.2826
14	0.4375	0.4242	0.4118	0.4000	0.3889	0.3784	0.3684	0.3590	0.3500	0.3415	0.3333	0.3256	0.3182	0.3111	0.3043
15	0.4688	0.4545	0.4412	0.4286	0.4167	0.4054	0.3947	0.3846	0.3750	0.3659	0.3571	0.3488	0.3409	0.3333	0.3261
16	0.5000	0.4848	0.4706	0.4571	0.4444	0.4324	0.4211	0.4103	0.4000	0.3902	0.3810	0.3721	0.3636	0.3556	0.3478
17	0.5313	0.5152	0.5000	0.4857	0.4722	0.4595	0.4474	0.4359	0.4250	0.4146	0.4048	0.3953	0.3864	0.3778	0.3696
18	0.5625	0.5455	0.5294	0.5143	0.5000	0.4865	0.4737	0.4615	0.4500	0.4390	0.4286	0.4186	0.4091	0.4000	0.3913
19	0.5938	0.5758	0.5588	0.5429	0.5278	0.5135	0.5000	0.4872	0.4750	0.4634	0.4524	0.4419	0.4318	0.4222	0.4130
20	0.6250	0.6061	0.5882	0.5714	0.5556	0.5405	0.5263	0.5128	0.5000	0.4878	0.4762	0.4651	0.4545	0.4444	0.4348
21	0.6563	0.6364	0.6176	0.6000	0.5833	0.5676	0.5526	0.5385	0.5250	0.5122	0.5000	0.4884	0.4773	0.4667	0.4565
22	0.6875	0.6667	0.6471	0.6286	0.6111	0.5946	0.5789	0.5641	0.5500	0.5366	0.5238	0.5116	0.5000	0.4889	0.4783
23	0.7188	0.6970	0.6765	0.6571	0.6389	0.6216	0.6053	0.5897	0.5750	0.5610	0.5476	0.5349	0.5227	0.5111	0.5000
24	0.7500	0.7273	0.7059	0.6857	0.6667	0.6486	0.6316	0.6154	0.6000	0.5854	0.5714	0.5581	0.5455	0.5333	0.5217
25	0.7813	0.7576	0.7353	0.7143	0.6944	0.6757	0.6579	0.6410	0.6250	0.6098	0.5952	0.5814	0.5682	0.5556	0.5435
26	0.8125	0.7879	0.7647	0.7429	0.7222	0.7027	0.6842	0.6667	0.6500	0.6341	0.6190	0.6047	0.5909	0.5778	0.5652
27	0.8438	0.8182	0.7941	0.7714	0.7500	0.7297	0.7105	0.6923	0.6750	0.6585	0.6429	0.6279	0.6136	0.6000	0.5870
28	0.8750	0.8485	0.8235	0.8000	0.7778	0.7568	0.7368	0.7179	0.7000	0.6829	0.6667	0.6512	0.6364	0.6222	0.6087
29	0.9063	0.8788	0.8529	0.8286	0.8056	0.7838	0.7632	0.7436	0.7250	0.7073	0.6905	0.6744	0.6591	0.6444	0.6304
30	0.9375	0.9091	0.8824	0.8571	0.8333	0.8108	0.7895	0.7692	0.7500	0.7317	0.7143	0.6977	0.6818	0.6667	0.6522
31	0.9688	0.9394	0.9118	0.8857	0.8611	0.8378	0.8158	0.7949	0.7750	0.7561	0.7381	0.7209	0.7045	0.6889	0.6739
32		0.9697	0.9412	0.9143	0.8889	0.8649	0.8421	0.8205	0.8000	0.7805	0.7619	0.7442	0.7273	0.7111	0.6957
33			0.9706	0.9429	0.9167	0.8919	0.8684	0.8462	0.8250	0.8049	0.7857	0.7674	0.7500	0.7333	0.7174
34				0.9714	0.9444	0.9189	0.8947	0.8718	0.8500	0.8293	0.8095	0.7907	0.7727	0.7556	0.7391
35					0.9722	0.9459	0.9211	0.8974	0.8750	0.8537	0.8333	0.8140	0.7955	0.7778	0.7609
36						0.9730	0.9474	0.9231	0.9000	0.8780	0.8571	0.8372	0.8182	0.8000	0.7826
37							0.9737	0.9487	0.9250	0.9024	0.8810	0.8605	0.8409	0.8222	0.8043
38								0.9744	0.9500	0.9268	0.9048	0.8837	0.8636	0.8444	0.8261
39									0.9750	0.9512	0.9286	0.9070	0.8864	0.8667	0.8478
40										0.9756	0.9524	0.9302	0.9091	0.8889	0.8696
41											0.9762	0.9535	0.9318	0.9111	0.8913
42												0.9767	0.9545	0.9333	0.9130
43													0.9773	0.9556	0.9348
44														0.9778	0.9565
45															0.9783

J.3 Expected Ranked Probability $<F_{i,n}>$ for $n = 45, \ldots, 60$

The equation for $<F_{i,n}>$ is (cf. Section 4.1.2.1, Eq. (4.15)):

$$< F_{i,n} >= \frac{i}{n+1}$$

								n								
i	46	47	48	49	50	51	52	53	54	55	56	57	58	59	60
1	0.0213	0.0208	0.0204	0.0200	0.0196	0.0192	0.0189	0.0185	0.0182	0.0179	0.0175	0.0172	0.0169	0.0167	0.0164
2	0.0426	0.0417	0.0408	0.0400	0.0392	0.0385	0.0377	0.0370	0.0364	0.0357	0.0351	0.0345	0.0339	0.0333	0.0328
3	0.0638	0.0625	0.0612	0.0600	0.0588	0.0577	0.0566	0.0556	0.0545	0.0536	0.0526	0.0517	0.0508	0.0500	0.0492
4	0.0851	0.0833	0.0816	0.0800	0.0784	0.0769	0.0755	0.0741	0.0727	0.0714	0.0702	0.0690	0.0678	0.0667	0.0656
5	0.1064	0.1042	0.1020	0.1000	0.0980	0.0962	0.0943	0.0926	0.0909	0.0893	0.0877	0.0862	0.0847	0.0833	0.0820
6	0.1277	0.1250	0.1224	0.1200	0.1176	0.1154	0.1132	0.1111	0.1091	0.1071	0.1053	0.1034	0.1017	0.1000	0.0984
7	0.1489	0.1458	0.1429	0.1400	0.1373	0.1346	0.1321	0.1296	0.1273	0.1250	0.1228	0.1207	0.1186	0.1167	0.1148
8	0.1702	0.1667	0.1633	0.1600	0.1569	0.1538	0.1509	0.1481	0.1455	0.1429	0.1404	0.1379	0.1356	0.1333	0.1311
9	0.1915	0.1875	0.1837	0.1800	0.1765	0.1731	0.1698	0.1667	0.1636	0.1607	0.1579	0.1552	0.1525	0.1500	0.1475
10	0.2128	0.2083	0.2041	0.2000	0.1961	0.1923	0.1887	0.1852	0.1818	0.1786	0.1754	0.1724	0.1695	0.1667	0.1639
11	0.2340	0.2292	0.2245	0.2200	0.2157	0.2115	0.2075	0.2037	0.2000	0.1964	0.1930	0.1897	0.1864	0.1833	0.1803
12	0.2553	0.2500	0.2449	0.2400	0.2353	0.2308	0.2264	0.2222	0.2182	0.2143	0.2105	0.2069	0.2034	0.2000	0.1967
13	0.2766	0.2708	0.2653	0.2600	0.2549	0.2500	0.2453	0.2407	0.2364	0.2321	0.2281	0.2241	0.2203	0.2167	0.2131
14	0.2979	0.2917	0.2857	0.2800	0.2745	0.2692	0.2642	0.2593	0.2545	0.2500	0.2456	0.2414	0.2373	0.2333	0.2295
15	0.3191	0.3125	0.3061	0.3000	0.2941	0.2885	0.2830	0.2778	0.2727	0.2679	0.2632	0.2586	0.2542	0.2500	0.2459
16	0.3404	0.3333	0.3265	0.3200	0.3137	0.3077	0.3019	0.2963	0.2909	0.2857	0.2807	0.2759	0.2712	0.2667	0.2623
17	0.3617	0.3542	0.3469	0.3400	0.3333	0.3269	0.3208	0.3148	0.3091	0.3036	0.2982	0.2931	0.2881	0.2833	0.2787
18	0.3830	0.3750	0.3673	0.3600	0.3529	0.3462	0.3396	0.3333	0.3273	0.3214	0.3158	0.3103	0.3051	0.3000	0.2951
19	0.4043	0.3958	0.3878	0.3800	0.3725	0.3654	0.3585	0.3519	0.3455	0.3393	0.3333	0.3276	0.3220	0.3167	0.3115
20	0.4255	0.4167	0.4082	0.4000	0.3922	0.3846	0.3774	0.3704	0.3636	0.3571	0.3509	0.3448	0.3390	0.3333	0.3279
21	0.4468	0.4375	0.4286	0.4200	0.4118	0.4038	0.3962	0.3889	0.3818	0.3750	0.3684	0.3621	0.3559	0.3500	0.3443
22	0.4681	0.4583	0.4490	0.4400	0.4314	0.4231	0.4151	0.4074	0.4000	0.3929	0.3860	0.3793	0.3729	0.3667	0.3607
23	0.4894	0.4792	0.4694	0.4600	0.4510	0.4423	0.4340	0.4259	0.4182	0.4107	0.4035	0.3966	0.3898	0.3833	0.3770
24	0.5106	0.5000	0.4898	0.4800	0.4706	0.4615	0.4528	0.4444	0.4364	0.4286	0.4211	0.4138	0.4068	0.4000	0.3934
25	0.5319	0.5208	0.5102	0.5000	0.4902	0.4808	0.4717	0.4630	0.4545	0.4464	0.4386	0.4310	0.4237	0.4167	0.4098
26	0.5532	0.5417	0.5306	0.5200	0.5098	0.5000	0.4906	0.4815	0.4727	0.4643	0.4561	0.4483	0.4407	0.4333	0.4262
27	0.5745	0.5625	0.5510	0.5400	0.5294	0.5192	0.5094	0.5000	0.4909	0.4821	0.4737	0.4655	0.4576	0.4500	0.4426
28	0.5957	0.5833	0.5714	0.5600	0.5490	0.5385	0.5283	0.5185	0.5091	0.5000	0.4912	0.4828	0.4746	0.4667	0.4590
29	0.6170	0.6042	0.5918	0.5800	0.5686	0.5577	0.5472	0.5370	0.5273	0.5179	0.5088	0.5000	0.4915	0.4833	0.4754
30	0.6383	0.6250	0.6122	0.6000	0.5882	0.5769	0.5660	0.5556	0.5455	0.5357	0.5263	0.5172	0.5085	0.5000	0.4918
31	0.6596	0.6458	0.6327	0.6200	0.6078	0.5962	0.5849	0.5741	0.5636	0.5536	0.5439	0.5345	0.5254	0.5167	0.5082
32	0.6809	0.6667	0.6531	0.6400	0.6275	0.6154	0.6038	0.5926	0.5818	0.5714	0.5614	0.5517	0.5424	0.5333	0.5246
33	0.7021	0.6875	0.6735	0.6600	0.6471	0.6346	0.6226	0.6111	0.6000	0.5893	0.5789	0.5690	0.5593	0.5500	0.5410
34	0.7234	0.7083	0.6939	0.6800	0.6667	0.6538	0.6415	0.6296	0.6182	0.6071	0.5965	0.5862	0.5763	0.5667	0.5574
35	0.7447	0.7292	0.7143	0.7000	0.6863	0.6731	0.6604	0.6481	0.6364	0.6250	0.6140	0.6034	0.5932	0.5833	0.5738
36	0.7660	0.7500	0.7347	0.7200	0.7059	0.6923	0.6792	0.6667	0.6545	0.6429	0.6316	0.6207	0.6102	0.6000	0.5902
37	0.7872	0.7708	0.7551	0.7400	0.7255	0.7115	0.6981	0.6852	0.6727	0.6607	0.6491	0.6379	0.6271	0.6167	0.6066
38	0.8085	0.7917	0.7755	0.7600	0.7451	0.7308	0.7170	0.7037	0.6909	0.6786	0.6667	0.6552	0.6441	0.6333	0.6230
39	0.8298	0.8125	0.7959	0.7800	0.7647	0.7500	0.7358	0.7222	0.7091	0.6964	0.6842	0.6724	0.6610	0.6500	0.6393
40	0.8511	0.8333	0.8163	0.8000	0.7843	0.7692	0.7547	0.7407	0.7273	0.7143	0.7018	0.6897	0.6780	0.6667	0.6557

								n							
i	46	47	48	49	50	51	52	53	54	55	56	57	58	59	60
41	0.8723	0.8542	0.8367	0.8200	0.8039	0.7885	0.7736	0.7593	0.7455	0.7321	0.7193	0.7069	0.6949	0.6833	0.6721
42	0.8936	0.8750	0.8571	0.8400	0.8235	0.8077	0.7925	0.7778	0.7636	0.7500	0.7368	0.7241	0.7119	0.7000	0.6885
43	0.9149	0.8958	0.8776	0.8600	0.8431	0.8269	0.8113	0.7963	0.7818	0.7679	0.7544	0.7414	0.7288	0.7167	0.7049
44	0.9362	0.9167	0.8980	0.8800	0.8627	0.8462	0.8302	0.8148	0.8000	0.7857	0.7719	0.7586	0.7458	0.7333	0.7213
45	0.9574	0.9375	0.9184	0.9000	0.8824	0.8654	0.8491	0.8333	0.8182	0.8036	0.7895	0.7759	0.7627	0.7500	0.7377
46	0.9787	0.9583	0.9388	0.9200	0.9020	0.8846	0.8679	0.8519	0.8364	0.8214	0.8070	0.7931	0.7797	0.7667	0.7541
47		0.9792	0.9592	0.9400	0.9216	0.9038	0.8868	0.8704	0.8545	0.8393	0.8246	0.8103	0.7966	0.7833	0.7705
48			0.9796	0.9600	0.9412	0.9231	0.9057	0.8889	0.8727	0.8571	0.8421	0.8276	0.8136	0.8000	0.7869
49				0.9800	0.9608	0.9423	0.9245	0.9074	0.8909	0.8750	0.8596	0.8448	0.8305	0.8167	0.8033
50					0.9804	0.9615	0.9434	0.9259	0.9091	0.8929	0.8772	0.8621	0.8475	0.8333	0.8197
51						0.9808	0.9623	0.9444	0.9273	0.9107	0.8947	0.8793	0.8644	0.8500	0.8361
52							0.9811	0.9630	0.9455	0.9286	0.9123	0.8966	0.8814	0.8667	0.8525
53								0.9815	0.9636	0.9464	0.9298	0.9138	0.8983	0.8833	0.8689
54									0.9818	0.9643	0.9474	0.9310	0.9153	0.9000	0.8852
55										0.9821	0.9649	0.9483	0.9322	0.9167	0.9016
56											0.9825	0.9655	0.9492	0.9333	0.9180
57												0.9828	0.9661	0.9500	0.9344
58													0.9831	0.9667	0.9508
59														0.9833	0.9672
60															0.9836

J.4 Median Ranked Probability $F_{M,i,n}$ for $n = 1, \ldots , 30$

The equation for $F_{M,i,n}$ is (cf. Section 4.1.2.1, Eq. (4.16)):

$$B(F_{M,i,n}; i, n + 1 - i) = \frac{\Gamma(n+1)}{\Gamma(i) \cdot \Gamma(n+1-i)} \int_0^{F_{M,i,n}} p^{i-1} \cdot (1-p)^{n-i} dp = \frac{1}{2}$$

i	n=1	2	3	4	5	6	7	8	9	10	11	12	13	14	15
1	0.5000	0.2929	0.2063	0.1591	0.1294	0.1091	0.0943	0.0830	0.0741	0.0670	0.0611	0.0561	0.0519	0.0483	0.0452
2		0.7071	0.5000	0.3857	0.3138	0.2644	0.2285	0.2011	0.1796	0.1623	0.1480	0.1360	0.1258	0.1170	0.1094
3			0.7937	0.6143	0.5000	0.4214	0.3641	0.3205	0.2862	0.2586	0.2358	0.2167	0.2004	0.1865	0.1743
4				0.8409	0.6862	0.5786	0.5000	0.4402	0.3931	0.3551	0.3238	0.2976	0.2753	0.2561	0.2394
5					0.8706	0.7356	0.6359	0.5598	0.5000	0.4517	0.4119	0.3785	0.3502	0.3258	0.3045
6						0.8909	0.7715	0.6795	0.6069	0.5483	0.5000	0.4595	0.4251	0.3954	0.3697
7							0.9057	0.7989	0.7138	0.6449	0.5881	0.5405	0.5000	0.4651	0.4348
8								0.9170	0.8204	0.7414	0.6762	0.6215	0.5749	0.5349	0.5000
9									0.9259	0.8377	0.7642	0.7024	0.6498	0.6046	0.5652
10										0.9330	0.8520	0.7833	0.7247	0.6742	0.6303
11											0.9389	0.8640	0.7996	0.7439	0.6955
12												0.9439	0.8742	0.8135	0.7606
13													0.9481	0.8830	0.8257
14														0.9517	0.8906
15															0.9548

i	16	17	18	19	20	21	22	23	24	25	26	27	28	29	30
1	0.0424	0.0400	0.0378	0.0358	0.0341	0.0325	0.0310	0.0297	0.0285	0.0273	0.0263	0.0253	0.0245	0.0236	0.0228
2	0.1027	0.0968	0.0915	0.0868	0.0825	0.0786	0.0751	0.0719	0.0690	0.0662	0.0637	0.0614	0.0592	0.0572	0.0553
3	0.1637	0.1542	0.1458	0.1383	0.1315	0.1253	0.1197	0.1146	0.1099	0.1055	0.1015	0.0978	0.0944	0.0911	0.0881
4	0.2247	0.2118	0.2002	0.1899	0.1805	0.1721	0.1644	0.1573	0.1509	0.1449	0.1394	0.1343	0.1296	0.1252	0.1210
5	0.2859	0.2694	0.2547	0.2415	0.2297	0.2189	0.2091	0.2001	0.1919	0.1843	0.1774	0.1709	0.1648	0.1592	0.1540
6	0.3471	0.3270	0.3092	0.2932	0.2788	0.2657	0.2538	0.2430	0.2330	0.2238	0.2153	0.2074	0.2001	0.1933	0.1869
7	0.4082	0.3847	0.3637	0.3449	0.3280	0.3126	0.2986	0.2858	0.2741	0.2632	0.2532	0.2440	0.2354	0.2274	0.2199
8	0.4694	0.4423	0.4182	0.3966	0.3771	0.3594	0.3433	0.3286	0.3151	0.3027	0.2912	0.2806	0.2707	0.2614	0.2528
9	0.5306	0.5000	0.4727	0.4483	0.4263	0.4063	0.3881	0.3715	0.3562	0.3422	0.3292	0.3171	0.3059	0.2955	0.2858
10	0.5918	0.5577	0.5273	0.5000	0.4754	0.4531	0.4329	0.4143	0.3973	0.3816	0.3671	0.3537	0.3412	0.3296	0.3187
11	0.6529	0.6153	0.5818	0.5517	0.5246	0.5000	0.4776	0.4572	0.4384	0.4211	0.4051	0.3903	0.3765	0.3637	0.3517
12	0.7141	0.6730	0.6363	0.6034	0.5737	0.5469	0.5224	0.5000	0.4795	0.4605	0.4431	0.4268	0.4118	0.3977	0.3846
13	0.7753	0.7306	0.6908	0.6551	0.6229	0.5937	0.5671	0.5428	0.5205	0.5000	0.4810	0.4634	0.4471	0.4318	0.4176
14	0.8363	0.7882	0.7453	0.7068	0.6720	0.6406	0.6119	0.5857	0.5616	0.5395	0.5190	0.5000	0.4824	0.4659	0.4506
15	0.8973	0.8458	0.7998	0.7585	0.7212	0.6874	0.6567	0.6285	0.6027	0.5789	0.5569	0.5366	0.5176	0.5000	0.4835
16	0.9576	0.9032	0.8542	0.8101	0.7703	0.7343	0.7014	0.6714	0.6438	0.6184	0.5949	0.5732	0.5529	0.5341	0.5165
17		0.9600	0.9085	0.8617	0.8195	0.7811	0.7462	0.7142	0.6849	0.6578	0.6329	0.6097	0.5882	0.5682	0.5494
18			0.9622	0.9132	0.8685	0.8279	0.7909	0.7570	0.7259	0.6973	0.6708	0.6463	0.6235	0.6023	0.5824
19				0.9642	0.9175	0.8747	0.8356	0.7999	0.7670	0.7368	0.7088	0.6829	0.6588	0.6363	0.6154
20					0.9659	0.9214	0.8803	0.8427	0.8081	0.7762	0.7468	0.7194	0.6941	0.6704	0.6483
21						0.9675	0.9249	0.8854	0.8491	0.8157	0.7847	0.7560	0.7293	0.7045	0.6813
22							0.9690	0.9281	0.8901	0.8551	0.8226	0.7926	0.7646	0.7386	0.7142
23								0.9703	0.9310	0.8945	0.8606	0.8291	0.7999	0.7726	0.7472
24									0.9715	0.9338	0.8985	0.8657	0.8352	0.8067	0.7801
25										0.9727	0.9363	0.9022	0.8704	0.8408	0.8131
26											0.9737	0.9386	0.9056	0.8748	0.8460
27												0.9747	0.9408	0.9089	0.8790
28													0.9755	0.9428	0.9119
29														0.9764	0.9447
30															0.9772

J.5 Median Ranked Probability $F_{M,i,n}$ for $n = 31, \ldots, 45$

$$B(F_{M,i,n}; i, n+1-i) = \frac{\Gamma(n+1)}{\Gamma(i) \cdot \Gamma(n+1-i)} \int_0^{F_{M,i,n}} p^{i-1} \cdot (1-p)^{n-i} dp = \frac{1}{2}$$

	n														
i	31	32	33	34	35	36	37	38	39	40	41	42	43	44	45
1	0.0221	0.0214	0.0208	0.0202	0.0196	0.0191	0.0186	0.0181	0.0176	0.0172	0.0168	0.0164	0.0160	0.0156	0.0153
2	0.0536	0.0519	0.0503	0.0489	0.0475	0.0462	0.0449	0.0438	0.0427	0.0416	0.0406	0.0396	0.0387	0.0379	0.0370
3	0.0853	0.0827	0.0802	0.0779	0.0757	0.0736	0.0716	0.0697	0.0680	0.0663	0.0647	0.0632	0.0617	0.0603	0.0590
4	0.1172	0.1136	0.1101	0.1069	0.1039	0.1011	0.0983	0.0958	0.0933	0.0910	0.0888	0.0867	0.0847	0.0828	0.0810
5	0.1491	0.1444	0.1401	0.1360	0.1322	0.1285	0.1251	0.1218	0.1187	0.1158	0.1130	0.1103	0.1078	0.1054	0.1030
6	0.1809	0.1753	0.1701	0.1651	0.1605	0.1560	0.1519	0.1479	0.1441	0.1406	0.1372	0.1339	0.1308	0.1279	0.1251
7	0.2128	0.2063	0.2001	0.1942	0.1887	0.1836	0.1786	0.1740	0.1696	0.1654	0.1614	0.1575	0.1539	0.1504	0.1471
8	0.2447	0.2372	0.2301	0.2234	0.2170	0.2111	0.2054	0.2001	0.1950	0.1901	0.1855	0.1812	0.1770	0.1730	0.1692
9	0.2766	0.2681	0.2600	0.2525	0.2453	0.2386	0.2322	0.2261	0.2204	0.2149	0.2097	0.2048	0.2000	0.1955	0.1912
10	0.3086	0.2990	0.2900	0.2816	0.2736	0.2661	0.2590	0.2522	0.2458	0.2397	0.2339	0.2284	0.2231	0.2181	0.2133
11	0.3405	0.3299	0.3200	0.3107	0.3019	0.2936	0.2857	0.2783	0.2712	0.2645	0.2581	0.2520	0.2462	0.2406	0.2353
12	0.3724	0.3609	0.3500	0.3398	0.3302	0.3211	0.3125	0.3044	0.2966	0.2893	0.2823	0.2756	0.2693	0.2632	0.2574
13	0.4043	0.3918	0.3800	0.3690	0.3585	0.3486	0.3393	0.3305	0.3221	0.3141	0.3065	0.2992	0.2923	0.2857	0.2794
14	0.4362	0.4227	0.4100	0.3981	0.3868	0.3762	0.3661	0.3565	0.3475	0.3389	0.3307	0.3229	0.3154	0.3083	0.3015
15	0.4681	0.4536	0.4400	0.4272	0.4151	0.4037	0.3929	0.3826	0.3729	0.3637	0.3549	0.3465	0.3385	0.3308	0.3235
16	0.5000	0.4845	0.4700	0.4563	0.4434	0.4312	0.4197	0.4087	0.3983	0.3884	0.3790	0.3701	0.3616	0.3534	0.3456
17	0.5319	0.5155	0.5000	0.4854	0.4717	0.4587	0.4464	0.4348	0.4237	0.4132	0.4032	0.3937	0.3846	0.3760	0.3677
18	0.5638	0.5464	0.5300	0.5146	0.5000	0.4862	0.4732	0.4609	0.4492	0.4380	0.4274	0.4173	0.4077	0.3985	0.3897
19	0.5957	0.5773	0.5600	0.5437	0.5283	0.5138	0.5000	0.4870	0.4746	0.4628	0.4516	0.4410	0.4308	0.4211	0.4118
20	0.6276	0.6082	0.5900	0.5728	0.5566	0.5413	0.5268	0.5130	0.5000	0.4876	0.4758	0.4646	0.4539	0.4436	0.4338
21	0.6595	0.6391	0.6200	0.6019	0.5849	0.5688	0.5536	0.5391	0.5254	0.5124	0.5000	0.4882	0.4769	0.4662	0.4559
22	0.6914	0.6701	0.6500	0.6310	0.6132	0.5963	0.5803	0.5652	0.5508	0.5372	0.5242	0.5118	0.5000	0.4887	0.4779
23	0.7234	0.7010	0.6800	0.6602	0.6415	0.6238	0.6071	0.5913	0.5763	0.5620	0.5484	0.5354	0.5231	0.5113	0.5000
24	0.7553	0.7319	0.7100	0.6893	0.6698	0.6514	0.6339	0.6174	0.6017	0.5868	0.5726	0.5590	0.5461	0.5338	0.5221
25	0.7872	0.7628	0.7400	0.7184	0.6981	0.6789	0.6607	0.6435	0.6271	0.6116	0.5968	0.5827	0.5692	0.5564	0.5441
26	0.8191	0.7937	0.7699	0.7475	0.7264	0.7064	0.6875	0.6695	0.6525	0.6363	0.6210	0.6063	0.5923	0.5789	0.5662
27	0.8509	0.8247	0.7999	0.7766	0.7547	0.7339	0.7143	0.6956	0.6779	0.6611	0.6451	0.6299	0.6154	0.6015	0.5882
28	0.8828	0.8556	0.8299	0.8058	0.7830	0.7614	0.7410	0.7217	0.7034	0.6859	0.6693	0.6535	0.6384	0.6240	0.6103
29	0.9147	0.8864	0.8599	0.8349	0.8113	0.7889	0.7678	0.7478	0.7288	0.7107	0.6935	0.6771	0.6615	0.6466	0.6323
30	0.9464	0.9173	0.8899	0.8640	0.8395	0.8164	0.7946	0.7739	0.7542	0.7355	0.7177	0.7008	0.6846	0.6692	0.6544
31	0.9779	0.9481	0.9198	0.8931	0.8678	0.8440	0.8214	0.7999	0.7796	0.7603	0.7419	0.7244	0.7077	0.6917	0.6765
32		0.9786	0.9497	0.9221	0.8961	0.8715	0.8481	0.8260	0.8050	0.7851	0.7661	0.7480	0.7307	0.7143	0.6985
33			0.9792	0.9511	0.9243	0.8989	0.8749	0.8521	0.8304	0.8099	0.7903	0.7716	0.7538	0.7368	0.7206
34				0.9798	0.9525	0.9264	0.9017	0.8782	0.8559	0.8346	0.8145	0.7952	0.7769	0.7594	0.7426
35					0.9804	0.9538	0.9284	0.9042	0.8813	0.8594	0.8386	0.8188	0.8000	0.7819	0.7647
36						0.9809	0.9551	0.9303	0.9067	0.8842	0.8628	0.8425	0.8230	0.8045	0.7867
37							0.9814	0.9562	0.9320	0.9090	0.8870	0.8661	0.8461	0.8270	0.8088
38								0.9819	0.9573	0.9337	0.9112	0.8897	0.8692	0.8496	0.8308
39									0.9824	0.9584	0.9353	0.9133	0.8922	0.8721	0.8529
40										0.9828	0.9594	0.9368	0.9153	0.8946	0.8749
41											0.9832	0.9604	0.9383	0.9172	0.8970
42												0.9836	0.9613	0.9397	0.9190
43													0.9840	0.9621	0.9410
44														0.9844	0.9630
45															0.9847

J.6 Median Ranked Probability $F_{M,i,n}$ for $n = 46, \ldots, 60$

$$B(F_{M,i,n}; i, n+1-i) = \frac{\Gamma(n+1)}{\Gamma(i) \cdot \Gamma(n+1-i)} \int_0^{F_{M,i,n}} p^{i-1} \cdot (1-p)^{n-i} dp = \frac{1}{2}$$

i	46	47	48	49	50	51	52	53	54	55	56	57	58	59	60
1	0.0150	0.0146	0.0143	0.0140	0.0138	0.0135	0.0132	0.0130	0.0128	0.0125	0.0123	0.0121	0.0119	0.0117	0.0115
2	0.0362	0.0355	0.0347	0.0340	0.0333	0.0327	0.0321	0.0315	0.0309	0.0303	0.0298	0.0293	0.0288	0.0283	0.0278
3	0.0577	0.0565	0.0553	0.0542	0.0531	0.0521	0.0511	0.0501	0.0492	0.0483	0.0475	0.0466	0.0458	0.0451	0.0443
4	0.0792	0.0776	0.0760	0.0744	0.0729	0.0715	0.0702	0.0688	0.0676	0.0664	0.0652	0.0640	0.0629	0.0619	0.0609
5	0.1008	0.0987	0.0966	0.0947	0.0928	0.0910	0.0892	0.0876	0.0860	0.0844	0.0829	0.0815	0.0801	0.0787	0.0774
6	0.1224	0.1198	0.1173	0.1149	0.1126	0.1105	0.1083	0.1063	0.1044	0.1025	0.1006	0.0989	0.0972	0.0956	0.0940
7	0.1439	0.1409	0.1380	0.1352	0.1325	0.1299	0.1274	0.1251	0.1227	0.1205	0.1184	0.1163	0.1143	0.1124	0.1105
8	0.1655	0.1620	0.1587	0.1555	0.1524	0.1494	0.1465	0.1438	0.1411	0.1386	0.1361	0.1338	0.1315	0.1293	0.1271
9	0.1871	0.1831	0.1794	0.1757	0.1722	0.1689	0.1656	0.1625	0.1595	0.1567	0.1539	0.1512	0.1486	0.1461	0.1437
10	0.2087	0.2043	0.2000	0.1960	0.1921	0.1883	0.1847	0.1813	0.1779	0.1747	0.1716	0.1686	0.1657	0.1630	0.1603
11	0.2302	0.2254	0.2207	0.2162	0.2119	0.2078	0.2039	0.2000	0.1963	0.1928	0.1894	0.1861	0.1829	0.1798	0.1768
12	0.2518	0.2465	0.2414	0.2365	0.2318	0.2273	0.2230	0.2188	0.2147	0.2109	0.2071	0.2035	0.2000	0.1967	0.1934
13	0.2734	0.2676	0.2621	0.2568	0.2517	0.2468	0.2421	0.2375	0.2332	0.2289	0.2249	0.2210	0.2172	0.2135	0.2100
14	0.2950	0.2888	0.2828	0.2770	0.2715	0.2663	0.2612	0.2563	0.2516	0.2470	0.2426	0.2384	0.2343	0.2304	0.2265
15	0.3166	0.3099	0.3035	0.2973	0.2914	0.2857	0.2803	0.2750	0.2700	0.2651	0.2604	0.2558	0.2514	0.2472	0.2431
16	0.3381	0.3310	0.3242	0.3176	0.3113	0.3052	0.2994	0.2938	0.2884	0.2831	0.2781	0.2733	0.2686	0.2641	0.2597
17	0.3597	0.3521	0.3448	0.3379	0.3311	0.3247	0.3185	0.3125	0.3068	0.3012	0.2959	0.2907	0.2857	0.2809	0.2763
18	0.3813	0.3733	0.3655	0.3581	0.3510	0.3442	0.3376	0.3313	0.3252	0.3193	0.3136	0.3082	0.3029	0.2978	0.2928
19	0.4029	0.3944	0.3862	0.3784	0.3709	0.3636	0.3567	0.3500	0.3436	0.3374	0.3314	0.3256	0.3200	0.3146	0.3094
20	0.4245	0.4155	0.4069	0.3987	0.3907	0.3831	0.3758	0.3688	0.3620	0.3554	0.3491	0.3430	0.3372	0.3315	0.3260
21	0.4460	0.4366	0.4276	0.4189	0.4106	0.4026	0.3949	0.3875	0.3804	0.3735	0.3669	0.3605	0.3543	0.3483	0.3425
22	0.4676	0.4577	0.4483	0.4392	0.4305	0.4221	0.4140	0.4063	0.3988	0.3916	0.3846	0.3779	0.3714	0.3652	0.3591
23	0.4892	0.4789	0.4690	0.4595	0.4503	0.4416	0.4331	0.4250	0.4172	0.4096	0.4024	0.3954	0.3886	0.3820	0.3757
24	0.5108	0.5000	0.4897	0.4797	0.4702	0.4610	0.4522	0.4438	0.4356	0.4277	0.4201	0.4128	0.4057	0.3989	0.3923
25	0.5324	0.5211	0.5103	0.5000	0.4901	0.4805	0.4713	0.4625	0.4540	0.4458	0.4379	0.4302	0.4229	0.4157	0.4088
26	0.5540	0.5423	0.5310	0.5203	0.5099	0.5000	0.4904	0.4813	0.4724	0.4639	0.4556	0.4477	0.4400	0.4326	0.4254
27	0.5755	0.5634	0.5517	0.5405	0.5298	0.5195	0.5096	0.5000	0.4908	0.4819	0.4734	0.4651	0.4571	0.4494	0.4420
28	0.5971	0.5845	0.5724	0.5608	0.5497	0.5390	0.5287	0.5187	0.5092	0.5000	0.4911	0.4826	0.4743	0.4663	0.4586
29	0.6187	0.6056	0.5931	0.5811	0.5695	0.5584	0.5478	0.5375	0.5276	0.5181	0.5089	0.5000	0.4914	0.4831	0.4751
30	0.6403	0.6267	0.6138	0.6013	0.5894	0.5779	0.5669	0.5562	0.5460	0.5361	0.5266	0.5174	0.5086	0.5000	0.4917
31	0.6619	0.6479	0.6345	0.6216	0.6093	0.5974	0.5860	0.5750	0.5644	0.5542	0.5444	0.5349	0.5257	0.5169	0.5083
32	0.6834	0.6690	0.6552	0.6419	0.6291	0.6169	0.6051	0.5937	0.5828	0.5723	0.5621	0.5523	0.5429	0.5337	0.5249
33	0.7050	0.6901	0.6758	0.6621	0.6490	0.6364	0.6242	0.6125	0.6012	0.5904	0.5799	0.5698	0.5600	0.5506	0.5414
34	0.7266	0.7112	0.6965	0.6824	0.6689	0.6558	0.6433	0.6312	0.6196	0.6084	0.5976	0.5872	0.5771	0.5674	0.5580
35	0.7482	0.7324	0.7172	0.7027	0.6887	0.6753	0.6624	0.6500	0.6380	0.6265	0.6154	0.6046	0.5943	0.5843	0.5746
36	0.7698	0.7535	0.7379	0.7230	0.7086	0.6948	0.6815	0.6687	0.6564	0.6446	0.6331	0.6221	0.6114	0.6011	0.5912
37	0.7913	0.7746	0.7586	0.7432	0.7285	0.7143	0.7006	0.6875	0.6748	0.6626	0.6509	0.6395	0.6286	0.6180	0.6077
38	0.8129	0.7957	0.7793	0.7635	0.7483	0.7337	0.7197	0.7062	0.6932	0.6807	0.6686	0.6570	0.6457	0.6348	0.6243
39	0.8345	0.8169	0.8000	0.7838	0.7682	0.7532	0.7388	0.7250	0.7116	0.6988	0.6864	0.6744	0.6628	0.6517	0.6409
40	0.8561	0.8380	0.8206	0.8040	0.7881	0.7727	0.7579	0.7437	0.7300	0.7169	0.7041	0.6918	0.6800	0.6685	0.6575
41	0.8776	0.8591	0.8413	0.8243	0.8079	0.7922	0.7770	0.7625	0.7484	0.7349	0.7219	0.7093	0.6971	0.6854	0.6740

							n								
i	46	47	48	49	50	51	52	53	54	55	56	57	58	59	60
42	0.8992	0.8802	0.8620	0.8445	0.8278	0.8117	0.7961	0.7812	0.7668	0.7530	0.7396	0.7267	0.7143	0.7022	0.6906
43	0.9208	0.9013	0.8827	0.8648	0.8476	0.8311	0.8153	0.8000	0.7853	0.7711	0.7574	0.7442	0.7314	0.7191	0.7072
44	0.9423	0.9224	0.9034	0.8851	0.8675	0.8506	0.8344	0.8187	0.8037	0.7891	0.7751	0.7616	0.7486	0.7359	0.7237
45	0.9638	0.9435	0.9240	0.9053	0.8874	0.8701	0.8535	0.8375	0.8221	0.8072	0.7929	0.7790	0.7657	0.7528	0.7403
46	0.9850	0.9645	0.9447	0.9256	0.9072	0.8895	0.8726	0.8562	0.8405	0.8253	0.8106	0.7965	0.7828	0.7696	0.7569
47		0.9854	0.9653	0.9458	0.9271	0.9090	0.8917	0.8749	0.8589	0.8433	0.8284	0.8139	0.8000	0.7865	0.7735
48			0.9857	0.9660	0.9469	0.9285	0.9108	0.8937	0.8773	0.8614	0.8461	0.8314	0.8171	0.8033	0.7900
49				0.9860	0.9667	0.9479	0.9298	0.9124	0.8956	0.8795	0.8639	0.8488	0.8343	0.8202	0.8066
50					0.9862	0.9673	0.9489	0.9312	0.9140	0.8975	0.8816	0.8662	0.8514	0.8370	0.8232
51						0.9865	0.9679	0.9499	0.9324	0.9156	0.8994	0.8837	0.8685	0.8539	0.8397
52							0.9868	0.9685	0.9508	0.9336	0.9171	0.9011	0.8857	0.8707	0.8563
53								0.9870	0.9691	0.9517	0.9348	0.9185	0.9028	0.8876	0.8729
54									0.9872	0.9697	0.9525	0.9360	0.9199	0.9044	0.8895
55										0.9875	0.9702	0.9534	0.9371	0.9213	0.9060
56											0.9877	0.9707	0.9542	0.9381	0.9226
57												0.9879	0.9712	0.9549	0.9391
58													0.9881	0.9717	0.9557
59														0.9883	0.9722
60															0.9885

J.7 Probability of Expected Ranked Weibull Plotting Position $F(<Z_{i,n}>)$ for $n = 1, \ldots, 30$

$$F(< Z_{i,n} >) = 1 - \exp(- \exp(< Z_{i,n} >)) \text{ with } Z_{i,n} = \ln(- \ln(1 - F_{i,n}))$$

Note: for plots of $Z = {}^{10}\log(-\ln(1 - F))$ against ${}^{10}\log(t)$ (cf. Section 5.4.1, Eq. (5.24)), $F(<Z_{i,n}>)$ is the same as with $Z = \ln(-\ln(1 - F))$.

								n								
i	1	2	3	4	5	6	7	8	9	10	11	12	13	14	15
1	0.4296	0.2448	0.1707	0.1310	0.1062	0.0893	0.0771	0.0678	0.0605	0.0546	0.0498	0.0457	0.0423	0.0393	0.0367
2		0.6747	0.4683	0.3583	0.2901	0.2438	0.2102	0.1847	0.1648	0.1487	0.1355	0.1245	0.1151	0.1070	0.1000
3			0.7762	0.5932	0.4797	0.4028	0.3471	0.3050	0.2720	0.2454	0.2236	0.2054	0.1899	0.1765	0.1650
4				0.8304	0.6714	0.5632	0.4851	0.4261	0.3799	0.3428	0.3123	0.2868	0.2651	0.2465	0.2303
5					0.8639	0.7247	0.6239	0.5478	0.4883	0.4405	0.4012	0.3684	0.3405	0.3166	0.2958
6						0.8865	0.7633	0.6699	0.5969	0.5384	0.4903	0.4502	0.4161	0.3868	0.3614
7							0.9027	0.7925	0.7060	0.6366	0.5796	0.5321	0.4918	0.4571	0.4271
8								0.9150	0.8154	0.7350	0.6691	0.6141	0.5675	0.5275	0.4928
9									0.9245	0.8337	0.7588	0.6964	0.6434	0.5980	0.5586
10										0.9322	0.8488	0.7788	0.7195	0.6686	0.6245
11											0.9384	0.8614	0.7957	0.7394	0.6905
12												0.9437	0.8721	0.8102	0.7566
13													0.9481	0.8812	0.8228
14														0.9519	0.8892
15															0.9551

								n								
i	16	17	18	19	20	21	22	23	24	25	26	27	28	29	30
1	0.0345	0.0325	0.0307	0.0291	0.0277	0.0264	0.0252	0.0241	0.0231	0.0222	0.0214	0.0206	0.0199	0.0192	0.0185
2	0.0938	0.0884	0.0836	0.0792	0.0753	0.0718	0.0686	0.0656	0.0629	0.0604	0.0581	0.0560	0.0540	0.0522	0.0504
3	0.1548	0.1458	0.1378	0.1307	0.1242	0.1184	0.1131	0.1082	0.1037	0.0996	0.0958	0.0923	0.0891	0.0860	0.0832
4	0.2161	0.2036	0.1924	0.1824	0.1734	0.1653	0.1578	0.1510	0.1448	0.1391	0.1338	0.1289	0.1243	0.1201	0.1161
5	0.2776	0.2615	0.2471	0.2343	0.2227	0.2122	0.2027	0.1940	0.1860	0.1786	0.1718	0.1655	0.1596	0.1542	0.1491
6	0.3391	0.3194	0.3019	0.2862	0.2721	0.2593	0.2476	0.2369	0.2272	0.2182	0.2098	0.2021	0.1950	0.1883	0.1821
7	0.4007	0.3775	0.3567	0.3382	0.3215	0.3063	0.2925	0.2799	0.2684	0.2577	0.2479	0.2388	0.2304	0.2225	0.2151
8	0.4624	0.4355	0.4116	0.3902	0.3709	0.3534	0.3375	0.3230	0.3096	0.2974	0.2860	0.2755	0.2658	0.2567	0.2482
9	0.5241	0.4936	0.4665	0.4422	0.4203	0.4005	0.3825	0.3660	0.3509	0.3370	0.3241	0.3122	0.3012	0.2909	0.2812
10	0.5859	0.5518	0.5215	0.4943	0.4698	0.4477	0.4275	0.4091	0.3922	0.3766	0.3623	0.3490	0.3366	0.3251	0.3143
11	0.6478	0.6100	0.5765	0.5464	0.5193	0.4948	0.4725	0.4522	0.4335	0.4163	0.4004	0.3857	0.3720	0.3593	0.3474
12	0.7097	0.6683	0.6315	0.5986	0.5689	0.5420	0.5176	0.4953	0.4748	0.4560	0.4386	0.4224	0.4075	0.3935	0.3805
13	0.7717	0.7267	0.6866	0.6507	0.6185	0.5892	0.5627	0.5384	0.5161	0.4956	0.4767	0.4592	0.4429	0.4278	0.4136
14	0.8339	0.7851	0.7418	0.7030	0.6681	0.6365	0.6078	0.5815	0.5575	0.5353	0.5149	0.4960	0.4784	0.4620	0.4467
15	0.8961	0.8437	0.7970	0.7553	0.7177	0.6838	0.6529	0.6247	0.5989	0.5751	0.5531	0.5328	0.5139	0.4963	0.4799
16	0.9580	0.9023	0.8523	0.8077	0.7675	0.7311	0.6981	0.6679	0.6402	0.6148	0.5913	0.5696	0.5493	0.5304	0.5129
17		0.9605	0.9077	0.8601	0.8173	0.7785	0.7433	0.7111	0.6817	0.6546	0.6295	0.6063	0.5848	0.5648	0.5461
18			0.9628	0.9126	0.8671	0.8259	0.7885	0.7544	0.7231	0.6943	0.6678	0.6432	0.6203	0.5990	0.5794
19				0.9648	0.9170	0.8735	0.8338	0.7977	0.7646	0.7341	0.7060	0.6799	0.6557	0.6327	0.6108
20					0.9666	0.9210	0.8792	0.8411	0.8061	0.7740	0.7443	0.7169	0.6914	0.6677	0.6454
21						0.9682	0.9246	0.8845	0.8477	0.8139	0.7826	0.7537	0.7270	0.7020	0.6774
22							0.9697	0.9279	0.8893	0.8538	0.8210	0.7907	0.7626	0.7363	0.7119
23								0.9710	0.9310	0.8937	0.8594	0.8276	0.7981	0.7705	0.7448
24									0.9723	0.9337	0.8979	0.8646	0.8338	0.8051	0.7783
25										0.9734	0.9363	0.9016	0.8695	0.8395	0.8117
26											0.9744	0.9387	0.9052	0.8740	0.8449
27												0.9754	0.9409	0.9085	0.8782
28													0.9763	0.9430	0.9115
29														0.9771	0.9449
30															0.9779

J.8 Probability of Expected Ranked Weibull Plotting Position $F(<Z_{i,n}>)$ for $n = 31, \ldots, 45$

$$F(< Z_{i,n} >) = 1 - \exp(- \exp(< Z_{i,n} >)) \text{ with } Z_{i,n} = \ln(- \ln(1 - F_{i,n}))$$

Note: for plots of $Z = {}^{10}\log(-\ln(1 - F))$ against ${}^{10}\log(t)$ (cf. Section 5.4.1, Eq. (5.24)), $F(<Z_{i,n}>)$ is the same as with $Z = \ln(-\ln(1 - F))$.

							n								
i	31	32	33	34	35	36	37	38	39	40	41	42	43	44	45
1	0.0179	0.0174	0.0169	0.0164	0.0159	0.0155	0.0151	0.0147	0.0143	0.0139	0.0136	0.0133	0.0130	0.0127	0.0124
2	0.0488	0.0473	0.0459	0.0445	0.0433	0.0421	0.0410	0.0399	0.0389	0.0379	0.0370	0.0361	0.0353	0.0345	0.0337
3	0.0805	0.0780	0.0757	0.0735	0.0714	0.0694	0.0675	0.0658	0.0641	0.0625	0.0610	0.0595	0.0582	0.0569	0.0556
4	0.1124	0.1089	0.1056	0.1025	0.0996	0.0969	0.0943	0.0918	0.0895	0.0872	0.0851	0.0831	0.0812	0.0794	0.0776
5	0.1443	0.1398	0.1356	0.1317	0.1279	0.1244	0.1211	0.1179	0.1149	0.1120	0.1093	0.1067	0.1043	0.1019	0.0996
6	0.1763	0.1708	0.1657	0.1608	0.1563	0.1519	0.1479	0.1440	0.1403	0.1368	0.1335	0.1304	0.1273	0.1245	0.1217
7	0.2082	0.2018	0.1957	0.1900	0.1846	0.1795	0.1747	0.1701	0.1658	0.1617	0.1577	0.1540	0.1504	0.1470	0.1438
8	0.2402	0.2328	0.2258	0.2192	0.2130	0.2071	0.2015	0.1962	0.1912	0.1865	0.1820	0.1777	0.1736	0.1696	0.1659
9	0.2722	0.2638	0.2558	0.2484	0.2413	0.2347	0.2284	0.2224	0.2167	0.2113	0.2062	0.2013	0.1967	0.1922	0.1880
10	0.3042	0.2948	0.2859	0.2776	0.2697	0.2623	0.2552	0.2485	0.2422	0.2362	0.2305	0.2250	0.2198	0.2148	0.2101
11	0.3363	0.3258	0.3160	0.3068	0.2981	0.2898	0.2821	0.2747	0.2677	0.2610	0.2547	0.2487	0.2429	0.2374	0.2322
12	0.3683	0.3569	0.3461	0.3360	0.3265	0.3174	0.3089	0.3008	0.2932	0.2859	0.2789	0.2723	0.2660	0.2600	0.2543
13	0.4003	0.3879	0.3762	0.3652	0.3549	0.3451	0.3358	0.3270	0.3187	0.3107	0.3032	0.2960	0.2892	0.2826	0.2764
14	0.4324	0.4189	0.4063	0.3945	0.3832	0.3727	0.3626	0.3532	0.3442	0.3356	0.3275	0.3197	0.3123	0.3052	0.2985
15	0.4644	0.4500	0.4364	0.4237	0.4116	0.4003	0.3895	0.3793	0.3697	0.3605	0.3517	0.3434	0.3354	0.3278	0.3206
16	0.4965	0.4811	0.4666	0.4529	0.4401	0.4279	0.4164	0.4055	0.3952	0.3853	0.3760	0.3671	0.3586	0.3505	0.3427
17	0.5285	0.5121	0.4967	0.4822	0.4685	0.4555	0.4433	0.4317	0.4207	0.4102	0.4002	0.3908	0.3817	0.3731	0.3648
18	0.5606	0.5432	0.5268	0.5114	0.4969	0.4831	0.4702	0.4578	0.4462	0.4351	0.4245	0.4144	0.4049	0.3957	0.3869
19	0.5927	0.5743	0.5570	0.5407	0.5253	0.5108	0.4970	0.4840	0.4717	0.4599	0.4488	0.4381	0.4280	0.4183	0.4091
20	0.6248	0.6054	0.5871	0.5699	0.5537	0.5384	0.5239	0.5102	0.4972	0.4848	0.4731	0.4618	0.4511	0.4409	0.4312
21	0.6569	0.6365	0.6173	0.5992	0.5822	0.5661	0.5508	0.5364	0.5227	0.5097	0.4973	0.4855	0.4743	0.4636	0.4533
22	0.6890	0.6676	0.6474	0.6285	0.6106	0.5937	0.5777	0.5626	0.5482	0.5346	0.5216	0.5092	0.4975	0.4862	0.4754
23	0.7211	0.6987	0.6776	0.6578	0.6390	0.6214	0.6046	0.5888	0.5738	0.5595	0.5459	0.5329	0.5206	0.5088	0.4976
24	0.7533	0.7298	0.7078	0.6871	0.6675	0.6490	0.6316	0.6150	0.5993	0.5844	0.5702	0.5567	0.5438	0.5315	0.5197
25	0.7855	0.7610	0.7380	0.7164	0.6960	0.6767	0.6585	0.6412	0.6248	0.6093	0.5945	0.5804	0.5669	0.5541	0.5418
26	0.8176	0.7922	0.7682	0.7457	0.7244	0.7044	0.6854	0.6674	0.6504	0.6342	0.6188	0.6041	0.5901	0.5767	0.5640
27	0.8499	0.8233	0.7984	0.7750	0.7529	0.7321	0.7123	0.6937	0.6759	0.6591	0.6431	0.6278	0.6133	0.5994	0.5861
28	0.8821	0.8546	0.8287	0.8044	0.7814	0.7598	0.7393	0.7199	0.7015	0.6840	0.6674	0.6515	0.6364	0.6220	0.6082
29	0.9144	0.8858	0.8590	0.8337	0.8099	0.7875	0.7662	0.7461	0.7271	0.7089	0.6917	0.6753	0.6596	0.6447	0.6304
30	0.9467	0.9171	0.8893	0.8631	0.8385	0.8152	0.7932	0.7724	0.7526	0.7339	0.7160	0.6990	0.6828	0.6673	0.6525
31	0.9787	0.9484	0.9196	0.8925	0.8670	0.8430	0.8202	0.7987	0.7782	0.7588	0.7403	0.7228	0.7060	0.6900	0.6747
32		0.9793	0.9499	0.9220	0.8956	0.8707	0.8472	0.8249	0.8038	0.7837	0.7647	0.7465	0.7292	0.7127	0.6969
33			0.9800	0.9514	0.9242	0.8985	0.8742	0.8512	0.8294	0.8087	0.7890	0.7703	0.7524	0.7353	0.7190
34				0.9806	0.9528	0.9263	0.9013	0.8775	0.8550	0.8337	0.8134	0.7940	0.7756	0.7580	0.7412
35					0.9811	0.9542	0.9283	0.9039	0.8807	0.8587	0.8377	0.8178	0.7988	0.7807	0.7634
36						0.9817	0.9554	0.9302	0.9064	0.8837	0.8621	0.8416	0.8220	0.8034	0.7856
37							0.9822	0.9566	0.9320	0.9087	0.8865	0.8654	0.8453	0.8261	0.8078
38								0.9827	0.9577	0.9337	0.9109	0.8892	0.8685	0.8488	0.8300
39									0.9831	0.9588	0.9354	0.9131	0.8918	0.8715	0.8522
40										0.9836	0.9598	0.9369	0.9151	0.8943	0.8744
41											0.9840	0.9608	0.9384	0.9170	0.8966
42												0.9844	0.9617	0.9398	0.9189
43													0.9847	0.9626	0.9412
44														0.9851	0.9634
45															0.9854

J.9 Probability of Expected Ranked Weibull Plotting Position $F(<Z_{i,n}>)$ for $n = 46, \ldots, 60$

$$F(< Z_{i,n} >) = 1 - \exp(- \exp(< Z_{i,n} >)) \text{ with } Z_{i,n} = \ln(- \ln(1 - F_{i,n}))$$

Note: for plots of $Z = {}^{10}\log(-\ln(1 - F))$ against ${}^{10}\log(t)$ (cf. Section 5.4.1, Eq. (5.24)), $F(<Z_{i,n}>)$ is the same as with $Z = \ln(-\ln(1 - F))$.

									n						
i	46	47	48	49	50	51	52	53	54	55	56	57	58	59	60
1	0.0121	0.0119	0.0116	0.0114	0.0112	0.0109	0.0107	0.0105	0.0103	0.0102	0.0100	0.0098	0.0096	0.0095	0.0093
2	0.0330	0.0323	0.0316	0.0310	0.0304	0.0298	0.0292	0.0287	0.0281	0.0276	0.0271	0.0267	0.0262	0.0258	0.0253
3	0.0544	0.0532	0.0521	0.0511	0.0501	0.0491	0.0482	0.0472	0.0464	0.0455	0.0447	0.0439	0.0432	0.0425	0.0418
4	0.0759	0.0743	0.0728	0.0713	0.0699	0.0685	0.0672	0.0659	0.0647	0.0636	0.0624	0.0613	0.0603	0.0593	0.0583
5	0.0975	0.0954	0.0935	0.0916	0.0897	0.0880	0.0863	0.0847	0.0831	0.0816	0.0802	0.0788	0.0774	0.0761	0.0748
6	0.1191	0.1166	0.1141	0.1118	0.1096	0.1075	0.1054	0.1034	0.1015	0.0997	0.0979	0.0962	0.0945	0.0930	0.0914
7	0.1407	0.1377	0.1349	0.1321	0.1295	0.1270	0.1245	0.1222	0.1199	0.1178	0.1157	0.1137	0.1117	0.1098	0.1080
8	0.1623	0.1589	0.1556	0.1524	0.1494	0.1465	0.1437	0.1410	0.1384	0.1359	0.1334	0.1311	0.1289	0.1267	0.1246
9	0.1839	0.1800	0.1763	0.1727	0.1693	0.1660	0.1628	0.1597	0.1568	0.1539	0.1512	0.1486	0.1460	0.1436	0.1412
10	0.2055	0.2012	0.1970	0.1930	0.1892	0.1855	0.1819	0.1785	0.1752	0.1720	0.1690	0.1660	0.1632	0.1604	0.1578
11	0.2271	0.2223	0.2177	0.2133	0.2091	0.2050	0.2011	0.1973	0.1936	0.1901	0.1868	0.1835	0.1803	0.1773	0.1744
12	0.2488	0.2435	0.2385	0.2336	0.2290	0.2245	0.2202	0.2161	0.2121	0.2082	0.2045	0.2010	0.1975	0.1942	0.1910
13	0.2704	0.2647	0.2592	0.2539	0.2489	0.2440	0.2393	0.2348	0.2305	0.2263	0.2223	0.2184	0.2147	0.2111	0.2075
14	0.2920	0.2858	0.2799	0.2742	0.2688	0.2635	0.2585	0.2536	0.2489	0.2444	0.2401	0.2359	0.2318	0.2279	0.2241
15	0.3137	0.3070	0.3007	0.2945	0.2887	0.2830	0.2776	0.2724	0.2674	0.2625	0.2579	0.2534	0.2490	0.2448	0.2407
16	0.3353	0.3282	0.3214	0.3149	0.3086	0.3026	0.2968	0.2912	0.2858	0.2806	0.2757	0.2708	0.2662	0.2617	0.2573
17	0.3569	0.3494	0.3421	0.3352	0.3285	0.3221	0.3159	0.3100	0.3043	0.2988	0.2934	0.2883	0.2834	0.2786	0.2740
18	0.3786	0.3706	0.3629	0.3555	0.3484	0.3416	0.3351	0.3288	0.3227	0.3169	0.3112	0.3058	0.3005	0.2955	0.2906
19	0.4002	0.3917	0.3836	0.3758	0.3683	0.3611	0.3542	0.3476	0.3412	0.3350	0.3290	0.3233	0.3177	0.3123	0.3072
20	0.4219	0.4129	0.4043	0.3961	0.3882	0.3807	0.3734	0.3664	0.3596	0.3531	0.3468	0.3407	0.3349	0.3292	0.3238
21	0.4435	0.4341	0.4251	0.4165	0.4082	0.4002	0.3925	0.3851	0.3780	0.3712	0.3646	0.3582	0.3521	0.3461	0.3404
22	0.4651	0.4553	0.4458	0.4368	0.4281	0.4197	0.4117	0.4039	0.3965	0.3893	0.3824	0.3757	0.3692	0.3630	0.3570
23	0.4868	0.4765	0.4666	0.4571	0.4480	0.4392	0.4308	0.4227	0.4149	0.4074	0.4002	0.3932	0.3864	0.3799	0.3736
24	0.5084	0.4977	0.4873	0.4774	0.4679	0.4588	0.4500	0.4415	0.4334	0.4255	0.4180	0.4107	0.4036	0.3968	0.3902
25	0.5301	0.5189	0.5081	0.4978	0.4878	0.4783	0.4691	0.4603	0.4518	0.4436	0.4358	0.4281	0.4208	0.4137	0.4068
26	0.5518	0.5401	0.5288	0.5181	0.5078	0.4978	0.4883	0.4791	0.4703	0.4618	0.4535	0.4456	0.4380	0.4306	0.4234

								n							
i	46	47	48	49	50	51	52	53	54	55	56	57	58	59	60
27	0.5734	0.5613	0.5496	0.5384	0.5277	0.5174	0.5075	0.4979	0.4887	0.4799	0.4713	0.4631	0.4551	0.4475	0.4400
28	0.5951	0.5825	0.5704	0.5588	0.5476	0.5369	0.5266	0.5167	0.5072	0.4980	0.4891	0.4806	0.4723	0.4643	0.4566
29	0.6167	0.6037	0.5911	0.5791	0.5676	0.5565	0.5458	0.5355	0.5256	0.5161	0.5069	0.4981	0.4895	0.4812	0.4732
30	0.6384	0.6249	0.6119	0.5994	0.5875	0.5760	0.5650	0.5543	0.5441	0.5342	0.5247	0.5156	0.5067	0.4981	0.4899
31	0.6601	0.6461	0.6327	0.6198	0.6074	0.5956	0.5841	0.5731	0.5626	0.5524	0.5425	0.5330	0.5239	0.5150	0.5065
32	0.6818	0.6673	0.6534	0.6401	0.6274	0.6151	0.6033	0.5920	0.5810	0.5705	0.5603	0.5505	0.5411	0.5319	0.5231
33	0.7034	0.6885	0.6742	0.6605	0.6473	0.6347	0.6225	0.6108	0.5995	0.5886	0.5781	0.5680	0.5583	0.5488	0.5397
34	0.7251	0.7097	0.6950	0.6808	0.6673	0.6542	0.6417	0.6296	0.6180	0.6068	0.5959	0.5855	0.5755	0.5657	0.5563
35	0.7468	0.7310	0.7158	0.7012	0.6872	0.6738	0.6608	0.6484	0.6364	0.6249	0.6138	0.6030	0.5926	0.5826	0.5729
36	0.7685	0.7522	0.7366	0.7216	0.7072	0.6933	0.6800	0.6672	0.6549	0.6430	0.6316	0.6205	0.6098	0.5995	0.5896
37	0.7902	0.7734	0.7573	0.7419	0.7271	0.7129	0.6992	0.6860	0.6734	0.6612	0.6494	0.6380	0.6270	0.6164	0.6062
38	0.8119	0.7947	0.7781	0.7623	0.7471	0.7324	0.7184	0.7049	0.6918	0.6793	0.6672	0.6555	0.6442	0.6333	0.6228
39	0.8336	0.8159	0.7989	0.7827	0.7670	0.7520	0.7376	0.7237	0.7103	0.6974	0.6850	0.6730	0.6614	0.6503	0.6394
40	0.8554	0.8372	0.8198	0.8030	0.7870	0.7716	0.7568	0.7425	0.7288	0.7156	0.7028	0.6905	0.6786	0.6672	0.6561
41	0.8771	0.8585	0.8406	0.8234	0.8070	0.7912	0.7760	0.7614	0.7473	0.7337	0.7206	0.7080	0.6958	0.6841	0.6727
42	0.8989	0.8797	0.8614	0.8438	0.8270	0.8108	0.7952	0.7802	0.7658	0.7519	0.7385	0.7255	0.7131	0.7010	0.6893
43	0.9207	0.9010	0.8823	0.8642	0.8470	0.8304	0.8144	0.7990	0.7843	0.7700	0.7563	0.7431	0.7303	0.7179	0.7060
44	0.9424	0.9224	0.9031	0.8847	0.8670	0.8500	0.8336	0.8179	0.8028	0.7882	0.7741	0.7606	0.7475	0.7348	0.7226
45	0.9642	0.9437	0.9240	0.9051	0.8870	0.8696	0.8528	0.8368	0.8213	0.8064	0.7920	0.7781	0.7647	0.7517	0.7392
46	0.9857	0.9650	0.9449	0.9255	0.9070	0.8892	0.8721	0.8556	0.8398	0.8245	0.8098	0.7956	0.7819	0.7687	0.7559
47		0.9861	0.9657	0.9460	0.9270	0.9088	0.8913	0.8745	0.8583	0.8427	0.8277	0.8131	0.7991	0.7856	0.7725
48			0.9864	0.9664	0.9471	0.9285	0.9106	0.8934	0.8768	0.8609	0.8455	0.8307	0.8164	0.8025	0.7892
49				0.9866	0.9671	0.9481	0.9298	0.9123	0.8954	0.8791	0.8634	0.8482	0.8336	0.8195	0.8058
50					0.9869	0.9678	0.9491	0.9312	0.9139	0.8973	0.8812	0.8658	0.8508	0.8364	0.8225
51						0.9872	0.9684	0.9501	0.9325	0.9155	0.8991	0.8833	0.8681	0.8534	0.8391
52							0.9874	0.9690	0.9510	0.9337	0.9170	0.9009	0.8853	0.8703	0.8558
53								0.9877	0.9696	0.9519	0.9349	0.9184	0.9026	0.8873	0.8725
54									0.9879	0.9701	0.9528	0.9360	0.9199	0.9042	0.8892
55										0.9881	0.9707	0.9536	0.9371	0.9212	0.9058
56											0.9883	0.9712	0.9544	0.9382	0.9225
57												0.9886	0.9717	0.9552	0.9392
58													0.9888	0.9722	0.9560
59														0.9890	0.9726
60															0.9891

J.10 Expected Ranked Weibull Plotting Position $<Z_{i,n}>$ for $n = 1, \ldots, 30$

$$< Z_{i,n} >= \left[-\gamma + i \binom{n}{i} \sum_{j=0}^{i-1} \binom{i-1}{j} \frac{(-1)^{i-j} \cdot \ln(n-j)}{n-j} \right] \text{ with } Z_{i,n} = \ln(-\ln(1 - F_{i,n}))$$

with $\gamma \approx 0.577215665$ the Euler constant.

Note: for plots of $Z = {}^{10}\log(-\ln(1 - F))$ against ${}^{10}\log(t)$ (cf. Section 5.4.1, Eq. (5.24)), multiply the $Z_{i,n}$ values below by ${}^{10}\log(e) = 0.434294482$.

i	1	2	3	4	5	6	7	8	9	10	11	12	13	14	15
								n							
1	−0.5772	−1.2704	−1.6758	−1.9635	−2.1867	−2.3690	−2.5231	−2.6567	−2.7744	−2.8798	−2.9751	−3.0621	−3.1422	−3.2163	−3.2853
2		0.1159	−0.4594	−0.8128	−1.0709	−1.2750	−1.4441	−1.5884	−1.7144	−1.8262	−1.9267	−2.0180	−2.1016	−2.1788	−2.2504
3			0.4036	−0.1061	−0.4256	−0.6627	−0.8525	−1.0111	−1.1475	−1.2672	−1.3739	−1.4703	−1.5581	−1.6387	−1.7133
4				0.5735	0.1069	−0.1884	−0.4097	−0.5882	−0.7383	−0.8681	−0.9825	−1.0849	−1.1776	−1.2623	−1.3404
5					0.6902	0.2545	−0.0224	−0.2312	−0.4005	−0.5436	−0.6678	−0.7777	−0.8763	−0.9659	−1.0478
6						0.7773	0.3653	0.1029	−0.0958	−0.2574	−0.3946	−0.5140	−0.6199	−0.7152	−0.8019
7							0.8460	0.4528	0.2022	0.0120	−0.1432	−0.2752	−0.3904	−0.4928	−0.5852
8								0.9021	0.5244	0.2837	0.1007	−0.0489	−0.1764	−0.2879	−0.3873
9									0.9493	0.5846	0.3523	0.1756	0.0308	−0.0928	−0.2010
10										0.9899	0.6362	0.4112	0.2399	0.0994	−0.0206
11											1.0252	0.6812	0.4626	0.2961	0.1595
12												1.0565	0.7209	0.5080	0.3458
13													1.0845	0.7564	0.5485
14														1.1097	0.7884
15															1.1327

i	16	17	18	19	20	21	22	23	24	25	26	27	28	29	30
								n							
1	−3.3498	−3.4104	−3.4676	−3.5217	−3.5729	−3.6217	−3.6683	−3.7127	−3.7553	−3.7961	−3.8353	−3.8731	−3.9094	−3.9445	−3.9784
2	−2.3172	−2.3798	−2.4387	−2.4944	−2.5471	−2.5971	−2.6448	−2.6903	−2.7338	−2.7755	−2.8156	−2.8541	−2.8911	−2.9269	−2.9614
3	−1.7827	−1.8475	−1.9084	−1.9658	−2.0200	−2.0715	−2.1204	−2.1671	−2.2116	−2.2543	−2.2952	−2.3344	−2.3723	−2.4087	−2.4438
4	−1.4126	−1.4800	−1.5431	−1.6024	−1.6584	−1.7113	−1.7616	−1.8095	−1.8552	−1.8988	−1.9407	−1.9808	−2.0194	−2.0566	−2.0924
5	−1.1235	−1.1937	−1.2592	−1.3207	−1.3786	−1.4332	−1.4850	−1.5342	−1.5811	−1.6259	−1.6687	−1.7098	−1.7493	−1.7872	−1.8237
6	−0.8814	−0.9549	−1.0233	−1.0872	−1.1471	−1.2037	−1.2571	−1.3078	−1.3560	−1.4020	−1.4459	−1.4880	−1.5284	−1.5672	−1.6045
7	−0.6693	−0.7467	−0.8182	−0.8849	−0.9472	−1.0058	−1.0611	−1.1135	−1.1631	−1.2104	−1.2556	−1.2987	−1.3400	−1.3797	−1.4179
8	−0.4770	−0.5588	−0.6342	−0.7040	−0.7691	−0.8300	−0.8874	−0.9415	−0.9928	−1.0415	−1.0880	−1.1323	−1.1747	−1.2153	−1.2544
9	−0.2976	−0.3849	−0.4647	−0.5382	−0.6064	−0.6700	−0.7297	−0.7858	−0.8389	−0.8892	−0.9371	−0.9827	−1.0263	−1.0680	−1.1080
10	−0.1259	−0.2200	−0.3051	−0.3830	−0.4548	−0.5215	−0.5838	−0.6423	−0.6974	−0.7495	−0.7989	−0.8459	−0.8907	−0.9336	−0.9746
11	0.0426	−0.0601	−0.1518	−0.2350	−0.3112	−0.3815	−0.4468	−0.5078	−0.5652	−0.6192	−0.6704	−0.7190	−0.7652	−0.8093	−0.8515
12	0.2126	0.0985	−0.0017	−0.0914	−0.1727	−0.2473	−0.3161	−0.3802	−0.4401	−0.4964	−0.5495	−0.5998	−0.6475	−0.6930	−0.7364
13	0.3902	0.2601	0.1487	0.0506	−0.0371	−0.1168	−0.1899	−0.2574	−0.3203	−0.3791	−0.4344	−0.4866	−0.5361	−0.5831	−0.6279
14	0.5851	0.4302	0.3030	0.1939	0.0979	0.0119	−0.0662	−0.1379	−0.2042	−0.2660	−0.3238	−0.3782	−0.4296	−0.4782	−0.5245
15	0.8174	0.6182	0.4666	0.3419	0.2350	0.1409	0.0565	−0.0201	−0.0905	−0.1557	−0.2164	−0.2733	−0.3268	−0.3774	−0.4253
16	1.1537	0.8440	0.6486	0.4998	0.3776	0.2727	0.1803	0.0974	0.0221	−0.0471	−0.1112	−0.1709	−0.2269	−0.2796	−0.3294
17		1.1731	0.8684	0.6765	0.5304	0.4103	0.3073	0.2166	0.1351	0.0610	−0.0071	−0.0701	−0.1289	−0.1841	−0.2360
18			1.1910	0.8910	0.7022	0.5587	0.4406	0.3394	0.2501	0.1700	0.0971	0.0300	−0.0321	−0.0900	−0.1443
19				1.2076	0.9120	0.7262	0.5849	0.4688	0.3691	0.2813	0.2024	0.1306	0.0645	0.0033	−0.0538
20					1.2232	0.9315	0.7485	0.6094	0.4950	0.3969	0.3103	0.2326	0.1618	0.0968	0.0364
21						1.2378	0.9498	0.7693	0.6322	0.5195	0.4228	0.3375	0.2609	0.1911	0.1269
22							1.2515	0.9670	0.7889	0.6537	0.5426	0.4472	0.3631	0.2875	0.2186
23								1.2644	0.9832	0.8073	0.6739	0.5642	0.4701	0.3871	0.3125
24									1.2767	0.9985	0.8248	0.6930	0.5847	0.4918	0.4098
25										1.2883	1.0130	0.8412	0.7110	0.6041	0.5123
26											1.2993	1.0267	0.8569	0.7281	0.6224
27												1.3098	1.0398	0.8717	0.7444
28													1.3198	1.0522	0.8858
29														1.3293	1.0641
30															1.3385

J.11 Expected Ranked Weibull Plotting Position $<Z_{i,n}>$ for $n = 31, \ldots, 45$

$$< Z_{i,n} >= \left[-\gamma + i \binom{n}{i} \sum_{j=0}^{i-1} \binom{i-1}{j} \frac{(-1)^{i-j} \cdot \ln(n-j)}{n-j} \right] \text{ with } Z_{i,n} = \ln(-\ln(1-F_{i,n}))$$

with $\gamma \approx 0.577215665$ the Euler constant.

Note: for plots of $Z = {}^{10}\log(-\ln(1-F))$ against ${}^{10}\log(t)$ (cf. Section 5.4.1, Eq. (5.24)), multiply the $Z_{i,n}$ values below by ${}^{10}\log(e) = 0.434294482$.

							n								
i	31	32	33	34	35	36	37	38	39	40	41	42	43	44	45
1	−4.0112	−4.0430	−4.0737	−4.1036	−4.1326	−4.1607	−4.1881	−4.2148	−4.2408	−4.2661	−4.2908	−4.3149	−4.3384	−4.3614	−4.3839
2	−2.9947	−3.0270	−3.0583	−3.0886	−3.1180	−3.1466	−3.1744	−3.2014	−3.2277	−3.2534	−3.2784	−3.3028	−3.3266	−3.3499	−3.3726
3	−2.4778	−2.5106	−2.5424	−2.5732	−2.6031	−2.6321	−2.6603	−2.6877	−2.7144	−2.7404	−2.7657	−2.7904	−2.8146	−2.8381	−2.8611
4	−2.1270	−2.1604	−2.1927	−2.2241	−2.2544	−2.2839	−2.3125	−2.3403	−2.3674	−2.3938	−2.4195	−2.4445	−2.4689	−2.4928	−2.5160
5	−1.8590	−1.8930	−1.9260	−1.9578	−1.9887	−2.0187	−2.0477	−2.0760	−2.1035	−2.1302	−2.1562	−2.1816	−2.2064	−2.2305	−2.2541
6	−1.6404	−1.6751	−1.7087	−1.7411	−1.7725	−1.8030	−1.8326	−1.8613	−1.8892	−1.9163	−1.9427	−1.9684	−1.9935	−2.0180	−2.0419
7	−1.4546	−1.4900	−1.5242	−1.5573	−1.5893	−1.6203	−1.6503	−1.6795	−1.7078	−1.7354	−1.7622	−1.7883	−1.8137	−1.8386	−1.8628
8	−1.2919	−1.3281	−1.3630	−1.3967	−1.4293	−1.4609	−1.4915	−1.5211	−1.5499	−1.5779	−1.6052	−1.6317	−1.6575	−1.6826	−1.7072
9	−1.1464	−1.1834	−1.2190	−1.2535	−1.2867	−1.3189	−1.3500	−1.3802	−1.4095	−1.4380	−1.4656	−1.4925	−1.5187	−1.5443	−1.5691
10	−1.0140	−1.0519	−1.0883	−1.1235	−1.1574	−1.1902	−1.2220	−1.2527	−1.2825	−1.3115	−1.3396	−1.3669	−1.3936	−1.4195	−1.4447
11	−0.8919	−0.9307	−0.9680	−1.0040	−1.0386	−1.0721	−1.1045	−1.1359	−1.1662	−1.1957	−1.2243	−1.2521	−1.2791	−1.3055	−1.3311
12	−0.7780	−0.8178	−0.8561	−0.8929	−0.9283	−0.9625	−0.9956	−1.0276	−1.0585	−1.0885	−1.1177	−1.1460	−1.1734	−1.2002	−1.2262
13	−0.6707	−0.7116	−0.7509	−0.7886	−0.8249	−0.8599	−0.8937	−0.9263	−0.9579	−0.9885	−1.0182	−1.0470	−1.0749	−1.1021	−1.1286
14	−0.5687	−0.6108	−0.6512	−0.6899	−0.7271	−0.7630	−0.7975	−0.8309	−0.8632	−0.8944	−0.9246	−0.9539	−0.9824	−1.0101	−1.0370
15	−0.4710	−0.5144	−0.5560	−0.5958	−0.6341	−0.6708	−0.7062	−0.7403	−0.7733	−0.8052	−0.8360	−0.8659	−0.8950	−0.9231	−0.9505
16	−0.3767	−0.4217	−0.4646	−0.5056	−0.5449	−0.5826	−0.6189	−0.6539	−0.6876	−0.7202	−0.7517	−0.7822	−0.8118	−0.8405	−0.8684
17	−0.2851	−0.3317	−0.3761	−0.4184	−0.4589	−0.4977	−0.5350	−0.5708	−0.6054	−0.6387	−0.6709	−0.7021	−0.7323	−0.7616	−0.7900
18	−0.1956	−0.2440	−0.2900	−0.3338	−0.3755	−0.4155	−0.4539	−0.4907	−0.5261	−0.5603	−0.5933	−0.6251	−0.6559	−0.6858	−0.7148
19	−0.1080	−0.1579	−0.2057	−0.2511	−0.2943	−0.3356	−0.3751	−0.4129	−0.4493	−0.4844	−0.5182	−0.5508	−0.5823	−0.6128	−0.6424
20	−0.0199	−0.0728	−0.1227	−0.1698	−0.2147	−0.2574	−0.2981	−0.3372	−0.3746	−0.4106	−0.4453	−0.4787	−0.5109	−0.5421	−0.5723
21	0.0674	0.0118	−0.0404	−0.0896	−0.1362	−0.1805	−0.2227	−0.2630	−0.3016	−0.3386	−0.3742	−0.4085	−0.4416	−0.4735	−0.5044
22	0.1553	0.0965	0.0416	−0.0099	−0.0586	−0.1046	−0.1484	−0.1901	−0.2299	−0.2681	−0.3047	−0.3400	−0.3739	−0.4066	−0.4382
23	0.2446	0.1820	0.1240	0.0697	0.0188	−0.0292	−0.0748	−0.1180	−0.1593	−0.1987	−0.2365	−0.2727	−0.3076	−0.3412	−0.3736
24	0.3361	0.2690	0.2072	0.1499	0.0963	0.0460	−0.0015	−0.0465	−0.0893	−0.1301	−0.1691	−0.2065	−0.2424	−0.2769	−0.3102
25	0.4313	0.3585	0.2922	0.2311	0.1745	0.1215	0.0717	0.0247	−0.0198	−0.0621	−0.1025	−0.1411	−0.1781	−0.2136	−0.2478
26	0.5317	0.4517	0.3797	0.3142	0.2538	0.1978	0.1454	0.0961	0.0497	0.0056	−0.0363	−0.0763	−0.1145	−0.1511	−0.1863
27	0.6399	0.5502	0.4710	0.3999	0.3351	0.2754	0.2199	0.1681	0.1194	0.0734	0.0298	−0.0117	−0.0513	−0.0891	−0.1254
28	0.7599	0.6565	0.5677	0.4895	0.4191	0.3550	0.2959	0.2410	0.1897	0.1415	0.0960	0.0528	0.0117	−0.0274	−0.0649
29	0.8993	0.7747	0.6723	0.5845	0.5071	0.4374	0.3740	0.3155	0.2612	0.2104	0.1627	0.1176	0.0748	0.0341	−0.0047
30	1.0755	0.9122	0.7888	0.6875	0.6005	0.5239	0.4550	0.3921	0.3342	0.2805	0.2302	0.1829	0.1382	0.0959	0.0556
31	1.3472	1.0863	0.9246	0.8023	0.7019	0.6159	0.5400	0.4717	0.4095	0.3522	0.2989	0.2491	0.2022	0.1580	0.1160
32		1.3556	1.0968	0.9364	0.8152	0.7158	0.6306	0.5554	0.4878	0.4261	0.3693	0.3166	0.2672	0.2208	0.1769
33			1.3637	1.1068	0.9478	0.8277	0.7291	0.6447	0.5702	0.5032	0.4421	0.3858	0.3335	0.2846	0.2386
34				1.3715	1.1164	0.9587	0.8396	0.7419	0.6582	0.5844	0.5180	0.4575	0.4017	0.3498	0.3014
35					1.3790	1.1257	0.9692	0.8511	0.7543	0.6712	0.5980	0.5322	0.4722	0.4169	0.3655
36						1.3863	1.1347	0.9793	0.8622	0.7661	0.6838	0.6112	0.5459	0.4864	0.4316
37							1.3932	1.1433	0.9891	0.8728	0.7776	0.6959	0.6239	0.5592	0.5001
38								1.4000	1.1516	0.9985	0.8831	0.7886	0.7076	0.6361	0.5719
39									1.4065	1.1597	1.0076	0.8931	0.7992	0.7189	0.6480
40										1.4129	1.1675	1.0165	0.9027	0.8096	0.7298
41											1.4190	1.1750	1.0250	0.9120	0.8195
42												1.4249	1.1824	1.0333	0.9210
43													1.4307	1.1895	1.0413
44														1.4363	1.1963
45															1.4418

J.12 Expected Ranked Weibull Plotting Position $<Z_{i,n}>$ for $n = 46, \ldots, 60$

$$< Z_{i,n} >= \left[-\gamma + i \binom{n}{i} \sum_{j=0}^{i-1} \binom{i-1}{j} \frac{(-1)^{i-j} \cdot \ln(n-j)}{n-j} \right] \text{ with } Z_{i,n} = \ln(-\ln(1-F_{i,n}))$$

with $\gamma \approx 0.577215665$ the Euler constant.

Note: for plots of $Z = {}^{10}\log(-\ln(1-F))$ against ${}^{10}\log(t)$ (cf. Section 5.4.1, Eq. (5.24)), multiply the $Z_{i,n}$ values below by ${}^{10}\log(e) = 0.434294482$.

i	n 46	47	48	49	50	51	52	53	54	55	56	57	58	59	60
1	−4.4059	−4.4274	−4.4484	−4.4690	−4.4892	−4.5090	−4.5285	−4.5475	−4.5662	−4.5845	−4.6026	−4.6203	−4.6377	−4.6548	−4.6716
2	−3.3948	−3.4166	−3.4379	−3.4587	−3.4791	−3.4991	−3.5187	−3.5380	−3.5568	−3.5753	−3.5935	−3.6114	−3.6289	−3.6462	−3.6631
3	−2.8836	−2.9056	−2.9271	−2.9482	−2.9688	−2.9890	−3.0088	−3.0282	−3.0473	−3.0660	−3.0844	−3.1024	−3.1201	−3.1375	−3.1546
4	−2.5388	−2.5611	−2.5828	−2.6041	−2.6250	−2.6454	−2.6654	−2.6850	−2.7043	−2.7232	−2.7417	−2.7599	−2.7778	−2.7953	−2.8126
5	−2.2771	−2.2996	−2.3217	−2.3432	−2.3643	−2.3849	−2.4052	−2.4250	−2.4444	−2.4635	−2.4822	−2.5006	−2.5186	−2.5364	−2.5538
6	−2.0652	−2.0880	−2.1103	−2.1321	−2.1534	−2.1743	−2.1947	−2.2148	−2.2344	−2.2537	−2.2726	−2.2912	−2.3094	−2.3273	−2.3448
7	−1.8864	−1.9095	−1.9320	−1.9541	−1.9757	−1.9968	−2.0175	−2.0377	−2.0576	−2.0771	−2.0962	−2.1149	−2.1333	−2.1514	−2.1691
8	−1.7311	−1.7545	−1.7773	−1.7997	−1.8215	−1.8429	−1.8638	−1.8843	−1.9044	−1.9241	−1.9434	−1.9623	−1.9809	−1.9991	−2.0170
9	−1.5934	−1.6171	−1.6403	−1.6629	−1.6850	−1.7066	−1.7278	−1.7485	−1.7688	−1.7887	−1.8083	−1.8274	−1.8462	−1.8646	−1.8827
10	−1.4693	−1.4934	−1.5168	−1.5397	−1.5621	−1.5840	−1.6055	−1.6264	−1.6470	−1.6671	−1.6868	−1.7062	−1.7252	−1.7438	−1.7621
11	−1.3561	−1.3804	−1.4042	−1.4275	−1.4501	−1.4723	−1.4940	−1.5153	−1.5361	−1.5564	−1.5764	−1.5959	−1.6151	−1.6339	−1.6524
12	−1.2516	−1.2763	−1.3004	−1.3240	−1.3470	−1.3695	−1.3915	−1.4130	−1.4340	−1.4546	−1.4748	−1.4946	−1.5140	−1.5330	−1.5517
13	−1.1544	−1.1795	−1.2039	−1.2278	−1.2512	−1.2739	−1.2962	−1.3180	−1.3393	−1.3602	−1.3806	−1.4006	−1.4202	−1.4395	−1.4583
14	−1.0632	−1.0887	−1.1135	−1.1378	−1.1615	−1.1846	−1.2071	−1.2292	−1.2508	−1.2719	−1.2926	−1.3128	−1.3327	−1.3522	−1.3712
15	−0.9771	−1.0031	−1.0283	−1.0529	−1.0769	−1.1004	−1.1233	−1.1457	−1.1675	−1.1889	−1.2099	−1.2304	−1.2505	−1.2702	−1.2895
16	−0.8955	−0.9218	−0.9475	−0.9725	−0.9969	−1.0207	−1.0439	−1.0666	−1.0888	−1.1105	−1.1317	−1.1524	−1.1728	−1.1927	−1.2122
17	−0.8176	−0.8444	−0.8705	−0.8959	−0.9207	−0.9449	−0.9684	−0.9915	−1.0139	−1.0359	−1.0574	−1.0785	−1.0991	−1.1192	−1.1390
18	−0.7429	−0.7702	−0.7968	−0.8226	−0.8478	−0.8724	−0.8963	−0.9197	−0.9425	−0.9648	−0.9866	−1.0079	−1.0288	−1.0492	−1.0692
19	−0.6710	−0.6989	−0.7259	−0.7523	−0.7779	−0.8028	−0.8272	−0.8509	−0.8740	−0.8967	−0.9188	−0.9404	−0.9615	−0.9822	−1.0025
20	−0.6016	−0.6300	−0.6576	−0.6844	−0.7105	−0.7358	−0.7606	−0.7847	−0.8082	−0.8312	−0.8536	−0.8755	−0.8970	−0.9180	−0.9385
21	−0.5343	−0.5633	−0.5914	−0.6187	−0.6453	−0.6711	−0.6963	−0.7208	−0.7447	−0.7680	−0.7908	−0.8130	−0.8348	−0.8561	−0.8769
22	−0.4688	−0.4984	−0.5271	−0.5550	−0.5821	−0.6084	−0.6340	−0.6589	−0.6832	−0.7069	−0.7301	−0.7527	−0.7747	−0.7963	−0.8174
23	−0.4049	−0.4351	−0.4645	−0.4929	−0.5205	−0.5474	−0.5734	−0.5988	−0.6236	−0.6477	−0.6712	−0.6941	−0.7165	−0.7384	−0.7598
24	−0.3422	−0.3733	−0.4033	−0.4323	−0.4605	−0.4879	−0.5145	−0.5403	−0.5655	−0.5900	−0.6139	−0.6373	−0.6600	−0.6823	−0.7040
25	−0.2807	−0.3125	−0.3433	−0.3730	−0.4018	−0.4297	−0.4569	−0.4832	−0.5089	−0.5338	−0.5582	−0.5819	−0.6050	−0.6276	−0.6497
26	−0.2201	−0.2528	−0.2843	−0.3147	−0.3442	−0.3727	−0.4004	−0.4273	−0.4535	−0.4789	−0.5037	−0.5278	−0.5513	−0.5743	−0.5967
27	−0.1603	−0.1938	−0.2261	−0.2573	−0.2875	−0.3167	−0.3450	−0.3725	−0.3992	−0.4251	−0.4503	−0.4749	−0.4988	−0.5222	−0.5450
28	−0.1009	−0.1354	−0.1686	−0.2007	−0.2316	−0.2615	−0.2905	−0.3186	−0.3458	−0.3723	−0.3980	−0.4230	−0.4474	−0.4712	−0.4943
29	−0.0418	−0.0774	−0.1117	−0.1446	−0.1764	−0.2071	−0.2367	−0.2654	−0.2933	−0.3203	−0.3466	−0.3721	−0.3969	−0.4211	−0.4447
30	0.0171	−0.0197	−0.0550	−0.0889	−0.1216	−0.1531	−0.1835	−0.2129	−0.2414	−0.2691	−0.2959	−0.3219	−0.3472	−0.3719	−0.3959
31	0.0761	0.0379	0.0015	−0.0335	−0.0672	−0.0996	−0.1308	−0.1610	−0.1902	−0.2184	−0.2458	−0.2724	−0.2983	−0.3234	−0.3479
32	0.1353	0.0957	0.0579	0.0218	−0.0129	−0.0463	−0.0784	−0.1094	−0.1393	−0.1683	−0.1963	−0.2235	−0.2499	−0.2756	−0.3005
33	0.1951	0.1539	0.1146	0.0772	0.0413	0.0069	−0.0262	−0.0581	−0.0888	−0.1185	−0.1472	−0.1751	−0.2021	−0.2283	−0.2537
34	0.2557	0.2126	0.1717	0.1328	0.0956	0.0601	0.0259	−0.0069	−0.0385	−0.0690	−0.0985	−0.1270	−0.1546	−0.1814	−0.2074
35	0.3175	0.2722	0.2295	0.1889	0.1503	0.1134	0.0781	0.0443	0.0117	−0.0197	−0.0499	−0.0792	−0.1075	−0.1349	−0.1615
36	0.3806	0.3330	0.2881	0.2457	0.2055	0.1672	0.1306	0.0956	0.0619	0.0296	−0.0015	−0.0316	−0.0606	−0.0887	−0.1159
37	0.4458	0.3952	0.3479	0.3034	0.2613	0.2214	0.1834	0.1471	0.1124	0.0790	0.0469	0.0160	−0.0138	−0.0426	−0.0705
38	0.5134	0.4594	0.4093	0.3623	0.3182	0.2765	0.2368	0.1991	0.1631	0.1286	0.0955	0.0636	0.0329	0.0033	−0.0253
39	0.5843	0.5262	0.4726	0.4229	0.3763	0.3325	0.2910	0.2517	0.2143	0.1785	0.1443	0.1114	0.0798	0.0493	0.0199
40	0.6594	0.5962	0.5385	0.4854	0.4360	0.3898	0.3463	0.3052	0.2661	0.2290	0.1935	0.1595	0.1268	0.0954	0.0651
41	0.7403	0.6705	0.6077	0.5505	0.4977	0.4487	0.4028	0.3596	0.3188	0.2801	0.2432	0.2079	0.1741	0.1417	0.1105
42	0.8292	0.7505	0.6812	0.6189	0.5620	0.5097	0.4610	0.4155	0.3726	0.3320	0.2936	0.2569	0.2219	0.1884	0.1562
43	0.9298	0.8386	0.7604	0.6916	0.6297	0.5733	0.5213	0.4729	0.4277	0.3851	0.3449	0.3067	0.2703	0.2355	0.2022
44	1.0491	0.9383	0.8476	0.7700	0.7017	0.6402	0.5842	0.5325	0.4845	0.4396	0.3973	0.3573	0.3194	0.2832	0.2487
45	1.2030	1.0566	0.9465	0.8565	0.7794	0.7114	0.6504	0.5947	0.5434	0.4957	0.4511	0.4091	0.3694	0.3317	0.2958

								n							
i	46	47	48	49	50	51	52	53	54	55	56	57	58	59	60
46	1.4471	1.2095	1.0639	0.9545	0.8650	0.7884	0.7209	0.6603	0.6050	0.5540	0.5066	0.4623	0.4206	0.3811	0.3437
47		1.4523	1.2159	1.0711	0.9623	0.8734	0.7972	0.7302	0.6699	0.6149	0.5643	0.5172	0.4732	0.4317	0.3925
48			1.4573	1.2220	1.0780	0.9699	0.8815	0.8058	0.7391	0.6792	0.6246	0.5743	0.5276	0.4838	0.4426
49				1.4622	1.2280	1.0848	0.9772	0.8893	0.8141	0.7479	0.6883	0.6341	0.5841	0.5376	0.4941
50					1.4670	1.2339	1.0914	0.9844	0.8970	0.8222	0.7564	0.6972	0.6433	0.5936	0.5474
51						1.4716	1.2396	1.0978	0.9914	0.9045	0.8301	0.7647	0.7058	0.6522	0.6028
52							1.4762	1.2451	1.1041	0.9982	0.9118	0.8378	0.7727	0.7142	0.6609
53								1.4806	1.2506	1.1102	1.0049	0.9189	0.8454	0.7806	0.7224
54									1.4850	1.2559	1.1161	1.0113	0.9258	0.8527	0.7883
55										1.4892	1.2610	1.1219	1.0177	0.9326	0.8598
56											1.4933	1.2661	1.1276	1.0239	0.9392
57												1.4974	1.2711	1.1332	1.0299
58													1.5014	1.2759	1.1386
59														1.5053	1.2806
60															1.5091

J.13 Weights for Linear Regression of Weibull-2 Data for $n = 1, \ldots, 30$

$$w_{i,n} = \frac{1}{<Z_{i,n}^2> - <Z_{i,n}>^2} \quad \text{with } Z_{i,n} = \ln(-\ln(1 - F_{i,n}))$$

Note: with $Z = {}^{10}\log(-\ln(1 - F))$ (cf. Section 5.4.1, Eq. (5.24)), multiply the $w_{i,n}$ values below by $({}^{10}\log(e))^{-2} = 5.30189811$. For weighted averages there is no need to adjust.

i	n														
	1	2	3	4	5	6	7	8	9	10	11	12	13	14	15
1	0.6079	0.6079	0.6079	0.6079	0.6079	0.6079	0.6079	0.6079	0.6079	0.6079	0.6079	0.6079	0.6079	0.6079	0.6079
2		1.4619	1.5186	1.5342	1.5407	1.5439	1.5458	1.5470	1.5478	1.5483	1.5487	1.5490	1.5493	1.5494	1.5496
3			2.2297	2.4066	2.4632	2.4885	2.5020	2.5100	2.5152	2.5188	2.5213	2.5232	2.5246	2.5257	2.5266
4				2.9068	3.2415	3.3600	3.4163	3.4476	3.4669	3.4796	3.4885	3.4949	3.4996	3.5033	3.5062
5					3.5104	4.0234	4.2192	4.3167	4.3728	4.4081	4.4319	4.4487	4.4610	4.4703	4.4775
6						4.0554	4.7571	5.0412	5.1881	5.2749	5.3308	5.3690	5.3963	5.4166	5.4321
7							4.5529	5.4481	5.8281	6.0308	6.1535	6.2339	6.2896	6.3300	6.3602
8								5.0111	6.1012	6.5825	6.8462	7.0089	7.1172	7.1934	7.2490
9									5.4362	6.7208	7.3070	7.6357	7.8420	7.9812	8.0801
10										5.8333	7.3103	8.0040	8.4007	8.6535	8.8262
11											6.2061	7.8730	8.6756	9.1427	9.4444
12												6.5578	8.4113	9.3238	9.8632
13													6.8907	8.9277	9.9503
14														7.2071	9.4240
15															7.5086

i	n														
	16	17	18	19	20	21	22	23	24	25	26	27	28	29	30
1	0.6079	0.6079	0.6079	0.6079	0.6079	0.6079	0.6079	0.6079	0.6079	0.6079	0.6079	0.6079	0.6079	0.6079	0.6079
2	1.5497	1.5498	1.5499	1.5500	1.5500	1.5501	1.5501	1.5502	1.5502	1.5502	1.5502	1.5503	1.5503	1.5503	1.5503
3	2.5273	2.5279	2.5284	2.5288	2.5291	2.5294	2.5296	2.5299	2.5300	2.5302	2.5304	2.5305	2.5306	2.5307	2.5308
4	3.5085	3.5103	3.5119	3.5131	3.5142	3.5151	3.5159	3.5166	3.5172	3.5177	3.5181	3.5185	3.5189	3.5192	3.5195
5	4.4832	4.4877	4.4915	4.4946	4.4971	4.4993	4.5012	4.5028	4.5042	4.5054	4.5064	4.5074	4.5082	4.5089	4.5096
6	5.4441	5.4537	5.4615	5.4679	5.4732	5.4776	5.4814	5.4846	5.4874	5.4898	5.4919	5.4938	5.4954	5.4969	5.4982
7	6.3835	6.4017	6.4164	6.4283	6.4381	6.4463	6.4532	6.4591	6.4641	6.4685	6.4723	6.4756	6.4785	6.4811	6.4835
8	7.2911	7.3237	7.3494	7.3702	7.3872	7.4012	7.4130	7.4230	7.4315	7.4388	7.4452	7.4507	7.4556	7.4599	7.4638
9	8.1532	8.2088	8.2522	8.2867	8.3147	8.3376	8.3567	8.3728	8.3864	8.3981	8.4082	8.4170	8.4247	8.4315	8.4374
10	8.9501	9.0424	9.1132	9.1688	9.2133	9.2494	9.2793	9.3042	9.3252	9.3432	9.3586	9.3719	9.3835	9.3937	9.4027
11	9.6528	9.8037	9.9170	10.0044	10.0734	10.1290	10.1743	10.2119	10.2434	10.2701	10.2929	10.3125	10.3296	10.3445	10.3575
12	10.2158	10.4618	10.6414	10.7771	10.8825	10.9662	11.0338	11.0892	11.1354	11.1742	11.2071	11.2354	11.2598	11.2810	11.2996
13	10.5633	10.9686	11.2537	11.4635	11.6231	11.7477	11.8471	11.9277	11.9942	12.0496	12.0964	12.1363	12.1705	12.2002	12.2260
14	10.5567	11.2445	11.7036	12.0294	12.2707	12.4553	12.6002	12.7163	12.8110	12.8892	12.9547	13.0101	13.0574	13.0982	13.1336
15	9.9020	11.1445	11.9077	12.4218	12.7894	13.0634	13.2742	13.4404	13.5741	13.6836	13.7743	13.8506	13.9152	13.9706	14.0185
16	7.7968	10.3630	11.7149	12.5540	13.1240	13.5344	13.8421	14.0800	14.2685	14.4207	14.5457	14.6497	14.7373	14.8119	14.8758
17		8.0728	10.8085	12.2690	13.1844	13.8110	14.2650	14.6073	14.8732	15.0847	15.2563	15.3975	15.5155	15.6151	15.7000
18			8.3377	11.2396	12.8078	13.7996	14.4834	14.9818	15.3594	15.6542	15.8895	16.0810	16.2393	16.3717	16.4839
19				8.5926	11.6573	13.3324	14.4006	15.1420	15.6853	16.0991	16.4233	16.6832	16.8953	17.0711	17.2186
20					8.8382	12.0625	13.8435	14.9880	15.7874	16.3762	16.8266	17.1809	17.4659	17.6993	17.8932
21						9.0752	12.4560	14.3419	15.5626	16.4201	17.0549	17.5424	17.9274	18.2381	18.4932
22							9.3043	12.8387	14.8283	16.1249	17.0407	17.7218	18.2470	18.6631	19.0000
23								9.5260	13.2110	15.3034	16.6756	17.6498	18.3775	18.9407	19.3884
24									9.7409	13.5738	15.7677	17.2152	18.2478	19.0223	19.6238
25										9.9493	13.9274	16.2218	17.7442	18.8352	19.6566
26											10.1518	14.2724	16.6662	18.2630	19.4124
27												10.3486	14.6093	17.1014	18.7722
28													10.5402	14.9385	17.5278
29														10.7267	15.2604
30															10.9085

J.14 Weights for Linear Regression of Weibull-2 Data for $n = 31, \ldots, 45$

$$w_{i,n} = \frac{1}{<Z_{i,n}^2> - <Z_{i,n}>^2} \text{ with } Z_{i,n} = \ln(-\ln(1 - F_{i,n}))$$

Note: with $Z = {}^{10}\log(-\ln(1 - F))$ (cf. Section 5.4.1, Eq. (5.24)), multiply the $w_{i,n}$ values below by $({}^{10}\log(e))^{-2} = 5.30189811$. For weighted averages there is no need to adjust.

								n							
i	31	32	33	34	35	36	37	38	39	40	41	42	43	44	45
1	0.6079	0.6079	0.6079	0.6079	0.6079	0.6079	0.6079	0.6079	0.6079	0.6079	0.6079	0.6079	0.6079	0.6079	0.6079
2	1.5503	1.5503	1.5504	1.5504	1.5504	1.5504	1.5504	1.5504	1.5504	1.5504	1.5504	1.5504	1.5504	1.5504	1.5504
3	2.5309	2.5310	2.5310	2.5311	2.5311	2.5312	2.5312	2.5313	2.5313	2.5314	2.5314	2.5314	2.5315	2.5315	2.5315
4	3.5198	3.5200	3.5202	3.5204	3.5206	3.5207	3.5209	3.5210	3.5211	3.5212	3.5213	3.5214	3.5215	3.5216	3.5217
5	4.5102	4.5107	4.5112	4.5116	4.5120	4.5124	4.5127	4.5130	4.5133	4.5136	4.5138	4.5140	4.5142	4.5144	4.5146
6	5.4994	5.5004	5.5013	5.5022	5.5030	5.5037	5.5043	5.5049	5.5055	5.5060	5.5065	5.5069	5.5073	5.5076	5.5080
7	6.4855	6.4874	6.4890	6.4906	6.4919	6.4932	6.4943	6.4954	6.4963	6.4972	6.4980	6.4988	6.4994	6.5001	6.5007
8	7.4672	7.4702	7.4730	7.4754	7.4777	7.4797	7.4816	7.4833	7.4848	7.4862	7.4875	7.4888	7.4899	7.4909	7.4919
9	8.4428	8.4475	8.4518	8.4556	8.4591	8.4622	8.4651	8.4677	8.4701	8.4723	8.4743	8.4761	8.4778	8.4794	8.4809
10	9.4107	9.4178	9.4241	9.4299	9.4350	9.4397	9.4439	9.4477	9.4513	9.4545	9.4574	9.4602	9.4627	9.4650	9.4671
11	10.3691	10.3794	10.3886	10.3968	10.4042	10.4108	10.4169	10.4224	10.4274	10.4320	10.4362	10.4401	10.4436	10.4469	10.4499
12	11.3159	11.3304	11.3433	11.3548	11.3652	11.3745	11.3829	11.3906	11.3975	11.4039	11.4097	11.4150	11.4199	11.4245	11.4286
13	12.2487	12.2687	12.2865	12.3023	12.3165	12.3292	12.3407	12.3511	12.3605	12.3691	12.3770	12.3842	12.3908	12.3969	12.4025
14	13.1645	13.1917	13.2158	13.2371	13.2562	13.2733	13.2887	13.3025	13.3151	13.3266	13.3371	13.3466	13.3554	13.3634	13.3709
15	14.0601	14.0965	14.1286	14.1570	14.1823	14.2049	14.2252	14.2434	14.2600	14.2750	14.2887	14.3012	14.3126	14.3231	14.3327
16	14.9312	14.9795	15.0218	15.0592	15.0923	15.1218	15.1483	15.1720	15.1935	15.2129	15.2306	15.2467	15.2614	15.2748	15.2872
17	15.7731	15.8365	15.8918	15.9405	15.9835	16.0217	16.0558	16.0863	16.1138	16.1387	16.1613	16.1818	16.2005	16.2176	16.2333
18	16.5797	16.6624	16.7343	16.7971	16.8525	16.9015	16.9450	16.9840	17.0190	17.0505	17.0790	17.1049	17.1285	17.1500	17.1697
19	17.3439	17.4512	17.5439	17.6247	17.6954	17.7578	17.8132	17.8624	17.9066	17.9462	17.9820	18.0144	18.0439	18.0707	18.0953
20	18.0564	18.1952	18.3144	18.4177	18.5077	18.5867	18.6566	18.7185	18.7739	18.8234	18.8680	18.9083	18.9449	18.9781	19.0085
21	18.7058	18.8851	19.0380	19.1696	19.2837	19.3834	19.4711	19.5487	19.6177	19.6793	19.7346	19.7844	19.8295	19.8704	19.9076
22	19.2774	19.5091	19.7051	19.8725	20.0168	20.1422	20.2520	20.3486	20.4343	20.5105	20.5787	20.6400	20.6953	20.7454	20.7908
23	19.7519	20.0520	20.3033	20.5163	20.6986	20.8561	20.9932	21.1133	21.2193	21.3134	21.3972	21.4723	21.5398	21.6008	21.6561
24	20.1035	20.4941	20.8174	21.0887	21.3191	21.5167	21.6877	21.8368	21.9677	22.0833	22.1860	22.2777	22.3599	22.4339	22.5008
25	20.2968	20.8088	21.2268	21.5736	21.8653	22.1135	22.3268	22.5117	22.6731	22.8150	22.9406	23.0522	23.1520	23.2416	23.3224
26	20.2809	20.9599	21.5046	21.9503	22.3210	22.6335	22.8998	23.1291	23.3281	23.5022	23.6555	23.7913	23.9122	24.0204	24.1176
27	19.9798	20.8955	21.6136	22.1912	22.6649	23.0598	23.3933	23.6781	23.9237	24.1373	24.3243	24.4892	24.6355	24.7659	24.8827
28	19.2721	20.5378	21.5007	22.2580	22.8687	23.3708	23.7902	24.1451	24.4487	24.7109	24.9392	25.1395	25.3163	25.4733	25.6134
29	17.9458	19.7631	21.0867	22.0968	22.8935	23.5376	24.0683	24.5124	24.8889	25.2115	25.4906	25.7340	25.9478	26.1367	26.3047
30	15.5753	18.3557	20.2455	21.6268	22.6842	23.5204	24.1980	24.7574	25.2266	25.6249	25.9669	26.2631	26.5218	26.7493	26.9507
31	11.0859	15.8836	18.7579	20.7198	22.1586	23.2632	24.1390	24.8501	25.4386	25.9329	26.3534	26.7149	27.0285	27.3028	27.5443
32		11.2591	16.1856	19.1528	21.1862	22.6822	23.8340	24.7494	25.4943	26.1119	26.6317	27.0745	27.4557	27.7870	28.0770
33			11.4282	16.4816	19.5407	21.6449	23.1980	24.3969	25.3519	26.1307	26.7776	27.3230	27.7883	28.1895	28.5386
34				11.5935	16.7717	19.9217	22.0964	23.7062	24.9521	25.9468	26.7596	27.4359	28.0070	28.4951	28.9165
35					11.7553	17.0564	20.2962	22.5408	24.2072	25.4999	26.5343	27.3811	28.0870	28.6839	29.1949
36						11.9135	17.3358	20.6645	22.9784	24.7010	26.0405	27.1145	27.9955	28.7310	29.3540
37							12.0685	17.6101	21.0267	23.4095	25.1880	26.5742	27.6878	28.6028	29.3681
38								12.2203	17.8795	21.3832	23.8342	25.6683	27.1010	28.2542	29.2034
39									12.3691	18.1443	21.7340	24.2528	26.1423	27.6213	28.8140
40										12.5150	18.4045	22.0795	24.6655	26.6100	28.1352
41											12.6582	18.6605	22.4197	25.0724	27.0716
42												12.7987	18.9122	22.7549	25.4737
43													12.9366	19.1600	23.0852
44														13.0721	19.4039
45															13.2053

J.15 Weights for Linear Regression of Weibull-2 Data for $n = 46, \ldots, 60$

$$w_{i,n} = \frac{1}{<Z_{i,n}^2> - <Z_{i,n}>^2} \quad \text{with } Z_{i,n} = \ln(-\ln(1 - F_{i,n}))$$

Note: with $Z = {}^{10}\log(-\ln(1 - F))$ (cf. Section 5.4.1, Eq. (5.24)), multiply the $w_{i,n}$ values below by $({}^{10}\log(e))^{-2} = 5.30189811$. For weighted averages there is no need to adjust.

							n								
i	46	47	48	49	50	51	52	53	54	55	56	57	58	59	60
1	0.6079	0.6079	0.6079	0.6079	0.6079	0.6079	0.6079	0.6079	0.6079	0.6079	0.6079	0.6079	0.6079	0.6079	0.6079
2	1.5504	1.5505	1.5505	1.5505	1.5505	1.5505	1.5505	1.5505	1.5505	1.5505	1.5505	1.5505	1.5505	1.5505	1.5505
3	2.5315	2.5316	2.5316	2.5316	2.5316	2.5316	2.5317	2.5317	2.5317	2.5317	2.5317	2.5317	2.5317	2.5318	2.5318
4	3.5218	3.5218	3.5219	3.5219	3.5220	3.5221	3.5221	3.5222	3.5222	3.5222	3.5223	3.5223	3.5224	3.5224	3.5224
5	4.5148	4.5149	4.5151	4.5152	4.5153	4.5154	4.5156	4.5157	4.5158	4.5159	4.5159	4.5160	4.5161	4.5162	4.5163
6	5.5083	5.5086	5.5089	5.5092	5.5094	5.5096	5.5098	5.5100	5.5102	5.5104	5.5106	5.5107	5.5109	5.5110	5.5112
7	6.5012	6.5018	6.5022	6.5027	6.5031	6.5035	6.5039	6.5042	6.5046	6.5049	6.5052	6.5055	6.5057	6.5060	6.5062
8	7.4928	7.4936	7.4944	7.4951	7.4958	7.4964	7.4970	7.4976	7.4981	7.4986	7.4991	7.4995	7.5000	7.5004	7.5007
9	8.4823	8.4835	8.4847	8.4858	8.4869	8.4878	8.4887	8.4896	8.4904	8.4911	8.4919	8.4925	8.4932	8.4938	8.4943
10	9.4691	9.4710	9.4727	9.4743	9.4758	9.4772	9.4785	9.4798	9.4809	9.4820	9.4831	9.4840	9.4850	9.4858	9.4867
11	10.4528	10.4554	10.4578	10.4601	10.4622	10.4642	10.4660	10.4678	10.4694	10.4709	10.4724	10.4737	10.4750	10.4762	10.4774
12	11.4325	11.4361	11.4394	11.4425	11.4454	11.4482	11.4507	11.4531	11.4553	11.4574	11.4593	11.4612	11.4629	11.4646	11.4661
13	12.4077	12.4126	12.4170	12.4212	12.4251	12.4287	12.4321	12.4352	12.4382	12.4410	12.4436	12.4461	12.4484	12.4506	12.4527
14	13.3777	13.3841	13.3900	13.3954	13.4005	13.4053	13.4097	13.4138	13.4177	13.4214	13.4248	13.4280	13.4310	13.4339	13.4366
15	14.3416	14.3499	14.3575	14.3646	14.3712	14.3773	14.3830	14.3883	14.3933	14.3980	14.4024	14.4066	14.4104	14.4141	14.4176
16	15.2986	15.3092	15.3189	15.3279	15.3363	15.3441	15.3514	15.3582	15.3645	15.3705	15.3760	15.3813	15.3862	15.3908	15.3952
17	16.2477	16.2610	16.2733	16.2847	16.2953	16.3051	16.3142	16.3228	16.3307	16.3382	16.3452	16.3517	16.3579	16.3636	16.3691
18	17.1878	17.2045	17.2199	17.2341	17.2473	17.2595	17.2708	17.2814	17.2913	17.3006	17.3092	17.3173	17.3249	17.3321	17.3388
19	18.1178	18.1385	18.1575	18.1751	18.1914	18.2064	18.2205	18.2335	18.2457	18.2570	18.2676	18.2776	18.2869	18.2957	18.3039
20	19.0362	19.0617	19.0851	19.1067	19.1266	19.1451	19.1622	19.1782	19.1930	19.2068	19.2198	19.2319	19.2432	19.2538	19.2638
21	19.9416	19.9728	20.0014	20.0277	20.0520	20.0744	20.0952	20.1146	20.1325	20.1493	20.1649	20.1795	20.1932	20.2060	20.2181
22	20.8323	20.8702	20.9049	20.9368	20.9662	20.9934	21.0185	21.0418	21.0634	21.0835	21.1023	21.1198	21.1363	21.1516	21.1660
23	21.7063	21.7522	21.7941	21.8326	21.8680	21.9007	21.9308	21.9587	21.9846	22.0087	22.0311	22.0521	22.0716	22.0899	22.1071
24	22.5616	22.6168	22.6673	22.7135	22.7559	22.7950	22.8310	22.8643	22.8952	22.9238	22.9505	22.9753	22.9985	23.0202	23.0405
25	23.3955	23.4619	23.5223	23.5776	23.6282	23.6748	23.7176	23.7572	23.7938	23.8278	23.8593	23.8887	23.9160	23.9416	23.9655
26	24.2053	24.2848	24.3570	24.4229	24.4831	24.5384	24.5892	24.6360	24.6793	24.7194	24.7565	24.7911	24.8233	24.8533	24.8813
27	24.9877	25.0827	25.1687	25.2470	25.3185	25.3839	25.4439	25.4992	25.5501	25.5973	25.6409	25.6814	25.7191	25.7543	25.7870
28	25.7390	25.8522	25.9545	26.0474	26.1319	26.2092	26.2799	26.3449	26.4047	26.4600	26.5111	26.5585	26.6025	26.6434	26.6816
29	26.4548	26.5895	26.7110	26.8209	26.9208	27.0118	27.0949	27.1712	27.2413	27.3059	27.3656	27.4208	27.4720	27.5197	27.5640
30	27.1299	27.2901	27.4342	27.5641	27.6819	27.7889	27.8865	27.9759	28.0578	28.1332	28.2027	28.2669	28.3264	28.3816	28.4330
31	27.7582	27.9489	28.1195	28.2731	28.4118	28.5375	28.6519	28.7563	28.8519	28.9397	29.0205	29.0950	29.1640	29.2279	29.2872
32	28.3327	28.5595	28.7618	28.9431	29.1063	29.2539	29.3878	29.5097	29.6211	29.7232	29.8169	29.9033	29.9830	30.0568	30.1253
33	28.8446	29.1147	29.3545	29.5687	29.7608	29.9339	30.0906	30.2328	30.3625	30.4810	30.5897	30.6896	30.7816	30.8667	30.9455
34	29.2835	29.6057	29.8904	30.1435	30.3697	30.5728	30.7560	30.9219	31.0727	31.2102	31.3360	31.4514	31.5576	31.6555	31.7461
35	29.6366	30.0219	30.3604	30.6599	30.9264	31.1648	31.3791	31.5726	31.7479	31.9074	32.0529	32.1862	32.3085	32.4211	32.5250

									n						
i	46	47	48	49	50	51	52	53	54	55	56	57	58	59	60
36	29.8879	30.3502	30.7538	31.1089	31.4234	31.7035	31.9543	32.1799	32.3838	32.5687	32.7370	32.8907	33.0315	33.1609	33.2801
37	30.0172	30.5743	31.0572	31.4794	31.8512	32.1808	32.4747	32.7380	32.9752	33.1896	33.3842	33.5615	33.7236	33.8721	34.0087
38	29.9985	30.6739	31.2543	31.7580	32.1988	32.5875	32.9324	33.2401	33.5162	33.7650	33.9901	34.1947	34.3811	34.5516	34.7080
39	29.7974	30.6223	31.3240	31.9278	32.4525	32.9121	33.3178	33.6781	34.0000	34.2889	34.5495	34.7855	35.0000	35.1957	35.3749
40	29.3674	30.3849	31.2398	31.9679	32.5952	33.1409	33.6195	34.0423	34.4182	34.7542	35.0562	35.3287	35.5757	35.8004	36.0055
41	28.6429	29.9145	30.9662	31.8510	32.6056	33.2565	33.8233	34.3210	34.7610	35.1526	35.5030	35.8181	36.1027	36.3608	36.5958
42	27.5274	29.1446	30.4554	31.5413	32.4560	33.2372	33.9118	34.4998	35.0167	35.4742	35.8816	36.2464	36.5748	36.8716	37.1410
43	25.8697	27.9775	29.6403	30.9904	32.1104	33.0552	33.8629	34.5612	35.1706	35.7068	36.1817	36.6051	36.9845	37.3263	37.6354
44	23.4108	26.2605	28.4221	30.1304	31.5196	32.6737	33.6484	34.4828	35.2050	35.8358	36.3913	36.8839	37.3233	37.7174	38.0727
45	19.6440	23.7319	26.6462	28.8613	30.6149	32.0431	33.2313	34.2360	35.0971	35.8431	36.4954	37.0704	37.5807	38.0362	38.4452
46	13.3362	19.8805	24.0485	27.0270	29.2953	31.0940	32.5611	33.7833	34.8180	35.7058	36.4757	37.1497	37.7442	38.2722	38.7440
47		13.4649	20.1135	24.3609	27.4030	29.7241	31.5678	33.0737	34.3298	35.3946	36.3091	37.1030	37.7985	38.4126	38.9586
48			13.5915	20.3431	24.6691	27.7744	30.1481	32.0364	33.5811	34.8710	35.9658	36.9070	37.7250	38.4422	39.0760
49				13.7161	20.5695	24.9732	28.1412	30.5672	32.5000	34.0833	35.4071	36.5318	37.4998	38.3417	39.0807
50					13.8387	20.7927	25.2735	28.5037	30.9816	32.9588	34.5805	35.9380	37.0927	38.0874	38.9534
51						13.9594	21.0128	25.5700	28.8619	31.3914	33.4127	35.0727	36.4640	37.6485	38.6700
52							14.0783	21.2299	25.8627	29.2160	31.7968	33.8620	35.5602	36.9851	38.1995
53								14.1954	21.4442	26.1519	29.5660	32.1978	34.3068	36.0430	37.5014
54									14.3108	21.6556	26.4376	29.9121	32.5946	34.7470	36.5212
55										14.4245	21.8643	26.7200	30.2543	32.9872	35.1830
56											14.5367	22.0704	26.9990	30.5928	33.3758
57												14.6472	22.2739	27.2747	30.9277
58													14.7563	22.4748	27.5474
59														14.8639	22.6734
60															14.9700

References

1 New York Herald (1880). *The electric light*, 6. New York Herald.

2 BSI. PAS 55-1 (2008). *Asset Management. Part 1: Specification for the Optimized Management of Physical Assets*. London: British Standards Institution.

3 ISO. 55000:2014 (corr. 2014-03) (2014). *Asset Management – Overview, Principles and Terminology*. Geneva, CH: International Standards Organization.

4 ISO. 55001:2014 (2014). *Asset Management – Requirements*. Geneva, CH: International Standards Organization.

5 ISO. 55002 (2014). *Asset Management – Guidelines on the Application of ISO 55001*. Geneva, CH: International Standards Organization.

6 The Institute of Asset Management (2015). *Asset Management – An Anatomy*. Bristol, UK: The Institute of Asset Management.

7 Burlando, A. (2010). IDEAS University of Oregon Economics Department Working Papers – The Impact of Transitory Income on Birth Weights: Evidence from a Blackout in Zanzibar. https://ideas.repec.org/p/ore/uoecwp/2010-1.html (accessed 2015).

8 van der Laan, R. (2013). *Foundation Rural Energy Services – Impact Assessment of Rural Electrification*. FRES http://www.fres.nl/en/docman/puclicaties/92-pwc-rapport/file.html (accessed 1 February 2015.

9 Van der Klauw, B. and Wang, L. (2004). Child Mortality in Rural India. World Bank Policy Research Working Paper 3281.

10 Ross, R. and van Schaik, N. (2003). PD diagnostics on cables and terminations for CBM. Proceedings ICPADM, Nagoya, Japan.

11 Ross, R. (2015). Evaluating methods for detecting end-of-life and non-compliance of asset populations. Proceedings of Cigré SCD1 Colloquium Trends in Technology, Materials, Testing and Diagnostics Applied to Electric Power Systems, Rio de Janeiro, Brazil.

12 Ross, R. (2017). Health index methodologies for decision-making on asset maintenance and replacement. Cigré 2017 Colloquium of Study Committees A3, B4 & D1, Winnipeg, Canada.

13 Dissado, L.A. and Fothergill, J.C. (1992). *Electrical Degradation and Breakdown in Polymers, Part IV*. London: Institution of Electrical Engineers.

14 Dakin, T.W. (1948). Electrical insulation deterioration treated as a chemical rate phenomenon. *Transactions of the American Institute of Electrical Engineers Part 1: Communication and Electronics* 67 (1): 113–122.

Reliability Analysis for Asset Management of Electric Power Grids, First Edition. Robert Ross.
© 2019 John Wiley & Sons Ltd. Published 2019 by John Wiley & Sons Ltd.
Companion website: www.wiley.com/go/ross/reliabilityanalysis

15 Jaccard, P. (1912). The distribution of the flora in the alpine zone. *New Phytologist* 11 (2): 37–50.

16 Ypma, P.A. and Ross, R. (2017). Determining the similarity between observed and expected ageing behavior. Proceedings of ICEMPE (1st International Conference on Electrical Materials and Power Equipment), Xi'an, China.

17 Fisher, R.A. and Tippett, L.H.C. (1928). Limiting forms of the frequency distribution of the largest or smallest member of a sample. *Mathematical Proceedings of the Cambridge Philosophical Society* 24 (2): 180–190.

18 Gnedenko, B.V. (1943). Sur la distribution limite du terme maximum d'une serie aleatoire. *Annals of Mathematics* 44 (3): 423–453.

19 Choi, K. (1994). On the medians of gamma distributions and an equation of Ramanujan. *Proceedings of the American Mathematical Society* 12 (1): 245–251.

20 de Moivre, A. (1718). *The Doctrine of Chances*. London: Millar.

21 de Laplace, P. (1812). *Théorie analytique des probabilités*. Paris: Ve. Courcier.

22 Brisbane, A. (1911). Speakers give sound advice. *Syracuse Post Standard* 28: 18.

23 Oresme, N. (1353). Tractatus de configurationibus qualitatum et motuum, France. p. I.iv.

24 Montanari, G., Fothergil, J., Hampton, N. et al. (2004). *IEEE Std 930-2004 Guide for the Statistical Analysis of Electrical Insulation Breakdown Data*. New York: The Institute of Electrical and Electronics Engineers.

25 Kaplan, E. and Meier, P. (1958). Nonparametric estimation from incomplete observations. *Journal of the American Statistical Association* 53: 457–481.

26 NIST/SEMATECH. (2013). e-Handbook of Statistical Methods. US Department of Commerce Agency NIST, 30 October 2013. http://www.itl.nist.gov/div898/handbook/toolaids/pff/apr.pdf (accessed 5 March 2016).

27 Ross, R. (1996). Bias and standard deviation due to Weibull parameter estimation for small data sets. *IEEE Transactions on Dielectrics and Electrical Insulation* 3 (1): 28–42.

28 Ross, R. (1994). Graphical methods for plotting and evaluating Weibull distributed data. Proceedings of the 4th International Conference on Properties and Applications of Dielectrics Materials, Brisbane, Australia.

29 IEC TC112. IEC 62539(E):2007 (2007). *Guide for the Statistical Analysis of Electrical Insulation Breakdown Data*, 2007-07. Geneva, CH: International Electrotechnical Committee.

30 Blom, G. (1958). *Statistical Estimates and Transformed Beta Variables*, 73 ff. New York: Wiley.

31 Duane, J. (1964). Learning curve approach to reliability monitoring. *IEEE Transactions on Aerospace* 2 (2): 536–566.

32 Smith, S.A. and Oren, S.S. (1980). Reliability growth of repairable systems. *Naval Research Logistics* 27 (4): 539–547.

33 Crow, L. (1974). Reliability analysis for complex, repairable systems. In: *Reliability and Biometry*, 379–410. Philadelphia, PA: SIAM.

34 Crow, L. (1975). On tracking reliability growth. Proceedings of 1975 Annual Reliability and Maintainability Symposium.

35 Crow, L. (2011). Reliability growth planning, analysis and management. Tutorial Notes 2011 Reliability and Maintainability Symposium.

36 Cramér, H. (1946). *Mathematical Methods of Statistics*. Princeton, NJ: Princeton University Press.

37 Rao, C.R. (1945). Information and the accuracy attainable in the estimation of statistical parameters. *Bulletin of the Calcutta Mathematical Society* 37: 81–89.

38 Fisher, R.A. (1922). On the mathematical foundations of theoretical statistics. *Philosophical Transactions of the Royal Society of London, Series A: Containing Papers of a Mathematical or Physical Character* 222: 309–368.

39 Fisher, R.A. (1925). Theory of statistical estimation. *Proceedings of the Cambridge Philosophical Society* 22 (5): 700–725.

40 Wikipedia (2016). *Maximum Likelihood Estimation*. Wikimedia Foundation, Inc. https://en.wikipedia.org/w/index.php?title=Maximum_likelihood_estimation&oldid=726117543(accessed 22 July 2016.

41 Thoman, D., Bain, L., and Antle, C. (1969). Inferences on the parameters of the Weibull distribution. *Technometrics* 11 (03): 445–460.

42 Ross, R. (1994). Formulas to describe the bias and standard deviation of the ML-estimated Weibull shape parameter. *IEEE Transactions on Dielectrics and Electrical Insulation* 1 (2): 247–253.

43 Ross, R. (1995). Comparison of methods to reduce the maximum likelihood bias on Weibull parameters. IEEE International Conference on Solid Dielectrics ICSD'95, Leicester.

44 CRC (1980). *CRC Handbook of Chemistry and Physics*, 60e (ed. R.C. Weast and M.J. Astle). Boca Raton, FL: CRC Press.

45 ReliaSoft Corporation. (n.d.). *ReliaWiki: Crow-AMSAA (NHPP)*. ReliaSoft Corporation. http://reliawiki.org/index.php?title=Crow-AMSAA_(NHPP) (accessed 3 January 2018).

46 Weibull, W. (1951). A statistical distribution of wide applicability. *Journal of Applied Mechanics* 18: 293–297.

47 Comerford, N. (2005). Crow/AMSAA reliability growth plots. Proceedings of 16th Annual Conference 2005 VANZ, Rotorua.

48 Meijden, M., Janssen, A., Ford, G. et al. (2006). *Asset Management of Transmission Systems and Associated Cigré Activities (TB309)*. Paris: Cigré.

49 Rijks, E., Sanchis, G., Ford, G. et al. (2010). *Transmission Asset Risk Management (TB422)*. Paris: Cigré.

50 US Department of Defense (1980). *Procedures for Performing a Failure Mode, Effects and Criticality Analysis*. Washington, D.C.: Department of Defense.

51 Bartnikas, R. and Srivastava, K. (2000). *Power and Communication Cables – Theory and Applications*, 331–426. New York: Wiley-IEEE Press.

52 Densley, R., Bartnikas, R., and Bernstein, B. (1994). Multiple stress aging of solid-dielectric extruded dry-cured insulation systems for power transmission cables. *IEEE Transactions on Power Delivery* 9 (1): 559–571.

53 Larsen, P. (1983). Dyeing methods used for detection of water-trees in extruded cable insulation. *Electra* 86: 53–59.

54 Ross, R., Smit, J.J., and Aukema, P. (1992). Staining of water trees with methylene blue explained. Proceedings of IEEE International Conference on Solid Dielectrics (ICSD), Sestri Levante, Italy.

55 Ross, R. (1998). Inception and propagation mechanisms of water treeing. *IEEE Transaction on Dielectrics and Electrical Insulation* 5 (5): 660–680.

56 Taylor, B. (1715). *Methodus Incrementorum Directa et Inversa*. London: Prostant apud Gul. Innys ad Insignia Principis in Coemeterio Paulino.

57 Ross, R. (1993). Unbiasing formulae for the Weibull shape parameter as estimated with the maximum likelihood method. Joint Conference 1993 International Workshop on Electrical Insulation & 25th Symposium on Electrical Insulating Materials, Nagoya, Japan.

58 Draper, N.R. and Smith, H. (1981). *Applied Regression Analysis*, 2e. New York: Wiley.

59 Gentle, J.E., Härdle, W.K., and Mori, Y. (2012). *Handbook of Computational Statistics – Concepts and Methods*, 2e (ed. J.E. Gentle, W.K. Härdle and Y. Mori). Berlin: Springer-Verlag.

60 Greenberger, M. (1961). An a priori determination of serial correlation in computer generated random numbers. *Mathematics of Computation* 15: 383–389.

61 Hull, T. and Dobell, A. (1962). Random number generators. *SIAM Review* 4 (3): 230–254.

62 Lehmer, D.H. (1949). Mathematical methods in large-scale computing units. Proceedings of a Second Symposium on Large-Scale Digital Calculating Machinery, Cambridge, MA.

63 Knuth, D. (1997). Seminumerical algorithms. In: *The Art of Computer Programming*, 3e. Reading, MA: Addison-Wesley.

64 Leemis, L.H. and Park, S.K. (2006). Section 2.2: Lehmer random number generators: implementation. In: *Discrete-Event Simulation: A First Course*. London: Pearson.

65 Jain, R. (2008). *Random-Number Generation*. Washington University of St. Louis http://www.cs.wustl.edu/~jain/cse567-08/ftp/k_26rng.pdf (accessed 28 February 2018.

66 van Asselt, R., Pijlgroms, R., Smeets, W. et al. (1991). Analyse en numerieke wiskunde Deel 3. In: *Wiskunde voor het hoger onderwijs*, 2e, 193–194. Culemborg, Netherlands: Educaboek bv.

67 Coveyou, R. and MacPherson, R. (1967). Fourier analysis of uniform random number generators. *Journal of the ACM* 14: 100–119.

68 National Bureau of Standards (1970). *Handbook of Mathematical Functions*, 9e (ed. M. Abramowitz and I.A. Stegun). New York: Dover Publications.

69 Råde, L. and Westergren, B. (1988). *BETA β Mathematics Handbook*. Lund: Studentlitteratur.

70 Ross, R. (1998). Reliability case studies. IEE Disucussion Meeting – Reliability and Survival: Failure Analysis (11 February 1998), London.

71 Mann, N.R., Schafer, R.E., and Singpurwalla, N.D. (1974). *Methods for Statistical Analysis of Reliability and Life Data*. New York: Wiley.

Index

a

abnormal 94
absorbing state 305, 306, 314
accelerated ageing 52–54, 322, 339, 348, 395, 397
accuracy 73, 195, 210
accurate 228
adjugate 376
adjusted
  plotting position 185, 203, 231, 238, 240, 254, 260, 261, 264
  rank 167, 231, 238, 253, 260
  ranking 164, 183, 185, 190, 203, 263
ageing
  conditions 50
  dose 49, 56, 58, 322, 349
  history 52
alert level 58
alternative hypothesis 407
AM 1, 158
AMSAA 197, 263, 327
analytical 259
  expression 193, 212, 260, 263
  solution 249
analytically correct 212
AND operation 280
approximation 212, 373, 379, 408
arithmetic mean 63
Arrhenius 339
as good as new 120, 318
asset management 2, 197, 211, 217, 242, 263
asymmetry 70
asymptote 128, 378
asymptotic 219

b

behaviour 153, 212, 263, 369
  distributions 153
autocorrelation 81
availability 291, 296, 298, 303–305, 312–314, 324, 353
average 71, 124, 224

bad luck 119
batch 31
bath tub model 36, 37, 40, 60, 115, 123, 353, 360
Bayes, Thomas 32, 41, 60
  theorem 42, 60
Bayesian
  approach 67, 286, 346, 351, 363
  mean 66
  statistics 41
Bernoulli trial 134–138, 153
best fit 174, 182, 192, 203, 344, 381, 383
beta 102
beta distribution 106–112, 152, 168, 174, 203, 233, 342, 347, 354, 399
  characteristic function 111
  moment generating function 111
bias 209, 217–220, 223, 242–248, 263, 378
  effect 220
  formulas 218, 243
bijective 97
binomial 102
  coefficient 136, 144, 149, 368, 369
binomial distribution 135–138, 143, 144, 150, 151, 153
  generalization 149
  moment generating function 139

Reliability Analysis for Asset Management of Electric Power Grids, First Edition. Robert Ross.
© 2019 John Wiley & Sons Ltd. Published 2019 by John Wiley & Sons Ltd.
Companion website: www.wiley.com/go/ross/reliabilityanalysis